Advanced Radiation Protection Dosimetry

Series in Medical Physics and Biomedical Engineering

Series Editors: John G. Webster, E. Russell Ritenour, Slavik Tabakov, and Kwan Hoong Ng

Recent books in the series:

Advance Radiation Protection Dosimetry
Shaheen A. Dewji and Nolan E. Hertel (Eds)

Clinical Radiotherapy Physics with MATLAB: A Problem-Solving Approach
Pavel Dvorak

Advances in Particle Therapy: A Multidisciplinary Approach
Manjit Dosanjh and Jacques Bernier (Eds)

Radiotherapy and Clinical Radiobiology of Head and Neck Cancer
Loredana G. Marcu, Iuliana Toma-Dasu, Alexandru Dasu, and Claes Mercke

Problems and Solutions in Medical Physics: Diagnostic Imaging Physics
Kwan Hoong Ng, Jeannie Hsiu Ding Wong, and Geoffrey D. Clarke

Advanced and Emerging Technologies in Radiation Oncology Physics
Siyong Kim and John W. Wong (Eds)

A Guide to Outcome Modeling in Radiotherapy and Oncology: Listening to the Data
Issam El Naqa (Ed)

Advanced MR Neuroimaging: From Theory to Clinical Practice
Ioannis Tsougos

Quantitative MRI of the Brain: Principles of Physical Measurement, Second Edition
Mara Cercignani, Nicholas G. Dowell, and Paul S. Tofts (Eds)

A Brief Survey of Quantitative EEG
Kaushik Majumdar

Handbook of X-ray Imaging: Physics and Technology
Paolo Russo (Ed)

Graphics Processing Unit-Based High-Performance Computing in Radiation Therapy
Xun Jia and Steve B. Jiang (Eds)

Targeted Muscle Reinnervation: A Neural Interface for Artificial Limbs
Todd A. Kuiken, Aimee E. Schultz Feuser, and Ann K. Barlow (Eds)

For more information about this series, please visit: https://www.crcpress.com/Series-in-Medical-Physics-and-Biomedical-Engineering/book-series/CHMEPHBIOENG

Advanced Radiation Protection Dosimetry

Edited by
Shaheen A. Dewji
and
Nolan E. Hertel

CRC Press
Taylor & Francis Group
Boca Raton London New York

CRC Press is an imprint of the
Taylor & Francis Group, an **informa** business

This manuscript has been authored by UT-Battelle, LLC under Contract No. DE-AC05-00OR22725 with the U.S. Department of Energy. The United States Government retains and the publisher, by accepting the article for publication, acknowledges that the United States Government retains a non-exclusive, paid-up, irrevocable, world-wide license to publish or reproduce the published form of this manuscript, or allow others to do so, for United States Government purposes. The Department of Energy will provide public access to these results of federally sponsored research in accordance with the DOE Public Access Plan (http://energy.gov/downloads/doe-public-access-plan).

CRC Press
Taylor & Francis Group
6000 Broken Sound Parkway NW, Suite 300
Boca Raton, FL 33487-2742

First issued in paperback 2020

© 2019 by Taylor & Francis Group, LLC
CRC Press is an imprint of Taylor & Francis Group, an Informa business

No claim to original U.S. Government works

ISBN-13: 978-1-4987-8543-3 (hbk)
ISBN-13: 978-0-367-78003-6 (pbk)

Library of Congress Cataloging-in-Publication Data

Names: Dewji, Shaheen A., editor. | Hertel, Nolan E., 1950- author.
Title: Advanced radiation protection dosimetry / edited by Shaheen A. Dewji, Nolan E. Hertel.
Other titles: Series in medical physics and biomedical engineering.
Description: Boca Raton, FL : CRC Press, Taylor & Francis Group, [2019] |
Series: Series in medical physics and biomedical engineering | Includes bibliographical references and index.
Identifiers: LCCN 2018048996| ISBN 9781498785433 (hbk ; alk. paper) | ISBN 1498785433 (hbk ; alk. paper) | ISBN 9780429055362 (eBook) | ISBN 0429055366 (eBook)
Subjects: LCSH: Radiation dosimetry. | Radiation--Safety measures.
Classification: LCC R906 .A38 2019 | DDC 612/.014480287--dc23
LC record available at https://lccn.loc.gov/2018048996

Visit the Taylor & Francis Web site at
http://www.taylorandfrancis.com

and the CRC Press Web site at
http://www.crcpress.com

To Pasquale and to Jennifer

Contents

About the Series, ix

The International Organization for Medical Physics, xi

Preface, xiii

Acknowledgments, xv

Editors, xvii

Contributors, xix

External Reviewers, xxv

CHAPTER 1 ▪ Introduction   1

    SHAHEEN A. DEWJI AND NOLAN E. HERTEL

CHAPTER 2 ▪ Fundamental Concepts and Quantities   11

    KEN G. VEINOT

CHAPTER 3 ▪ Evolution of Radiation Protection Guidance in
the United States   79

    RONALD L. KATHREN

CHAPTER 4 ▪ Radiation Detection and Measurement   123

    JOSEPH C. MCDONALD

CHAPTER 5 ▪ Reference Individuals Defined for External and
Internal Radiation Dosimetry   169

    WESLEY E. BOLCH

CHAPTER 6 ▪ Biokinetic Models   215

    RICH LEGGETT

CHAPTER 7 ▪ Dosimetric Models   307

    JOHN R. FORD, JR. AND JOHN W. POSTON, SR.

CHAPTER 8 ■ Dose Coefficients   335

NOLAN E. HERTEL AND DEREK JOKISCH

CHAPTER 9 ■ Cancer Risk Coefficients   395

DAVID PAWEL

CHAPTER 10 ■ Interpretation of Bioassay Results to Assess the Intake of Radionuclides   417

DAVID MCLAUGHLIN

Index   465

About the Series

The *Series in Medical Physics and Biomedical Engineering* describes the applications of physical sciences, engineering, and mathematics in medicine and clinical research. The series seeks (but is not restricted to) publications in the following topics:

- Artificial organs
- Assistive technology
- Bioinformatics
- Bioinstrumentation
- Biomaterials
- Biomechanics
- Biomedical engineering
- Clinical engineering
- Imaging
- Implants
- Medical computing and mathematics
- Medical/surgical devices
- Patient monitoring
- Physiological measurement
- Prosthetics
- Radiation protection, health physics, and dosimetry
- Regulatory issues
- Rehabilitation engineering
- Sports medicine
- Systems physiology
- Telemedicine
- Tissue engineering
- Treatment

The International Organization for Medical Physics

THE INTERNATIONAL ORGANIZATION FOR Medical Physics (IOMP) represents over 18,000 medical physicists worldwide and has a membership of 80 national and 6 regional organizations, together with a number of corporate members. Individual medical physicists of all national member organizations are also automatically members.

The mission of the IOMP is to advance medical physics practice worldwide by disseminating scientific and technical information, fostering the educational and professional development of medical physics, and promoting the highest quality medical physics services for patients.

A World Congress on Medical Physics and Biomedical Engineering is held every three years in cooperation with the International Federation for Medical and Biological Engineering (IFMBE) and the International Union for Physics and Engineering Sciences in Medicine (IUPESM). A regionally based international conference, the International Congress of Medical Physics (ICMP) is held between World Congresses. The IOMP also sponsors international conferences, workshops, and courses.

The IOMP has several programmes to assist medical physicists in developing countries. The joint IOMP Library Programme supports 75 active libraries in 43 developing countries, and the Used Equipment Programme coordinates equipment donations. The Travel Assistance Programme provides a limited number of grants to enable physicists to attend the World Congresses.

The IOMP co-sponsors the *Journal of Applied Clinical Medical Physics*. The IOMP publishes, twice a year, an electronic bulletin, *Medical Physics World*. The IOMP also publishes e-Zine, an electronic newsletter, about six times a year. The IOMP has an agreement with Taylor & Francis for the publication of the *Medical Physics and Biomedical Engineering* series of textbooks. IOMP members receive a discount.

The IOMP collaborates with international organizations, such as the World Health Organization (WHO), the International Atomic Energy Agency (IAEA), and other international professional bodies, such as the International Radiation Protection Association (IRPA) and the International Commission on Radiological Protection (ICRP), to promote the development of medical physics and the safe use of radiation and medical devices.

Guidance on education, training, and professional development of medical physicists is issued by the IOMP, which is collaborating with other professional organizations in the development of a professional certification system for medical physicists that can be implemented on a global basis.

The IOMP website (www.iomp.org) contains information on all the activities of the IOMP, policy statements 1 and 2, and the *IOMP: Review and Way Forward* which outlines all the activities of IOMP and plans for the future.

Preface

ALTHOUGH MANY RADIATION PROTECTION scientists and engineers employ dose coefficients computed from the methodologies presented herein, not many are aware of the details of the origin of those dose coefficients. The methodologies presented in this book are used in the preparation of dose coefficients for regulatory limits for releases from nuclear facilities under normal conditions; the determination of dose limits to members of the public; and to determine emergency response actions when there is a nuclear facility incident (e.g., Fukushima) or a nuclear security incident (e.g., improvised nuclear device or radiological dispersal device). The techniques presented in the book can further be employed to determine radiation doses to people given nuclear medicine treatments and inadvertent exposure to occupation workers.

The book is the first of its kind in over 40 years to address the topic of radiation protection dosimetry in intimate detail, and is intended to form a comprehensive summary of the current state-of-the-art computational dosimetry techniques, with the overarching goal of capturing the high-level knowledge used to generate fundamental radiation protection dosimetry quantities. Topics presented in the scope of this text include advanced radiation dosimetry concepts and regulatory applications considering both external and internal pathways. This book may be seen as a book written at the depth of the Fitzgerald, Brownell, and Mahoney book entitled *Mathematical Theory of Radiation Dosimetry* published by Gordon and Breach Science Publishers in 1967, with the exception that the material is presented in an alternative arrangement and the latest international computational dosimetry methods are elaborated.

The topics are presented in a logical order, rather than in the historical order of the development of dosimetry fundamental concepts and applications. The methods presented are largely based on, or were used, to compute dose coefficients for the latest International Commission on Radiation Protection (ICRP) and International Commission on Radiation Units and Measurements (ICRU) publications and recommendations. It is hoped that these methods are elucidated in more practical terms than may be found in the ICRP/ICRU publications and that the concepts presented are relevant to an international audience. For completeness, we have included a chapter on radiation detection and measurement, which serves to link the computational dosimetry presented to the measurement of dose. Notably, the content in this book contains the most up-to-date computational dosimetry models, where each chapter is authored by an esteemed (if not singular) subject matter expert in that field of study, and consolidated into an edited volume to form this book. The authors investigate both the origins and methodologies of dose coefficient calculations. The book covers all the methods used in modern radiation protection dosimetry and will be of great benefit to the radiation protection community and to graduate radiation protection programs.

Shaheen A. Dewji
College Station, Texas

Nolan E. Hertel
Atlanta, Georgia

Acknowledgments

The editors would like to thank the contributors for their time and commitment in developing the content for this book, as well as the reviewers, who also undertook notable labors to enhance the rigor of the content presented here. The editors would finally like to thank those in the office of Nuclear Regulatory Research, especially Tanya Palmateer-Oxenberg, in the United States Nuclear Regulatory Commission for their support in the actuation of this critical knowledge preservation project.

Additional thanks are given to Liz Dallas and Clay Easterly from Easterly Scientific Inc.; Diane Kosier, Greg Zimmerman, and Michael Johnson from Oak Ridge National Laboratory; and Rebecca Tadesse, Sami Sherbini, and Mohammed Saba from the U.S. Nuclear Regulatory Commission, for their administrative support during the preparation of the manuscript for this book.

Editors

Shaheen A. Dewji, PhD, is an Assistant Professor in the Department of Nuclear Engineering at Texas A&M University and is a Faculty Fellow of the Center for Nuclear Security Science & Policy Initiatives. Having established the Radiological Engineering, Detection, and Dosimetry (RED²) Laboratory at Texas A&M University, Dr. Dewji's research has focused on harnessing both computational capabilities in radiation transport modeling, and experimental measurements using radiation detection for applications in radiation protection, dosimetry, health physics, and nuclear materials accounting. Her research thrust in computational dosimetry investigates the development of dose coefficients using age/sex-specific anthropomorphic computational phantoms and radionuclide biokinetic models for occupational nuclear workers, members of the public, nuclear medicine, space, defense, and emergency response. In her prior role as a Radiological R&D Staff Scientist in the Center for Radiation Protection Knowledge at Oak Ridge National Laboratory, Dr. Dewji's research investigated nuclear medicine patient-release criteria using bespoke biokinetic models of adult patients. Her research also focused on the computation of external dose coefficients in an update of the Environmental Protection Agency's Federal Guidance Report (FGR) series on external exposure to radionuclides in environmental media, resulting in the publication of FGR 15; and in the computation of specific absorbed fraction data for inhalation/ingestion internal dose coefficients (FGR 16). She is the recipient of the 2018 Health Physics Society Elda E. Anderson Award, and the 2015 ORNL Science Serving Society Award.

Dr. Dewji completed her master's and PhD degrees in nuclear and radiological engineering at the Georgia Institute of Technology in Atlanta. She received her bachelor of science in physics from the University of British Columbia.

Nolan E. Hertel, PhD, is a Professor of Nuclear and Radiological Engineering at the Georgia Institute of Technology and holds a Joint Faculty Appointment in the Center for Radiation Protection Knowledge at the Oak Ridge National Laboratory. He received the Distinguished Scientific Achievement Award from the Health Physics Society in 2016 and the Rockwell Lifetime Achievement Award from the Radiation Protection and Shielding Division of the American Nuclear Society in 2018. Dr. Hertel is a recognized expert in radiation protection, shielding, detection, transport, and dosimetry; and has been actively engaged in nuclear engineering education and research as a university professor and private consultant for 39 years. He is a licensed professional engineer in the State of Georgia. He was a co-author of ICRP 116, the EPA Federal Guidance Report 15, and is the co-chair of the ICRU Report Committee on operational quantities. He received his BS and MS in degrees in nuclear engineering from Texas A&M University, and his PhD from the University of Illinois at Urbana-Champaign.

Contributors

 **Wesley E. Bolch** is a Professor of Biomedical Engineering and Medical Physics in the J. Crayton Pruitt Family Department of Biomedical Engineering at the University of Florida (UF). He serves as Director of ALRADS—the Advanced Laboratory for Radiation Dosimetry Studies at UF. He has been certified by the American Board of Health Physics since 1994, and licensed in Radiological Health Engineering by the Texas Board of Professional Engineers since 1992. In 2011, Mr. Bolch was elected fellow of both the Health Physics Society (HPS) and the American Association of Physicists in Medicine (AAPM). He has been a member of the Society of Nuclear Medicine's Medical Internal Radiation Dose (MIRD) Committee since 1993, a member of the National Council on Radiation Protection and Measurements (NCRP) since 2005, a member of Committee 2 of the International Commission on Radiological Protection (ICRP) since 2005, and a member of the U.S. delegation of the United Nations Scientific Committee on the Effects of Atomic Radiation (UNSCEAR) since 2015. He has published over 210 peer-reviewed journal articles, co-authored/edited 16 books/book chapters, and served as author on two NCRP Reports, six ICRP publications, and two MIRD monographs. He is the recipient of the 2014 Distinguished Scientific Achievement Award by the Health Physics Society acknowledging outstanding contributions to the science and technology of radiation safety.

John Ford is an Associate Professor in the Department of Nuclear Engineering at Texas A&M University in College Station, Texas. He has been teaching radiation protection, instrumentation, internal dose assessment, Monte Carlo transport calculations, and radiation biology for over 20 years. Mr. Ford has research experience in radiation biology, health, and medical physics, participating in research projects funded by DOE, NASA, and NSBRI.

Derek Jokisch serves as a Professor of Physics and Chair of the Department of Physics and Engineering at Francis Marion University in Florence, South Carolina. Since 2014, he has held a Joint Faculty Appointment in the Center for Radiation Protection Knowledge at Oak Ridge National Laboratory. He earned his BA degree in nuclear engineering from the University of Illinois, and his master's and doctoral degrees in health physics from the University of Florida. He is currently serving as a member of Committee 2 on Doses from Radiation Exposure for the International Commission on Radiological Protection.

Ronald L. Kathren is a Professor Emeritus and retired Director of the United States Transuranium and Uranium Registries at Washington State University at Tri-Cities (WSUTC). His broad health physics interests include uranium and the transuranium elements, history of the radiological sciences, radiation dosimetry and standards, and environmental radioactivity. He holds degrees from UCLA and the University of Pittsburgh, is a licensed professional engineer and board certified by both the American Board of Health Physics and the American Academy of Environmental Engineers. In addition to serving on numerous national scientific committees and standards working groups, he is author or co-author of nearly 200 papers in the peer-reviewed scientific literature and several books, including *Ionizing Radiation: Tumorigenic and Tumoricidal Effects*, *Radioactivity in the Environment*, *Radiation Protection*, and *The Plutonium Story*. His many honors include both the Elda Anderson and Distinguished Scientific Achievement Awards of the Health Physics Society; the Arthur Humm Award of the National Registry of Radiation Protection Technologists; the Hartman Orator and Medalist of the Radiology Centennial; and he was named to the Distinguished Public Health Alumni of the University of Pittsburgh. He is a past president of both the Health Physics Society and the American Academy of Health Physics. He helped to start the historical Radiological and Allied Sciences book collection at WSUTC with a donation of his personal library of 3400 volumes and endowed a scholarship for science and history of science students at that institution. He and his wife Susan are the parents of two children and reside in Richland, Washington.

Rich Leggett is a Research Scientist in the Environmental Sciences Division at Oak Ridge National Laboratory (ORNL). He received his PhD in mathematics from the University of Kentucky in 1972, and taught mathematics at Ruhr University in Bochum, Germany, and the University of Tennessee before joining the Health Physics Division at ORNL in 1976. In 1979, he published a foundational mathematical theorem that remains one of the most frequently cited tools for establishing the existence of multiple solutions of boundary value problems arising in biology, chemistry, and physics. Since the early 1980s, his main research interest has been physiological systems modeling, with applications to the biokinetics, dosimetry, and risk analysis of radionuclides and chemical toxins. He is a member of Committee 2 of the International Commission on Radiological Protection (ICRP) and the ICRP Task Group on Internal Dose Coefficients (IDC). His physiological systems models of the human circulation, skeleton, and alimentary transport, and his biokinetic models for over 50 elements have been adopted by the ICRP as international standards for derivation of radiation dose estimates and interpretation of bioassay data. He is the author of ICRP Publication 70, *Basic Anatomical and Physiological Data for Use in Radiological Protection: The Skeleton*, and co-author of a number of other ICRP reports, including a series on doses to members of the public from intake of radionuclides (1989–1996), the Reference Person document (2002), and the Human Alimentary Tract Model (2006). In 1995, he was named ORNL Author of the Year for the paper, "An Age-Specific Kinetic Model of Lead Metabolism in Humans."

Joseph C.McDonald is an Emeritus Laboratory Fellow, retired from the Pacific Northwest National Laboratory. He has worked for 50 years in the field of radiological physics, and has served on national and international standards writing committees. His work includes 150 publications, scientific society awards, patents, and university level teaching. He is a Fellow of the American Physical Society, and a Fellow of the Health Physics Society, and was Editor-in-Chief of the journal *Radiation Protection Dosimetry*. Mr. McDonald served as president of the Health Physics Society's Accelerator Section, president of the Council on Ionizing Radiation Measurements and Standards (CIRMS), and as vice-chairman of the Panel on Reference Nuclear Data for the National Nuclear Data Center. He served on report committees for the International Electrotechnical Commission, the American National Standards Institute (ANSI), and the Health Physics Society Standards Committee. He was chairman of the International Commission on Radiation Measurements and Units (ICRU) Report 76, *Measurement Quality Assurance for Ionizing Radiation Dosimetry*, a member of the ICRU Committee for Report 66, *Determination of Operational Dose Equivalent Quantities for Neutrons*, and Chairman of the International Organization for Standardization (ISO) Working Group on Neutron Reference Radiations. He was co-author of a book entitled *Dosimetry for Radiation*

Processing. He received an R&D 100 Award and holds patents for the development of Optically Stimulated Luminescent Dosimetry. Mr. McDonald has also served on the faculty of the Cornell University Medical College, the University of California at Los Angeles, Washington State University, Tri-Cities, and the Columbia Basin College. He currently resides in Pasco, Washington.

David McLaughlin has served the majority of his 30+ year professional career as a practicing internal dosimetrist for UT-Battelle at the Oak Ridge National Laboratory. Mr. McLaughlin is a graduate of the University of Lowell holding both a BS in radiological health physics, and an MS in radiological sciences and protection. He is certified by the American Board of Health Physics and has served on the American Academy of Health Physics panel of examiners for the Part II certification exam. He has served on various ANSI working groups, was an advisor to the Department of Energy Expert Group on Internal Dosimetry, and has collaborated with the Radiation Emergency Assistance Center/Training Site (REAC/TS) following radiological events.

David J. Pawel a member of the (U.S.) National Council on Radiation Protection and Measurements, is an expert on assessing health risks from environmental exposures to ionizing radiation. At the EPA, his responsibilities include the evaluation of radioepidemiological literature and implementation of the Agency's radiogenic cancer risk models, and he is co-author of journal articles and technical reports on EPA radiation risk models and projections for the U.S. population. Prior experience includes the evaluation of radiation risks to the atomic bomb survivors as a research scientist at the Radiation Effects Research Foundation (RERF) in Japan from 1992–1994. In 2003, Mr. Pawel was awarded a Gilbert W. Beebe Fellowship, which enabled him to revisit RERF to study methods to improve cancer-specific radiogenic risk estimates. He is a recipient of EPA's Science Achievement Award for his work on risk assessment methodology. Dr. Pawel served as co-chair for the Conference on Radiation and Health, a highly acclaimed biennial meeting which brings together researchers of various disciplines to discuss their latest findings on radiogenic health effects. He has also served as a member of the U.S. delegation to the United Nations Scientific Committee on the Effects of Atomic Radiation (UNSCEAR).

John W. Poston, Sr., is a Professor of nuclear engineering at Texas A&M University and an Associate Director of the Nuclear Power Institute. He is a BS graduate of Lynchburg College in Lynchburg, Virginia, and received the MS degree and a PhD degree in nuclear engineering from the Georgia Institute of Technology in Atlanta. Before coming to Texas A&M, he was employed as an experimental reactor physicist with Babcock & Wilcox in Lynchburg, a health physicist and section head in the Health Physics Division at Oak Ridge National Laboratory, Tennessee, and an associate professor at Georgia Tech. His research and interests include radiation detection, internal dosimetry, mathematical phantoms, and, more recently, response to terrorist activities involving radiation/radioactivity. He is an elected fellow of the Health Physics Society, the American Nuclear Society, and the American Association for the Advancement of Science. In addition, he served as vice president of the National Council on Radiation Protection and Measurements (NCRP) for seven years and was elected a distinguished emeritus member of the NCRP in 2002.

Ken G. Veinot is the Senior Health Physicist at the Y-12 National Security Complex in Oak Ridge, Tennessee. After receiving his doctorate from the Georgia Institute of Technology, he began his career at Y-12 working in both the internal and external dosimetry programs. He was the operations manager of the TLD processing center and the technical lead for the external dosimetry program. Following this, he served as the instrumentation and technical programs lead, and now functions as the senior HP for the Radiological Control program providing technical guidance to all areas of health physics at the site. He has been involved with the U.S. Department of Energy Laboratory Accreditation Program for over 15 years, where he serves as a lead assessor. He also works for the Center for Radiation Protection Knowledge at the Oak Ridge National Laboratory, has authored or co-authored a number of peer-reviewed articles, and has served on ICRP and ICRU committees. He received his certification by the American Board of Health Physics in 2004. He and his wife are parents of a son and daughter and live in the Knoxville, Tennessee area.

External Reviewers

Michael Bellamy, Oak Ridge National Laboratory

Mike Boyd, U.S. Environmental Protection Agency

Keith Eckerman, Oak Ridge National Laboratory (Ret.)

Mauritius Hiller, Oak Ridge National Laboratory

Vincent Holahan, U.S. Nuclear Regulatory Commission

Phil Jalbert, U.S. Environmental Protection Agency

Cynthia Jones, U.S. Nuclear Regulatory Commission

Choonsik Lee, National Cancer Institute, National Institutes of Health

Ryan Manger, University of California, San Diego

Minh-Thuy Nguyen, U.S. Nuclear Regulatory Commission

Cailin O'Connell, Texas A&M University

Tanya Palmateer-Oxenberg, U.S. Nuclear Regulatory Commission

Charles Potter, Sandia National Laboratories

Jerome Puskin, U.S. Environmental Protection Agency (Ret.)

Casper Sun, U.S. Nuclear Regulatory Commission

Jon Walsh, U.S. Environmental Protection Agency

CHAPTER 1

Introduction

Shaheen A. Dewji and Nolan E. Hertel

CONTENTS

1.1	Regulation of Radiation Dose	2
	1.1.1 Reports of the ICRU	2
	1.1.2 Reports of the ICRP	3
	1.1.3 Reports and Commentaries of the NCRP	3
	1.1.4 U.S. Regulations	3
	1.1.4.1 Environmental Protection Agency (EPA)	3
	1.1.4.2 Nuclear Regulatory Commission (NRC)	4
	1.1.4.3 Standards and Guidelines	4
	1.1.5 International Committees and Organizations	4
	1.1.5.1 Other Governmental and Non-Governmental Organizations	5
1.2	Radiation Protection Professional Societies	5
1.3	Synopsis of the Book	5
	References	7

THE ABILITY TO ACCURATELY quantify radiation and its potential health effects remains the driver for ensuring its safe and secure use of nuclear technologies. Managing the benefits and detriments of radiation dates back to the discovery of X-rays by Wilhelm Röntgen, whose work using penetrating radiation on his wife's hand underscored the effects of radiation on the human body (Röntgen 1895, 1896). Regulation has since evolved for protecting occupational workers in nuclear facilities or handling nuclear material, as well as exposure limits for members of the public pertaining to the nuclear fuel cycle, nuclear medicine, emergency response, national defense, and space exploration.

The consequences of radiation exposure have led to the development of the scientific field of radiation dosimetry. Radiation dosimetry addresses how ionizing radiation interacts with matter (i.e., within tissues and organs of the human body) and the effects of energy deposited. Ionizing radiation is characterized by the ability to excite and ionize interacting atoms in matter. The ionization energy required to cause a valence electron to escape from an atom, hence causing ionization, ranges from 4 to 25 eV, thus requiring energies in excess of this range to be classified as ionizing radiation. Damage caused by ionizing radiation is the damage that could occur by the incident wave (i.e., X-ray, gamma rays) or particle (alpha, beta, neutron) breaking up a molecule. In the context of radiation dosimetry, the target molecule of primary interest is deoxyribonucleic acid (DNA) in exposed organs or tissues. If the organ or tissue region is small and the energy deposited by ionizing radiation is low, then the risk of delayed effects, such as cancer, arguably remains low; however, if damage accumulates successively over a prolonged period of time (i.e., chronic exposure), or if a high-energy field interacts with tissues in a short period of time (i.e., acute exposure), cancerous and non-cancerous effects can be more serious. Radiation effects can be classified as deterministic (e.g., cataracts, radiation sickness) or stochastic (e.g., hereditary, cancer, non-cancer).

1.1 REGULATION OF RADIATION DOSE

The philosophy for dose regulation and radiation protection has revolved around the principle of "As Low as Reasonably Achievable" (ALARA). In keeping with the ALARA principle, the benefits of activities involving radiation must balance the risk of detrimental effects. The growth of the use of ionizing radiation has necessitated the creation of organizations and standards for developing the scientific and technical basis of the safe application of ionizing radiation. The International Commission on Radiation Units and Measurements (ICRU) established in 1925, and the International Commission on Radiological Protection (ICRP) established shortly thereafter in 1928 are the core institutions that provide recommendations to the international community on radiation protection, with the fundamental models in internal and external dosimetry informing the guidance. Within the United States, the National Council on Radiation Protection and Measurements (NCRP) was chartered by the U.S. Congress in 1964 to collect, analyze, develop, and disseminate information and recommendations in the public interest regarding radiation protection and measurements (NCRP 2015). Federal government regulations and guidance in the United States are provided primarily by the U.S. Nuclear Regulatory Commission and the U.S. Environmental Protection Agency, each of which reflects to varying degrees some level of scientific application, radiation protection guidance, and policy recommendations from the ICRP/ICRU/NCRP regarding exposure limits.

Recommendations relevant to radiation doses for human exposure to ionizing radiation, are provided in a plurality of reports, regulations, and standards, for the interested reader.*

1.1.1 Reports of the ICRU

The International Commission on Radiation Units and Measurements (ICRU) establishes international standards for radiation units and measurement (accessible via: https://icru.org).

- ICRU Report 57/ICRP Publication 74: Conversion Coefficients for Use in Radiological Protection Against External Radiation (1998)

- ICRU Report 85a: Fundamental Quantities and Units for Ionizing Radiation (2011)

---

* The list provided is not exhaustive, but is current at the time of publication of this work. Interested readers are encouraged to consult the latest ICRP, ICRU, and NCRP for the latest publications and recommendations.

1.1.2 Reports of the ICRP

The International Commission on Radiological Protection (ICRP) serves as the world-wide organization that forms the basis for radiological protection standards, legislation, guidelines, programs, and practice (accessible via: http://www.icrp.org).

- ICRP Publication 89: Basic Anatomical and Physiological Data for Use in Radiological Protection—Reference Values (2002)

- ICRP Publication 103: Recommendations of the International Commission on Radiological Protection (2007) [as an update to ICRP Publication 30 (1979), (1980), (1981), (1988), and ICRP Publication 60 (1991)]

- ICRP Publication 107: Nuclear Decay Data for Dosimetric Calculations (2008)

- ICRP Publication 110: Adult Reference Computational Phantoms (2009)

- ICRP Publication 116: Conversion Coefficients for Radiological Protection Quantities for External Radiation Exposures (2010)

- ICRP Publication 133: The ICRP Computational Framework for Internal Dose Assessment for Reference Adults: Specific Absorbed Fractions (2016a)

- ICRP Publication 130/134/137: Occupational Intakes of Radionuclides: Parts 1–3 (2015), (2016b), (2018)

1.1.3 Reports and Commentaries of the NCRP

The National Council on Radiation Protection and Measurements (NCRP) supports radiation protection in the U.S. by providing independent scientific analysis, information, and recommendations that represent the consensus of leading scientists (accessible via: https://ncrponline.org).

- NCRP Report 116: Limitation of Exposure to Ionizing Radiation (1993)

- NCRP Statement 10: Recent Applications of the NCRP Public Dose Limit Recommendation for Ionizing Radiation (2004)

1.1.4 U.S. Regulations

1.1.4.1 Environmental Protection Agency (EPA)

The EPA Federal Guidance Report (FGR) series is employed by federal and state agencies in developing radiation protection regulations and standards to protect the American public from harmful effects of radiation (accessible via: https://www.epa.gov/radiation/federal-guidance-radiation-protection).

- Federal Guidance Report 13: Cancer Risk Coefficients for Environmental Exposure to Radionuclides: Updates and Supplements (Eckerman et al. 1999)

- Federal Guidance Report 15: External Exposure to Radionuclides in Air, Water and Soil (Bellamy et al. 2018) [as an update to Federal Guidance Report 12 (Eckerman and Ryman 1993)]*

1.1.4.2 Nuclear Regulatory Commission (NRC)

Title 10 of the Code of Federal Regulation (CFR) details the requirements binding on all persons and organizations who receive a license from NRC to use nuclear materials or operate nuclear facilities† (accessible via: https://www.nrc.gov/reading-rm/doc-collections/cfr/).

- Code of Federal Regulations, Title 10, Part 20 (10CFR20): Standards for Protection Against Radiation (U.S. Nuclear Regulatory Commission 1993)
- Code of Federal Regulations, Title 10, Part 50 (10CFR50): Domestic Licensing of Production and Utilization Facilities (U.S. Nuclear Regulatory Commission 2004)

1.1.4.3 Standards and Guidelines

- The American National Standards Institute (ANSI) is the premier source for timely, relevant, and actionable information on national, regional, and international standards, including publications on radiation sources, detectors, instrumentation, and operations (accessible via: https://www.ansi.org).

1.1.5 International Committees and Organizations

A multitude of organizations—national, international, and non-governmental—are active in radiation protection and dosimetry activities and regulation. For the reader's benefit, we list those frequently referenced in this book; we include additional organizations in Section 1.1.5.1.

- The Committee on Medical Internal Radiation Dose (MIRD) Committee was established under the Society of Nuclear Medicine and Molecular Imaging to develop standard methods, models, assumptions, and mathematical schema for assessing internal radiation doses from administered radiopharmaceuticals (accessible via: http://www.snmmi.org).
- The United Nations Scientific Committee on the Effects of Atomic Radiation (UNSCEAR), was established in 1955 by the General Assembly of the United Nations with the mandate to assess and report levels and effects of exposure to ionizing radiation. Governments and organizations consult with the Committee's findings as the scientific basis for evaluating radiation risk and for establishing protection guidelines (accessible via: http://www.unscear.org).

---

* Recommendations under FGR15 are based primarily on recommendations of ICRP Publication 103 (2007).
† Recommendations under 10CFR20 are primarily based on recommendations of ICRP Publication 26 (1977).

- The International Atomic Energy Agency (IAEA) is the world's central intergovernmental forum for scientific and technical co-operation in the nuclear field. Its objectives focus on the safe, secure, and peaceful uses of nuclear science and technology, contributing to international peace and security under the United Nations (accessible via: https://www.iaea.org).

1.1.5.1 Other Governmental and Non-Governmental Organizations

- There are a number of other international and organizations in various countries; a subset of these organizations are listed here for their contributions to recent international recommendations non-governmental:

  - The CONCERT-European Joint Program for the Integration of Radiation Protection Research operates as an umbrella structure for the research initiatives jointly launched by the radiation protection research platforms MELODI, ALLIANCE, NERIS, and EURADOS in the European Union (accessible via: http://www.concert-h2020.eu).

  - The Japanese Atomic Energy Agency national nuclear regulatory agency in Japan (accessible via: https://www.jaea.go.jp/english/).

1.2 RADIATION PROTECTION PROFESSIONAL SOCIETIES

Readers are strongly encouraged to investigate and engage in the activities of professional societies engaged in radiation protection and dosimetry efforts. A plurality of professional societies and organizations undertake key activities in the development of radiation protection technical and policy recommendations including, but not limited to, the following:

- Health Physics Society (HPS)—accessible via: http://hps.org/

- American Academy of Health Physics (AAHP)—accessible via: https://www.hps1.org/aahp/

- American Nuclear Society (ANS)—accessible via: http://www.ans.org

- Radiation Research Society (RRS)—accessible via: https://www.radres.org

- International Radiation Protection Association (IRPA) and its member organizations—accessible via: http://www.irpa.net

1.3 SYNOPSIS OF THE BOOK

This book elaborates on foundational concepts in radiation protection and dosimetry, focusing on the historical evolution of regulation and guidance, scientific models in radiation dosimetry, radiation measurement of exposure and uptake, and applications of these models in evaluating radiation exposure/uptake risk.

Extensive, but not exclusive, use is made of the International System of Units (SI) in this book, due to the historical use of traditional units in the United States. Readers are encouraged to review the challenges and recommendations associated with the adoption of SI in the workshop proceedings published by the National Academies of Sciences, Engineering, and Medicine entitled "Adopting the International System of Units for Radiation Measurements in the United States: Proceedings of a Workshop" (2017).

This book commences with a discussion of fundamental physics concepts and definitions in radiation protection and dosimetry in Chapter 2. An overview of quantities and units begins the chapter, followed by reviews of atomic structure, radioactive decay, a condensed history of atomic models and their development, interaction of radiation with matter, dosimetric terminology, and finally a summary of radiation protection quantities, including operational and protection quantities. This chapter is intended to provide introductory background information for the following chapters and can be used as a reference for all radiation protection practitioners and students.

The book takes a novel and in-depth historical review in Chapter 3, which traces, in linear historical fashion, the key events that led to the adoption of radiation protection guidance, beginning with the discovery of X-rays and radioactivity. Early efforts at protective guidance were initially delayed by the belief that radiation was not harmful, and thus protection was not required. Recognition of the hazard, drawing upon experience of the radium dial painters and the Manhattan Project, coupled with the definition and acceptance of a unit for radiation exposure, facilitated the formation of protective guidance, both nationally and internationally by scientific committees. Discussion of transition from a dose-based to a risk-based model of radiation regulation, that is, the inception of the linear no-threshold philosophy, is investigated in a historical context.

Devices used for the detection and measurement of external sources of ionizing radiation, along with methods for their calibration and testing, are described in Chapter 4. The two basic classes of radiation detection and measurement devices—active (e.g., ion-chamber, proportional counters, Geiger–Müller counters, and scintillation detectors) and passive (e.g., thermoluminescent detectors) powered devices are described. The chapter concludes with a discussion on radiological calibrations that are traceable to national standards for radiation detection and measurement devices, and personal dosimeters. This chapter provides an foundational snapshot of measurement devices that provide a link to the remainder of the book which largely addresses computational dosimetry techniques.

The discussion following in Chapters 5–8 addresses the scientific models in radiation dosimetry employing reference phantoms and biokinetic models, leading to dosimetric models and the methods used in the computation of dose coefficients. Chapter 5 focuses on defining the history, concepts, and practical applications of the Reference Individual, as defined by the ICRP. Reference individuals discussed include the newborn, 1-year-old, 5-year-old, 10-year-old, 15-year-old, and adults, for both males and females. This is followed by a discussion of the anatomical aspects of the ICRP Reference Individual in the context of stylized, voxel, and hybrid forms of computational phantom models. This discussion further addresses physiological aspects of the ICRP Reference Individual, focusing on metabolic rates. Comparisons are drawn to parallel efforts to define reference individuals in populations outside the ICRP definition.

The evolution of the physiological and anatomical models prepare the reader for a subsequent discussion on the models of radionuclide inhalation, ingestion, and systemic biokinetics, which are addressed next in Chapter 6. Biokinetic models are used to predict the time-dependent distribution, retention, and excretion of substances that enter the body through inhalation, ingestion, wounds, intact skin, or direct injection into blood. This chapter reviews the history of biokinetic models used in radiation protection or nuclear medicine to predict the time-dependent behavior of radionuclides in the human body. Further discussion follows on the ICRP's latest biokinetic modeling system for workers and members of the public involving recycling models for all radionuclides and increased realism in the treatment of radioactive progeny produced in the body by radioactive decay.

The development of methods and mathematical models used for the calculation of absorbed dose in human tissues due to internal or external radiation exposures is addressed in Chapter 7. The sources of nuclear decay data, development of the source and target tissue concepts, and the methods used to calculate the absorbed fraction of energy in a particular tissue due to a range of radiations are discussed.

Chapter 8 integrates the models of Chapters 5–7 in the discussion of computational dosimetry approaches to generate dose coefficients for both external and internal dosimetry. The general concept of a dose coefficient for converting fluence to dose is discussed, followed by a brief presentation of the radiation transport methods that have been used to compute dose coefficients. The methods used to produce dose coefficients for ingestion and inhalation are covered with discussions of special quantities used in that approach, followed by a discussion of the computation of dose coefficients for external irradiation environmental fields. Examples of dose coefficients are provided for five different intakes of radionuclides.

The final sections of the book in Chapters 9 and 10 focus on the applications of the models discussed in Chapters 5–8. The evaluation of cancer risk from exposure to individual radionuclides is discussed in Chapter 9 for internal and external exposures. Chapter 9 provides details on the application of cancer risk coefficients, their limitations, and how they are computed for a given population and exposure pathway. The last section of Chapter 9 includes examples to provide further insight on the proper use of risk coefficients.

The final section of the book, Chapter 10, focuses on the theory and practice of interpreting biokinetic models introduced in Chapter 6 through the use of bioassay measurements to estimate the intake of radionuclides following known or suspected intakes. Mathematical solutions for acute and chronic exposures are developed for both open and closed multicompartmental catenary linked systems. The effect of recycling between compartments is also examined. The concepts of retention and excretion functions (and fractions) are introduced and various bioassay fitting techniques are discussed. Four case studies are provided to facilitate application of the models and enlighten the reader.

REFERENCES

Bellamy, M. B., S. A. Dewji, R. W. Leggett, M. M. Hiller, K. G. Veinot, R. P. Manger, K. F. Eckerman et al. 2018. *Federal Guidance Report No. 15: External Exposure to Radionuclides in Air, Water, and Soil*. Washington, DC: U.S. Environmental Protection Agency.

Eckerman, K. F., R. W. Leggett, C. B. Nelson, J. S. Pushkin, and A. C. B. Richardson. 1999. *Federal Guidance Report No. 13: Cancer Risk Coefficients for Environmental Exposure to Radionuclides.* Environmental Protection Agency (ed). Washington, DC: U.S. Environmental Protection Agency.

Eckerman, K. F., and J. C. Ryman. 1993. *Federal Guidance Report No. 12: External Exposure to Radionuclides in Air, Water, and Soil.* Washington, DC: U.S. Environmental Protection Agency.

International Commission on Radiation Units and Measurements. 2011. "ICRU Report 85a: Fundamental Quantities and Units for Ionizing Radiation." *J ICRU* 11 (11): 1–137.

International Commission on Radiation Units and Measurements. 1998. *ICRU Report 57: Conversion Coefficients for Use in Radiological Protection against External Radiation.* Bethesda, MD: International Commission on Radiation Units and Measurements.

International Commission on Radiological Protection. 1977. ICRP Publication 26: Recommendations of the ICRP. *Ann ICRP* 1 (3).

International Commission on Radiological Protection. 1979. ICRP Publication 30 (Part 1): Limits for Intakes of Radionuclides by Workers. *Ann ICRP* 2 (3–4).

International Commission on Radiological Protection. 1980. ICRP Publication 30 (Part 2): Limits for Intakes of Radionuclides by Workers. *Ann ICRP* 4 (3–4).

International Commission on Radiological Protection. 1981. ICRP Publication 30 (Part 3): Limits for Intakes of Radionuclides by Workers. *Ann ICRP* 6 (2–3).

International Commission on Radiological Protection. 1988. ICRP Publication 30 (Part 4): Limits for Intakes of Radionuclides by Workers. *Ann ICRP* 19 (4).

International Commission on Radiological Protection. 1991. "ICRP Publication 60: 1990 Recommendations of the International Commission on Radiological Protection." *Ann ICRP* 21 (1–3):1–167.

International Commission on Radiological Protection. 2002. "ICRP Publication 89: Basic Anatomical and Physiological Data for Use in Radiological Protection—Reference Values." *Ann ICRP* 32 (3–4):1–277.

International Commission on Radiological Protection. 2007. "ICRP Publication 103: The 2007 Recommendations of the International Commission on Radiological Protection." *Ann ICRP* 37 (2–4).

International Commission on Radiological Protection. 2008. "ICRP Publication 107: Nuclear Decay Data for Dosimetric Calculations." *Ann ICRP* 38 (3):1–26.

International Commission on Radiological Protection. 2009. "ICRP Publication 110: Adult Reference Computational Phantoms." *Ann ICRP* 39 (2):1–165.

International Commission on Radiological Protection. 2010. "ICRP Publication 116: Conversion Coefficients for Radiological Protection Quantities for External Radiation Exposures." *Ann ICRP* 40 (2–5):1–257.

International Commission on Radiological Protection. 2015. "ICRP Publication 130: Occupational Intakes of Radionuclides: Part 1." *Ann ICRP* 44 (2):1–188.

International Commission on Radiological Protection. 2016a. "ICRP Publication 133: The ICRP Computational Framework for Internal Dose Assessment for Reference Adults: Specific Absorbed Fractions." *Ann ICRP* 45 (2):1–74.

International Commission on Radiological Protection. 2016b. "ICRP Publication 134: Occupational Intakes of Radionuclides: Part 2." *Ann ICRP* 45 (3–4):1–352.

International Commission on Radiological Protection. 2018. "ICRP Publication 137: Occupational Intakes of Radionuclides: Part 3." *Ann ICRP* 46 (3–4):1–486.

National Academies of Sciences, Engineering, and Medicine. 2017. *Adopting the International System of Units for Radiation Measurements in the United States: Proceedings of a Workshop.* Washington, DC: National Academies Press.

National Council on Radiation Protection and Measurements. 1993. *NCRP Report 116: Limitation of Exposure to Ionizing Radiation.* Bethesda, MD: NCRP.

National Council on Radiation Protection and Measurements. 2004. *NCRP Statement 10: Recent Applications of the NCRP Public Dose Limit for Ionizing Radiation.* Bethesda, MD: NCRP.

National Council on Radiation Protection and Measurements. 2015. "NCRP Mission". Accessed August 2017. https://icru.org/icru-at-a-glance-pdf/uncategorised/icru-at-a-glance.

Röntgen, Wilhelm Conrad. 1895. 'Über eine neue Art von Strahlen' *Sitsungsberichte der Physikalisch-medicinischen Gesellschaft zu Würzburg* 9: 132–141.

Röntgen, Wilhelm Conrad. 1896. "On a New Kind of Rays." *Science* 3 (59): 227–231.

U.S. Nuclear Regulatory Commission. 1993. *Code of Federal Regulations Title 10, Part 20: Standards for Protection against Radiation.* Washington, DC: Government Printing Office.

U.S. Nuclear Regulatory Commission. 2004. *Code of Federal Regulations Title 10, Part 50—Domestic Licensing of Production and Utilization Facilities.* Washington, DC: Government Publishing Office.

CHAPTER 2

Fundamental Concepts and Quantities

Ken G. Veinot

CONTENTS

2.1 International Standard Units   13
    2.1.1 Quantities and Units   13
    2.1.2 The International System of Units (SI) and the Corresponding System of Quantities   13
    2.1.3 SI Base Units   14
       2.1.3.1 Units with Special Names and Symbols   14
    2.1.4 Traditional Units for Radiation Protection   15
2.2 Atomic Structure   16
    2.2.1 Proton   16
    2.2.2 Neutron   16
    2.2.3 Electron   16
    2.2.4 The Nucleus   17
       2.2.4.1 Binding Energy   17
       2.2.4.2 Mass Defect   18
       2.2.4.3 Binding Energy per Nucleon   18
    2.2.5 Liquid Drop Model   21
    2.2.6 Electron Orbital Structure   22
       2.2.6.1 Bohr Model   22
       2.2.6.2 Sommerfeld Model   25
    2.2.7 Excitation   25
    2.2.8 Ionization   25
2.3 Nuclear Reactions   26
    2.3.1 Absorption   26
    2.3.2 Fission   27
    2.3.3 Fusion   30
    2.3.4 Elastic Scatter   30
    2.3.5 Inelastic Scatter   31

2.4	Radioactive Decay	31
	2.4.1 Alpha Decay	31
	2.4.2 Beta Decay	34
	2.4.3 Positron Decay	35
	2.4.4 Electron Capture	35
	2.4.5 Internal Conversion	36
	2.4.6 Spontaneous Fission	36
	2.4.7 Gamma Emissions	37
	2.4.8 Isomeric Transitions (Metastable Energy States)	37
	2.4.9 Radioactive Decay Law	37
	2.4.10 Radioactive Half-Life and Decay Constant	38
	2.4.11 Production and Decay	39
	2.4.12 Specific Activity	40
	2.4.13 Serial Decay	40
	2.4.14 Secular Equilibrium	42
	2.4.15 Transient Equilibrium	43
	2.4.16 Branching Ratios	43
2.5	Interaction of Radiation with Matter	44
	2.5.1 Electrons	44
	2.5.1.1 Soft Collisions	44
	2.5.1.2 Hard Collisions	44
	2.5.1.3 Bremsstrahlung	45
	2.5.2 Photons	45
	2.5.2.1 Photoelectric Effect	46
	2.5.2.2 Compton Scatter	47
	2.5.2.3 Pair Production	48
	2.5.2.4 Photonuclear Reactions	49
	2.5.3 Neutrons	49
	2.5.3.1 Cross Sections	49
	2.5.3.2 Absorption	50
	2.5.3.3 Elastic Scatter	51
	2.5.3.4 Inelastic Scatter	51
	2.5.4 Heavy Charged Particles	52
	2.5.5 Range	52
	2.5.6 Attenuation Coefficients	52
	2.5.7 Mass Energy Transfer Coefficient	53
	2.5.8 Mass Energy Absorption Coefficient	54
2.6	Radiation Protection Quantities and Units	54
	2.6.1 Kerma	54
	2.6.2 Absorbed Dose	56
	2.6.2.1 Charged Particle Equilibrium	57
	2.6.3 Exposure–Dose Relationship	57
	2.6.4 Linear Energy Transfer and Quality Factor	59

2.7   Protection Quantities   61
    2.7.1   ICRP Publication 26   62
    2.7.2   ICRP Publication 60   63
    2.7.3   ICRP Publication 103   64
2.8   Operational Quantities   67
    2.8.1   Ambient Dose Equivalent   67
    2.8.2   Personal Dose Equivalent   68
References   70
Definitions   71

A‌LTHOUGH IT IS ASSUMED that the reader is familiar with the concepts of the atom, radioactive decay, and other radiation protection fundamentals, a few important points of interest are covered in this chapter. It provides an introduction to units and quantities, gives brief descriptions of nuclear and atomic principles, describes the development of various atomic models, the mechanisms and mathematics of radioactive decay, and the interaction of radiation with matter, and concludes by covering the principles of radiation dosimetry, including some of the various quantities that may be of interest.

2.1 INTERNATIONAL STANDARD UNITS

The modern metric system of measurement is the International System of Units (SI). The International Bureau of Weights and Measures (BIPM) was established in 1875 with the task of ensuring the unification of measurements via the fundamental standards and scales used for the measurement of the principal physical quantities, as well as to maintain international prototypes. The BIPM publishes an SI brochure (Bureau International des Poids et Mesures (BIPM) 2006) that describes and defines the various quantities and units used in the SI. The National Institute of Standards and Technology (NIST) is tasked with implementing the SI system in the United States, and NIST Special Publication (SP) 330 (Taylor and Thompson 2008) is the U.S. version of the BIPM SI brochure.

2.1.1 Quantities and Units

In order to be meaningful, the value of a quantity must be expressed using both a value and a unit. The unit provides a reference and magnitude for the value. For a particular quantity, various units may be used, thus it is imperative that a consistent set of reference units be defined. Units should be chosen so that they are readily available, constant, and easy to realize with high accuracy. It is convenient to choose definitions for a small number of units termed "base units," and then to define units for all other quantities that can be derived from these base units. Measurements of quantities should be traceable to a national or international standard.

2.1.2 The International System of Units (SI) and the Corresponding System of Quantities

The base quantities used in the SI are length, mass, time, electric current, thermodynamic temperature, amount of substance, and luminous intensity. The corresponding base units are the meter, the kilogram, the second, the ampere, the kelvin, the mole, and the candela.

2.1.3 SI Base Units

Physical quantities are organized in a system of dimensions, with each of the seven base quantities used in the SI having its own dimension. All other quantities are derived quantities: they can be written in terms of the base quantities, and their dimensions are products of powers of the dimensions of the base quantities.

The formal definitions of the SI base units are (Taylor and Thompson 2008):

- Unit of length (meter): The meter is the length of the path travelled by light in a vacuum during a time interval of $1/299,792,458$ of a second. The original international prototype of the meter, constructed of a platinum-iridium alloy, is maintained at the BIPM under conditions specified in 1889 when it was originally adopted.

- Unit of mass (kilogram): The international prototype of the kilogram, a rod made of a platinum-iridium alloy, is kept at the BIPM under specific conditions and the kilogram is equal to the mass of this standard. Because of the accumulation of contaminants on its surface, the reference mass is taken to be the mass of the standard immediately after cleaning.

- Unit of time (second): The unit of time, the second, is the duration of 9,192,631,770 periods of the radiation corresponding to the transition between two hyperfine levels of the ground state of the ^{133}Cs atom at rest at a temperature of 0 K.

- Unit of electric current (ampere): The ampere is defined as that constant current which, if maintained in two straight parallel conductors of infinite length, of negligible circular cross section, and placed one meter apart in vacuum, would produce between these conductors a force equal to 2×10^{-7} Newton per meter of length.

- Unit of thermodynamic temperature (kelvin): The kelvin, the unit of thermodynamic temperature, is the fraction $1/273.16$ of the thermodynamic temperature of the triple point of reference water.

- Unit of amount of substance (mole): The unit of amount of substance is called the mole and is defined by specifying the mass of carbon 12 that constitutes one mole of carbon 12 atoms. By international agreement this was fixed at 0.012 kg, i.e., 12 g.

- Unit of luminous intensity (candela): The unit of luminous intensity, the candela, is the luminous intensity, in a given direction, of a source that emits monochromatic radiation of frequency 540×10^{12} hertz and that has a radiant intensity in that direction of $1/683$ watt per steradian. The base quantities and units of the SI, and their symbols, are listed in Table 2.1 along with the unit name and unit symbol.

2.1.3.1 Units with Special Names and Symbols

Derived units are products of powers of base units. Coherent derived units are products of powers of base units that include no numerical factor other than one. The base and

TABLE 2.1   SI Base Units

Base Quantity		SI Base Unit	
Name	Symbol	Name	Symbol
length	l; x; r, etc.	meter	m
mass	m	kilogram	kg
time, duration	t	second	s
electric current	I; i	ampere	A
thermodynamic temperature	T	kelvin	K
amount of substance	n	mole	mol
luminous intensity	Iv	candela	cd

TABLE 2.2   Example Derived Units in the SI with Special Names and Symbols

	SI Coherent Derived Unit			
Derived Quantity	Name	Symbol	Expressed in Terms of Other SI Units	Expressed in Terms of SI Base Units
Pressure, stress	Pascal	Pa	$N\,m^{-2}$	$m^{-1}\,kg\,s^{-2}$
Energy, work	Joule	J	$N\,m$	$m^2\,kg\,s^{-2}$
Power, radiant flux	Watt	W	$J\,s^{-1}$	$m^2\,kg\,s^{-3}$
Electric charge	Coulomb	C		$s\,A$
Radionuclide activity	Becquerel	Bq		s^{-1}
Absorbed dose	Gray	Gy	$J\,kg^{-1}$	$m^2\,s^{-2}$
Energy imparted	Gray	Gy	$J\,kg^{-1}$	$m^2\,s^{-2}$
Kerma	Gray	Gy	$J\,kg^{-1}$	$m^2\,s^{-2}$
Dose equivalent	Sievert	Sv	$J\,kg^{-1}$	$m^2\,s^{-2}$
Ambient dose equivalent	Sievert	Sv	$J\,kg^{-1}$	$m^2\,s^{-2}$
Directional dose equivalent	Sievert	Sv	$J\,kg^{-1}$	$m^2\,s^{-2}$
Personal dose equivalent	Sievert	Sv	$J\,kg^{-1}$	$m^2\,s^{-2}$

coherent derived units of the SI form a coherent set designated the set of coherent SI units. For convenience, certain derived units have been given special names and symbols, and examples are given in Table 2.2. Examples of SI coherent derived units relevant to radiation protection are given in Table 2.3.

2.1.4 Traditional Units for Radiation Protection

Since radiation protection has evolved over the past century, many of the associated units have been modified from their original form, and historical units are still in use in some countries or in references. Their relationship to the SI is a matter of conversion and some examples are listed in Table 2.4.

TABLE 2.3   Examples of SI Coherent Derived Units

	SI Coherent Derived Unit		
Derived Quantity	**Name**	**Symbol**	**Expressed in Terms of SI Base Units**
Specific energy	joule per kilogram	$J\,kg^{-1}$	$m^{-1}\,s^{-2}$
Energy density	joule per cubic meter	$J\,m^{-3}$	$kg\,m^{-1}\,s^{-2}$
Exposure (x- and γ-rays)	coulomb per kilogram	$C\,kg^{-1}$	$s\,A\,kg^{-1}$
Absorbed dose rate	gray per second	$Gy\,s^{-1}$	$m^{-2}\,s^{-3}$

TABLE 2.4   Relationship between Traditional Units and SI Units

Quantity	**Traditional Unit**	**Traditional Symbol**	**Equivalent SI Unit**
Exposure	Roentgen	R	$2.58 \times 10^{-4}\,C\,kg^{-1}$
Absorbed dose	Rad	rad	$0.01\,Gy$
Dose equivalent	Rem	rem	$0.01\,Sv$
Activity	Curie	Ci	$3.7 \times 10^{10}\,Bq$

2.2 ATOMIC STRUCTURE

2.2.1 Proton

The nucleus of an atom is composed of protons and neutrons. Protons are formed by the combination of two up quarks and one down quark, with each up quark having a unit charge of +2/3 and the down quark having a −1/3 unit charge, yielding a net unit charge for the proton of +1 (1 unit charge = 1.602×10^{-19} Coulombs). The proton has a rest mass of 1.673×10^{-27} kg (Baum, Knox, and Miller 2002) or, using the mass-energy relationship $E = mc^2$, 938.272 MeV/c^2 where one atomic mass unit (AMU) is taken to be 931.5 MeV/c^2. The number of protons in an atom determines its atomic number (Z) and, therefore, the chemical properties of charge-neutral elements.

2.2.2 Neutron

Neutrons are composed of two down quarks and one up quark and thus have a neutral charge. Neutrons have a rest mass of 1.675×10^{-27} kg (Baum, Knox, and Miller 2002) (939.571 MeV/c^2) which is slightly larger than a proton. Elements can contain different numbers of neutrons to form isotopes. For example, hydrogen most commonly contains only one proton in its nucleus, but its isotopes may include one neutron (deuteron) or two (triton). Free neutrons decay to a proton and electron with a half-life of about 10 minutes.

2.2.3 Electron

Electrons are fundamental particles having a unit charge of −1 and a mass of 9.109×10^{-31} kg (0.511 MeV/c^2) and they orbit the nucleus at energy levels that vary depending on the number of electrons. For electrically neutral atoms, the Z number is equal to the number of electrons. The structure of each electron orbit is a defined pattern that has undergone significant study and consists of shells that correspond to various discrete energy states. The electrons may move between shells via excitation and de-excitation—a process to be expanded upon in later sections.

2.2.4 The Nucleus

The atomic nucleus consists of neutrons and protons that are tightly bound and in motion. The atomic number, Z, of a nucleus is the number of protons in the nucleus, and (for charge neutral atoms) the number of electrons orbiting the nucleus. The integer atomic mass, A, is the total number of protons and neutrons in the nucleus, and it represents the mass of an atom in atomic mass units (AMU). The neutron number of a nucleus, N, is numerically equal to $(A - Z)$. The elements are determined based on the number of protons present in the nucleus.

A nuclide refers to a species of atomic nuclei with a specific Z and A value. If X is the chemical symbol for a nuclide, then it is written as $^A_Z X$, and occasionally as $^A_Z X^N$. Often the subscript Z is not written, as the chemical symbol corresponds uniquely to atoms with a given atomic number, and because it is found from $N = A - Z$, the value of N is also commonly omitted.

Isotopes are defined to be nuclides having the same Z number, but different A numbers (e.g., ^{57}Co, ^{59}Co, and ^{60}Co are isotopes of cobalt). Most of the chemical properties of neutral isotopes are the same, but the nuclear properties can be vastly different. Some elements have only one stable naturally occurring isotope, for example, ^{23}Na, ^{27}Al, ^{9}Be, ^{103}Rh, and ^{19}F, while others have many naturally occurring stable isotopes, for example, mercury ($Z = 80$) has seven (^{196}Hg, ^{198}Hg, ^{199}Hg, ^{200}Hg, ^{201}Hg, ^{202}Hg, and ^{204}Hg). Nuclides with $Z > 83$ have no naturally occurring stable isotopes.

Isotones are defined as nuclides having the same number of neutrons, but different numbers of total nucleons or integer atomic masses (e.g., $^{36}_{16}$S and $^{37}_{17}$Cl). Isobars are all nuclides having the same number of nucleons or integer atomic masses, but different numbers of protons, for example, $^{36}_{16}$S and $^{36}_{18}$Ar.

2.2.4.1 Binding Energy

Since protons have a +1 charge, Coulombic repulsion effects, which has a $1/r^2$ dependence, must be overcome for nuclei with $Z \geq 1$ to remain bound. This is achieved via the exchange of mesons that have an effective range on the order of 1–2 F (1 F = 10^{-15} m), and this force acts on both protons and neutrons. At distances $< \sim 0.5$ F, the nuclear force becomes repulsive. Thus, as additional protons are incorporated in the nucleus, additional (electrically neutral) neutrons may be required to allow for sufficient π-meson exchange between nucleons to overcome the Coulombic repulsion. This exchange is termed the strong nuclear force, and it has a property known as saturability, which allows a single nucleon to only interact with a finite number of other nearby nucleons. The radius of a proton is on the order of one F, while a nucleus has a radius on the order of one to a few F, depending on its composition. Various mesons are involved in the exchange to maintain charge parity, depending on whether the exchange occurs between protons, neutrons, or neutron-proton pairs. This exchange is manifested in the form of binding energy.

The binding energy of nucleons may be found from the mass energy of the nuclide. The mass energy ($m_N c^2$) is the atomic mass energy ($m_A c^2$) less the mass energy of the Z electrons and the electron binding energy:

$$M_N c^2 = M_A c^2 - Z M_e c^2 + \sum_{i=1}^{Z} B_i \tag{2.1}$$

where B_i is the binding energy of the ith electron, which is quite small compared to the nuclear energies and can be ignored (Krane 1988). The nuclear binding energy (*BE*) is the difference in the total mass of the nucleus compared to the sum of the masses of the constituent number of protons and neutrons (this is a result of the mass-energy equivalence):

$$BE = \left\{ Z m_p + N m_n - \left[m\left(^A X \right) - Z m_e \right] \right\} c^2 \tag{2.2}$$

For neutral atoms Equation (2.2) becomes

$$BE = \left\{ Z m\left(^1 H \right) + N m_n - m\left(^A X \right) \right\} c^2 \tag{2.3}$$

where ^1H is the mass of the hydrogen atom.

2.2.4.2 Mass Defect

The mass defect of an atom is a result of the binding energies required to overcome the repulsive effects of the Coulombic forces of the protons, and is also a consequence of the mass-energy relationship. Thus, observed atomic masses are less than the sum of their constituent neutron and proton masses. For example, the mass of $^4_2 He$ is $4.002603u$, and the masses of the proton (hydrogen) and neutron are $1.007825u$ and $1.008665u$, respectively (Baum, Knox, and Miller 2002). The mass defect is:

$$\Delta M = Z_{^1 H} M_{^1 H} + N_{^1_0 n} M_{^1_0 n} - M_{^A_Z X}$$

$$\Delta M = 2\left(1.00783u \right) + 2\left(1.00866u \right) - 4.002603u$$

$$\Delta M = 0.030377u = 28.3 \text{ MeV} \tag{2.4}$$

When dispersed nucleons combine to form a nucleus, the energy of the resultant system must decrease by ΔE, to account for the binding energy of the nucleus. For $^4_2 He$, the decrease in energy is 28.3 MeV or $0.030377u$, which is equivalent to a mass of about 5×10^{-29} kg. Examples of atomic masses and mass defects are provided in Table 2.5 (Baum, Knox, and Miller 2002; Krane 1988).

2.2.4.3 Binding Energy per Nucleon

The binding energy per nucleon *BE/A* is the quotient of the total binding energy of the nucleus and the number of nucleons. For $^4_2 He$, it is simply $28.3 \text{ MeV} / 4 = 7.075 \text{ MeV}$. The *BE/A* is telling, in that it gives a measure of the propensity of a nucleus to divide into

TABLE 2.5 Atomic Masses and Mass Defects (the Sum of the Mass of the Constituent Protons and Neutrons Are Also Listed)

	Z	A	Observed Atomic Mass (MeV/c^2)	Mass of Constituents (MeV/c^2)	Mass Defect[a] (MeV)
^1_0n	0	1	939.571	939.571	–
^1_1H	1	1	938.789	938.789	–
^2_1H	1	2	1876.136	1878.360	2.22
^3_1H	1	3	2809.450	2817.932	8.48
^3_2He	2	3	2809.431	2817.149	7.72
^4_2He	2	4	3728.425	3756.721	28.30
$^{13}_7N$	7	13	12114.846	12208.952	94.11
$^{14}_7N$	7	14	13043.863	13148.523	104.66
$^{15}_8O$	8	15	13975.355	14087.312	111.96
$^{16}_8O$	8	16	14899.263	15026.884	127.62

[a] The listed mass defect assumes $1u = 931.5\ MeV/c^2$.

separate components or to join with other nucleons to form heavier nuclei. Values of the binding energies per nucleon are shown in Figure 2.1 for $A = 1$ to $A = 227$. The range for BE/A is between about 7.5 and 8.5 for most of the nuclides, with ^{62}Ni having the highest value of 8.795 MeV/nucleon, followed by ^{58}Fe (8.792 MeV/A) and ^{56}Fe (8.790 MeV/A).

A change in the nuclear structure that drives the number of nucleons toward the maxima of the binding energy curve will result in a release of energy as a consequence of the change in the initial and final masses of the nuclei. For example, two nuclides to the left of the maxima that combine to form a nuclide with higher binding energy will produce excess energy through the fusion process. A nuclide to the right of the maxima that splits to form two (or more) nuclei will release energy via the fission process. The fission and fusion processes will be discussed later.

Of interest are the energies required to remove a proton or a neutron from the nucleus, which can be determined by comparing the binding energies of the initial nuclide to that of the nuclide with the proton or neutron removed. For protons, this is termed the proton separation energy, S_p:

$$S_p = B\left(^A_Z X^N\right) - B\left(^{A-1}_{Z-1} X^N\right)$$

$$S_p = \left[m\left(^{A-1}_{Z-1}X^N\right) - m\left(^A_Z X^N\right) + m\left(^1H\right) \right]c^2 \tag{2.5}$$

Similarly, the neutron separation energy, S_n, can be found from:

$$S_n = B\left(^A_Z X^N\right) - B\left(^{A-1}_Z X^{N-1}\right)$$

$$S_n = \left[m\left(^{A-1}_Z X^{N-1}\right) - m\left(^A_Z X^N\right) + m_n \right]c^2 \tag{2.6}$$

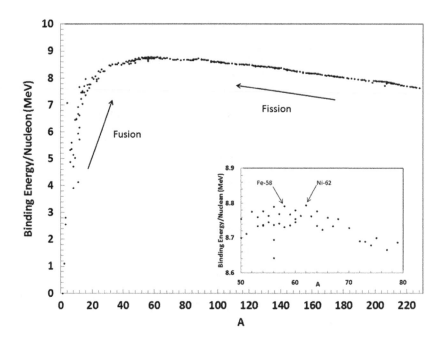

FIGURE 2.1   Average binding energy per nucleon for $A = 1$ to 224. *Inset plot shows details around the most stable nuclei.*

For ^{4}He the neutron separation energy is:

$$S_n = \left[m\left({}^{A-1}_{Z}X^{N-1} \right) - m\left({}^{A}_{Z}X^{N} \right) + m_n \right] c^2$$

$$S_n = \left[m\left({}^{3}_{2}He^3 \right) - m\left({}^{4}_{2}He^2 \right) + m_n \right] c^2$$

$$S_n = \left[2809.431 - 3728.425 + 939.571 \right] c^2$$

$$S_n = 20.58 \text{ MeV} \tag{2.7}$$

And the proton separation energy is:

$$S_p = \left[m\left({}^{A-1}_{Z-1}X^{N} \right) - m\left({}^{A}_{Z}X^{N} \right) + m_p \right] c^2$$

$$S_p = \left[m\left({}^{3}_{1}H^2 \right) - m\left({}^{4}_{2}He^2 \right) + m_p \right] c^2$$

$$S_p = \left[2809.450 - 3728.425 + 938.789 \right] c^2$$

$$S_p = 19.81 \text{ MeV} \tag{2.8}$$

Mass defect calculations are a convenient means to determine Q-values for various reactions. As an example, to calculate the energy of neutrons produced by a deuterium-tritium (D-T) generator, the mass defect method may be applied. Assuming negligible initial kinetic energies of the deuteron and triton, the total Q-value of the reaction is:

$$\,^2_1H + \,^3_1H \rightarrow \,^4_2H + \,^1_0n + Q$$

$$1876.136 \text{ MeV} + 2809.450 \text{ MeV} = 3728.425 \text{ MeV} + 939.571 \text{ MeV} + Q$$

$$4685.586 \text{ MeV} = 4667.996 \text{ MeV} + Q$$

$$17.59 \text{ MeV} = Q \qquad\qquad (2.9)$$

This reaction Q-value is shared in the form of kinetic energy between the resultant ^4He nucleus and the neutron and is found using conservation of energy methods yielding 3.52 MeV for the ^4He atom and 14.07 MeV for the neutron. For high energies, relativistic effects may also need to be considered.

2.2.5 Liquid Drop Model

If it is assumed that the interior mass densities of nuclides are approximately the same and their total binding energies are approximately proportional to their masses (i.e., $\Delta E/A$ ≈ constant), then the masses of the nuclei should have characteristics similar to that of a liquid drop. The analogy is that for liquids, the interior densities of drops are the same and the heat of vaporization is proportional to the drop's mass (Krane 1988). This approach was one of the first used to describe atomic nuclei, and resulted in reasonable predictions of observed nuclear properties. The liquid drop model simulates the impacts on binding energy from the volume of the nucleus, surface effects, Coulombic repulsion, asymmetry effects (which accounts for the observed tendency of nuclei to have $Z = N$), and a pairing term that incorporates the phenomena that more stable nuclei tend to have even Z and N numbers.

The liquid drop model provides an approximate description of the behavior of nuclei with regard to its mass (or binding energy). As discussed earlier, the binding energy is a measure of the stability of the atom. An examination of the binding energy per nucleon values indicate that certain values of N and Z have significantly higher binding energies than nearby nuclides and thus are unusually stable. These values of Z and/or N, are termed "magic numbers," and include:

$$Z \text{ and } / \text{ or } N = 2, 8, 20, 28, 50, 82, 126$$

The binding energies per nucleon for nuclides having magic Z and/or N are higher and as a result are more stable, and this is especially true for nuclides where both Z and N are equal to a magic number. The effect is even more pronounced if a measure of stability more sensitive than the BE/A is used, such as the neutron separation energy discussed previously. Consider, for example, the neutron separation energy of ^{16}O which has both $Z = N = 8$ and a $BE/A = 7.976$:

$$S_n = \left[m\left(\,^{A-1}_{Z}X^{N-1} \right) - m\left(\,^A_Z X^N \right) + m_n \right] c^2$$

$$S_n = \left[m\left({}^{15}_{8}O^7 \right) - m\left({}^{16}_{8}O^8 \right) + m_n \right] c^2$$

$$S_n = \left[13975.355 - 14899.263 + 939.571 \right] c^2$$

$$S_n = 15.66 \, \text{MeV} \tag{2.10}$$

While for ${}^{14}_{7}N$ (which has a similar BE/A of 7.476) the S_n is

$$S_n = \left[m\left({}^{A-1}_{Z}X^{N-1} \right) - m\left({}^{A}_{Z}X^N \right) + m_n \right] c^2$$

$$S_n = \left[m\left({}^{13}_{7}N^6 \right) - m\left({}^{14}_{7}N^7 \right) + m_n \right] c^2$$

$$S_n = \left[12114.846 - 13043.863 + 939.571 \right] c^2$$

$$S_n = 10.55 \, \text{MeV} \tag{2.11}$$

Shell models of the nucleus, roughly analogous to those used for electron orbits which will be discussed in the next section, attempt to account for the combinatorial structure of the nucleons and explain the appearance of the so-called magic numbers. However, contradictions to the electron model are evident. Potential energy supplied by the positively charged nucleus is felt by the negatively charged electrons as they orbit the nucleus. Also, the electrons move in their orbital shells without a high probability of collision, since these electron orbits are several orders of magnitude larger than the nucleus. Whereas the electron orbital field has a low density since its radius is relatively large (on the order of 10^{-10} m), the nuclear radius is on the order of 10^{-15} m. However, empirical evidence for the shell structure of the nucleus is found by observing the alpha emission energies of nuclides whose daughter has neutron numbers near $N = 126$ (alpha decay will be discussed in more detail in Section 2.4.1). As an example, Figure 2.2 shows the average alpha particle energies for isotopes of polonium. Note the dramatic peak in decay energy for daughter $N = 126$ (a magic number), which is indicative of an underlying shell structure.

2.2.6 Electron Orbital Structure

2.2.6.1 Bohr Model

The manner in which electrons orbit the nucleus of the atom is commonly represented as being analogous to the planetary orbits about the sun. Although this is an inaccurate model, early representations of atomic structure using this model predicted observed results surprisingly well. Physicists recognized that such planetary models would inevitably lead to the electrons losing energy in their orbits and eventually plunging into the nucleus. This model also failed to explain the observed discrete spectral lines from atomic nuclei as the orbits decayed. Under the planetary model, these lines would be continuous rather than discrete if any radii orbit were allowed, or if the electron continuously emitted energy as a result of centripetal forces associated with their orbital motion about

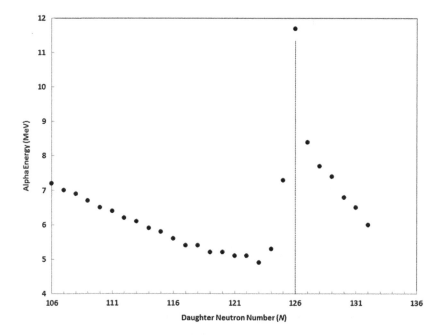

FIGURE 2.2 Alpha particle energy for isotopes of polonium. Alpha particle energy is plotted as a function of the daughter N number.

the nucleus. Niels Bohr made the first attempt to explain these discrete spectra for the hydrogen atom by assuming a quantization model of orbital energy levels. The Bohr model of the atom assumed circular electron orbits, with negligible electron mass compared to the nucleus, and the nucleus remaining fixed in space (Krane 1988; Eisberg and Resnick 1985). Using hydrogen, since it has the simplest atomic structure, the condition of mechanical stability of the electron is described by such classical mechanics as

$$\frac{1}{4\pi\epsilon_o} \frac{Ze^2}{r^2} = m\frac{v^2}{r} \tag{2.12}$$

where $1/4\pi\epsilon_o$ is Coulomb's law constant equal to $8.988x\times10^9$ N·m^2·C^{-2}, Z is the charge of the nucleus, e is the magnitude of electron charge, r is the orbital radius, m is the electron rest mass, and v is the speed of the electron in orbit. The left side of Equation (2.12) is the Coulombic force acting on the electron, while the right side represents the centripetal force of the electron in its orbit. The potential energy for an electron at a distance r from the nucleus is given by

$$V = -\int_r^\infty \frac{Ze^2}{4\pi\epsilon_o r^2} dr = -\frac{Ze^2}{4\pi\epsilon_o r} \tag{2.13}$$

and using Equation (2.12), the kinetic energy is

$$K = \frac{1}{2}mv^2 = \frac{Ze^2}{4\pi\epsilon_o 2r} \tag{2.14}$$

The total energy of the system is the sum of the potential and kinetic energies

$$E = K + V = \frac{Ze^2}{4\pi\epsilon_o 2r} \tag{2.15}$$

By applying a quantization, $mvr = n\hbar$, to Equation (2.15) the orbital radius, r, is

$$r = 4\pi\epsilon_o \frac{n^2\hbar^2}{mZe^2} \tag{2.16}$$

Bohr surmised that this quantization of orbital angular momentum would lead to a quantization of total energy:

$$E = -\frac{mZ^2e^4}{(4\pi\epsilon_o)^2 2\hbar^2} \frac{1}{n^2} \tag{2.17}$$

where n is the orbital number (Eisberg and Resnick 1985). For $n = 1$, this represents the energy required to completely remove the electron from the ground state to infinity, and this is termed the ionization energy. Other values of n (e.g., 2, 3, 4, etc.) represent various energy levels of the electron orbit. Finding the differences in these various energy levels showed that this model predicted the spectral lines observed via spectrometry. The total ionization energy for hydrogen from its ground state is:

$$E = -\frac{mZ^2e^4}{(4\pi\epsilon_o)^2 2\hbar^2} \frac{1}{n^2}$$

$$E = -2.177 * 10^{-18} \, J = -13.59 \text{ eV} \tag{2.18}$$

Additional orbital energies can be found for $n = 2$ (–3.39 eV), $n = 3$ (–1.51 eV), and so on, and these energies correspond closely to the observed spectral lines for hydrogen.

Electrons need not move from a higher orbit directly to the ground state. Instead, they can move between integer states (e.g., from 5 to 3, 3 to 2, etc.), emitting discrete energy photons as they do. Since the orbits are quantized, the electrons change orbits in discrete steps rather than through continuous energy loss. For electrons that move between orbital states, the differences between these states correspond to an observed frequency of radiation emitted that equals the difference in the final (n_f) and initial (n_i) orbits by the relationship

$$v = \frac{E_i - E_f}{h} \tag{2.19}$$

which is a consequence of Einstein's postulate $E = hv$. Conversely, increasing the orbital energy requires the supply of energy at least equal to the difference in potential between the orbits.

2.2.6.2 Sommerfeld Model

The Bohr model of the atom employed a simplified atomic model. Two significant assumptions were that the mass of the nucleus was infinitely large compared to the mass of the electron, and the nucleus remained stationary. To account for the mass of the nucleus, a planetary model type correction was made taking the form

$$\mu = \frac{mM}{m+M} \tag{2.20}$$

where m is the electron mass and M is the mass of the nucleus (Eisberg and Resnick 1985).

As opposed to the requirement that electrons traverse in circular orbits, the Sommerfeld model incorporated elliptical orbital models as a means to explain the fine structure observed in spectra. Despite the elliptical orbital path, the total electron energy is the same as the circular orbit. Sommerfeld's equation is

$$E = -\frac{\mu Z^2 e^4}{\left(4\pi\varepsilon_o\right)2n^2\hbar^2}\left[1+\frac{\alpha^2 Z^2}{n}\left(\frac{1}{n_\theta}-\frac{3}{4n}\right)\right] \tag{2.21}$$

where n is termed the principal quantum number, n_θ is the azimuthal quantum number, and α is the fine structure constant equal to approximately 1/137 (Eisberg and Resnick 1985).

2.2.7 Excitation

If an orbital electron absorbs energy, it may be excited to an upper energy region, but still be bound by the nucleus. The electron then decays to the ground state releasing the excess energy as a photon. The emitted photon energy is the difference between the electron excited state and the ground state which leads to the spectroscopic lines commonly observed for various elements. Additional subsequent emissions are possible since the emitted photons may excite electrons in other orbital shells. As these electrons de-excite, photons are emitted. As an example, consider the spectral emissions from hydrogen shown in Figure 2.3. The wavelengths corresponding to the various de-excitations from n_i to n_f are depicted in Figure 2.3 and may be obtained from Equation (2.17) and the relation $E = h\nu$. Note that, as a consequence of the quantization model, only discrete energy levels are allowed.

2.2.8 Ionization

The energy associated with an electron in the ground state in a hydrogen atom is −13.6 eV. As discussed earlier, if sufficient energy is imparted to this electron, it can move to excited orbital states. If energy greater than the ground state (e.g., 13.6 eV for hydrogen) is imparted to the electron, it is removed completely from the bound state ($n = \infty$). This process is termed ionization and results in a free electron and a positively charged nucleus. It is important to note that the amount of energy that can be absorbed is essentially limitless and excess energy greater than the ionization energy is manifested in the form of kinetic energy of the electron. For electromagnetic radiation $E = h\nu$, and this corresponds to a

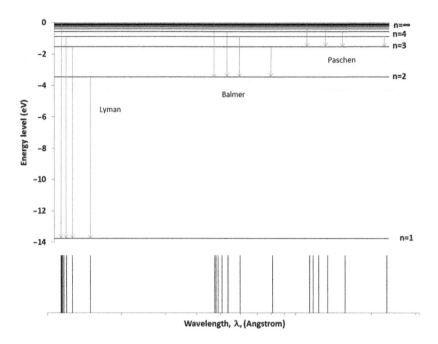

FIGURE 2.3   Spectral emission lines from de-excitation of selected hydrogen energy levels.

minimum frequency (or conversely, maximum wavelength, since $c = v / \lambda$) required to produce ionization.

Each element will have different excitation and ionization levels according to its orbital structure. Thus, de-excitation between orbitals produces photon emissions that are characteristic of the orbital structure and, consequently, of the element. These emissions are termed characteristic X-rays and are routinely used in the field of gamma spectroscopy. If an inner shell electron is ionized, outer shell electrons can de-excite to replace the empty orbit and, in the process, emit these characteristic X-rays. A cascade effect may occur resulting in multiple photon emissions.

2.3 NUCLEAR REACTIONS

2.3.1 Absorption

Nuclear reactions involve the absorption of an incoming particle by a nucleus. This absorption produces a compound nucleus that subsequently breaks up to yield some final set of particles. A nuclear reaction can be thought of in the following manner: a projectile a, usually the lighter particle, interacts with the target, which is another particle or nucleus, X. The compound nucleus is formed and breaks up very rapidly (typically on the order of 10^{-14} seconds) into two reaction products, b and Y. The particles a and X are commonly referred to as the initial constellation, and b and Y as the reactants in the final constellation. The reaction may be written in the form $X(a,b)Y$.

$$a + X \rightarrow \left[\text{Compound Nucleus} \right] \rightarrow Y + b \qquad (2.22)$$

Since energy is either required for the reaction to occur (endothermic), or is released (exothermic) as a result of total mass changes, Equation (2.22) is amended to

$$a + X \rightarrow \left[\text{Compound Nucleus} \right] \rightarrow Y + b + Q \tag{2.23}$$

where Q can be negative or positive depending on the reaction energy requirements.

Nuclear reactions follow the laws of conservation:

- The total number of nucleons (total A) is conserved.
- The total charge (total Z) is conserved.
- The total energy (including changes in total mass) is conserved.
- The total momentum is conserved.

A common source of neutrons for use in the laboratory involves an alpha particle reacting with a ^{9}Be nucleus to create a neutron and a ^{12}C nucleus, as shown below. Note that a gamma ray can be emitted, so it is also shown in the final constellation. The reaction is

$$ {}^9_4\text{Be} + {}^4_2\text{He} \rightarrow \left[{}^{13}_6\text{C}^* \right] \rightarrow {}^{12}_6\text{C} + {}^1_0\text{n} + \gamma \tag{2.24}$$

which can also be written as

$$ {}^9_4\text{Be}(\alpha, n) {}^{12}_6\text{C} \tag{2.25}$$

In actuality, there are multiple reactions that can occur from this alpha particle bombardment of beryllium, so, in a reaction, the endpoints or final constellation can be multifaceted. Consider, for example, the possible reactions of neutron absorption by ${}^{14}_7\text{N}$:

$$ {}^1_0\text{n} + {}^{14}_7\text{N} \rightarrow \left[{}^{15}_7\text{N} \right] \rightarrow \begin{Bmatrix} {}^{11}_5\text{B} + {}^4_2\text{He} \\ {}^{13}_6\text{C} + {}^2_1\text{H} \\ {}^{14}_6\text{C} + {}^1_1\text{H} \\ {}^{14}_7\text{N} + {}^1_0\text{n} \\ {}^{13}_7\text{N} + 2{}^1_0\text{n} \\ {}^{15}_7\text{N} + \gamma \end{Bmatrix} \tag{2.26}$$

2.3.2 Fission

In the fission process, an atomic nucleus, usually a heavy nucleus with an atomic weight greater than 200 AMUs, splits into at least two smaller nuclei. These smaller nuclei are referred to as fission products. In addition to the fission products, neutrons are emitted, along with photons created from de-excitation of nuclear states or other reactions and decays. These initial radiations consisting of neutrons, photons, and fission products are

TABLE 2.6   Atomic Masses of Fission Products

Radionuclide	Atomic Mass (AMU)
$^{235}_{92}\text{U}$	235.0439
$^{91}_{38}\text{Sr}$	90.9102
$^{143}_{54}\text{Xe}$	142.9273
^{1}n	1.0087

referred to as prompt radiations. Since the fission products are heavy and strongly ionized, their ranges in material (including air) are very short and are of no real consequence from an external dosimetry standpoint, but can be significant if internalized.

The fission process results in a large release of energy in the form of kinetic energies of the neutrons and fission products. The mass defect of the $^{235}_{92}\text{U}$ atom is about 1783 MeV or 7.59 MeV/nucleon. Typical fission products of the $^{235}_{92}\text{U}$ $(n, \text{fission})$ reaction are $^{91}_{38}\text{Sr}$ and $^{143}_{54}\text{Xe}$ plus two neutrons. The masses of each of the products are listed in Table 2.6 (Baum, Knox, and Miller 2002).

The mass defect from the reaction in Equation (2.27) is ~184 MeV.

$$^{235}_{92}\text{U} + ^{1}_{0}\text{n} \rightarrow ^{91}_{38}\text{Sr} + ^{143}_{54}\text{Xe} + 2^{1}_{0}\text{n} \tag{2.27}$$

This excess energy of 184 MeV is shared between all the fission products (including the neutrons) primarily in the form of kinetic energy. Of special interest are the associated energies of the neutrons. Since the kinetic energy is shared among the various fission fragments as well as the neutrons, the neutrons exhibit an energy spectrum. The neutron energy spectrum for the fission process can be well described by a Maxwell–Boltzmann distribution:

$$\chi(E) = \frac{2}{\sqrt{\pi}} T^{-\frac{3}{2}} E^{\frac{1}{2}} e^{\frac{-E}{T}} \tag{2.28}$$

where T is the temperature and E is the neutron energy. This method ensures that the total neutron fluence has an energy distribution so that

$$\int_{0}^{\infty} \chi(E)dE = 1 \tag{2.29}$$

Temperature values for common fission reactions range from 1.25 to 1.5 with the ^{235}U fission being well fit with a temperature of 1.29 MeV. For comparison, the spontaneous fission of ^{252}Cf (a common neutron calibration source), has a temperature of 1.42 MeV. Neutron fission spectra for ^{235}U, ^{252}Cf, and ^{239}Pu ($T = 1.333$ MeV) are shown in Figure 2.4. These spectra are based on the neutron mission probabilities per fission and do not include effects from moderation as the neutrons traverse a critical assembly. The most probable neutron

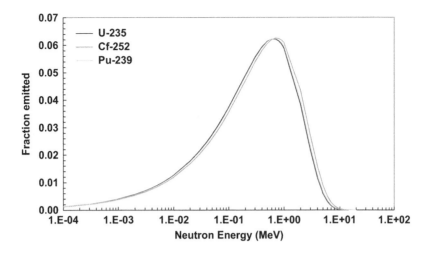

FIGURE 2.4   Neutron fission spectra for ^{235}U, ^{252}Cf, and ^{239}Pu. Neutron energy spectral distributions shown are based on Equation (2.30) and temperatures are quoted in the text.

emission energy can be found by taking the first derivative of Equation (2.28) with respect to E, and setting the result equal to zero to find the maxima:

$$\frac{d\chi(E)}{dE} = \frac{d}{dE}\left[\frac{2}{\sqrt{\pi}}T^{-\frac{3}{2}}E^{\frac{1}{2}}e^{\frac{-E}{T}}\right] = 0 \tag{2.30}$$

$$E = \frac{1}{2}T \tag{2.31}$$

The most probable energies for ^{235}U, ^{239}Pu, and ^{252}Cf are 0.65 MeV, 0.67 MeV, and 0.71 MeV, respectively.

The average energy, $\bar{E}$, can be found by integrating Equation (2.28):

$$\bar{E} = \int E\chi(E)dE \tag{2.32}$$

This integration yields average energies for ^{235}U, ^{252}Cf, and ^{239}Pu of 1.94 MeV, 2.0 MeV, and 2.13 MeV, respectively.

Regarding neutron spectra, the wide range of energies commonly encountered (from 10^{-9} to tens and hundreds of MeV for many workplaces, and even higher for accelerator facilities and cosmic sources) can create the appearance of discontinuities when plotting the spectra. These effects are a result of energy binning, which are often chosen for calculation or measurement convenience. For this reason, the use of lethargy fluence, U, is used. The lethargy fluence is the fluence within an energy region divided by the logarithmic (base 10 or natural log) difference in the energy bin width:

$$U = \frac{\varphi(E_i)}{\ln(E_i) - \ln(E_{i-1})} \tag{2.33}$$

where $\varphi(E_i)$ is the fluence of particles within the energy region from E_{i-1} to E_i.

2.3.3 Fusion

Recalling the discussion in Section 2.2.4.1 and referring to Figure 2.1, it can be seen that heavy elements can fission to produce more stable nuclei. Conversely, lighter elements (e.g., $Z < \sim26$) can combine to form heavier elements through the fusion process. This process is most important for light nuclei such as hydrogen and helium whose low BE/A provides for a more energy-favorable process. As with fission, the resulting masses following this reaction have a lower BE/A. The most obvious example of the fusion process is the conversion of hydrogen to helium in stars. Although several chains can occur, for main-sequence stars such as our sun, the Proton-Proton (PP I) chain dominates (Krane 1988, Zeilik, Gregory, and Smith 1992):

$$^{1}H + {}^{1}H \rightarrow {}^{2}H + e^{+} + \nu \quad Q = 0.42 \text{ MeV} \tag{2.34}$$

$$^{2}H + {}^{1}H \rightarrow {}^{3}He + \gamma \quad Q = 5.5 \text{ MeV} \tag{2.35}$$

$$^{3}He + {}^{3}He \rightarrow {}^{4}He + {}^{1}H \quad Q = 12.9 \text{ MeV} \tag{2.36}$$

The first two steps occur twice so the total energy released in the entire process is roughly 25 MeV. For the first reaction to occur the proton must be converted to a neutron and a positron. This process has a low probability of occurring. This low probability, along with the resistance to proton-proton combination (due to Coulombic repulsion), results in a slow reaction rate for the first step of the *PPI* chain. Other reactions in the solar heating cycle occur depending on, among other things, the size, age, and temperature of the star. The carbon–nitrogen–oxygen (CNO) cycle, for example, involves the fusion of carbon and nitrogen with hydrogen. Once hydrogen fuel is exhausted, fusion with helium and eventually higher-Z elements can occur, depending on the stellar mass and temperature. These fusion chains progress until significant formation of iron occurs. Since iron has a high BE/A (refer to Figure 2.1), the fusion process is no longer energy-favorable and the energy production stops, ultimately leading to star death.

2.3.4 Elastic Scatter

As particles traverse materials, they may interact with either bound electrons or the nuclei. Neglecting Coulombic repulsion effects, after interacting, the incident particle is scattered at some angle, and in the process transfers some of its initial energy to the scattering object. When a neutron interacts with an atomic nucleus and undergoes a scatter event without first being absorbed, the reaction is termed "elastic scattering," and is analogous to the reaction of billiard balls colliding. While these reactions can occur with charged particles as well, Coulombic effects must be considered, and usually prevent pure elastic processes. For the non-relativistic case, the kinematics for neutrons are solved using conservation of energy and momentum principles. Using these relationships, initial neutron energy, E_{o}, initial velocity V_{o}, and assuming the target nucleus (M_{N}) is initially at rest

$$\frac{1}{2}M_{n}V_{o}^{2} = \frac{1}{2}M_{n}V_{n}^{2} + M_{N}V_{N}^{2} \tag{2.37}$$

and

$$M_n V_o = M_n V_n + M_N V_N \qquad (2.38)$$

the maximum fraction of energy transferred, Q_{max}, to the nucleus via an elastic scatter reaction is:

$$Q_{max} = \frac{4 M_n M_N E_o}{\left(M_n + M_N \right)^2} \qquad (2.39)$$

From Equation (2.39), it is seen that the maximum energy transfer is produced by light nuclei (Knoll 2010; Attix 2008), and for this reason compounds containing light nuclei (e.g., H_2O) are the preferred choice for shielding neutrons.

2.3.5 Inelastic Scatter

In some cases, the incident particle does not scatter off the target nucleus, but instead is first absorbed by the nucleus, then re-emitted. In conjunction with this re-emission, some other reactions or emissions may occur, including, for example, fission of the nucleus or the emission of gamma rays. Since these other processes require energy, the incident particle is re-emitted at a lower energy. This process is termed "inelastic scatter." As a rule, the likelihood of an inelastic scatter reaction occurring is dependent on the incident particle energy and, when compared to elastic scatter, is much less probable (Knoll 2010; Attix 2008).

2.4 RADIOACTIVE DECAY

When a nucleus is in an energy state that is not the lowest possible for a system with a given number of nucleons, nuclear decays or transformations can occur. Multiple transformative processes are possible. A brief description of those most applicable to dosimetric applications is provided in this section.

2.4.1 Alpha Decay

For nuclei with $Z > 82$, a primary mode of decay is via the emission of an alpha particle. This is an energetically favorable process since the mass of the daughter and the mass of the alpha particle are less than the mass of the parent as a result of the decrease in the Coulombic energy. This energy difference is manifested in the form of kinetic energy of the alpha particle and recoil energy of the nucleus. The alpha decay process is highly energetic, with typical decay energies ranging between about 4 MeV and 9 MeV. The alpha decay equation is

$$^{A}_{Z}X_N \rightarrow \ ^{A-4}_{Z-2}X_N + \ ^{4}_{2}He_2 + Q \qquad (2.40)$$

Using mass defect analysis, it can be shown that alpha decay is one of the few methods that can energetically occur. Consider, for example, emissions of other particles

($^1_0$n, $^2_0$H, etc.) from the $^{212}_{84}$Po nucleus and the resulting mass defects. For alpha decay, the mass defect is

$$\Delta M = M\left(^{212}_{84}\text{Po}\right) - \left[M\left(^{208}_{82}\text{Pb}\right) + M\left(^4_2\text{He}\right)\right]$$

$$\Delta M = 211.98887u - \left(207.97665u + 4.002603u\right)$$

$$\Delta M = 0.009617u = 8.96 \text{ MeV} \tag{2.41}$$

By conservation of momentum the alpha particle energy is found to be 8.8 MeV. Since ΔM is positive, the decay is energetically possible. For other particles, however, this is not true. The decay of $^{212}_{84}$Po by proton emission is not energetically allowed, since $\Delta M \leq 0$:

$$\Delta M = M\left(^{212}_{84}\text{Po}\right) - \left[M\left(^{211}_{83}\text{Bi}\right) + M\left(^1_1\text{H}\right)\right]$$

$$\Delta M = 211.98887u - \left(210.98727u + 1.007825u\right)$$

$$\Delta M = -0.006225u = -5.8 \text{ MeV} \tag{2.42}$$

The process of alpha decay is of interest since the total potential energy acting on an alpha particle includes the nuclear force and Coulombic repulsion effects. In fact, according to classical physics models, alpha decay could not occur. The Coulombic repulsion felt by an alpha particle (with a +2 charge) that is in contact with a $^{212}_{84}$Po nucleus is

$$V_o = \frac{2Ze^2}{4\pi\varepsilon_o r'} \tag{2.43}$$

where r' is the charge density half value radius a and has a value of about 8×10^{-15} m (Eisberg and Resnick 1985). The Coulombic potential increases as the alpha particle approaches the nucleus until it reaches a maximum of 30 MeV per Equation (2.43) with the alpha particle and nucleus touching (Eisberg and Resnick 1985). Inside the nucleus, the alpha particle participates in the nuclear binding energy exchange and is trapped by this much stronger force (see Figure 2.5) (Krane 1988; Eisberg and Resnick 1985). Prior to decay, the alpha particle is contained within the nucleus and is maintained in that state with a 30 MeV barrier to escape. Since alpha particle energies are observed to be well below this value, the emission process should not occur. However, through the use of quantum mechanical models developed by Schrödinger, the decay was found to be possible via barrier penetration processes termed "tunneling." This process is a consequence of the wavelike and statistical nature of matter. The probability that an alpha particle will tunnel through the barrier is dependent on the configuration of the nucleus. The emitted alpha particle energy is related to a nuclide's decay time (half-life)—higher energy alpha particles tend to originate in nuclei having shorter half-lives. This is a quantum mechanical effect where the probability of tunneling is higher for higher energy emissions. Figure 2.6 shows alpha particle energies emitted from isotopes of thorium ($Z = 90$)

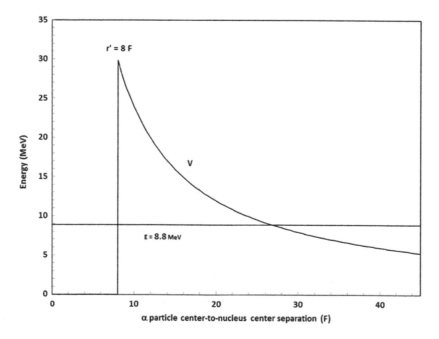

FIGURE 2.5   Representation of the Coulombic and total barrier energy potentials acting on an alpha particle and a $^{212}_{84}$Po nucleus. Also shown is the observed alpha particle emission energy.

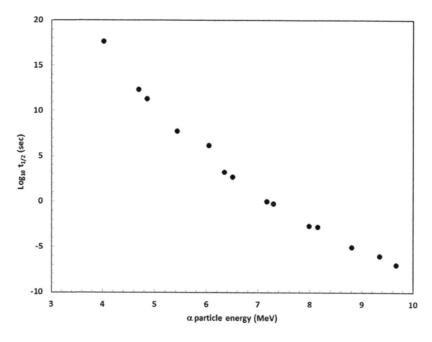

FIGURE 2.6   Alpha emission energies compared to the half-life of thorium isotopes.

and the isotope's half-life ($\log_{10}$ in seconds). Note the correlation between decay half time and alpha particle energy.

Following alpha emission, the nucleus may be left in an excited state, and additional transitions may occur resulting in the emission of photons. These photon emissions will be discussed in Section 2.4.7.

2.4.2 Beta Decay

The term beta decay is used to describe three processes—electron emission, positron (anti-electron) emission, or electron capture. The latter two processes will be discussed in later sections. Beta decay involves the transition of a neutron into a proton-electron pair, and results in an increase by one of the Z number:

$$_{Z}^{A}X^{N} \rightarrow {_{Z+1}^{A}}X^{N-1} + e^{-} \tag{2.44}$$

If Equation (2.44) were complete, then it would be expected, from mass defect calculations, that the electron would be emitted mono-energetically. In fact, the emitted electron has a distribution in energy from near zero to the maximum energy predicted from the mass defect calculation. This phenomenon gave rise to the theory that a second particle, the neutrino, was emitted simultaneously from the nucleus and shared the kinetic energy. Later, this was revised to the antineutrino, $\bar{\nu}$, which, like the neutrino, has no charge, spin $= 1/2$, and a very small mass (i.e., on the order of eV) (Krane 1988). The antineutrino shares a portion of the kinetic energy with the electron and, since the fraction of energy carried by the electron and antineutrino will vary with each decay, the observed electron energy is in the form of a spectrum. Thus, the equation (Equation 2.44) is rewritten as

$$_{Z}^{A}X^{N} \rightarrow {_{Z+1}^{A}}X^{N-1} + e^{-} + \bar{\nu} \tag{2.45}$$

Consider the decay of ^{90}Sr :

$$^{90}\text{Sr} \rightarrow {^{90}\text{Y}} + e^{-} + \bar{\nu} + Q \tag{2.46}$$

The energy of this decay can be determined from the mass defect equation under the simplifying assumptions that electron binding energy effects are negligible.

$$Q = \left\{ \left[M\left(_{Z}^{A}X\right) - Z\left(M_{e^{-}}\right) \right] - \left[M\left(_{Z+1}^{A}X\right) - (Z+1)\left(M_{e^{-}}\right) \right] - \left(M_{e^{-}}\right) \right\} c^{2} \tag{2.47}$$

The electron masses cancel and

$$Q = \left\{ \left[M\left(_{Z}^{A}X\right) \right] - \left[M\left(_{Z+1}^{A}X\right) \right] \right\} c^{2}$$

$$Q = \left\{ 89.907738 - 89.907152 \right\} c^{2}$$

$$Q = 0.546 \text{ MeV} \tag{2.48}$$

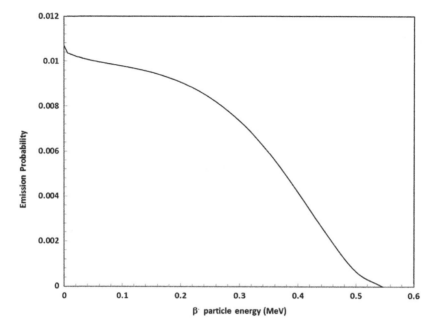

FIGURE 2.7   Beta spectrum from ^{90}Sr decay. The maximum electron energy is 0.546 MeV.

The total energy available to the electron and antineutrino in this decay is 0.546 MeV. The observed decay spectrum for this process is shown in Figure 2.7 (Eckerman and Sjoreen 2014).

2.4.3 Positron Decay

For nuclei that are unstable as a result of having too few neutrons, a proton may be converted to a neutron and positron via the positron decay process. Since the positron is identical to an electron, except with a +1 charge, the decay process must include a neutrino:

$$^A_Z X^N \rightarrow {}^A_{Z-1}X^{N+1} + e^+ + \nu \tag{2.49}$$

The resultant nucleus has one fewer proton and, to maintain charge parity, an electron is emitted. The positron decay energy is

$$E = \left[M_Z - M_{Z-1} - 2m_e \right] c^2 \tag{2.50}$$

where m_e is the mass of the electron or positron. For this reason, the initial mass must be greater than the final atomic mass by 2 × 0.511 MeV = 1.022 MeV for the reaction to be energetically possible.

2.4.4 Electron Capture

Per Equation (2.50), the rest mass of the final atom must be at least $2 \cdot m_e$ greater than the initial atomic mass for the decay to be energetically possible. If this is not the case, and the nucleus is neutron-deficient, the capturing of an orbital electron allows for an alternate decay process:

$$e^- + {}_Z^A X \rightarrow {}_{Z-1}^A X + \nu \tag{2.51}$$

This process is termed electron capture and results in the transformation of a proton into a neutron. Thus, the Z number is decreased by one. Since the result is the same as positron decay, it is a competing process for nuclei with high Z/A ratios, and is a more common decay mode. The innermost (e.g., K and L shell) electrons are most likely to be captured and, as a result, characteristic X-rays from orbital cascade may result as the shells are re-filled. The energy equation for electron capture is

$$E = \left[M_Z + m_e - M_{Z-1} \right] c^2 - BE_{e^-} \tag{2.52}$$

where BE_{e^-} is the binding energy of the captured orbital electron.

2.4.5 Internal Conversion

In the quantum mechanical model, there is a non-zero probability that an inner shell electron may penetrate the nucleus. This electron may then absorb excess energy from an excited nuclear state and be emitted. Since the excited nuclear states are themselves discrete and no neutrino is involved, the emitted electron appears as a single energy equal to the energy difference between the excited nuclear state and the binding energy of the electron of the daughter. Thus, the internal conversion of an electron is a competing decay mode to gamma emission, which will be discussed in Section 2.4.7. Following the internal conversion process, a cascade of electrons fills the empty shell, resulting in characteristic X-ray emissions. Although internal conversion most often occurs with K-shell electrons, other shells may contribute. Thus, there are a number of possible energies from the internal conversion process. The nuclear structure is unchanged, and there is no change in the atomic number (or the atomic mass number).

2.4.6 Spontaneous Fission

The emission of alpha particles, as discussed in Section 2.4.1, is energetically favorable, since the BE/A of the ${}_2^4 He$ atom is high, leading to both a decrease in the Coulombic force of the nucleus and higher BE/A of the daughter (Eisberg and Resnick 1985). A similar effect would be seen for emission of other particles, such as ${}_6^{12}C$ or other higher Z nuclei. Emissions of larger nuclei are referred to as spontaneous fission, and become a relevant decay mode in elements with $Z \geq \sim 92$, and significant in elements with $Z > 100$. The emission of neutrons often accompanies this decay. Typically, nuclei that undergo spontaneous fission do so for some fraction of the total decays, as alpha decay is a competing process. For example, ${}^{252}Cf$ decays via both spontaneous fission and alpha decay, with spontaneous fission occurring in about 3% of the decays (Eckerman and Endo 2007).

TABLE 2.7   Alpha and Gamma Decay Energies for ^{214}Po and Intensities

α Energy (MeV)	Decay Energy (keV)	Intensity (%)	γ Energy (keV)	Intensity (%)
7.687	7834	99.9	800	1×10^{-2}
6.902	7034	0.01	298	5×10^{-5}
6.610	6736	0.00005	87	4×10^{-8}

2.4.7 Gamma Emissions

Following radioactive decay, the nucleus may be left in an excited energy state—see the discussion on nuclear shell models from Section 2.2.5 and Figure 2.2. When ^{214}Po undergoes alpha decay to ^{210}Pb (which is itself unstable), photons are emitted from the nucleus. These photons are referred to as gamma emissions since they originate in the nucleus, whereas X-rays originate in the electron orbital shells. The alpha energies and intensities for ^{214}Po decay are listed in Table 2.7 along with the three most intense photon emissions that accompany the alpha emissions (Hubbell and Seltzer 1995). The listed decay energy was determined from the observed alpha particle energy and conservation of momentum. If the 7834 keV decay energy represents the transition to the ground state of the nucleus, then the difference between the 7034 keV and the 7834 keV decays (~800 keV) should appear as an emitted photon energy (Krane 1988; Eisberg and Resnick 1985). Similarly, the difference in the next two decay energies (6736 keV and 7034 keV) is 298 keV, and this photon energy is also observed. Other photon energies represent transitions between various nuclear energy states. Since the gamma emissions are characteristic of the shell structure of the nucleus, each atom will decay with specific energies. These characteristic emission energies are routinely used to identify the radioactive nucleus via gamma spectroscopy.

2.4.8 Isomeric Transitions (Metastable Energy States)

Gamma decay from excited nuclear states typically occurs with a half-life on the order of 10^{-9} seconds or less. Isotopes with longer gamma decay times are said to be in isomeric or metastable energy states. These isomers are denoted with an "m" in the atomic number, such as ^{99m}Tc or ^{110m}Ag .

2.4.9 Radioactive Decay Law

Radioactive decay is governed by energy perturbations in the nucleus. These perturbations are probabilistic, and based on quantum theory (Krane 1988). As a result, statistical methods may be employed to describe the decay process. The probability of a nucleus undergoing decay is

$$-\frac{dN/dt}{N} \tag{2.53}$$

where N is the number of radionuclides present at time t (Fitzgerald, Mahoney, and Brownell 1967). The quantity $-(dN/dt)/N$ is assumed to be proportional to a constant, provided the number of nuclei is large enough. This constant of proportionality is termed the decay constant, λ, and has units of inverse time (e.g., s^{-1}).

The decay constant represents the probability that an unstable nucleus will undergo a decay in the time interval $t + dt$. Using this constant, Equation (2.53) can be rewritten as:

$$\lambda = -\frac{dN/dt}{N} \tag{2.54}$$

Using an initial condition that $N = N_o$ at $t = 0$, Equation (2.54) can be integrated to yield the exponential law of radioactive decay:

$$N(t) = N_o e^{-\lambda t} \tag{2.55}$$

2.4.10 Radioactive Half-Life and Decay Constant

The radioactive half-life, $t_{1/2}$, is defined as the time for half the initial number of nuclei, N_o to decay, and is found by setting $N(t)/N_o = 1/2$ in Equation (2.55) and solving for $t_{1/2}$:

$$t_{1/2} = \frac{0.693}{\lambda} \tag{2.56}$$

The mean lifetime, τ, is the average time a nucleus will survive before decaying. If the number of nuclei that decay in the interval t and $t + dt$ is $|dN/dt|dt$, then the mean lifetime is found from (Fitzgerald, Mahoney, and Brownell 1967)

$$\tau = \frac{\int_0^\infty t|dN/dt|dt}{\int_0^\infty |dN/dt|dt} \tag{2.57}$$

which reduces to the inverse of the decay constant:

$$\tau = \frac{1}{\lambda} \tag{2.58}$$

The mean lifetime of particles assumes the particle is at rest. For particles that are traveling at a significant fraction of the speed of light, c, relativistic effects on time dilation (and length contraction) must be considered. These effects may become significant, for example, in particle accelerators and in cosmological studies.

The decay constant, λ, is the probability a nuclide will decay, and for a given number of parent nuclei, N, at time t. The activity, A, is the product of the decay probability per nuclide and the total number of nuclides:

$$A = \lambda N \tag{2.59}$$

Differentiating Equation (2.55) and using the relation in Equation (2.54), the activity as a function of time is found:

$$\frac{d}{dt}\big(N(t)\big) = \frac{d}{dt}\big(N_o e^{-\lambda t}\big)$$

(2.60)

and

$$\lambda = -\frac{\dfrac{dN}{dt}}{N}$$

(2.61)

$$\lambda N(t) = \lambda N_o e^{-\lambda t}$$

(2.62)

$$A(t) = A_o e^{-\lambda t}$$

(2.63)

The SI unit of activity is the Becquerel (Bq), which has units of decays per second. The conventional unit is the Curie (Ci) where $1\ \mathrm{Ci} = 3.7 \times 10^{10}\ \mathrm{Bq}$.

2.4.11 Production and Decay

Certain absorption reactions that were discussed in Section 2.3.1 can produce radioactive atoms, which may then decay themselves. A common example is the production of unstable nuclei following bombardment by neutrons. For this case, the number of atoms transformed, N, by an irradiation of neutrons is

$$N = \phi\sigma n$$

(2.64)

where ϕ is the neutron fluence, σ is the neutron absorption cross section (to be discussed later), and n is the number of target atoms. This transformation equation is applied for a single fluence of neutrons and does not consider the decay of unstable product nuclei. Thus, the activity at a time t can be found by determining the rate of production of transformed atoms minus the decay rate of transformed atoms using the relation from Equation (2.59)

$$\frac{dN}{dt} = \phi\sigma n - \lambda N$$

(2.65)

Equation (2.65) can be re-arranged (Fitzgerald, Mahoney, and Brownell 1967) and transformed by $e^{\lambda t}$:

$$e^{\lambda t}\frac{dN}{dt} + \lambda N e^{\lambda t} = \phi\sigma n\big(e^{\lambda t}\big)$$

(2.66)

The derivative of $Ne^{\lambda t}$ with respect to t is found:

$$\frac{d}{dt}\big(Ne^{\lambda t}\big) = N\left[\frac{d}{dt}\big(e^{\lambda t}\big)\right] + \left[\frac{d}{dt}\big(N\big)\right]e^{\lambda t}$$

(2.67)

$$\frac{d}{dt}\left(Ne^{\lambda t}\right) = \lambda Ne^{\lambda t} + e^{\lambda t}\frac{dN}{dt} \tag{2.68}$$

The right side of this equation is identical to the left side of Equation (2.66) which can now be rewritten as

$$\frac{d}{dt}\left(Ne^{\lambda t}\right) = \phi\sigma ne^{\lambda t} \tag{2.69}$$

This can be integrated with respect to t to yield

$$\int\frac{d}{dt}\left(Ne^{\lambda t}\right) = \int\phi\sigma ne^{\lambda t}dt \tag{2.70}$$

$$N\lambda = \phi\sigma n + Ce^{-\lambda t} \tag{2.71}$$

$N\lambda$ is the activity as a function of time. Using an initial condition that $A_o = 0$ at $t = 0$, the constant of integration is $C = -\phi\sigma n$. Finally, the constant of integration is substituted into Equation (2.71) to yield:

$$N\lambda = \phi\sigma n\left(1 - e^{-\lambda t}\right) \tag{2.72}$$

2.4.12 Specific Activity

Often the concentration of activity, instead of the total activity, is desired. Since activity can be defined as the product of the number of radioactive nuclei, N, and the decay probability, λ, the concentration of activity can be defined in units of N per gram:

$$SA = N\lambda = \frac{\lambda N_A}{A} \tag{2.73}$$

where SA is the specific activity, N_A is Avogadro's number (6.022×10^{23} atoms/mole), and A is the atomic weight in units of grams/mole.

2.4.13 Serial Decay

Equation (2.55) defines the number of daughter atoms produced from a parent decaying with decay constant of λ. This equation can be used to determine the number of daughter atoms that may in turn themselves decay if the daughter nuclide is not stable. If the daughter decays with constant λ_2, the rate of change of the number of daughter nuclei at a time t is the number of daughter nuclei produced from parent decay less the number of daughters that have decayed (Fitzgerald, Mahoney, and Brownell 1967):

$$\frac{dN_2}{dt} = \lambda_1 N_1 - \lambda_2 N_2 \tag{2.74}$$

Similarly, the rate of change of a third daughter, N_3, is proportional to the production rate of N_2 or $\lambda_2 N_2 = dN_3 / dt$, assuming the third daughter is stable.

The net change in the daughter nuclei is found by letting $N_1(t) = N_1(0)$. Equation (2.55) becomes

$$N_1(t) = N_1(0)e^{-\lambda_1 t} \tag{2.75}$$

and Equation (2.74) can be written as

$$\frac{dN_2}{dt} + \lambda_2 N_2 = \lambda_1 N_1(0)e^{-\lambda_1 t} \tag{2.76}$$

Applying the factor $e^{\lambda_2 t}$

$$\frac{dN_2}{dt}e^{\lambda_2 t} + \lambda_2 N_2 e^{\lambda_2 t} = \lambda_1 N_1(0)e^{(\lambda_2 - \lambda_1)t} \tag{2.77}$$

since

$$\frac{d}{dt}\left(N_2 e^{\lambda_2 t}\right) = e^{\lambda_2 t}\frac{dN_2}{dt} + \lambda_2 N_2 e^{\lambda_2 t} \tag{2.78}$$

Equation (2.77) can be rewritten as

$$\frac{d}{dt}\left(N_2 e^{\lambda_2 t}\right) = \lambda_1 N_1(0)e^{(\lambda_2 - \lambda_1)t} \tag{2.79}$$

Integrating this with respect to t yields

$$N_2 e^{\lambda_2 t} = \frac{\lambda_1}{\lambda_2 - \lambda_1}N_1(0)e^{(\lambda_2 - \lambda_1)t} + C \tag{2.80}$$

and for the initial condition $N_2 = N_2(0)$ at $t = 0$

$$C = N_2(0) - \frac{\lambda_1}{\lambda_2 - \lambda_1}N_1(0) \tag{2.81}$$

and $N_2(t)$ is

$$N_2(t) = N_2(0)e^{-\lambda_2 t} + \frac{\lambda_1}{\lambda_2 - \lambda_1}N_1(0)\left(e^{-\lambda_1 t} - e^{-\lambda_2 t}\right) \tag{2.82}$$

For the condition $N_2(0) = 0$, Equation (2.82) reduces to

$$N_2(t) = \frac{\lambda_1}{\lambda_2 - \lambda_1}N_1(0)\left(e^{-\lambda_1 t} - e^{-\lambda_2 t}\right) \tag{2.83}$$

(Fitzgerald, Mahoney, and Brownell 1967). Equation (2.74) can be generalized to include multiple generations of unstable progeny (e.g., N_3, N_4, N_5, etc.). Each successive generation is populated by the preceding parent, so Equation (2.74) takes the general form

$$\frac{dN_i}{dt} = \lambda_{i-1} N_{i-1} - \lambda_i N_i \qquad (2.84)$$

The general solution to this system is termed the Bateman equation (Krane 1988):

$$N_n(t) = N_1(0)\left(C_1 e^{-\lambda_1 t} + C_2 e^{-\lambda_2 t} + C_3 e^{-\lambda_3 t} + C_n e^{-\lambda_n t}\right) \qquad (2.85)$$

where

$$C_1 = \frac{\lambda_1 \lambda_2 \lambda_3 \cdots \lambda_{n-1}}{(\lambda_2 - \lambda_1)(\lambda_3 - \lambda_1)(\lambda_4 - \lambda_1)\cdots(\lambda_n - \lambda_1)}$$

$$C_2 = \frac{\lambda_1 \lambda_2 \lambda_3 \cdots \lambda_{n-1}}{(\lambda_1 - \lambda_2)(\lambda_3 - \lambda_2)(\lambda_4 - \lambda_2)\cdots(\lambda_n - \lambda_2)}$$

$$C_3 = \frac{\lambda_1 \lambda_2 \lambda_3 \cdots \lambda_{n-1}}{(\lambda_1 - \lambda_3)(\lambda_2 - \lambda_3)(\lambda_4 - \lambda_3)\cdots(\lambda_n - \lambda_3)}$$

$$C_n = \frac{\lambda_1 \lambda_2 \lambda_3 \cdots \lambda_{n-1}}{(\lambda_1 - \lambda_n)(\lambda_2 - \lambda_n)(\lambda_3 - \lambda_n)\cdots(\lambda_m - \lambda_n)} \qquad (2.86)$$

and where λ_m is used to indicate that the term $n = n$ is omitted in the last term of the denominator of C_n.

2.4.14 Secular Equilibrium

For the special case of serial decay where the daughter half-life is significantly shorter than the parent ($t_{1/2,1} \gg t_{2/2,}$), or, equivalently, the decay constant of the parent is much shorter than the daughter's ($\lambda_1 \ll \lambda_2$), Equation (2.83) reduces to

$$N_2(t) \cong N_1(0)\frac{\lambda_1}{\lambda_2}\left(1 - e^{-\lambda_2 t}\right) \qquad (2.87)$$

As t becomes large, the activity of the daughter approaches that of the parent, and the daughter is said to be in secular equilibrium with the parent:

$$\lambda_2 N_2 = \lambda_1 N_1 \qquad (2.88)$$

This is illustrated in Figure 2.8 for the case $t_{1/2,2} = 0.001(t_{1/2,1})$.

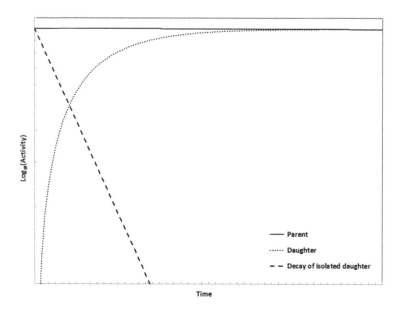

FIGURE 2.8 Secular equilibrium. Here the daughter half-life is 1/1000 that of the parent. Also shown is the decay of a separated daughter.

2.4.15 Transient Equilibrium

If the daughter's half-life is less than the parent's, but not in the extreme, as in secular equilibrium, the daughter will eventually establish transient equilibrium and decay at the same rate as it is produced by the parent. The decay equation for transient equilibrium is found from Equation (2.83) under the assumption that for times sufficiently large $e^{-\lambda_2 t}$ is negligible compared to $e^{-\lambda_1 t}$ and becomes

$$N_2(t) \cong N_1(0) \frac{\lambda_1}{\lambda_2 - \lambda_1} e^{-\lambda_1 t} \tag{2.89}$$

An example of transient equilibrium is shown in Figure 2.9 using a parent half-life that is five times that of the daughter.

2.4.16 Branching Ratios

When a nuclide decays, it may do so in more than one decay mode, or decay to different energy states. Isotopes of bismuth, for example, decay via electron capture, beta, positron, and alpha decay. The competing decay modes vary in frequency depending on the particular isotope. The frequency of a specific decay mode is given in terms of its branching ratio. Thus, for ^{211}Bi, which decays by either beta or alpha decay, the branching ratio identifies the probability of decay via each mode. The beta decay occurs 0.28% of the time, so the beta branching ratio is said to be 0.0028. Conversely, the alpha branching ratio is $1 - 0.0028 = 0.9972$. Other competing decay processes often involve metastable energy states, as discussed in Section 2.4.8. For example, ^{99}Mo undergoes β^- decay with a half-life

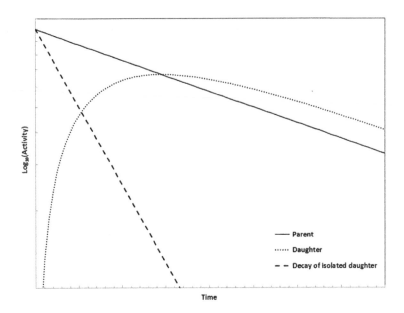

FIGURE 2.9   Transient equilibrium. Here the daughter half-life is 1/5 that of the parent. Also shown is the decay of a separated daughter.

of about 66 hours (Baum, Knox, and Miller 2002), thus increasing the Z number by one to form ^{99}Tc, which then decays to ^{99}Ru with a half-life of 2.13×10^5 years. However, only about 12% of the ^{99}Mo atoms decay directly to ^{99}Tc. The remaining 88% decay to form the metastable ^{99m}Tc, which then decays via gamma emission to ^{99}Tc. The ^{99}Tc has a half-life of about 6 hours. Thus, the branching ratio of ^{99}Mo to ^{99}Tc is 12%. The decay chain for ^{99}Mo is shown in Figure 2.10.

2.5 INTERACTION OF RADIATION WITH MATTER

2.5.1 Electrons

2.5.1.1 Soft Collisions

When charged particles traverse a material, they may interact with bound electrons via Coulombic interactions. Provided the incident particle does not directly encounter an orbiting electron or the atomic nucleus, the particle will continue along its original path with little deviation, and a small fraction of its energy will be transferred to the orbiting electrons. This may lead to excitation or ionization of the orbital electron. In the case of electrons, the energy transfer during a soft collision is small, but due to the relatively large radius of the electron cloud, the probability of this type of interaction is high and these so-called soft collisions may account for about half the total energy transfer (Knoll 2010; Attix 2008).

2.5.1.2 Hard Collisions

As discussed earlier, large fractions of energy can be transferred with scatter interactions. In the event that an incident electron approaches an atom within the radius of the electron

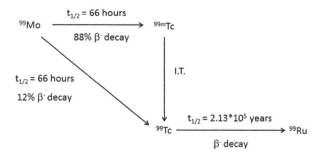

FIGURE 2.10   Decay chain graphic for ^{99}Mo.

cloud, the probability that it will interact directly with an orbital electron is significant. Since the orbital binding energy of the electron is typically very small compared with incident electron energies, the orbital electron can be treated as free. In this case, Equation (2.39) holds and the maximum energy transfer is high. The orbital electron may then be ionized, resulting in a delta ray, and this delta ray may then undergo additional interactions and create other ionizations along its path. Ionized inner-shell electrons will also result in subsequent decays of the electron shell, including the production of characteristic X-rays.

2.5.1.3 Bremsstrahlung

When a charged particle undergoes a change in its velocity, it loses kinetic energy. This energy loss results in an emitted photon termed bremsstrahlung. The amount of energy loss depends on the net change in kinetic energy of the charged particle, and the interaction probability is proportional to the square of the Z number of the medium through which the electron passes. In practical terms, the energy loss from bremsstrahlung becomes significant at higher electron energies.

The total energy loss for electrons is a combination of collisional and radiative losses. At lower energies, the losses from collisional processes dominate, with radiative losses becoming more significant with increasing energy. The relative energy loss, dE/dx, can be approximated by

$$\frac{\left(dE/dx\right)_r}{\left(dE/dx\right)_c} \cong \frac{E+mc^2}{mc^2}\frac{Z}{1600} \tag{2.90}$$

where $\left(dE/dx\right)_r$ is the radiative energy loss, $\left(dE/dx\right)_c$ the energy loss due to collisional processes, E is the incident electron energy, and Z is the atomic number of the medium (Knoll 2010; Attix 2008).

2.5.2 Photons

Photons are uncharged and thus are not affected by Coulombic fields. Photons instead transfer energy to electrons or, at higher energies, to the atomic nuclei. Although photons are massless, momentum is carried according to wavelength:

$$p = \frac{h}{v} \tag{2.91}$$

where h is Planck's constant. The photon energy is also associated with its wavelength:

$$E = h\nu \tag{2.92}$$

Thus, electromagnetic radiation is quantized.

There are three primary interactions that result in ionization: the photoelectric effect, Compton scattering, and pair production. The probability of a photon undergoing one of these processes is dependent on the photon energy ($E = h\nu$) and the atomic number (Z) of the medium with which the photon interacts. At energies below about 0.5 MeV, the photoelectric effect dominates, while in the region between one-half MeV and several MeV, Compton scattering dominates, and at higher energies ($E > 10$ MeV) pair production dominates. These energy regions are also dependent on the interaction medium. For low Z materials (air, water, etc.), the Compton scatter interaction dominates from about 20 keV up to approximately 20 MeV, and this region gradually narrows as the Z number increases (Attix 2008).

2.5.2.1 Photoelectric Effect

In the photoelectric (PE) effect, an orbital electron absorbs an incident photon and is ionized. The electron must be bound in order to satisfy conservation of energy and momentum requirements. Since the energy of the photon is characterized by its wavelength (or, alternatively, its frequency) per Equation (2.92) there is a limiting value below which the photoelectric effect cannot take occur. This minimum energy, per Equation (2.92), has a corresponding minimum frequency. The minimum energy required to produce photoelectrons is termed the work function, w. Using this work function and the incident photon energy, the kinetic energy of the ionized electron can be found:

$$K = h\nu - w \tag{2.93}$$

Thus, each material will have a minimum frequency below which the photoelectric effect will not occur. The probability that a photoelectron will be produced is dependent on the incident photon energy and the Z number of the absorber, and below about 100 keV the cross section, σ_τ is

$$\sigma_\tau \propto \frac{Z^4}{E^3} = \frac{Z^4}{(h\nu)^3} \tag{2.94}$$

In Figure 2.11, an incident photon of energy quanta $h\nu$ interacts with an inner shell electron of potential energy E_b. The photoelectric effect is not energetically possible unless $h\nu > E_b$. However, the smaller $h\nu$, the more likely an interaction will occur (as long as $h\nu > E_b$). In the case of a photoelectric event, the photon ceases to exist. The kinetic energy given to the electron, independent of the scattering angle θ, is

$$T = h\nu - E_b - T_a = h\nu - E_b \tag{2.95}$$

The kinetic energy T_a given to the recoiling atom is almost zero.

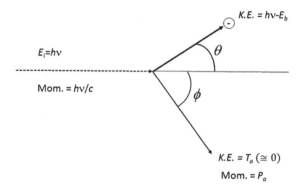

FIGURE 2.11   Kinematics of photoelectric events.

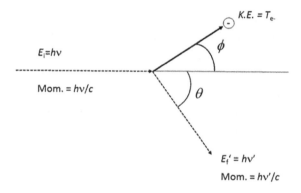

FIGURE 2.12   Kinematics of Compton scatter events.

The electron leaves the interaction at an angle θ with momentum p. Since the photon was totally absorbed, no scatter photon contributes to momentum conservation (this is accomplished by the recoiling atom).

A common phenomenon of plots of mass attenuation coefficients, especially for higher Z materials, is the appearance of a K and L edge. For instance, in the case of lead, this K-edge appears at 88-keV. Below this energy, the K-shell electrons can't participate in the PE effect because of their binding energy requirements. Only the L, M, and higher-shell electrons can participate (Krane 1988; Knoll 2010; Attix 2008).

2.5.2.2 Compton Scatter

Figure 2.12 depicts a photon with energy $h\nu$ and forward momentum $h\nu/c$ colliding with a stationary electron. Following the interaction, the electron has kinetic energy T and momentum p at an angle ϕ. The photon is scattered through an angle θ and has lost some energy to the electron, so its new energy is represented as $h\nu'$ and it now has momentum $h\nu'/c$. By conservation of energy and momentum, the energy of the redirected photon can be found:

$$hv' = \frac{hv}{1 + \dfrac{hv}{m_o c^2 (1 - \cos\theta)}} \tag{2.96}$$

where $m_o c^2$ is the rest mass of the electron (0.511 MeV) and hv and hv' are expressed in terms of MeV. The difference $hv - hv'$ represents the energy transferred to the electron.

Note that the amount of energy transferred to the electron is dependent on the angle through which the photon scatters. While it is impossible to predict the scatter angle for an individual photon, the Klein–Nishina equation allows for prediction of the average angular scattering distributions for various incident photon energies. These are described in terms of the differential scatter cross section, $d\sigma / d\omega$:

$$\frac{d\sigma}{d\omega} = Zr_o^2 \left(\frac{1}{1+\alpha(1-\cos\theta)} \right)^2 \left(\frac{1+\cos^2\theta}{2} \right) \left(1 + \frac{\alpha^2(1-\cos\theta)^2}{(1+\cos^2\theta)[1+\alpha(1-\cos\theta)]} \right) \quad (2.97)$$

A polar plot of the fraction of photons scattered through various angles as predicted by Equation (2.97) for various energies is shown in Figure 2.13. It can be seen that the fraction of photons scattered in the forward direction (to the right in Figure 2.13) increases with increasing incident energy (Krane 1988; Knoll 2010; Attix 2008).

2.5.2.3 Pair Production

Pair production is an absorption phenomenon in which a photon's kinetic energy is converted entirely into equivalent masses, namely an electron and positron. The production of both a positron and electron maintains charge parity, and both particles share the difference of the photon's original energy and the rest mass energy (2×0.511 MeV) in the form of kinetic energy. This can be expressed as

$$hv = E_- + E_+ = \left(m_o c^2 + K_- \right) + \left(m_o c^2 + K_+ \right) = K_- + K_+ + 2m_o c^2 \quad (2.98)$$

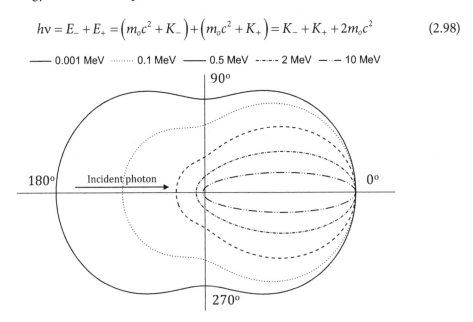

FIGURE 2.13 Polar plot of fraction of photons scattered through various angles as predicted by the Klein–Nishina equation for selected incident energies.

where E_+ and E_- are the total relativistic energies and K_+ and K_- are the positron and electron kinetic energies. The pair production process can only occur in a Coulombic field, most likely near the nucleus, in order to allow for conservation of energy and momentum. Since both a positron and electron are created, the minimum photon energy required for this process to occur is twice the electron rest mass. The resulting electron may deposit its energy locally while a portion of the positron energy may be spent in Coulombic interactions. Eventually, the positron slows and recombines with an available electron, creating two 0.511 MeV annihilation photons (Krane 1988; Knoll 2010; Attix 2008).

2.5.2.4 Photonuclear Reactions
Provided a photon has sufficient energy, it is possible for it to interact directly with a nucleus and supply it with sufficient excitation energy to cause the emission of a neutron. The free neutron shares the excess energy with the recoil nuclei. A common target for this reaction is ^9Be:

$$^9_4Be + h\nu \rightarrow \,^8_4Be + \,^1_0n \quad Q = -1.67\,\text{MeV} \tag{2.99}$$

Thus, $h\nu$ must be greater than 1.67 MeV for this reaction to be energetically possible.

The production rates for photonuclear reactions are generally low. Reactions in other materials are possible as well, and of particular interest are photoneutrons produced in high-energy medical accelerators. Materials such as tungsten have threshold energies on the order of several MeV and photoneutrons can provide additional shielding considerations for facilities employing accelerators capable of producing photons above this threshold energy, such as high-energy medical accelerators (Veinot et al. 1998).

2.5.3 Neutrons

2.5.3.1 Cross Sections
Neutrons interact with nuclei through scatter (elastic and inelastic) or absorption reactions. In many of these reactions, the probability of interaction per unit pathlength in a medium varies as a function of the neutron energy. Total interaction probabilities are described in terms of the cross section and are commonly given in units of barns (one barn = 10^{-24} cm^2). The microscopic cross section, σ, is the probability of a certain type of interaction occurring per nucleus. For a total number of atoms in a volume of interest, the microscopic cross section may be converted to a macroscopic cross section:

$$\Sigma = N\sigma \tag{2.100}$$

The macroscopic cross section is used as an indication of the interaction probability for a particular mass of material, takes into account the shape and dimensions of the material, and has units of cm^{-1}. Many cross sections have an energy dependence that decreases by the square root of the incident neutron energy (e.g., $1/\sqrt{E} = 1/v$). The neutron mean free path, λ is the inverse of the macroscopic cross section and describes the average path length a neutron will travel in a material before undergoing a reaction. The number of total reactions in a volume can be determined by combining the probability of a reaction occurring with the neutron fluence, ϕ. The fluence is a measure of particle density and is typically given for a specific energy. For a given position, r, the neutron reaction rate, R, is given by

$$R = \int_{E_{min}}^{E_{max}} \phi(E,r)\Sigma(E)dE \tag{2.101}$$

where E_{min} and E_{max} are the minimum and maximum neutron energies in the field at r.

2.5.3.2 Absorption

At low energies neutrons can be absorbed by nuclei, as discussed in Section 2.4.11. For neutrons with energies up to 10 MeV, the primary mode of interaction in tissue is through either absorption (for thermal neutrons) or elastic scatter (for higher neutron energies) reactions. In each case, charged particles are produced that typically deposit their energy near the reaction site. The number of reaction products and the charged particle spectrum depend on the reaction type (absorption or scatter), as well as the incident neutron energy.

Two absorption reactions are responsible for the largest dose contributions in tissue. The first is the $^1H(n,\gamma)^2H$ reaction, where the photon energy is 2.2 MeV and equals the binding energy of the deuteron. The second is the $^{14}N(n,p)^{14}C$ reaction with a Q-value of 0.62 MeV. This energy is shared by the proton and the recoiling ^{14}C atom (the proton receives 0.58 MeV).

Other absorption reactions are possible and vary depending on the target nuclei and incident neutron energy. A number of absorption reactions are important in the areas of radiation dosimetry and detection including:

- $^3He(n,p)^3H$—This reaction is commonly used for neutron detectors.

- $^6Li(n,\alpha)^4He$—^6Li-enriched LiF is a common material used in thermoluminescent detectors for neutron dosimetry (Veinot and Hertel 2001; Veinot and Hertel 2005b).

- $^{10}B(n,\alpha)^7Li$—Boron-10 detectors are commonly used for neutron detection.

- $^{14}N(n,p)^{14}C$—An important reaction for neutron dosimetry and for carbon-dating.

- $^{23}Na(n,\gamma)^{24}Na$—Since the body contains sodium, this reaction is used as a screening tool for persons irradiated in a nuclear criticality (Veinot, Gose, and Bogard 2009).

Finally, neutrons can produce fission reactions leading to large releases of energy as a result of the binding energies discussed earlier. In addition, at higher energies it is possible for multiple neutrons to be emitted by a nucleus following absorption of a neutron. These are referred to as (n,xn) reactions. These reactions are covered in more detail in Chapter 4.

2.5.3.3 Elastic Scatter

Since neutrons are uncharged and unaffected by the Coulombic fields of atoms, they frequently interact directly with the nucleus. For energetic neutrons, a common interaction is elastic scattering, much like a collision between billiard balls. In this reaction, momentum and kinetic energy are conserved and, assuming the neutron's velocity does not require relativistic corrections, classical mechanics may be used.

In scatter reactions, neutrons collide with atomic nuclei and are redirected after transferring a portion of their energy to the scattering atom. The incident neutron recoils on average 180 degrees and the fractional energy transferred, *f*, is determined by the mass of the scattering element:

$$f = \frac{2M}{(M+1)^2} \tag{2.102}$$

where M is the atomic mass of the scattering element. Using Equation (2.102), the average fractional energy transfer from elastic scatter is

$$E_f = E_i \frac{2M_1 M_n}{(M_1 + M_n)^2} \tag{2.103}$$

Values of E_f are 0.5 for hydrogen ($M = 1$), 0.142 for carbon ($M = 12$), 0.124 for nitrogen ($M = 14$), and 0.111 for oxygen ($M = 16$). The dose from elastic scatter of neutrons in tissue is dominated by hydrogen because of the high average energy transfer value, but contributions from other elements including carbon, oxygen, and nitrogen are also significant, particularly at high neutron energies where resonances in the scatter cross sections increase reaction probabilities.

The probabilities of one of the two reactions (absorption or scatter) occurring are dependent on the incident neutron energy. The absorbed dose to the tissue from incident neutrons is also determined by the reaction type, since the resultant particles and photons will deposit their energy according to their energies and ranges (stopping powers). Since the reaction type, reaction rate, and energy deposition after reaction are energy-dependent, the absorbed dose per incident neutron also exhibits a strong energy dependence (Veinot and Hertel 2005a). This effect is further complicated by the wide range of neutron energies typically encountered that span from fractions of eV to tens of MeV (Krane 1988; Knoll 2010; Attix 2008).

2.5.3.4 Inelastic Scatter

In some cases, neutrons may be absorbed by a nucleus and re-emitted at a different energy. This inelastic scatter process includes a change in the state of the nuclei and the re-emission of the neutron is often accompanied by a gamma photon. Inelastic scatter reactions generally have small cross sections at low neutron energies and become more important as the neutron energies reach various reaction thresholds.

2.5.4 Heavy Charged Particles

Heavy charged particles are atomic nuclei, including protons. These particles are characterized by their large size and net charge and thus have extremely short ranges in materials. The interaction mechanisms are Coulombic and primarily involve interactions with electrons, although they may involve direct interactions with other atomic nuclei. Because of their masses, the particles carry a large amount of momentum, and as they traverse a medium, they interact with a number of electrons and transfer energy in the process. The amount of energy transferred per interaction is small, but because of the large number of electrons, the slowing down process is continuous. As the particle continues to lose energy through the Coulombic interactions, the energy loss rate eventually increases until the particle is stopped. As the particle slows significantly, its interaction time with the electrons increases, further enhancing the loss rate. Some electrons may become bound to the incident nuclei, thereby reducing its charge (Krane 1988; Knoll 2010; Attix 2008).

2.5.5 Range

The concept of a particle range is somewhat dependent on the particle in question. For photons and neutrons, the average ranges can be defined in terms of the mean free paths of the particles and are the average distance travelled before an interaction occurs. Alternatively, the maximum straight-line distance of a field of particles can be found. For charged particles, effects such as straggling can complicate the definition of a particle range. In the case of heavy charged particles, the influence of straggling is minimal, particularly at high energies, owing to the large particle mass and momentum. For electrons, however, the particle path is rarely straight for any significant distance.

Heavy charged particle ranges are often given in terms of the half-value layer of an absorber. That is, the range, R is defined as

$$R = \frac{I}{I_o} = 0.5 \tag{2.104}$$

where I is the initial field intensity and I_o is the intensity at some depth in the absorber. Alternatively, the half-value range can be extrapolated to zero and this value given as the range.

In the case of electrons, the impact of straggling can be considerable. Additionally, since the electrons have a small mass, they can be scattered at large angles, further reducing observed flux through a moderator.

2.5.6 Attenuation Coefficients

The primary modes of photon interaction are the photoelectric effect, Compton scatter, and pair production as discussed in Sections 2.5.2.1–2.5.2.3. The summation of these interaction mechanisms describes the total attenuation coefficient, μ, which is the total probability that a photon will undergo an interaction. For a beam of photons, the decrease in intensity of photons having initial energy E is defined in terms of the attenuation coefficient as

$$\frac{\phi(E)}{\phi(E)_o} = e^{-\mu x} \tag{2.105}$$

where x is the thickness of the absorber medium. The mean free path, λ, is the inverse of the attenuation coefficient:

$$\lambda = \frac{1}{\mu} = \frac{\displaystyle\int_0^\infty x e^{-\mu x} dx}{\displaystyle\int_0^\infty e^{-\mu x} dx} \tag{2.106}$$

The linear attenuation coefficient has units of inverse distance. To account for the material density, the mass attenuation coefficient is more commonly used. This is simply the linear attenuation coefficient divided by the material density, μ/ρ, and has units of $cm^2 \cdot g^{-1}$.

The total mass attenuation coefficient is the sum of the coefficients for the various interaction mechanisms. In the case of photons, the total mass attenuation coefficient (μ/ρ) is the sum of the photoelectric effect (τ/ρ), the Compton effect (σ/ρ), and pair production (κ/ρ) coefficients (neglecting photonuclear effects) (Hubbell and Seltzer 1995).

2.5.7 Mass Energy Transfer Coefficient

For photons, the quantity of interest in radiation protection is the amount of energy transferred to electrons. In the case of the photoelectric effect, a bound electron absorbs the photon. Some energy is required to free the electron (the binding or ionization energy), B. The excess energy transferred by the incident photon is manifested in the form of kinetic energy of the electron, T. This kinetic energy is

$$T = E_\gamma - B = h\nu - B \tag{2.107}$$

The vacancy created by the ionized electron results in additional photon emissions and Auger electrons. If the average energy of these subsequent emissions is δ, then the total fraction of energy transferred to the electron produced in the photoelectric event is $1 - \delta/E_\gamma$. The mass energy transfer coefficient, μ_{tr}/ρ, accounts for these losses. The mass energy transfer coefficient can also be defined as the quotient of the fraction of incident radiant energy that is transferred to kinetic energy of charged particles (dR_{tr}/R) by the traversed distance, dl within a material having density, ρ:

$$\frac{\mu_{tr}}{\rho} = \frac{dR_{tr}}{R\rho dl} \tag{2.108}$$

Using the fractional energy losses due to other losses, δ, the mass energy transfer coefficient for the photoelectric effect is:

$$\frac{\mu_{tr}}{\rho} = \frac{\tau}{\rho}\left(1 - \frac{\delta}{E_\gamma}\right)$$

(2.109)

Radiative losses are not accounted for in Equation (2.109).

Similarly, for Compton scatter events, if the fraction of the incident photon energy that is converted to kinetic energy of the electrons produced in these events is $\bar{T}$, then the mass energy transfer coefficient is

$$\frac{\mu_{tr}}{\rho} = \frac{\sigma}{\rho}\frac{\bar{T}}{E_\gamma}$$

(2.110)

For pair production, the rest masses of the electron and positron are $2mc^2$, and the mass energy transfer coefficient is

$$\frac{\mu_{tr}}{\rho} = \frac{\kappa}{\rho}\left(1 - \frac{2mc^2}{E_\gamma}\right)$$

(2.111)

Combining these three modes of interactions, the total mass energy transfer coefficient is

$$\frac{\mu_{tr}}{\rho} = \frac{\tau}{\rho}\left(1 - \frac{\delta}{E_\gamma}\right) + \frac{\sigma}{\rho}\frac{\bar{T}}{E_\gamma} + \frac{\kappa}{\rho}\left(1 - \frac{2mc^2}{E_\gamma}\right)$$

(2.112)

(Krane 1988; Knoll 2010; Attix 2008).

2.5.8 Mass Energy Absorption Coefficient

As mentioned above, the mass energy transfer coefficients do not account for losses due to radiative processes. If g is the energy of liberated charged particles lost in radiative processes, then the mass energy absorption coefficient is

$$\frac{\mu_{en}}{\rho} = \left(1 - g\right)\frac{\mu_{tr}}{\rho}$$

(2.113)

2.6 RADIATION PROTECTION QUANTITIES AND UNITS

2.6.1 Kerma

There are two important fundamental quantities related to energy deposition within the body. The first, termed the kerma (K), is a non-stochastic quantity relevant to kinetic energy exchange via uncharged radiations. The kerma (Kinetic Energy Released per unit Mass) corresponds to the kinetic energy transferred to charged particles within a volume and is defined as the difference of the radiant energy entering a volume and that leaving a volume, with the exception of radiant energy leaving the volume that was produced within the volume itself. The kerma can be formally defined in terms of the energy transferred, ε_{tr}, as

$$\varepsilon_{tr} = R_{\text{in}} - R_{\text{out}} + \Sigma Q \tag{2.114}$$

where R_{in} and R_{out} are the radiant energies entering and leaving the volume and $\pounds Q$ is the change in rest mass in the volume (Krane 1988; Knoll 2010; Attix 2008). The outgoing radiant energy in Equation (2.114) does not include energy created in the volume and the rest mass changes can be positive or negative. When mass is converted to energy in the volume ΣQ is positive and, conversely, negative when energy is converted to mass.

The kerma is defined as the quotient of $d\varepsilon_{tr}/dm$ where $d\varepsilon_{tr}$ is the mean sum of the initial kinetic energies of all the charged particles liberated in a mass dm of a material by the uncharged particles incident on dm:

$$K = \frac{d\varepsilon_{tr}}{dm} \tag{2.115}$$

For photons, the kerma is given in terms of the energy fluence, Ψ, and the mass energy transfer coefficient, μ_{tr}/ρ, as

$$K = \int_E \Psi(E) \left(\frac{\mu_{tr}}{\rho} \right)_E dE \tag{2.116}$$

while for neutrons it is typically given in terms of the particle fluence:

$$K = \int_E \phi(E) \left(\frac{\mu_{tr}}{\rho} \right)_E E \, dE \tag{2.117}$$

In practice, so-called kerma coefficients are determined based on the material composition and neutron energy. These kerma coefficients, F_n, are defined as

$$F_n = \left(\frac{\mu_{tr}}{\rho} \right)_E (E) \tag{2.118}$$

An average value for F_n for a neutron spectrum can be found from

$$\left(\bar{F}_n \right)_{\phi(E)} = \frac{K}{\phi} = \frac{\int_E \phi(E)(F_n)_E \, dE}{\int_E \phi(E) dE} \tag{2.119}$$

(Krane 1988; Knoll 2010; Attix 2008).

Following the kinetic energy transfer to the charged particles (primarily electrons and positrons in the case of photons and nuclei in the case of neutrons), the charged particles themselves will lose energy through interactions in the volume. These energy losses may be in the form of radiative losses (bremsstrahlung in the case of electrons or annihilation in the case of positrons, for example) or through Coulombic or other collisional-type interactions. The kerma can be divided into two components based on the energy loss methods of the charged particles: collisional kerma and radiative kerma. These are denoted K_c and K_r, respectively.

The collisional kerma excludes the radiative losses by the liberated charged particles. For a given fluence, ϕ, of (uncharged) particles having energy E the K_c is

$$K_c = \phi E \frac{\mu_{en}}{\rho} = \phi E \left(\frac{\mu_{tr}}{\rho} \right)(1-g) \tag{2.120}$$

where μ_{en}/ρ is the mass energy absorption coefficient and g is the fraction of energy loss from radiative processes.

The collisional kerma can be expressed in terms of the distribution, ϕ_E, of the uncharged particle fluence with respect to energy as

$$K_c = \int \phi_E E \frac{\mu_{en}}{\rho} dE = \int \phi_E E \left(\frac{\mu_{tr}}{\rho} \right)(1-g)dE = K(1-\bar{g}) \tag{2.121}$$

where $\bar{g}$ is the mean value of g averaged over the distribution of the kerma with respect to the electron energy (ICRP 2006).

2.6.2 Absorbed Dose

The absorbed dose can be defined in a similar manner as kerma. In Equation (2.114), the energy transferred was defined. A quantity that accounts for both charged and uncharged radiant energies entering and leaving a volume is termed the energy imparted and is given by

$$\varepsilon = \left(R_{in} \right)_u - \left(R_{out} \right)_u + \left(R_{in} \right)_c - \left(R_{out} \right)_c + \Sigma Q \tag{2.122}$$

where $\left(R_{in} \right)_u$ and $\left(R_{out} \right)_u$ are the uncharged radiant energies entering and leaving the volume and $\left(R_{in} \right)_c$ and $\left(R_{out} \right)_c$ are the charged radiant energies entering and leaving the volume and ΣQ is as given in Equation (2.114) (ICRP 1998b; Knoll 2010). The absorbed dose can then be defined in terms of ε as

$$D = \frac{d\varepsilon}{dm} \tag{2.123}$$

Thus, the absorbed dose includes contributions from both charged and uncharged radiations.

2.6.2.1 Charged Particle Equilibrium

Charged particle equilibrium (CPE) refers to the condition where the number of charged particles, their energies, and their directions are constant throughout a volume. In other words, the distribution of charged-particle energy radiance does not vary within the volume. Consequently, the sums of the energies (excluding rest energies) of the charged particles entering and leaving the volume are equal (ICRP 2010). The numerical value of kerma approaches that of the absorbed dose to the degree that charged particle equilibrium exists, that radiative losses are negligible, and that the kinetic energy of the uncharged particles is large compared with the binding energy of the liberated charged particles. Thus, kerma is sometimes used as an approximation to the absorbed dose.

For computational dosimetry applications, the calculation of absorbed dose requires that charged particles be tracked which can significantly slow the process. For this reason, many calculations of doses are performed using the kerma approximation whereby it is assumed that when uncharged radiations undergo interactions, the subsequent energy exchange is deposited locally. This assumption eliminates the necessity to track secondary charged particles and is a reasonable approximation, provided charged particle equilibrium is established at the point of interest. In practice, CPE is lost in tissue at different energies depending on the depth at which the measurement is taken and intermediary effects such as buildup along the particle path before interaction with the medium of interest (e.g., the inclusion of air surrounding volume of tissue). In the case of photons, for example, CPE conditions are maintained up to about 3 MeV at a depth of 1 cm in tissue, while at the 3 mm depth, CPE conditions are met up to around 1.5 MeV. At the 0.007 cm depth used to determine the dose to the sensitive layer of skin, CPE is lost for photons with energies above about 200 keV (Veinot and Hertel 2010). Interfaces where changes in material composition or densities occur can also impact CPE since these may alter the production and range of charged particles.

2.6.3 Exposure–Dose Relationship

The quantities exposure and (absorbed) dose are of particular interest in radiation protection. From a practical standpoint, exposure is often one of the more commonly measured quantities in operational settings. While the exposure is defined as the ratio dQ/dm, it is dependent on the average energy required to create an ion pair in air, $\overline{W}$, termed the average or mean work function. There are, however, important subtleties that deserve mention. In fact, the exposure is more closely related to the collisional kerma of incident photons. As photons interact with the air, electrons are generated through the various interaction mechanisms. These electrons have initial kinetic energies following the interactions and undergo further interactions experiencing energy losses through either radiative (e.g., bremsstrahlung) or collisional processes as they traverse the medium. By definition, ionization resulting from energy losses from radiative processes are not included in the exposure quantity. Thus, the factor g representing the fraction of energy loss from radiative processes is incorporated as it was for the collisional kerma (Knoll 2010).

The kinetic energy of the electrons is denoted E_k, then the total kinetic energy loss from collisions is given by $E_k(1-g)$ and for n total electrons the total collisional energy loss is

$$\sum_n E_{k_n}(1-g_n)$$ (2.124)

Each electron can create more than one ion pair as it traverses the volume of interest. If N is the number of ion pairs formed by the electron and r_n the fraction created through radiative interactions, then $N(1-r_n)$ is the number of ion pairs created through collisional processes. Summing for n electrons

$$\sum_n N_n(1-r_n)$$ (2.125)

is the total number of ion pairs formed through collisional reactions. The mean energy expended in a gas per ion pair formed can then be found for large n:

$$\bar{W} = \frac{\sum_n E_{k_n}(1-g_n)}{\sum_n N_n(1-r_n)}$$ (2.126)

For dry air, a value of $\bar{W} = 33.97$ J/C is used (Knoll 2010).

For personnel measuring exposure (e.g., roentgen) in air, it is common practice in operational settings to record the results in units of dose or dose rate (Rad or Rad/hr) and sometimes in units of dose equivalent (rem or rem/hr) without realizing the significance of the quantities. Thus, it is of interest to investigate the correlation of these quantities and examine the differences in the values. As will be discussed in the sections to follow, correlating a measurement in air using a hand-held instrument (such as an ionization chamber) to dose in tissue is not possible without additional knowledge such as, for example, the photon energy spectrum at the point of measurement. However, given its common practice in operational radiation protection environments, it is important to understand the relationship between exposure and dose.

The exposure in air from a fluence of photons is found from:

$$X = \sum_i \phi E \left(\frac{\mu_{en}}{\rho}\right)_{E_{air}} \left(\bar{W}\right)_{air}$$ (2.127)

where i represents the ith photon having energy E. For example, consider a fluence of 10^9 photons/cm^2 with energy of 0.5 MeV, $\bar{W} = 33.97$ J/C, and mass energy absorption coefficient (μ_{en}/ρ) equal to 0.02966. The exposure is

$$X = \sum_i \phi E \left(\frac{\mu_{en}}{\rho}\right)_{E_{air}} \left(\bar{W}\right)_{air}$$

$$X = 7.00 * 10^{-5} \, C/kg = 0.271 \, R \qquad (2.128)$$

To find the dose in air, the work function is omitted to obtain the dose units of J/kg (Gy):

$$D_{air} = \sum_{i} \phi E \left(\frac{\mu_{en}}{\rho} \right)_{E_{air}}$$

$$D_{air} = 2.38 * 10^{-3} \, J/kg = 0.238 \, Rad \qquad (2.129)$$

Thus, for 500 keV photons, the ratio of dose in air to exposure is $0.238/0.271 = 0.88$, which is reasonably close to unity.

The dose in tissue from the same fluence, neglecting scatter, buildup, and other (important) effects, is found using the mass energy absorption coefficient for 0.5 MeV photons in tissue:

$$D_{tissue} = \sum_{i} \phi E \left(\frac{\mu_{en}}{\rho} \right)_{E_{tissue}}$$

$$D_{tissue} = 2.62 \times 10^{-3} \, J/kg = 0.262 \, Rad \qquad (2.130)$$

The ratio of dose in tissue (in units of rad) to exposure in air (in units of roentgen) is $0.262/0.271 = 0.97$.

A quick examination of the calculations above shows that the ratio of dose in tissue to exposure in air is simply the ratio of the mass energy absorption coefficients. These coefficients (Hubbell and Seltzer 1995) are energy-dependent, and Figure 2.14 shows the ratios of the μ_{en}/ρ values for tissue to air and water to air. Since the ratios are relatively constant between about 200 keV and 2 MeV, the ratio of exposure in air to dose in tissue determined using the method described above will remain close to 0.97. This provides a good approximation to tissue dose using relatively simple measurement methods for photon energies commonly encountered in many facilities and environments.

2.6.4 Linear Energy Transfer and Quality Factor

The linear energy transfer, L or LET, is the average linear rate of energy loss of a charged particle as it passes through a medium. That is, the LET represents the radiation energy lost per unit length of the path and is given by

$$L = \frac{dE}{dl} \qquad (2.131)$$

where dE is the mean energy lost by a charged particle from collisions with electrons over a distance dl in matter (ICRP 2007).

The quality factor, Q, is introduced in radiation protection to account for the differing biological effectiveness of radiation. It is determined using a quality factor function, $Q(L)$,

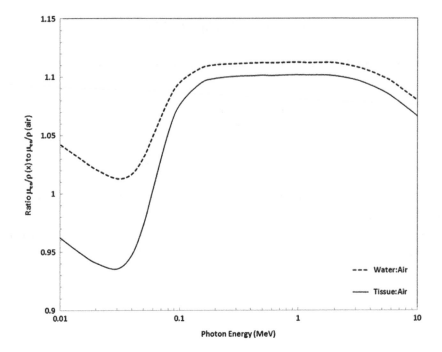

FIGURE 2.14 Ratios of mass energy absorption coefficients, μ_{en}/ρ, for air, water, and tissue.

which characterizes the biological effectiveness of a charged particle with an unrestricted LET, L, in water at a point of interest relative to the effectiveness of a reference radiation at this point (ICRU 2001).

In general, photons, X-rays, and gamma rays, with no energy specified, are used as reference radiation to which other radiations are compared. In the International Commission on Radiological Protection (ICRP) Publication 21 (ICRP 1973), the quality factor was specified for specific values of L. This $Q(L)-L$ relationship was revised in ICRP Publication 60 to reflect the relationship in Table 2.8 (ICRP 1991).

Figure 2.15 compares the quality factor versus LET relationship from ICRP Publication 21 and ICRP Publication 60. The quality factor versus LET relationship did not change between ICRP Publication 60 and ICRP Publication 103 (ICRP 1991; ICRU 1998a). The quality factor at a point in tissue is given by:

$$Q = \frac{1}{D} \int_{L=0}^{L=\infty} Q(L) D_L dL \tag{2.132}$$

where D is the absorbed dose in tissue, D_L is the distribution of D in unrestricted linear energy transfer L, and $Q(L)$ is the corresponding quality factor. The integration is performed over D_L, due to all charged particles, excluding their secondary electrons (ICRP 1991, 2007).

TABLE 2.8   Quality Factor Q(L) as a Function of Unrestricted LET (L)

L (keV μm^{-1})	Q(L)
<10	1
10–100	$0.32L - 2.2$
>100	$300/\sqrt{L}$

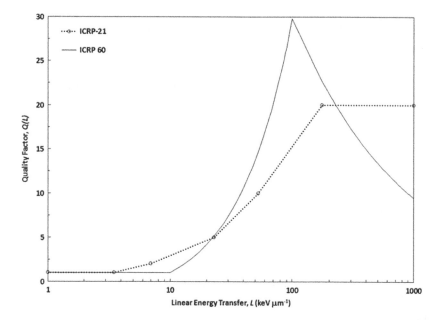

FIGURE 2.15   ICRP Publication 21 and ICRP Publication 60 $Q(L) - L$ relationships (ICRP 1973, 1991).

More specifically, the above is termed the "mean quality factor." When the integration is performed over all radiations and energies, the effective quality factor $\bar{Q}$ is found and is simply the ratio of dose equivalent to absorbed dose:

$$\bar{Q} = \frac{\int_L Q(L)\left(\dfrac{dD}{dL}\right)dL}{\int_L \left(\dfrac{dD}{dL}\right)dL} \tag{2.133}$$

or simply

$$\bar{Q} = \frac{H}{D} \tag{2.134}$$

2.7 PROTECTION QUANTITIES

In radiation protection, the primary aim is to protect the workers and members of the public. This protection must be balanced with associated costs and workplace efficiencies, particularly when doses received are low and the return on investment for protection methods reaches the point of diminishing returns. There are two primary categories of quantities that are used to limit exposures to persons: protection quantities and operational quantities. In broad terms, these may be described as risk-based limiting regulatory quantities (in the case of protection quantities) and those quantities that are used in operational settings and are measurable (the operational quantities). The protection quantities are intended to relate overall risk and are developed using humanoid phantoms meant to represent reference persons. These reference phantoms have undergone numerous changes as knowledge and capabilities have increased, allowing for more precise computation of dose to specific organs. The evolution of phantom designs and the associated dose coefficients will be discussed in more detail in Chapters 5 and 7, but a review of the evolution of the protection quantities will be covered briefly here.

2.7.1 ICRP Publication 26

In 1977, the ICRP released its recommendations in its Publication 26 (ICRP 1977). Included in this release was the concept of the protection quantity effective dose equivalent, H_E. The foundation for the effective dose equivalent was a risk-based approach to radiation protection. That is, various types and energies of radiations that delivered doses to specific organs and radiations were assigned factors that accounted for the radiation sensitivities of those organs and the effectiveness of various radiations at producing damage to cells. The computation of the effective dose equivalent involves calculating the average dose over an organ or tissue from a radiation. This organ dose is then coupled with a quality factor determined from the radiation type and (where appropriate) energy spectrum averaged over the organ or tissue. In the case of photons and electrons, the quality factor is taken to be unity, while for neutrons, the quality factor for a given organ or tissue and radiation, $Q_{T,R}$, is based on the quality factor versus linear energy transfer, $Q(L)-L$ function defined by the ICRP and the International Commission on Radiation Units and Measurements (ICRU). In this use, the linear energy transfer is related to the relative biological effectiveness of the radiation.

The mean organ dose equivalent, H_T, is found by averaging over an entire organ or tissue and applying the energy-dependent quality factor based on the radiation type and energy within the organ or tissue:

$$H_T = \iint_{m \, L} Q(L) D_L \frac{dL \, dm}{m} \tag{2.135}$$

where D_L is the distribution of absorbed dose in unrestricted linear energy transfer (LET), and the integral ranges over LET and the mass, m, of the organ (ICRP 1977).

For the specific case of neutrons having energy E, the mean quality factor, Q_n, is determined based on the LET of the produced radiations in a similar form as Equation (2.132), but stated more explicitly:

$$Q_n\left(E_n\right) = \frac{1}{D} \int_{L_{min}}^{L_{max}} Q\left(L\right) D_L\left(L, E_n\right) dL \tag{2.136}$$

Each organ and tissue is assigned a risk weighting factor, w_T, and the effective dose equivalent is found by summing over all organs and tissues the product of the dose equivalent delivered to that organ or tissue and its respective weighting factor:

$$H_E = \sum_T H_T w_T \tag{2.137}$$

The organ weighting factors were assigned based on their stochastic risk of developing cancers with the gonads being the highest of the six single organs assigned weighting factors. Also included were remainder organs that consisted of the five other organs not specifically assigned a weighting factor that received the highest dose. The organs and their respective weighting factors are listed in Table 2.9. Note that the remainder organ assigns a weighting factor of 0.06 to each of the five most highly irradiated organs not specifically assigned a weighting factor. Since these organs can change depending on the radiation, energy, and orientation of the radiation field, the effective dose equivalent is not additive.

2.7.2 ICRP Publication 60

The 1990 recommendations of the ICRP, released as its Publication 60, included a number of changes to the definition of the protection quantity (ICRP 1991). The protection quantity was renamed effective dose, E, and included a revised set of organs to be used during its calculation. Further, the modifier used to account for radiation quality was changed from the quality factor to the radiation weighting factor, w_R. Whereas the quality factor (for neutrons) under ICRP Publication 26 was determined based on the spectrum present in the organ or tissue, under ICRP Publication 60 the radiation weighting factor is determined based on the radiation type and energy spectrum incident on the body or phantom (ICRP 1977, 1991). Radiation weighted absorbed dose in the organ or tissue is referred to as the equivalent dose, H_T and is found from

$$H_T = \sum_R w_R D_{T,R} \tag{2.138}$$

where $D_{T,R}$ is the absorbed dose averaged over the organ or tissue T from radiation R. The protection quantity effective dose, E, is given by

$$E = \sum_T w_T H_T \tag{2.139}$$

TABLE 2.9   Organ and Tissue Weighting Factors under ICRP Publication 26 (ICRP 1977), ICRP Publication 60, and ICRP Publication 103 (ICRP 2007)

Organ/Tissue	ICRP Publication 26	ICRP Publication 60	ICRP Publication 103
Bone surface	0.03	0.01	0.01
Bladder	–	0.05	0.04
Brain	–	–	0.01
Breast	0.15	0.05	0.12
Colon	–	0.12	0.12
Gonads	0.25	0.20	0.08
Liver	–	0.05	0.04
Lungs	0.12	0.12	0.12
Esophagus	–	0.05	0.04
Red bone marrow	0.12	0.12	0.12
Salivary glands	–	–	0.01
Skin	–	0.01	0.01
Stomach	–	0.12	0.12
Thyroid	0.03	0.05	0.04
Remainder	0.30[a]	0.05[b]	0.12[c]

[a] The five most highly irradiated other organs and tissues.
[b] Adrenals, brain, upper large intestine, small intestine, kidneys, muscle, pancreas, spleen, thymus, and uterus.
[c] Adrenals, extrathoracic tissue, gallbladder, heart, kidneys, lymphatic nodes, muscle, oral mucosa, pancreas, prostate (male), small intestine, spleen, thymus, uterus/cervix (female).

The ICRP Publication 60 organ and tissue weighting factors are listed in Table 2.9. The approach of ICRP Publication 60 represents a somewhat more straightforward computational approach compared to the ICRP Publication 26 method, since the radiation weighting is based on the spectrum incident on the phantom instead of requiring the averaging over each organ and tissue.

The continuous function for the radiation weighting factor provided in ICRP Publication 60 (to serve as an approximation) is

$$w_R = 5 + 17e^{-\frac{(\ln(2E))^2}{6}} \qquad (2.140)$$

where E is the neutron energy in MeV.

2.7.3 ICRP Publication 103

The protection quantities defined under the ICRP Publication 103 recommendations were further modified in 2007 (ICRP 2007). The most significant changes include the introduction of male and female phantoms developed using high-resolution tomographic data of humans, as compared to previous phantoms (ICRP 2009), which were mathematical and androgynous, revisions to the recommended radiation weighting factors, and the number of organs specifically included in the definition of effective dose. The calculation of organ doses based on these tomographic phantoms are coupled with revised organ and tissue weighting factors with more organs assigned specific risk weighting factors. The w_T values are specified for each organ, with male and female reproductive organs included as part of the remainder component. Since the w_T values are unchanging, the ICRP Publication 103 effective dose is additive (ICRP 2007). Other changes include modifications to the radiation weighting factors for neutrons, as well as the inclusion of recommended w_R values for other particles. As in ICRP Publication 60, the w_R value is based on the radiation spectrum and type that is incident on the phantom (ICRP 1991). The organ and tissue weighting factors are provided in Table 2.9 and the radiation weighting factors of ICRP Publication 60 and ICRP Publication 103 are listed in Table 2.10.

As in ICRP Publication 60, the equivalent dose to each organ and tissue is determined according to Equation (2.138). The effective dose is determined by first finding the organ equivalent doses in both the male and female phantoms, averaging them, applying the appropriate value of w_T, and summing over all organs and tissues (ICRP 1991):

$$E = \sum_T w_T \left[\frac{H_T^M + H_T^F}{2} \right] \tag{2.141}$$

where H_T^M and H_T^F are the equivalent doses to organ or tissue T in the male and female phantoms, respectively.

The radiation weighting factors given in ICRP Publication 103 for neutrons are

$$w_R = \begin{cases} 2.5 + 18.2e^{-\frac{\left[\ln(E_n)\right]^2}{6}} & E_n < 1\,\text{MeV} \\[2em] 5 + 17.0e^{-\frac{\left[\ln(2E_n)\right]^2}{6}} & 1\,\text{MeV} \le E_n \le 50\,\text{MeV} \\[2em] 2.5 + 3.25e^{-\frac{\left[\ln(0.04E_n)\right]^2}{6}} & E_n > 50\,\text{MeV} \end{cases} \tag{2.142}$$

The ICRP Publication 60 and ICRP Publication 103 radiation weighting factors are compared in Figure 2.16.

The calculation method for determining effective dose is summarized in Figure 2.17 for the ICRP Publication 103 recommendations.

TABLE 2.10   Radiation Weighting Factors under ICRP Publication 60 (ICRP 1991) and ICRP Publication 103 (ICRP 2007)

Organ/Tissue	ICRP Publication 60	ICRP Publication 103
Photons	1	1
Electrons and muons	1	1
Protons and charged pions	5	2
Alpha particles, fission fragments, and heavy ions	20	20
Neutrons[a] < 10 keV	5	
10 keV to 100 keV	10	
>100 keV to 2 MeV	20	Continuous Function
>2 MeV to 20 MeV	10	
>20 MeV	5	

[a] A continuous function was provided to serve as an approximation.

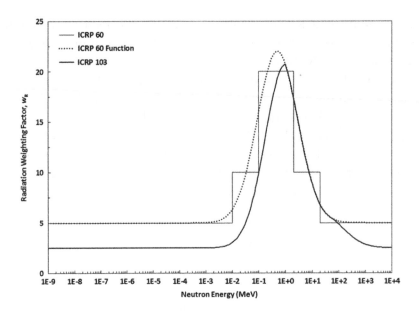

FIGURE 2.16 Radiation weighting factors given in ICRP Publication 60 and ICRP Publication 103 (ICRP 2007).

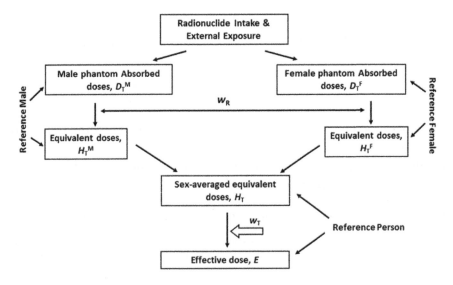

FIGURE 2.17 ICRP Publication 103 method for determining equivalent doses and effective doses (ICRP 2007).

In conjunction with the release of new recommendations, ICRU and/or ICRP task groups perform calculations of dose coefficients to be used in accordance with the revised recommendations. These reports, including ICRU Report 57 (ICRU 1998a) and ICRP Publication 74 (ICRP 1996) which followed the ICRP Publication 60 (ICRP 1991) guidelines, were updated in ICRP Publication 116 (ICRP 2010) according to guidance contained in ICRP Publication 103. These dose coefficients convert particle fluence as a function of energy to dosimetric quantities including organ equivalent dose and effective dose. A more thorough discussion of these dose coefficients including their method of calculation is contained in Chapters 7 and 8.

The lens of the eye is not an organ included in the effective dose, however, there has recently been increased concern of the effects of irradiation to the lens of the eye, which has led to the reduction in annual limits (ICRP 2012). Specific phantoms have been developed (Behrens 2012; Behrens and Dietze 2010) that provide more detail, especially of the sensitive regions of the eye.

The control of the exposure of local skin and the lens of the eye is intended to restrict tissue reactions that occur above a threshold dose. For evaluating tissue reactions, the mean absorbed dose is assessed, weighted by a specific relative biological effectiveness (RBE) for high-LET radiation. The absorbed dose to local skin is defined as the absorbed dose averaged over 1 cm^2 of the skin anywhere on the surface of the body. The specific annual dose limits recommended for the skin apply to the local skin dose at the most highly irradiated area of the skin.

2.8 OPERATIONAL QUANTITIES

Protection quantities are used to define dose limits for both workers and members of the public and are based on complex humanoid phantom models requiring the determination of absorbed doses to specific organs in the body. They are not point quantities and are not measurable. Thus, they are not appropriate for the calibration of either instruments or personnel dosimeters. To address this restriction, a second class of quantities, termed operational quantities, have been developed by the ICRU. These operational quantities are intended to be both measurable and representative of the corresponding protection quantity. It is important to note that operational quantities have been developed for both internal and external exposure. For external exposure, these quantities apply to dose (and dose rate) instrument calibrations and to personnel dosimeters. For internal exposures, the operational quantities are for measurements used to estimate intakes of radionuclides by personnel, including air or body concentrations. The following discussion will concentrate on the quantities used for external radiations.

For the calculation of operational quantities, specific phantoms are employed. Whereas the protection quantities use the radiation weighting factor to account for radiation effectiveness, the operational quantities apply the quality factor (see Section 2.6.4).

2.8.1 Ambient Dose Equivalent

For area monitoring, the operational quantity used to estimate the protection quantities for measurements and calibration of instruments should be a point quantity that provides a single value at that point for a given radiation field and should also be independent of the directional distribution of the radiation field. For the control of dose to personnel, the three primary protection quantities are the effective dose, which is analogous to the whole-body dose, the dose to the lens of the eye, and the dose to the skin. The operational quantity recommended by the ICRU to be used to approximate these is termed the "ambient dose." For the control of effective dose, the ambient dose equivalent, denoted as $H^*(10)$, is used. The ambient dose equivalent is defined as the dose equivalent that would be produced by the corresponding expanded and aligned field in the ICRU sphere at a depth of 10 mm on the radius opposing the direction of the aligned field. A second quantity is the directional dose equivalent, $H'(d,\Omega)$, which is used for assessing the dose to the skin and the extremities (hands, wrists, and feet), as well as the dose to the lens of the eye. The directional dose equivalent at a point in a radiation field is the dose equivalent that would be produced by the corresponding expanded field in the ICRU sphere at a depth, d, on a radius in a specified direction, Ω. For the lens of the eye, the depth, d, is taken to be 3 mm, while for skin $d = 0.07$ mm. In both $H^*(10)$ and $H'(d,\Omega)$, the sphere consists of ICRU tissue substitute which has a density of $1\,\mathrm{g \cdot cm^{-3}}$ and a mass composition of 76.2% oxygen, 11.1% carbon, 10.1% hydrogen, and 2.6% nitrogen. This material represents a significant limitation in that it cannot be manufactured but can be simulated in calculations. Both $H^*(10)$ and $H'(d,\Omega)$ retain their character of a point quantity and the property of additivity by using a fixed depth in the definition of each. Regarding the specifications of aligned and expanded fields, an expanded radiation field is defined as a field in which the fluence and its direction and energy distributions have the same value throughout the volume of interest as in the actual field at the point of reference. An expanded field is one such that the entire sphere is exposed to a homogeneous radiation field with the same fluence, energy distribution, and direction distribution as in the point of interest of the real radiation field (ICRU and Measurements 1998a, 2001). The quality factor is used to account for radiation quality.

2.8.2 Personal Dose Equivalent

The operational quantity used for the calibration of personnel dosimeters is termed the "personal dose equivalent," $H_p(d,\alpha)$ where α indicates the angle of incidence to the body. The personal dose equivalent is defined for the body and can be taken for different locations and depths. In order to maintain consistency, the location and depth need to be specified. As with the ambient and directional dose equivalent, depths are typically taken to be 10 mm for penetrating radiations, 3 mm for the lens of the eye, and 0.07 mm for the skin. A special case for the trunk of the body is often assumed since that is the most common location used for personnel dosimetry. In the case of the trunk, the ICRU specifies a 30 cm × 30 cm × 15 cm thick phantom constructed of ICRU tissue substitute and denotes the corresponding quantity as $H_{p,slab}(d,\alpha)$.

In practice, dosimeters are calibrated using phantoms that may be constructed completely of polymethyl-methacrylate (PMMA) or of a PMMA box filled with water.

The personal dose equivalent is defined for particles of type i as:

$$H_{p,i} = \int H_{p,i}(E,\alpha) \frac{d\phi_i}{dE} dEo \qquad (2.143)$$

where $d\phi_i / dE$ is the fluence of particles i with kinetic energies in the interval dE around E.

Although the personal dose equivalent provides a convenient means to calculate dose quantities at various depths and perform dosimeter calibrations, it is a limited approximation for extenuating cases such as dose to the extremities (i.e., the wrist and fingers) and to the lens of the eye. There is now strong interest in better modeling of the eye lens (Behrens 2012) in the assessment of doses to the eye lens and in the operational quantity $H_p(3)$.

Individual dosimeters for monitoring the dose to the lens of the eye should be designed and calibrated in terms of this quantity on an appropriate phantom. Right-circular cross section cylindrical phantoms or slab phantoms have been proposed (Behrens 2012; Behrens and Dietze 2010), though the first is more appropriate for tests of angular dependence of response, especially at large angles. The ICRU has given recommendations on phantoms, composed of ICRU four-element tissue, for calibration procedures of individual dosimeters to be worn on the trunk (ICRU 1992). The calibration of dosimeters for the determination of the operational quantities requires that it be placed on a phantom that provides a reasonable estimation of the backscatter properties of that part of the body on which it is to be worn. Several types of phantoms can be adequate and have been considered. The latest ICRU working groups are recommending that, for dosimeters to be worn on the finger, a right-circular cylinder phantom of diameter 19 mm and length 300 mm be used. For dosimeters to be worn on the wrist or ankle, a right-circular cylinder phantom of diameter 73 mm and length 300 mm has been proposed. These are similar to phantoms specified in ANSI N13.32 (ANSI 2008), specifically those referred to as the rod and pillar phantoms. Past designs have included aluminum inserts within the cylinder to serve as a surrogate for bone.

For the lens of the eye, a right-circular cylinder phantom composed of ICRU 4-element tissue of diameter 200 mm and height of 200 mm is recommended. When calibrating dosimeters on a phantom made of a material that is a surrogate for ICRU tissue, corrections may be necessary to account for differences in backscatter properties between ICRU tissue and the surrogate material. These differences will, of course, be dependent on the radiation types and energies.

Similar to the ambient and directional dose equivalent, the quality factor is currently used to account for radiation quality since it is a useful approximation to the RBE of various radiations. However, since it is generally applicable to whole-body irradiations that include sensitive and blood-forming regions, it is not completely appropriate for the modification to absorbed doses in the extremities. No guidance has been adopted on applications of modifying factors for these conditions, although several possibilities are being considered, including the use of specific RBE values or the adoption of the radiation weighting factor used for the protection quantities.

The appropriate operational quantity to be used in the various external exposure scenarios are listed in Table 2.11.

TABLE 2.11   Scheme of Operational Quantities Used for Dose Monitoring in External Exposure Situations

Task	Area Monitoring	Personnel Monitoring
Control of effective dose	Ambient dose equivalent $H^*(10)$	Personal dose equivalent $H_p(10)$
Control of dose to the lens of the eye	Directional dose equivalent $H'(3,\Omega)$	Personal dose equivalent $H_p(3)$
Control of dose to skin, hands, and feet	Directional dose equivalent $H'(0.07,\Omega)$	Personal dose equivalent $H_p(0.07)$

Note that under ICRP Publication 26 the term "dose equivalent" is used for both operational and protection quantities, while under ICRP Publication 60 and ICRP Publication 103, the protection quantities utilize "equivalent dose" for organs and tissues and "dose equivalent" for operational quantities.

Thus, it is important to clearly state which quantity, and under which guidance, is being determined or reported. Further complicating matters is the duplicated terminology of ICRP Publication 60 and ICRP Publication 103—since all three recommendations are being used by various agencies throughout the world, the possibility of confusion is significant and users must be aware of this and ensure the correct quantities are employed.

REFERENCES

American National Standards Institute. 2008. American National Standard: Performance Testing of Extremity Dosimeters. In *American National Standard ANSI/HPS N13.32*. McLean: Health Physics Society.

Attix, Frank Herbert. 2008. *Introduction to Radiological Physics and Radiation Dosimetry*, New York, NY: John Wiley & Sons.

Baum, Edward M., Harold D. Knox, and Thomas R. Miller. 2002. *Nuclides and Isotopes: Chart of the Nuclides*. 16th ed. New York, NY: Knolls Atomic Power Laboratory.

Behrens, R. 2012. "On the Operational Quantity Hp (3) for Eye Lens Dosimetry." *Journal of Radiological Protection* 32 (4):455.

Behrens, R., and G. Dietze. 2010. "Dose Conversion Coefficients for Photon Exposure of the Human Eye Lens." *Physics in Medicine and Biology* 56 (2):415.

Bureau International des Poids et Mesures (BIPM). 2006. "Le Système International D'Unites (SI)." Sèvres, France: Bureau International des Poids et Mesures.

Eckerman, K. F., and A. Endo. 2007. *MIRD Radionuclide Data and Decay Schemes*. 2nd ed. Reston, VA: Society of Nuclear Medicine.

Eckerman, K. F., and A. L. Sjoreen. 2014. *Radiological Toolbox V. 3.0.0*. Washington, DC: Oak Ridge National Laboratory.

Eisberg, R., and R. Resnick. 1985. *Quantum Physics of Atoms, Molecules, Solids, Nuclei, and Particles*. New York, NY: John Wiley and Sons.

Fitzgerald, Joseph J., F. J. Mahoney, and Gordon L. Brownell. 1967. *Mathematical Theory of Radiation Dosimetry*. New York, NY: Gordon and Breach Science Publishers, Inc.

Hubbell, John H., and Stephen M. Seltzer. 1995. *Tables of X-Ray Mass Attenuation Coefficients and Mass Energy-Absorption Coefficients 1 keV to 20 MeV for Elements Z= 1 to 92 and 48 Additional Substances of Dosimetric Interest*. Gaithersburg, MD: National Inst. of Standards and Technology-PL, Ionizing Radiation Div.

International Commission on Radiation Units and Measurements. 1992. ICRU Report 47: Measurement of Dose Equivalents from External Photon and Electron Radiations. *Journal of the International Commission on Radiation Units and Measurements* 24 (2).

International Commission on Radiation Units and Measurements. 1998a. ICRU Report 57: Conversion Coefficients for Use in Radiological Protection against External Radiation. *Journal of the International Commission on Radiation Units and Measurements* 29 (2).

International Commission on Radiation Units and Measurements. 1998b. ICRU Report 60: Fundamental Quantities and Units for Ionizing Radiation. *Journal of the International Commission on Radiation Units and Measurements* 31 (1).

International Commission on Radiation Units and Measurements. 2001. ICRU Report 66: Determination of Operational Dose Equivalent Quantities for Neutrons. *Journal of the International Commission on Radiation Units and Measurements* 1 (3).

International Commission on Radiation Units and Measurements. 2018. "Operational Quantities for External Radiation Exposure." *Journal of the International Commission on Radiation Units and Measurements*. In Draft.

International Commission on Radiological Protection. 1973. ICRP Publication 21: Data for Protection against Ionizing from External Sources: Supplement to ICRP Publication 15.

International Commission on Radiological Protection. 1977. ICRP Publication 26: Recommendations of the ICRP. *Annals of the ICRP* 1 (3).

International Commission on Radiological Protection. 1991. ICRP Publication 60: 1990 Recommendations of the ICRP. *Annals of the ICRP* 21 (1–3).

International Commission on Radiological Protection. 1996. ICRP Publication 74: Conversion Coefficients for Use in Radiological Protection against External Radiation. *Annals of the ICRP* 26 (3–4).

International Commission on Radiological Protection. 2007. ICRP Publication 103: The 2007 Recommendations of the International Commission on Radiological Protection. *Annals of the ICRP* 37 (2–4).

International Commission on Radiological Protection. 2009. ICRP Publication 110: Adult Reference Computational Phantoms. *Annals of the ICRP* 39 (2).

International Commission on Radiological Protection. 2010. ICRP Publication 116: Conversion Coefficients for Radiological Protection Quantities for External Radiation Exposures. *Annals of the ICRP* 40 (2–5).

International Commission on Radiological Protection. 2012. ICRP Publication 118: 2012 ICRP Statement on Tissue Reactions / Early and Late Effects of Radiation in Normal Tissues and Organs - Threshold Doses for Tissue Reactions in a Radiation Protection Context. *Annals of the ICRP* 41 (1–2).

Knoll, Glenn F. 2010. *Radiation Detection and Measurement*. 4th ed. Hoboken, NJ: John Wiley & Sons.

Krane, Kenneth S. 1988. *Introductory Nuclear Physics*. New York, NY: John Wiley & Sons.

Taylor, Barry N., and Ambler Thompson. 2008. *National Institute of Standards and Technology Special Publication 330: The International System of Units (SI)*. Gaithersburg, MD: U.S. Department of Commerce, National Institute of Standards and Technology.

Veinot, K. G., and N. E. Hertel. 2001. "Measured and Calculated Angular Responses of Panasonic Ud-809 Thermoluminescece Dosemeters to Neutrons." *Radiation Protection Dosimetry* 95 (1):25–30.

Veinot, K. G., and N. E. Hertel. 2005a. "Effective Quality Factors for Neutrons Based on the Revised ICRP/ICRU Recommendations." *Radiation Protection Dosimetry* 115 (1–4):536–541.

Veinot, K. G., and N. E. Hertel. 2005b. "Response of Harshaw Neutron Thermoluminescence Dosemeters in Terms of the Revised ICRP/ICRU Recommendations." *Radiation Protection Dosimetry* 113 (4):442–448.

Veinot, K. G., and N. E. Hertel. 2010. "Personal Dose Equivalent Conversion Coefficients for Photons to 1 GeV." *Radiation Protection Dosimetry* 145 (1):28–35.

Veinot, K. G., B. T. Gose, and J. S. Bogard. 2009. "Use of Portable Gamma Spectrometers for Identifying Persons Exposed in a Nuclear Criticality Event." *Nuclear Technology* 168 (1):17–20.

Veinot, K. G., N. E. Hertel, K. W. Brooks, and J. E. Sweezy. 1998. "Multisphere Neutron Spectra Measurements near a High Energy Medical Accelerator." *Health Physics* 75 (3):285–290.

Zeilik, M., S. A. Gregory, and E. V. P. Smith. 1992. *Introductory Astronomy and Astrophysics*. Orlando, FL: Harcourt Brace Jovanovich.

DEFINITIONS

The following definitions were taken from ICRU Report 85a (2011), Attix (2008), ICRP Publication 103, and ICRP Publication 92.

Absorbed Dose The quotient of the mean energy imparted, $d\bar{\epsilon}$, to a mass, dm. Absorbed dose has units of $J \cdot kg^{-1}$ or the special unit Gray, Gy.

$$D = \frac{d\bar{\epsilon}}{dm}$$

Absorbed Dose Rate The quotient of dD by dt, where dD is the increment of the absorbed dose in the time interval dt. The units of absorbed dose rate are $J\,kg^{-1}\,s^{-1}$.

$$\dot{D} = \frac{dD}{dt}$$

Charged Particle Equilibrium Charged particle equilibrium is established in a volume of interest when the energies, numbers, and directions of the charged particles are constant throughout this volume. This is equivalent to saying that the distribution of charged-particle energy radiance does not vary within the volume. In particular, the sums of the energies (excluding rest energies) of the charged particles entering and leaving the volume are equal.

Collisional Kerma A quantity related to the kerma, the collision or collisional kerma has been used as an approximation to absorbed dose when radiative losses are not negligible. The collisional kerma, K_C, excludes the radiative losses by the liberated charged particles, and for a fluence, ϕ, of uncharged particles of energy E in a specified material is given by

$$K_C = \int \phi_E E \frac{\mu_{en}}{\rho} dE = \int \phi_E E \frac{\mu_{tr}}{\rho}(1-g)dE = k(1-\bar{g})$$

where $\bar{g}$ is the mean value of g over the distribution of the kerma with respect to the electron energy. The expression of collision kerma in terms of the product of the kerma and a radiative-loss correction factor evaluated for the same material as the kerma suggests that one can refer to a value of collision kerma or collision-kerma rate for a specified material at a point in free space, or inside a different material.

Cross Section (Macroscopic) The product of the microscopic cross section, σ and the number of nuclei per unit volume are N. The macroscopic cross section has units of length^{-1}.

$$\Sigma = \sigma N$$

Cross Section (Microscopic) The microscopic cross section, σ, is the quotient of the probability, P, of an interaction occurring in a particle fluence, $\varnothing$. The units of σ are area^2 with the special unit barns where $1\,barn = 10^{-28}\,m^2$.

$$\sigma = \frac{P}{\phi}$$

A full description of an interaction process requires, inter alia, knowledge of the distributions of cross sections in terms of energy and direction of all emergent particles from the interaction. Such distributions, sometimes called "differential cross sections," are obtained by differentiations of r with respect to energy and solid angle.

Dose Equivalent The dose equivalent, H, is the product of Q and D at a point in tissue, where D is the absorbed dose and Q is the quality factor, at that point. With D_L being the distribution of the dose D in linear energy transfer L in water, and $Q(L)$ being the quality factor as a function of L in water:

$$H = QD = \int Q(L) D_L dL$$

The unit of dose equivalent is $J\,kg^{-1}$. The special name for the unit of dose equivalent is sievert (Sv).

Effective Dose The tissue weighted sum of equivalent doses in all specified organs and tissues of the body, given by the expression:

$$E = \sum_T w_T \sum_R w_R D_{T,R} = \sum_T w_T H_T$$

where H_T is the equivalent dose in an organ or tissue T, $D_{T,R}$ is the mean absorbed dose in an organ or tissue T from radiation of type R, and w_T is the tissue weighting factor. The sum is performed over organs and tissues considered to be sensitive to the induction of stochastic effects. The unit of effective dose $J \cdot kg^{-1}$, and its special name is sievert (Sv).

Energy Fluence The quotient of dR by da where dR is the radiant energy incident on a sphere of cross-sectional area da. The units of energy fluence are $J\,m^{-2}$.

$$\Psi = \frac{dR}{da}$$

Energy Fluence Rate The energy flux density is defined as the energy fluence rate at a time dt. The units of energy fluence are $J\,m^2\,s^{-1}$. The energy fluence rate is sometimes referred to as the energy flux or energy flux density.

$$\dot{\psi} = \frac{d\Psi}{dt} = \frac{d}{dt}\left(\frac{dR}{da}\right)$$

Energy Imparted The sum of all energy depositions, ϵ_i, in a volume. Energy imparted, ϵ, has units of Joules or eV.

$$\epsilon = \sum_i \epsilon_i$$

Energy Transferred The sum of the radiant energy, $R_{i,u}$ of uncharged particles entering a volume and the changes of rest energy ΣQ, of particles and nuclei that occur in the volume less the radiant energy, of uncharged particles exiting the volume except that which

originated from radiative losses of kinetic energy by charged particles in the volume, $R_{o,u}^n$.
The energy transferred, ϵ_{tr} is typically given in units of $cm^2\ g^{-1}$.

$$\epsilon_{tr} = R_{i,u} - R_{o,u}^n + \Sigma Q$$

Equivalent Dose The equivalent dose in an organ or tissue T is given by

$$H_T = \sum_{TR} w_R D_{T,R}$$

where $D_{T,R}$ is the mean absorbed dose from radiation of type R in the specified organ or tissue T and w_R is the radiation weighting factor. The unit of equivalent dose is $J \cdot kg^{-1}$, and its special name is sievert (Sv).

Exposure The exposure, X, is the quotient of the total charge of ions produced in air, dQ, when all the charged particles are stopped in a mass dm of air. Exposure has units of $C\ kg^{-1}$ or legacy units roentgen, R, where $1\,R = 2.58 \times 10^{-4}\ C\ kg^{-1}$.

$$X = \frac{dQ}{dm}$$

More specifically, the exposure is

$$X = \frac{e}{\overline{W}} \int \phi_E E \frac{\mu_{tr}}{\rho} (1-g) dE$$

where $\overline{W}$ is the work function equal to the average energy required to create an ion pair, e is the elemental charge, g is the fraction of the energy of liberated charged particles lost in radiative processes, μ_{tr}/ρ is the mass energy transfer coefficient, and ϕ_E is the fluence of photons having energy E.

Fluence The quotient of dN/dA where dN is the number of particles incident upon a small sphere of cross-sectional area da. In calculations, fluence is often alternatively expressed in terms of the path length, dl of trajectories of particles passing through a small volume dV. The fluence, ϕ, has units of particles m^{-2}.

$$\phi = \frac{dN}{da} = \frac{dl}{dV}$$

The distributions, ϕ_E of the fluence with respect to energy are given by

$$\phi_E = \frac{d\phi}{dE}$$

where $d\phi$ is the increment of fluence in the energy interval between E and $E+dE$. In certain circumstances, quantities involving the differential solid angle, $d\Omega$, are required. The complete representation of the double differential of fluence can be written $\phi_{E,\Omega}(E,\Omega)$.

Fluence Rate The quotient $d\phi/dt$, where $d\phi$ is the increment of the fluence in the time interval dt. The fluence rate, $\dot{\phi}$, has units particles m^{-2} s^{-1}. The fluence rate is sometimes called the flux or flux density.

$$\dot{\phi} = \frac{d\phi}{dt}$$

Kerma The quotient of the sum of the initial kinetic energies of charged particles liberated by uncharged particles, dE_{tr}, and the mass, dm, of the material. Kerma has units of J kg^{-1} and equivalent special units Gray, Gy.

$$K = \frac{dE_{tr}}{dm}$$

Kerma can also be expressed in terms of the distribution of uncharged particle fluence with respect to energy:

$$K = \int \phi_E E \frac{\mu_{tr}}{\rho} dE$$

where μ_{tr}/ρ is the mass energy transfer coefficient with respect to energy, E. The kerma consists of contributions from radiative and collisional processes.

Kerma Rate The kerma rate, $\dot{K}$, is the kerma as a function of time. Kerma rate has units J kg^{-1} s^{-1} or Gy s^{-1}.

$$\dot{K} = \frac{dK}{dt}$$

Lineal Energy The lineal energy, y, is the quotient of the energy imparted to matter in a volume by a single energy deposition event, ϵ_s, and the mean chord length, $\bar{l}$, of that volume:

$$y = \frac{\epsilon_s}{\bar{l}}$$

Linear Energy Transfer The quotient of dE_Δ by dL, where dE_Δ is the mean energy lost by the charged particles due to electronic interactions in traversing a distance dL, minus the mean sum of the kinetic energies in excess of Δ of all the electrons released by the charged particles. The unit of linear energy transfer is J m^{-1}. E_Δ may be expressed in eV, and hence L_Δ may be expressed in eV m^{-1}, or some convenient multiples or submultiples, such as keV $\cdot$ μm^{-1}. If Δ is expressed in eV, $L100$ is understood to be the linear energy transfer for an energy cutoff of 100 eV. If no energy cutoff is imposed, the unrestricted linear energy transfer, L_∞, is equal to the linear electronic stopping power, S_{el}, and may be denoted simply as L.

Mass Attenuation Coefficient The quotient of the fraction of particles that undergo interactions $\left(dN\!/\!N\right)$ by the traversed distance, dl within a material having density, ρ. The mass attenuation coefficient, $\mu\!/\!\rho$, has units $m^2\ kg^{-1}$.

$$\frac{\mu}{\rho} = \frac{dN}{N\rho dl}$$

Mass Energy Absorption Coefficient The product of unity less the fraction of the energy of liberated charged particles lost in radiative processes $\left(g\right)$ and the mass energy transfer coefficient. The mass energy absorption coefficient, $\mu_{en}\!/\!\rho$, has units $m^2\ kg^{-1}$.

$$\frac{\mu_{en}}{\rho} = \left(1-g\right)\frac{\mu_{tr}}{\rho}$$

Mass Energy Transfer Coefficient The quotient of the fraction of incident radiant energy that is transferred to kinetic energy of charged particles $dR_{tr}\!/\!R$ by the traversed distance, dl within a material having density, ρ. The mass energy transfer coefficient, $\mu_{tr}\!/\!\rho$, has units $m^2\ kg^{-1}$.

$$\frac{\mu_{tr}}{\rho} = \frac{dR_{tr}}{R\rho dl}$$

Mean Absorbed Dose The mean absorbed dose, D_T in a specified organ or tissue T, is given by:

$$D_T = \frac{1}{m_T}\int_{m_T} D\,dm$$

where m_T is the mass of the organ or tissue, and D is the absorbed dose in the mass element dm. The mean absorbed dose, D_T, equals the ratio of the mean energy imparted to the organ or tissue, $\bar{\epsilon}_T$, and m_T, the mass of the organ or tissue, thus:

$$D_T = \frac{\bar{\epsilon}_T}{m_T},$$

The unit of mean absorbed dose is $J\ kg^{-1}$, and its special name is gray $\left(Gy\right)$. The mean absorbed dose in an organ is sometimes termed organ dose.

Mean Energy Imparted The sum of the radiant energy, R_{in} of all charged and uncharged particles entering a volume and the changes of rest energy ΣQ, of particles and nuclei that occur in the volume less the radiant energy, R_{out} of all charged and uncharged particles exiting the volume. The mean energy imparted, $\bar{\epsilon}$ has units of Joules or eV.

$$\bar{\epsilon} = R_{in} - R_{out} + \sum Q$$

Mean Work Function The mean work function, W, also termed the mean energy expended in a gas per ion pair formed, is the quotient of the initial kinetic energy of a charged particle, E, and the mean total liberated charge (of either sign), N. The units of mean work function are J.

$$W = \frac{E}{N}$$

Particle Radiance The quotient of $d\dot{\phi}$ by $d\Omega$ where $d\dot{\phi}$ is the fluence rate of particles propagating within a solid angle $d\Omega$ around a specified direction. The unit of particle radiance is $m^{-2}s^{-1}sr^{-1}$.

$$\dot{\phi}_\Omega = \frac{d\dot{\phi}}{d\Omega}$$

The distributions of particle radiance with respect to energy is given by

$$\dot{\phi}_{\Omega,E} = \frac{d\dot{\phi}_\Omega}{dE}$$

where $d\dot{\phi}_\Omega$ is the particle radiance for particles of energy between E and $E + dE$. The quantity $\dot{\phi}_{\Omega,E}$ is sometimes termed angular flux or phase flux in radiation-transport theory.

Quality Factor (Effective) The effective quality factor, Q_{eff} is the ratio of dose equivalent, H to absorbed dose, D.

$$Q_{\text{eff}} = \frac{H}{D}$$

Quality Factor The quality factor, Q, is used to account for the different biological effectiveness of radiation. It is defined by a quality factor function, $Q(L)$, which characterizes the biological effectiveness of a charged particle with an unrestricted linear energy transfer, L, in water at a point of interest relative to the effectiveness of a reference radiation at this point. Generally, photons, X-rays, and gamma rays, with no energy specified, are used as reference radiation. Values for the quality factor, $Q(L)$, are given by

$$Q(L) = \begin{cases} 1 & L/(keV/\mu m) < 10 \\ 0.32/keV/\mu m - 2.2 & 10 \le L(keV/\mu m) \le 100 \\ 300/\sqrt{L\,keV/..m)} & L/(keV/\mu m) > 100 \end{cases}$$

Instead of tissue, the value of L for the charged particle at the point of interest is given by the data for water. Obviously, photons above about 10 keV and electrons (low-LET radiation with $L < 10\ keV\ \mu m^{-1}$) are weighted by $Q = 1$. Both simplifications are seen to be an approximation sufficient for usual radiation protection applications. The quality factor Q at a point in tissue is then given by:

$$Q = \frac{1}{D} \int\limits_{L=0}^{L=\infty} Q(L) D_L \, dL$$

where D is the absorbed dose in tissue and D_L is the distribution of D in unrestricted linear energy transfer L, and $Q(L)$ is the corresponding quality factor. The integration is to be performed over D_L, due to all charged particles, excluding their secondary electrons.

Radiation Weighting Factor A dimensionless factor by which the organ or tissue absorbed dose is multiplied to reflect the relative biological effectiveness of high-LET radiations compared with photon radiations. It is used to derive the equivalent dose from the mean absorbed dose in an organ or tissue.

Relative Biological Effectiveness The ratio of absorbed doses of two different types of radiation that produce the same effect in a tissue.

Stopping Power The expectation value of the rate of energy loss per unit path length by a charged particle. The units of stopping power are $J \, m^{-1}$. Stopping power consists of two components: collisional and radiative. Collisional stopping power is a result of hard (e.g., knock-on) and soft (Coulombic) interactions, while radiative stopping power is a result of radiative interactions, including bremsstrahlung. The mass stopping power is the stopping power divided by the density of the absorber material.

Tissue Weighting Factor The factor by which the equivalent dose in an organ or tissue T is weighted to represent the relative contribution of that organ or tissue to overall radiation detriment from stochastic effects. It is defined such that:

$$\sum_T w_T = 1$$

CHAPTER 3

Evolution of Radiation Protection Guidance in the United States

Ronald L. Kathren

CONTENTS

3.1 Introduction 80
3.2 Toward Speaking the Same Language 81
    3.2.1 Legal Bases of Radiation Protection Guidance 81
    3.2.2 Standards and Standards Setting Bodies 82
    3.2.3 Licensure and Certification 82
3.3 The Pioneer Era (1895–1905) and the Discovery
    of Ionizing Radiation 83
    3.3.1 The Discovery of X-Rays 83
    3.3.2 Discovery of Radioactivity 84
3.4 The Protection Pioneers 85
    3.4.1 Recognition of the Hazard 86
        3.4.1.1 Early Reports of Injury 86
        3.4.1.2 Protective Measures 88
        3.4.1.3 Additional Impetus 90
3.5 Quiescence (1905–1925) 91
    3.5.1 Status of Protection Standards 92
    3.5.2 First Efforts 93
3.6 Maturation: Standards Come of Age (1925–1950) 94
    3.6.1 The Tolerance Dose 95
    3.6.2 The Roentgen 95
    3.6.3 Organizing for Radiation Protection: The ICRU, ICRP, and NCRP 96
    3.6.4 Radium Rears Its Ugly Head 97
    3.6.5 The World at War 98
    3.6.6 The Immediate Postwar Period 99
3.7 Interlude 101

3.8 An Explosion of Radiation Regulations and Standards: Part I   102
    3.8.1 The 1950s: A Watershed   102
    3.8.2 From Tolerance Dose to the LNT Paradigm   105
3.9 An Explosion of Radiation Regulations and Standards: Part II   107
    3.9.1 Proliferation of Regulations and Standards   107
    3.9.2 The ICRP   108
    3.9.3 The NCRP   110
    3.9.4 The Federal Radiation Council   111
    3.9.5 Electronics Products   112
    3.9.6 The Environmental Protection Agency   112
    3.9.7 The AEC Fissions: The NRC and DOE   113
    3.9.8 Other Regulatory Agencies   115
3.10 Into the New Millennium   115
    3.10.1 Site Remediation   116
    3.10.2 Some Concluding Remarks   117
References   118

3.1 INTRODUCTION

It is perhaps axiomatic that knowledge of the past, that is, history, is essential to understanding the present, and to provide a window, however small, into the future as the present moves forward in time. The course of the evolution of radiation protection guidance devolves into two distinct eras separated by World War II. During the first era, which lasted about half a century, the concern was primarily with X-ray protection, and, to a much lesser extent, protection from radium. Efforts at radiation protection moved slowly, almost glacially so, and were largely concentrated on simple solutions to specific problems. There were few organized efforts at developing protective guidance and methods, and what little research was done was relatively simple and unsophisticated, and largely devoted to improvements in X-ray generating apparatus and to the applications of both X-rays and, to a much lesser extent, radium in medicine. There was virtually no organized research with regard to the biological effects of ionizing radiation, or to the broader consideration of economic or environmental impacts. Little guidance was established in the form of standards and there were essentially no regulatory requirements. Progress in this regard was generally quite slow.

The second era was the antithesis of the first. The development of the atomic bomb and the creation of vast quantities of numerous radioactive species, and the associated newly declassified and well-organized research and development efforts of the Manhattan Project produced an explosion of research and knowledge that brought with it a flurry of activity that continues to this day and has resulted in a plethora of legislation, regulations, and standards. So intensive and productive has this effort been, based largely on the results obtained from research, that, using legislation and voluntary standards as the metric, the pace has been rather frenetic.

3.2 TOWARD SPEAKING THE SAME LANGUAGE

In the 120+ years since the discovery of ionizing radiation, a hierarchy of guidance for the protection of people and the environment from the potentially harmful effects of exposure has evolved. When describing guidance for control of ionizing radiation, the terms "regulations" and "standards" are often used generically and interchangeably, and frequently incorrectly by radiation protection and other personnel in casual and informal conversation. Nonetheless, both the speaker and the listener may fully understand what is stated and meant, for in the context of routine daily usage, words such as regulation and standard as applied to radiation protection have acquired the status of shoptalk, or even jargon. In more formal communications, such as peer-reviewed scientific and technical reports, textbooks and histories, and various legal documents, including licenses and investigative reports, such usage, although it often appears, is unacceptable. The terms regulation and standards, along with a host of other similarly used words, have specific and precise meanings and should be used at least in the formal sense to communicate radiation protection guidance, thereby avoiding ambiguity, misunderstandings, and miscommunications.

3.2.1 Legal Bases of Radiation Protection Guidance

At the top of the legal hierarchy is constitutional law, which deals primarily with the relationships of government to the citizenry and which, for all practical purposes, is far removed from the specific aspects of radiation protection guidance. At the second level of the hierarchy is statutory law, which in its broad general sense includes the body of rules, requirements, principles, standards, and specification of conduct incumbent upon the radiation protection practitioner as prescribed by a duly constituted and empowered legislative body or administrative body to whom such powers have been delegated by the legislative body, and case law which is law generated by decisions of the courts and various administrative bodies in the absence of applicable statutory law. Statutory law is thus typically the province of elected legislative bodies such as the U.S. Congress, the legislatures of the various states, or lower level political subdivisions such as counties, incorporated cities, and certain specially constituted bodies or districts. Laws can be very detailed, lengthy, and inclusive, but typically when technical matters are involved, laws are quite broad, leaving the task of issuing specific requirements in the form of regulations to boards and commissions, and specified governmental agencies.

Regulations constitute the third level of the hierarchy, and are frequently quite specific and detailed. Although technically not laws, regulations are typically derivatives of statutory laws and as such have the force of law. Regulations are promulgated by governmental departments and various governmental agencies, boards, and commissions to carry out the intent of the laws passed by a legislative body. Like laws, compliance with regulations is mandatory, and violation of regulations may subject the violator to monetary or other penalties.

Adjuncts to the regulations are regulatory guides which are issued by some agencies to provide clarification and assistance in carrying out regulatory requirements. Regulatory guides do not have the force of law, and compliance with them is therefore not mandatory, although it may be highly desirable.

3.2.2 Standards and Standards Setting Bodies

In addition to the mandatory requirements imposed by laws and regulations, there are standards of various types that provide radiation protection guidance, and these lie at the next lower level (fourth) of the guidance hierarchy. There are many different types and levels of standards, which are often referred to as "voluntary standards" since, except in a practical sense, these are not mandatory. Voluntary standards may be put forth by governmental agencies such as the National Institute of Standards and Technology (NIST) and may be in addition to regulatory standards issued by that body. Other voluntary standards may be set out by quasi-legal bodies, scientific and technical organizations, and committees and organizations created for that purpose. These include the federally chartered National Council on Radiation Protection and Measurements (NCRP), the International Commission on Radiological Protection (ICRP), the American National Standards Institute (ANSI), and the International Electrotechnical Commission (IEC). Unlike laws and regulations, standards are not mandatory but are recommendations, and as such are often incorporated into regulations. However, until this is done, the recommendations by such bodies are purely voluntary guidance. However, from a practical viewpoint, the recommendations set forth in standards put forth by recognized organizations such as those mentioned above are generally universally accepted, for a product that goes against what has become conventionally accepted, for example, which direction a screw must be turned to drive it in, may spell economic disaster.

In theory at least, any organization or individual can publish a standard, and the seemingly ordinary term "standard," widely used in radiation protection, can have multiple meanings in everyday usage. Often the specific meaning is obscure, and may not be entirely clear or obvious, even in the context of usage. The word standard, as commonly used by radiation protection personnel, has several meanings: it may be used to mean a specified desirable or required level of quality such as that promulgated in a document by a standards setting organization or regulatory body; it may mean the normal or accepted way in which things are done, as in a standard practice; it may mean a recognition of particular merit or acceptance; or it may mean that an agreed upon or contractual requirement was met. Consider, for example, the use of the word standard in the following sentence: "The measurement was not up to standard." The meaning of the word standard is not clear from the context, and more information is needed to determine which of the several specific meanings given above applies.

3.2.3 Licensure and Certification

In the broad general sense, licensure refers to the granting of permission to carry out certain acts that would otherwise not be permitted. In the more restrictive sense, licensure, as applied to radiation protection guidance, refers to the granting of permission by a governmental body to perform activities which, in the absence of a license, would be illegal. Thus, licensure is analogous to law, and indeed is codified in various compendia of laws. Licensure applies to business entities and other organizations, as well as individuals. Thus, a business may be permitted to operate if it meets specific qualifications which are provided for by law or regulation. Similarly, an individual who meets certain specific qualifications may be granted a license by the state or some other regulatory body to practice a particular profession; engineers may be licensed by the state to practice a particular branch of engineering and thus may perform actions such as signing off on engineering documents and are legally permitted to use the designation of "Professional Engineer" (PE).

Certification is in a sense analogous to standards and may be granted to an individual who has met certain specific educational and other requirements, often including the passage of an examination that indicates that the person so certified has achieved a specified level of competence. Certification is a voluntary process and typically carries with it no specific legal authority, although certain facility licenses and standards may specify or require certified individuals on the staff. The primary certifying body for radiation protection professionals (health physicists) is the American Board of Health Physics which began operations in 1960, and certifies radiation safety professionals with appropriate higher education and operational experience who successfully pass a rigorous two-part written examination. Those who qualify are eligible for membership in the American Academy of Health Physics. The National Registry of Radiation Protection Technologists (NRRPT) provides certification at the technologist level. Both bodies are private organizations and have no direct connection with any governmental agency. In Britain, the Society for Radiological Protection has also established a professional board certification program.

3.3 THE PIONEER ERA (1895–1905) AND THE DISCOVERY OF IONIZING RADIATION

The nineteenth century was marked by numerous scientific and technical advances along a broad technical front. Such diverse developments as photography, anesthesia, the steam engine, and ships powered by the burning of coal, the discovery of radio waves, and the industrial revolution had brought comforts of living and the knowledge of science to a level that, in the minds of many Americans at least, including some who were highly educated, indicated that there was little left to discover or create. But as the nineteenth century drew to a close, two parallel scientific discoveries made within a few weeks of each other changed that viewpoint, and excited both the scientific and the lay world, opening a whole new and hitherto unknown area of physics and creating the need for radiation protection guidance. These were the discoveries of X-rays and radioactivity.

3.3.1 The Discovery of X-Rays

The discovery of X-rays by Wilhelm Konrad Röntgen has been documented in detail (Glasser 1993), yet still bears at least brief mention here. On November 8, 1895, Röntgen, then a 50-year-old physics professor and rector at the University of Würzburg in Germany, was continuing his studies of electrical discharges in vacuum tubes. He had repeated the studies of other experimenters, and on this gray Friday afternoon, suspecting that cathode rays produced in electrically charged vacuum tubes could penetrate the glass walls of the tube, he covered a thick glass-walled Hittorf tube with black cardboard to hide the fluorescence produced in the tube walls when it was energized, completely darkened his laboratory, and activated the tube. He was surprised to see a pale green fluorescent glow coming from a cardboard screen coated with barium platinocyanide on a table some distance away from the energized tube that far exceeded the range of cathode rays in air. Röntgen clearly deduced that something else, likely some kind of unknown ray, was producing the effect and began a comprehensive examination of the properties of these hitherto undiscovered rays.

For the next seven weeks, Röntgen was fully occupied by his studies of the mysterious rays, virtually barricading himself in his laboratory and working at a feverish pace. Meals brought to him by his wife were left outside the laboratory door, and often went untouched, so intense was his effort to study these mysterious rays which penetrated solid matter, induced fluorescence in a variety of materials, fogged photographic plates, and, perhaps most significantly, were unaffected by a strong magnetic field, which definitively differentiated them from cathode rays. On December 28, 1895, he gave the handwritten preliminary report of his findings entitled "Über eine neue Art von Strahlen" (On a new kind of rays) to the Secretary of the Physikalisch-Medizinische Gesellschaft von Würzburg (Physical-Medical Society of Würzburg), and it was published in the Society's *Proceedings* a few days later (Röntgen 1895). On January 1, 1896, he sent reprints of his paper along with samples of X-ray pictures he had taken, including one of his wife's hand which clearly showed the outline of her bones and her large wedding ring to several colleagues, one of whom, Franz Exner, released the news of the discovery to the Viennese press which announced the discovery to the world on January 5, 1896.

Few scientific discoveries have provoked such worldwide study and interest as the discovery of X-rays. Röntgen's paper was quickly translated and republished in English in the widely respected British scientific journal *Nature* (Röntgen 1896). Experimenters all over the world—scientists and lay persons alike—were fascinated by these heretofore unknown penetrating rays, and began studies immediately, developing new and more powerful generating apparatus and ancillary devices, studying the properties and effects of these mysterious rays, and developing applications for them. Medical and commercial applications in medicine were immediately realized, but relatively little research was done into the biological effects of the new rays. In what has been characterized as "the remarkable year of 1896," no than 1,044 scientific and technical papers were published (Webster 1995), along with no than five books in the English language.

Within the first month after the discovery, physics professor M. I. Pupin of Columbia University made the first diagnostic radiograph, providing a Dr. Bull of New York with an X-ray photo showing numerous pieces of shot in the hand of a patient who had suffered a gunshot wound, thereby facilitating removal. Commercial X-ray photographers opened businesses selling radiographs they had made, or, for a fee, demonstrating to members of the public the bones in their hand. People lined up at county and state fairs for the same purpose, and numerous stories of miraculous and often bizarre bogus effects appeared in the news media of the day, further whetting the public's interest. Even scientists were not immune to the X-ray craze; one scientist published a paper claiming to have projected the image of a bone on the brain of a dog, causing it to salivate.

3.3.2 Discovery of Radioactivity

Shortly after the discovery of X-rays, the French physicist Henri Becquerel, following up on a suggestion by Henri Poincaré that X-rays might be also be produced by ordinary phosphorescent materials, began a systematic study of various phosphorescent materials in an effort to determine whether the luminescence induced by sunlight might contain X-rays as well. Believing that the mineral would be activated by sunlight, he initially placed a quantity of potassium uranyl sulfate, a compound known to be highly fluorescent, on a photographic plate and exposed the mineral to sunlight for several days. When developed, the photographic plate showed a clear image of the potassium uranyl sulfate salt, an observation he promptly reported to the Paris Academy of Science on February 4, 1896.

Continuing his experiments, Becquerel prepared a similar arrangement later in the month, but due to cloudy weather did not expose the combination to sunlight, but instead stored it in a dark drawer in his laboratory. When he developed the photographic plate a few days later, he was amazed to find the same dark spots on them, even more intensely darkened than before, even though there had been no exposure to sunlight. Becquerel correctly concluded that this unexpectedly strong effect was attributable to some kind of penetrating rays emitted by the uranium salt itself, reporting his extraordinary observation and conclusion to the Paris Academy of Science on March 2, 1896, only two months after Röntgen's bombshell report of the discovery of X-rays. But unlike the immense response engendered by the discovery of X-rays, the expression of interest in Becquerel's discovery was small and largely restricted to a fairly small number of physicists and chemists within the scientific community. Only about a dozen papers on radioactivity were published in the scientific literature during the year following the discovery.

Despite the relatively limited interest, Becquerel's discovery marked the beginning of a host of research devoted to the study of natural radioactivity. Perhaps most notable are the studies of the Curies, which led to the discovery of two new elements, polonium and radium, in 1898, and to the use of ^{226}Ra to treat cancers early in the twentieth century. Radium was considered a miracle element and along with its daughter radon was considered to have great therapeutic value, as well as other beneficial applications, some quite bizarre, including a radium-bearing fertilizer claimed to improve plant growth and health. Once again, although radium was known to destroy tissue, scant attention was paid to potential adverse effects from its unrestricted use. And, it is of interest to note that both Becquerel and Pierre Curie, were unknowing victims of the effects of radioactivity, each suffering a skin erythema on the abdomen from small samples of radioactive substances carried in their vest pockets.

3.4 THE PROTECTION PIONEERS

Given the near frenetic interest in studying and applying X-rays, coupled with the low energies and high X-ray output of tubes operated with no shielding and operators located nearby, it is not surprising that X-ray injuries should soon appear. There were, of course, no regulations or safety standards governing X-rays, and experimenters and users of the equipment were free to operate as they wished. The two-decade period from 1895 to 1915 has been dubbed "The Era of Protection Pioneers" in the history published by the Radiology Centennial to commemorate the 100th anniversary of the discovery. During this period, the hazardous nature of X-rays came to be recognized and fully accepted, protection standards made their appearance, and crude but effective protective techniques were developed by what has been termed "a small cadre of protection pioneers" (Kathren and Brodsky 1996), aided by new discoveries in physics leading to a greater understanding of both X-rays and radioactivity.

3.4.1 Recognition of the Hazard

At least initially, and for the first few years after the discovery, there was the more or less general belief that X-rays were biologically harmless. The basis for this belief was buttressed by the fact that X-rays were undetectable by the human senses and could also easily pass through solid matter (Kathren 1962). Soon reports of injuries somehow associated with the use of X-rays began to appear in the literature. And, given the fact that the nature of X-rays was not well understood, other explanations such as static charge seemed a more rational explanation of the cause of the adverse effects. Among the earliest reports stating that X-rays were safe was one by an X-ray experimenter who attested to the safety of X-rays generated with the static machine (Frei 1896). But there was no general agreement as to the cause of the injuries noted among the physicians and technicians working with X-rays, and those investigating the injuries were divided into two camps. One believed that other factors, such as techniques or the character of the generating equipment were responsible for the erythema and epilation that were sometimes associated with X-ray exposures, and noted that dermatological effects were not always seen. The other camp was convinced that it was the X-rays themselves that caused the effects, but could offer no explanation of the mechanism.

3.4.1.1 Early Reports of Injury

On March 3, 1896, what may well have been the first published accounts of injury associated with the use of X-rays were published in the respected British journal *Nature* (Edison 1896; Morton 1896). No less a personage than Thomas Edison, and the prominent New York pioneering radiologist William Morton independently reported adverse effects on the eyes associated with X-rays and fluorescent screens. Edison, who had ceased his studies with X-rays only two months after he began them, recommended against the continued use of X-rays, citing a causal effect with eye irritation. In the same month, John Daniel, an experimenter at Vanderbilt University, cautioned about skin burns and epilation which he himself had incurred in his research with X-rays (Daniel 1896a,b). Another apparent X-ray injury was also reported in March, in which case a physician attempting to obtain a lateral X-ray picture of his brain had exposed his head for a full hour at a distance of only a half inch from the tube, which resulted in epilation at the entry portal some three weeks later (Brecher and Brecher 1969). The skin dose was estimated as 4 Gy, certainly well above the threshold for skin erythema (Webster 1995).

In November 1896, less than a year after Röntgen announced his epic discovery to the world, pre-eminent British-born American electrical engineer Elihu Thomson performed an experiment to determine the potential deleterious effects of X-rays. Thomson, who later would become known as one of the founders of the General Electric Company and as the inventor of the arc lamp, exposed the little finger of his left hand to the direct radiation from an X-ray tube daily over a period of a few weeks. For the first week, there was no observable effect, after which he was able to describe in exquisite detail the erythema, swelling, stiffness, and excruciating pain that he endured before terminating the experiment and issuing a strong warning against overexposure (Thomson 1896a,b). As for the cause of the injury, Thomson put forth the hypothesis that the injury was attributable to a chemical effect within the tissues and noted "there is evidently a point beyond which exposure cannot go without causing serious damage" (Thomson 1896a).

Although acceptance of the fact that X-ray exposures could cause injury was growing, there seemed to be "a kind of euphoria within the medical community regarding X-ray dangers" (Kathren and Brodsky 1996). The deniers of the hazards of X-rays, although continuing to decline in numbers, nonetheless tenaciously held to their belief. While not denying that deleterious effects were observed, various alternative explanations other than X-ray exposure were put forth to account for the adverse effects that were observed, while maintaining that X-ray exposure was not the cause. By 1900, the shift was apparent, accelerated to some extent by continuing reports of injuries among those exposed to X-rays, converting some to the belief that the effects were indeed caused by the X-rays themselves. One such convert was the prominent radiologist Charles Lester Leonard, who in 1898 had definitively stated "The X-ray per se is incapable of injuring the tissues of the patient," ascribing the effects noted to interference with the nutrition of the afflicted part by induced static charges (Leonard 1898), later changed his view as more evidence indicated that X-rays were in fact the cause of untoward skin effects. Although by about 1900 the prevailing view accepted by most physicians and others who worked with X-rays, was that X-rays could prove hazardous, the controversy continued for several more years, despite a report in the medical literature of an X-ray induced death (Anonymous 1901), and, perhaps more significantly, the classic and prescient experiment of the Boston dentist William Rollins who, in 1901, demonstrated experimentally that X-rays could kill the higher forms of animal life (Rollins 1901b). As late as 1907, a book written by a prominent pioneering Philadelphia radiologist Mihran Kassabian, himself a victim of exposure to X-rays in 1910 (Brown 1936), listed no less than seven alternatives to X-rays as potentially the cause of erythema and other adverse skin effects, in addition to X-rays (Kassabian 1907).

Although the reports of skin burns, epilation, and other dermatologic injuries likely played a paramount role in recognizing X-ray hazards, reports of fatalities, public opinion, and potential legal ramifications also played a role. While it was generally felt that X-rays were nonlethal, a brief 1897 article in *Medical News* noted that an attorney defending a man charged with murder alleged that the actual cause of death was exposure to X-rays administered to the victim in an effort to locate the bullet in his head. The fact that *Medical News* chose to publish this item could indicate underlying doubts about the power of X-rays, but more likely it was published as a curiosity about an overzealous attorney. In January 1901, and somewhat more soberly, but still anonymously, came another brief report alleging a death attributable to overexposure to X-rays that appeared in as a news item in a respected peer-reviewed American medical journal (Anonymous 1901). Neither of these reports received much attention, for it was commonly accepted lore that X-rays were not lethal. In an even more telling report in that same journal the following month, William Herbert Rollins (Rollins 1901a,b) described fatality and miscarriage in guinea pigs experimentally exposed to X-rays. While this research got scant notice, it did establish unequivocally that exposure to X-rays could in fact kill the higher forms of life.

Recognition of the hazards of radiation was largely confined to consideration of X-rays. X-rays were widely used and easily produced with fairly simple apparatus, and the results of excessive exposure were prominently displayed. Radioactivity was certainly an interesting phenomenon, but other than a few naturally occurring radioactive substances that had been discovered, it was hard to come by and required more complex knowledge and skills than X-rays. Radium was the exception, and, like X-rays not considered harmful, but rather quite the opposite, and was treated as a panacea for another quarter of a century.

3.4.1.2 Protective Measures

Along with efforts to determine the cause of the dermatologic effects in some patients and X-ray workers, protective measures were proposed by some prescient X-ray pioneers. Since the actual causative agent of the effects, which occurred only sporadically, was not known, and little was known about the physics of the X-rays, many of these recommendations were little more than blind guesses. The problem was further exacerbated by the rapidly evolving improvements in techniques, plates, tubes, and generating apparatus. Suggestions to interpose screens or filters of various types between the X-ray tube and the patient to intercept static charges or other minute particles which many believed to be the cause of the X-ray burns were made. One such device was the proposal of Charles Lester Leonard in late 1897 who inserted a grounded metal conductor between the patient and the X-ray tube to eliminate so-called static burns (Leonard 1898). But even though the basis was incorrect, the screens were effective, for the X-ray spectra produced by the early unfiltered tubes contained a large fraction of low-energy X-ray photons which deposited most of their energy in the superficial layers of tissues, hence reducing the skin dose.

The earliest safety measures were the rather obvious limitation of exposure time and the admonition of Nikola Tesla to not get too close to the X-ray tube (Tesla 1896). But the premier radiation protection pioneer was a shy and reserved Harvard-educated Boston dentist, William Herbert Rollins, who richly merits the appellation of "the father of health physics" (Kathren 1964). Working alone and in his home, which also served as his laboratory, over the course of about 12 years following the discoveries of X-rays and radioactivity, Rollins published more than 200 individual contributions relating to X-rays, most of which were compiled into a book he published privately (Rollins 1904). All in all, his contributions to radiation protection and radiological science are far too numerous to mention. In addition to developing protective devices and designing improvements to X-ray tubes and generating apparatus, he was a true Renaissance man, making contributions in dentistry, horticulture, genetics, physics, and biology. At the time of his death in 1929 at the age of 77, he was actively involved in research on radio communications and television. Rollins research was self-financed—he spent more than $30,000 of his own funds, approximately one-million 2017 dollars, based on the Consumer Price Index. Although he could have gained numerous patents, and presumably some wealth, for his X-ray tube development work alone, he never patented anything, choosing instead to publish his work openly as a contribution to mankind.

Perhaps recognizing something about X-rays that others had not seen, Rollins referred to X-rays as X-Light in his numerous writings. The first of his more than 200 "Notes on X-Light" was published in mid-1896 and described an intraoral fluoroscope for dental application (Rollins 1901a,b). For the protection of both the operator and the patient, this device had non-radiable walls and leaded glass over the fluorescent screen. Over the subsequent years, Rollins addressed virtually every principal technique of diagnostic X-ray safety, often being the first to call attention to a particular source of unnecessary exposure and putting forth a means to reduce that exposure. In many respects, despite his inherent shyness, he was a virtually singular, yet quite influential voice calling for protective measures and methods to improve radiographic quality.

Among his numerous contributions to radiation safety, the following items are particularly noteworthy. Concluding that scattered radiation was of no value for radiography and actually increased the dose to the patient without concomitant benefit, he proposed the use of collimation to restrict beam size and reduce scatter, which also had the benefit of improving radiographic quality (Rollins 1898,1899). He created the non-radiable tube housings and promoted their use in conjunction with beam collimation (Rollins 1900, 1902a,b).

Arguably, Rollins' most significant contributions lay in the area of radiation biology. In 1901, he carried out a well-designed simple experiment in which two guinea pigs, kept in grounded Faraday cages to obviate effects from electric fields or static charges, were exposed to X-rays for two hours each day (Rollins 1901b). On the eighth day, one of the experimental animals died; the second died three days later. This was the first unequivocal demonstration of the potentially lethal nature of X-rays on otherwise healthy mammals, clearly demonstrating that X-rays were not, as many still maintained, harmless. Regrettably, little notice was taken of this experiment, which was followed in the subsequent issue of the journal by a description of the X-ray induced death of a fetus from X-ray exposure of a pregnant guinea pig (Rollins 1901a), leading him to express concern about exposure of pregnant women from pelvimetry or routine X-ray exams, a prescient concern that was not fully considered in regulations and standards for more than half a century. Finally, Rollins concluded his trilogy of guinea pig experiments with a third paper in which he discussed deep tissue effects noting that X-rays, which had already been applied to treatment of superficial lesions, might be useful for therapy of deep tissue cancers (Rollins 1902c).

The prescient Rollins was concerned by the wide range of potential hazards from the use of X-rays, and described many more firsts or near firsts in his privately printed book (Rollins 1904). His concerns included the generation of toxic gases from the operation of X-ray apparatus (Rollins 1901b), and he was among the first to recognize the cataractogenic potential of radiation exposure to the eyes, recommending that fluoroscopists wear leaded glasses with lenses one centimeter thick (Rollins 1903). His protection recommendations extended to radium, whose use in medicine he advocated only a year after the confirmation of its existence (Rollins 1903). The shy, prescient genius had laid down virtually every protection principle that would serve as the basis for regulations and standards in the coming decades.

In addition to Rollins, others proposed various protective measures, augmenting but sometimes reiterating or rediscovering what had already been put forth by Rollins.

Notable in this regard were Charles Lester Leonard, who reported that a grounded conductor between the X-ray tube and the patient would eliminate X-ray burns, which he attributed to the static charge associated with the generation of X-rays (1897). Elihu Thomson described an aluminum tube housing to protect against the so-called "soft" X-rays (1898), and Francis Williams recommended that fluoroscopists dark-adapt their eyes prior to examinations to reduce patient exposures (1898).

3.4.1.3 Additional Impetus

By 1900, and certainly by 1905, there was general acceptance that the cause of X-ray injuries had been identified and the means of protection proposed. The earlier prevailing attitude had shifted and the general belief now among physicians, scientists, the media, and others working with X-rays, as well as the general public was that X-rays were indeed the cause of adverse effects, and that precautions were needed to ensure their safe use in medicine. The change was largely attributable to a better understanding of the scientific aspects, particularly the physics, of X-rays and radioactivity; the large and growing body of publications in the medical and scientific literature describing untoward effects of operational exposure to X-rays, including deaths among some of the early workers, and research results; public and media pressure; threats of lawsuits with concomitant large monetary awards to the plaintiffs; and greater sophistication and less tolerance in the public attitudes regarding potential hazards. Still, there was not yet complete accord, and although rapidly dwindling in numbers and influence, a minority still stuck to the belief that X-rays were harmless. But there was also a modicum of pressure from the public and the media as evidenced by the efforts of a New York newspaper reporter named John Dennis, among others. Dennis proposed a state commission consisting of two physicians and an electrician to control licensure of X-ray operators to ensure competency, and further declared that injury to a subject by a radiographer was a criminal act.

No small impetus to the change in attitude regarding providing some sort of protection standards was given by a report in the prestigious and widely read *Journal of the American Medical Association* of a monetary award of $10,000, equivalent to about $300,000 in 2017 dollars, to a plaintiff who was judged to have suffered diagnostic X-ray burns, and by delayed deaths among the early X-ray workers attributable to their earlier exposures (Brown 1936). The first of these "X-ray martyrs" was Clarence Dally, Edison's assistant, who had suffered large exposures in 1896 and who subsequently died in 1904.

Along with recognition of the problem came a number of suggested stopgaps made by some notable protection pioneers who, despite a lack of knowledge of the nature of X-rays, recognized some means to minimize the untoward effects. However, still more information regarding the physical nature of the X-rays was needed, including a practical, universally accepted means of quantification, specified limitations on exposure, improvements in equipment, technique, and operations, as well as consensus standards of practice. Most of the progress came from the medical applications, for this was where most problems were seen. Radium was in short supply and expensive, and moreover had developed an aura of mystique and harmlessness, and in fact was promoted as beneficial to health and wellbeing, as well as having commercial value as an essential component of luminous paint. Hence the appearance of radium waters, sold as a tonic, about 1910, and patent medicines and other products containing or allegedly containing radium or uranium ores were beginning to appear along with claims of their miraculous curative, tonic, or other beneficial properties.

3.5 QUIESCENCE (1905–1925)

The stage was now set. The hazardous nature, of X-rays at least, had been identified and means to control the hazard proposed, and the frenetic early years after the discovery were at an end. Granted, much more needed to be done, including the establishment of a standard means of quantification, as well as monitoring of exposures, but certainly establishment of some level of acceptable protective standards, if not actual legislative controls, could still be developed with an eye toward modification as new information became available. Regrettably, this was not to be the case. Indeed, the years 1905 to 1925 were dubbed the "Dormant Era" by Kathren and Ziemer in their minihistory of the first 50 years of radiation protection (1980), during which the establishment of formal radiation safety guidance, whether by industry, professional organizations, or governmental bodies was very slow and very limited. During this period of so-called dormancy, there was, however, great progress made in applications, equipment, and basic scientific knowledge important to the preparation of meaningful and workable protections. Among these were proposals for a radiation unit based on ionization proposed by British physicist C.E.S. Phillips at the American Roentgen Ray Society (ARRS) meeting along with a measurement device by S.J.M. Allen of Cincinnati (Portmann 1933), and the introduction of the concept of half-value layer by Dutch physicist Theodore Christen in 1912 (Portmann 1933), both important underpinnings of meaningful standards for dose and dose measurements. Also put forward during this quiescent period for protection standards was the international radium standard and the Curie unit.

The urgency, and indeed the controversies, that existed during the pioneer years was much less, and it was generally accepted that overexposure to X-rays could result in somatic injuries such as skin burns and epilation. Exposures, and hence injuries, were much reduced, attributable to improvements in technique, but likely more so by improvements in generating apparatus and techniques to produce better quality radiographs. These included shorter exposure times, higher voltages, faster X-ray plates and films, increased distance from the tube to the patient, and limitation of beam size, among many others, and especially the development of the hot cathode tungsten target X-ray tube by William D. Coolidge in 1913. Experimenters and developers of equipment also experienced a reduction in exposures. And, along with the reductions in exposure came much increased usage and applications of X-rays in both diagnosis and therapy, and the two together, perhaps subtly, reduced the urgency and concern with respect to protection among X-ray users. But clearly, there was a reduction in reports of injuries, and indeed the number of reported X-ray induced injuries might have been exaggerated by reports of the same injury in multiple sources. In 1902, Ernest Amory Codman, a Boston-based radiologist affiliated with the Harvard Medical School and the Massachusetts General Hospital, examined reports of X-ray injury from around the world and found fewer than 200, with only 88 of these in the United States (Codman 1902). Of the American cases, fully 55% or 62.5% had occurred during 1896. Only a single case had been reported in 1901, the last year in Codman's study.

3.5.1 Status of Protection Standards

To a great degree, the establishment of formal generally accepted radiation protection standards was hampered by the times. Companies, individuals, and professions were by and large free to practice as they saw fit, typically unhampered by governmental requirements, broad consensus standards, or even general guidance. Licensure was nonexistent; anyone could own and operate X-ray apparatus or commercially use radium. Industry-wide safety standards and standards committees were rare, and, indeed, had really only begun to come into existence. A major step in the development of voluntary safety standards was the formation of the National Fire Protection Association in 1896, which was founded to prevent fires and write codes and standards, a clear safety purpose. Another group formed for safety reasons was the American Society for Testing and Materials (now ASTM International) created out of concern for railway safety in 1898. In an effort to reduce costly and dangerous breaks in rails, the ASTM developed a standard for the steel used in rails that was quickly adopted.

The X-ray safety problem was considered the purview of the medical profession, which was the primary user of X-rays, and thus would address the problem internally within the profession. A major positive step was the formation of the Roentgen Society of the United States in 1900, formally renamed the American Roentgen Ray Society (ARRS) two years later. The organization held annual conferences and published the transactions of these meetings until 1908, including not only documentation of the formal presentations, but of the informal (and sometimes somewhat contentious) discussions that took place. Most importantly, from the standpoint of protection, was the establishment of a Committee on Standards that included protection standards in its purview, but unfortunately was not particularly active in this area.

One hindrance to the development of protection standards was the lack of a commonly accepted quantity and unit for dose, and in particular one that could be applicable to the X-radiation and gamma radiation from radium. Early dosimeters were based on color changes induced in various chemical formulations, and as such were not always intercomparable. In 1905, Milton Franklin of Philadelphia proposed measurement based on ionization, and although this had considerable appeal, it was more than two decades later before such an ionization-based quantity was precisely defined and adopted as a standard. Without a standard quantity and associated unit, dose could not be consistently measured nor compared obviating meaningful comparisons to a great extent.

Another hindrance was the fascination of the lay public with the wonders of X-rays and radium and the virtually complete lack of concern with respect to potential radiation hazards. The widespread usage of X-rays in medical diagnosis—a painless procedure without apparent injury to the patient—supported the belief in the harmless nature of X-rays. X-ray baths were proposed for treatment of various diseases, including tuberculosis, widespread in the early years of the twentieth century, and as a depilatory by legitimate medical practitioners. According to Dr. Louis Harris, New York City Commissioner of Health, the latter use resulted in "countless cases" of young women with facial disfigurement. In particular, radium was touted as a panacea, encouraging belief in its (largely imaginary) special tonic and therapeutic properties, and found its way into all sorts of consumer products, as well as being applied legitimately in medical practices for numerous ailments, including such diverse conditions as baldness, hearing loss, and improving sexual potency. Radium waters and patient medicines were numerous and common, and radium found its way into or claimed to be included in numerous consumer products, including cosmetics, toothpastes, mouthwashes, hair tonics, and even candy bars. Legitimate medical uses included radium belts designed to be worn over affected organs and, somewhat ironically, "standard" radium preparations for treating by injection or oral administration subacute and chronic joint and muscular conditions, high blood pressure, nephritis, and the simple and pernicious anemias (Kathren 1978; Mould 1993; Weart and Weart 2009). However, there was an apparent unrecognized and inexplicable disconnection, in that the tissue destructive properties of higher doses of the same X-rays and radium touted to be harmless or even beneficial were used to treat both malignant and benign neoplasms.

3.5.2 First Efforts

During this relatively quiescent period of dormancy, there were two largely abortive efforts at establishing protection standards. In 1913, the Deutsche Röntgengesellschaft (German Roentgen Society) put forth a set of radiation protection recommendations from an organized body (Taylor 1979). Then, in June 1915, the British Roentgen Society (BRS), at the urging of Sidney Russ, unanimously adopted what has been identified as the first organized step toward radiation protection (Taylor 1971), passing a resolution calling for the adoption of stringent rules and directing that this be done expeditiously. Within five months, a comprehensive series of proposals for radiation protection were developed, dealing mostly with X-rays, but including a brief section on radium. Russ continued his efforts, and the following year the BRS passed a strongly worded resolution hoping (unsuccessfully) to influence legislation (Russ 1916). Taylor (1979) notes that because of World War I and general human indifference, this was essentially the last mention of radiation protection for five years (Taylor 1971). These historic documents are both reproduced verbatim in Taylor (1979).

The United States came somewhat late to the game, following the British effort by five years. In September 1920, led by the efforts of Detroit radiologist Preston M. Hickey, the ARRS established its Roentgen Ray Protection Committee, the first standing committee on X-ray protection. The was not accomplished without a struggle, for many radiologists, who were the leading and majority force within the ARRS, did not support the establishment of such a standing committee. To his credit, Hickey ultimately prevailed. The following year, the British issued a comprehensive formal set of recommendations and formally established the British X-Ray and Radium Protection Committee under the leadership of Professor Sydney Russ. The British recommendations introduced the concept of limited exposure time for persons working with X-rays and radium to no more than seven hours per day with Sundays and two half-days per week off (Kathren and Brodsky 1996; Williams and Ell 1986). Other countries, notably Sweden, Holland, Germany, the Soviet Union, and France also quickly adopted radiation protection recommendations and standards patterned after those of the British, which stood essentially unchanged for the next decade. And, in the United States, 1922 saw the adoption of radiation protection rules by the ARRS, patterned after those of the British.

As has been pointed out above, standards do not have the mandatory force of law or regulations, and hence the term "voluntary standards" is sometimes used. Implementation of the recommendations and guidance in these standards is therefore up to the individual and is often resisted for various reasons including cost, inconvenience, and the desire to be free from oversight. The British met with this potential lack of acceptance by a unique strategy. With encouragement and approval from the British committee, implementation of their recommendations was delegated to the National Physical Laboratory (NPL), which included the power of inspection and approval of radiology facilities. This strictly voluntary program was very well received by the radiological community. Where inspections revealed program inadequacies, recommendations would be made by the NPL, and when the program met standards, a certificate of approval was issued by the NPL (Taylor 1933).

3.6 MATURATION: STANDARDS COME OF AGE (1925–1950)

The quarter century between 1925 and 1950 was marked by a number of important changes with respect to radiation protection guidance. Not only was the standards process crystallized both on a domestic and international level, but knowledge and experience gained through the tragedy of the radium dial painters and the World War II Manhattan District effort to build nuclear explosives served to promote protection standards with more general applicability than just for use in medical applications.

Three decades after the discovery of X-rays and radioactivity (i.e., radium), radiology had become a recognized medical specialty. In addition to practitioners of the healing arts, literally hundreds of thousands of patients were exposed each year. It was left to the manufacturers of X-ray apparatus and the individual practitioner to adopt such measures as deemed appropriate, and while skin burns and other overt somatic effects in patients were by and large a thing of the past, there were subtle latent effects, including early deaths appearing among physicians, dentists, and others who worked with X-rays. In his moving tribute, Brown (1936) lists twenty X-ray pioneers who suffered early deaths as a result of their pioneering efforts with X-rays. This is but a small fraction of those whose deaths were hastened by X-ray exposure.

A testament to the effectiveness of the relatively few and simple protection standards and methods developed during the years 1925 to 1950 is clearly demonstrated by examining the mortality and morbidity of radiologists. In an illuminating epidemiological study of the mortality of British radiologists, Berrington and colleagues (2001) examined the causes of mortality in British radiologists over a period of 100 years. Compared to other British physicians of the same period, the standard mortality ratio (SMR) for all cancers in radiologists practicing from 1897 to 1920 was 75% greater than their physician peers, while those practicing from 1955 onward had a slightly lower SMR. The annual dose to the radiologists of that era was estimated at more than 1 Sv. The SMR for all cancers for radiologists in the intermediate years between 1920 and 1950 did not differ significantly from that of their non-radiologist peers (Berrington et al. 2001). Similar findings have been seen in mortality studies of American radiologists (Matanoski et al. 1975). The obvious conclusion derived from these and numerous other studies is that the X-ray exposure to those early British workers, all of whom predated the 1921 British standard, was the cause of their excess of cancer SMR (Berrington et al. 2001; Brenner and Hall 2003), underscoring the validity of even simple protective methods, at least in radiology.

3.6.1 The Tolerance Dose

The concept of a tolerance dose was introduced by the German-born American physicist Arthur Mutscheller, and led to a time of great progress with respect to radiation protection standards and the beginnings of legislation (Mutscheller 1925). The tolerance dose was the quantity of (X-ray) exposure that a person could experience continuously, or at intervals, with no demonstrable ill effect or damage to the blood or reproductive organs. From measurements and observations that radiologists and radiographers in the New York City area were free of observable deleterious X-ray effects, Mutscheller concluded that the doses they incurred were safe, and from this he derived a permissible tolerance dose of 1/100 of the skin erythema dose (SED) in a 30-day period, roughly equivalent to 2 mSv/day or about 720 mSv annually as expressed in modern units (Mutscheller 1925).

Also, inherent in the tolerance dose was the notion of recovery, viz., that in the event of overexposure, a person, given time, could recover and return to normal. Since there was no commonly accepted, or even defined, physical dose quantity or unit, initially at least, the tolerance dose was expressed in units of SED. There was no specific value for SED, which was dependent upon the characteristics of the exposing radiation, exposure conditions including rate and fractionation, and the individual susceptibility of the person exposed. Nonetheless, the tolerance dose concept was to serve as the basis for radiation protection standards and recommendations for dose limits for the next quarter century.

Although precedence goes to Mutscheller, other dose limit recommendations were proposed at about the same time, also in terms of the SED, and at a somewhat lower level, viz., one-tenth of the SED annually. But a proposal with great significance from the standpoint of regulatory authority was that of the Dutch Board of Health which proposed a limit of 1 SED per 90,000 working hours (Kaye 1927). Assuming the SED to be about 6 Sv, this equates to 6.67 µSv per hour, or about 4 mSv annually, quite a step down from Mutscheller, and more in keeping with current legal limits.

3.6.2 The Roentgen

A major milestone in the development of protection standards generally, and exposure limitation standards specifically, was the definition of the quantity of exposure and its measurement unit, the roentgen. This was a product of an international committee of physicists appointed at the Second International Congress on Radiology held in Stockholm in 1928. This committee was the start of what is now the ICRP (Bushong 1995; Taylor 1979).

The roentgen, a unit based on ionization in air, was defined as the quantity of X-radiation or gamma radiation that would produce in 1 cm^3 of air at standard temperature and pressure one electrostatic unit of charge. Over the years, this basic definition of the roentgen underwent a number of minor modifications, but essentially remained the same, until it was declared obsolete with the adoption of the Système Internationale (SI) in the latter half of the twentieth century. Although there is now no official unit for exposure in its special sense as applied to X-radiation and gamma radiation, exposure in the special sense can still be described in terms of Coulomb per kg of air which can readily be converted to the earlier definition through simple unitary conversion.

The roentgen was a physical unit, based on air ionization, and hence did not have any of the biological or other limitations of the SED. The importance of this unit to radiation protection was immediately realized, and indeed cannot be understated, for doses could now be measured in a consistent fashion and unambiguously specified and compared. Thus, the development of ionization chamber instrumentation to measure radiation in terms of, or convertible to, the roentgen was spurred on, as was personnel monitoring with film badges to replace periodic blood counts. Mutscheller reaffirmed his recommended permissible or tolerance dose of 1% of SED per month based on additional measurements made by himself and other investigators in 1928 and 1934 (Mutscheller 1928, 1934). Recognizing the dependence of the SED on the energy of the exposing radiation, his 1934 recommendation specified a limit of 3.4 R per month for low-energy radiation and 7.5 R month, roughly 900 mSv annually, for the more penetrating rays (Mutscheller 1928, 1934).

3.6.3 Organizing for Radiation Protection: The ICRU, ICRP, and NCRP

At the First International Congress on Radiology, held in 1925, the International X-ray Unit Committee (IXRUC) was created. Formation of this body, which in 1950 morphed into the International Commission on Radiological Units (ICRU), reflected the importance and need for a consensus radiation measurement unit and associated measurement techniques as fundamental, not only to diagnostic and therapeutic applications of radiation, but, albeit secondarily, also an essential underpinning of exposure limitation and other protective standards. Thus, the IXRUC devoted its efforts primarily to measurement techniques and instrumentation and attempting to resolve and explain the differences between measurements made by different investigators. For example, there were large differences between the French and German measurements of erythema dose, puzzling differences that could not be explained at the time, but now are understood as being related to wall effects (Taylor 1989).

The Second International Congress on Radiology, held in 1928, saw the formation of the International Committee on X-ray and Radium Protection (ICXRP), the forerunner of the modern day ICRP). This initially five-member body consisted of four physicists, one each from Great Britain, Germany, Sweden, and the United States, plus a physician from Great Britain, and was chaired by physicist G.W.C. Kaye of Great Britain. The American member was Lauriston Taylor who, upon his return to the United States, lobbied the various American radiological societies to form an American counterpart of the ICXRP in the form of a single national committee devoted to radiation protection. Taylor's efforts led to the formation of the Advisory Committee on X-ray and Radium Protection (ACXRP), the forerunner of the current NCRP in 1929. Both committees began their efforts quickly. The international body issued its first report in 1929, but curiously made no mention of a permissible dose or dose limit in it or in its subsequent report dated 1931 (International Congress of Radiology 1929; Bushong 1995; Taylor 1971).

Both the ACXRP and ICXRP proposed levels for what was termed the "tolerance dose" of 0.2 R d^{-1} (~2 mSv d^{-1}), the ACXRP in 1931 and the ICXRP in 1934 (Kathren and Ziemer 1980). Also in 1931, the League of Nations published a detailed and comprehensive report which set an exposure limit of 10^{-5}R s^{-1} for 8 hours, equivalent to a daily dose of about 2.5 mSv (Wintz and Rump 1931). Unfortunately, this excellent report never gained much traction, as the League of Nations was soon to fail, and the Wintz and Rump report faded into obscurity (Grigg 1965). In 1936, the ACXRP halved its proposed tolerance dose limit to 0.1 R d^{-1}, beginning what has been the downward slide in permissible dose levels that continues to this day, and has been well and beautifully chronicled since 1947 for both committees and American federal agencies by Jones, and for earlier years by Bushong (Jones 2005; Bushong 1995).

The early history of these two independent non-governmental standards bodies has been amply documented by Taylor and others in numerous publications and need not be repeated here (Taylor 1958a,b, 1971, 1979; Brodsky, Kathren, and Willis 1995; Bushong 1995). However, suffice to say, from rather modest beginnings, these two organizations have become the premier and most influential non-governmental radiation protection standards setting bodies in the world. Many of the recommendations of these voluntary, scientifically based standards setting bodies, in particular, but not limited to, proposed dose limits, have been incorporated into legislation internationally. These recommendations generally lead the legislative actions by several years.

3.6.4 Radium Rears Its Ugly Head

Additional impetus was given to the need for protective standards from a totally unexpected quarter. For nearly three decades, radium activated phosphors had been a component of self-luminous compounds and paints used primarily for watch, clock, and instrument dials, and radium dial painting was a flourishing industry. The bubble burst in 1924, when in a brief footnote to an article in the *Journal of the American Dental Association*, Theodore Blum, a New York dentist noted that an unusual number of his patients—all young women and radium dial painters—were affected with a particular syndrome of the jaw. This led Harrison Martland, medical examiner for nearby Essex County, New Jersey, to examine the mortality of these young female dial painters which resulted in a classic series of studies over the next 20 years, and the establishment of standards for internal emitters and airborne radioactivity largely through the pioneering studies of Robley D. Evans and his colleagues at the Massachusetts Institute of Technology (Evans 1933, 1974, 1981; Rowland 1994). Once again, radium captured the public fancy, but this time the experience of the dial painters shook the public confidence in the benign nature of radium, as did the revelations of radiation-induced injuries and death revealed in lawsuits brought against the U.S. Radium Corporation beginning in the 1920s. Interestingly, the last of the dial painters, Mrs. Mae Keene, died in 2014 at the age of 107. During her long lifetime, she suffered from two different cancers, but fortunately not the untreatable osteosarcoma that was the fate and death bringer for many of her coworkers.

Recognition of the need for standards and regulations to protect the general public was furthered as a result of the death in 1932 of the wealthy and prominent Pittsburgh industrialist Eban Byers, a consumer of large quantities of the radium-containing patent medicine Radithor. This unhappy event further eroded public confidence in the alleged beneficial nature of radium; Byers was a former U.S. Amateur Golf champion and a well-known and liked figure. The experiences of the dial painters and Byers led down a whole new and hitherto unconsidered protection path: limitation of radioactivity ingested within the body or inhaled via contaminated air. This path would be extremely important in a few years as a result of the American effort to build an atomic bomb. Flowing directly from the dial painter experience, and just in time for the atomic bomb development effort, were two standards pertaining to radioactivity in the body. One was the permissible amount of residual radium in the body considered safe, set at 0.1 µCi (3700 Bq) (Evans 1981; NBS 1941). The other was a standard for the maximum permissible concentration of thoron in air, set at 10^{-11} Ci L^{-1} (3.7×10^{-5} Bq m^{-3}) (Evans and Goodman 1940). These were the first true standards based on radioactivity in the body.

Over the years, the saga of the dial painters has spawned a number of books and numerous articles in the lay literature and resulted in a large amount of scientific study. Human experience with radium from a scientific perspective has been well-documented in the excellent book by Rowland (1994), published as the dial painter studies were wrapping up.

3.6.5 The World at War

World War II provided the greatest motivation for the development of radiation protection guidance, but because of military security requirements, very little appeared to have been done, as secrecy was maintained. From the outset, the effort to develop and build an atomic bomb brought with it concerns about the safety of the creation of vast amounts radioelements, many of which are highly radiotoxic, were never before found on earth, and were therefore outside of human experience. Protection for workers as well as the environment were important considerations and accordingly a large-scale research program was undertaken to study the nature of these new radioelements, not only from a physical standpoint but also to understand their biological aspects and thus gain the knowledge necessary to ensure the safety from radiation hazards consistent with the needs of a nation at war. The activities carried out during this period in the Manhattan Project have been documented in numerous publications, but most notably in the concise single volume by Barton C. Hacker and in the epic 2,000 page tome by J. Newell Stannard (Hacker 1987; Stannard and Baalman Jr. 1988). Unlike the relatively simple situation of external exposure, as was the case with X-rays, there were numerous fission products, isotopes of many elements that would be produced, all with differing radioactive decay properties, decay chains, measurement techniques, and biological characteristics vital to understanding and setting protective standards. And, given the heightened awareness of the potential dangers of radionuclides taken into the body because of the radium dial painter experience, there was special concern about plutonium, which was being produced, in immense quantities and, by analogy with radium, promised to be particularly hazardous on several levels.

The exigencies of wartime, on the one hand, slowed the development of voluntary standards of all kinds, but, on the other hand, accentuated the need and importance for such standards. With the new knowledge being gained in the Manhattan Engineer District

(MED), radiation protection becomes much more complex and demanding, requiring the development of new instrumentation and techniques, or as Jones so succinctly put it, "a more complicated task" (Jones 2005). Numerous standards were developed by the MED, but these were typically not formalized, in part because of security, and therefore should be considered as internal operating standards. Several, however, were of fundamental and lasting impact, and led to the formation of what might well be termed the era of radiation protection standards.

3.6.6 The Immediate Postwar Period

In the half decade following the end of World War II, a cornucopia of scientific and programmatic data flowed out of the MED and set the stage for future legislative and voluntary standards. Large gains were made with respect to radiation protection guidance, for with the end of the war and the easing of security restrictions, the scientific advances during the war years could be applied to civilian projects. These applications had begun within a year of the war's end through civilian use of radionuclides obtained from the fission process. There were few, if any, regulations to control civilian usage, and also few standards and guides, although these were on the way.

Recognizing its now much broader concern than X-rays and radium, the ACXRP, which had been largely dormant during the World War II years, became reinvigorated. Its name was changed to the National Committee on Radiation Protection operating out of the National Bureau of Standards (NBS), but with no statutory responsibility for radiation protection. It later adopted its current name, NCRP. In 1948, at its first formal meeting, the NCRP proposed a reduction in the permissible dose standard from 1 mSv (0.1 R) per day, comparable to about 300 mSv annually, to 3 mSv (0.3 R) per week, comparable to 150 mSv annually. The proposed limits were adopted in principle. The assumption was that the blood-forming organs were the most radiosensitive tissue. Higher limits were proposed for exposures limited to the skin, the hands and feet, persons over the age of 45, and accidental exposures. In addition, in the case of other radiations such as neutrons and alpha particles, the R unit would be modified by a multiplier known as the Relative Biological Effectiveness, to account for their more significant dose effect (Taylor 1971).

The full and complete recommendations, published as NCRP Report 17 in NBS Handbook 59, did not appear until 1954 (NBS 1954), and in subtle fashion introduced the recommended permissible dose concept, discarding the term "tolerance dose" and replacing it with "permissible dose," a semantic change that recognized the possibility that, for some effects, specifically genetic mutations, there might well be no threshold for the absence of a deleterious effect. This was strengthened in later publications, and indeed was the harbinger of the change from tolerance dose to a risk-based no-threshold model of response to radiation. Even prior to the formal publication, these recommendations had a very large influence on the establishment of protection standards and legislation within the United States and worldwide.

Standards and guides had been developed within MED for plutonium-based on animal studies carried out during the war, and in 1947 the great British–American radiological and health physicist Herbert M. Parker described the standard setting and operational radiation protection programs and limits within the MED. This was initially was published as a report and reprinted in 1980 as a landmark paper in a special edition of the journal *Health Physics* (Parker 1948). From a practical as well as a scientific standpoint, the evolution of an unambiguous system of radiological units, elegant in its simplicity, was a major contribution to the formation of radiation protection regulations and standards, much of which would have otherwise been impossible or meaningless. As early as 1937, the adequacy of the roentgen had been questioned by Gioaccino Failla, but during MED days it was recognized the inadequacies were far more serious than had been previously thought. The roentgen unit was inadequate to characterize doses from radiations other than photons, and indeed was defined only in terms of X-radiation and gamma radiation in air. It remained for Herbert M. Parker, then at Oak Ridge, to come up with a solution to the problem. In 1943, while at Oak Ridge, Parker proposed a purely physical unit based on energy deposition in matter that could be applied to any ionizing radiation—photons, beta, alpha, and neutron—or to mixed radiations (Parker 1950). He called this new unit the "rep," the acronym for radiation equivalent physical. Later this unit, which at one time was also known as the parker, was slightly revised and renamed the rad, the unit for the quantity absorbed dose. Parker also proposed a biological unit based on the rep to account for the different biological effectiveness of various radiations. This he named the "rem," the acronym for radiation equivalent man (Kathren et al. 1986). These two units and their associated quantities are the direct ancestral beginnings of the gray and the sievert, and of course trace their origins to the now obsolete roentgen.

Parker also established an air concentration limit for plutonium—the first so-called Maximum Permissible Concentration in Air—based on the dose to the lung tissue in 1944 in a report that was declassified in 1947. The permissible level that Parker derived is very close to the Derived Air Concentration value specified in standards and regulations today (Kathren et al. 1986).

The work of Parker, then at Hanford, with respect to standards for radioactivity within the body, was greatly enlarged by his now equivalent and former employee at Oak Ridge, Karl Z. Morgan, who prepared a comprehensive and classified report of maximum permissible amounts in the body and permissible concentrations in air and water for many radionuclides (Morgan 1947). The original report, drafted in 1945 but not declassified until 1954, served as the basis for the early standards regarding internal emitters, that is, radioactivity within the body, addressing to a great extent the concern raised by the dial painters who had ingested radium.

Also during the early postwar years, two important small conferences were held. These were the First and Second Tripartite Conference on Internal Dosimetry. Representative scientists from the United States, Canada, and Great Britain met in 1949 and 1950, and from these conferences emerged the basic system for handling radionuclides put forth by the NCRP and ICRP for at least the next two decades. The so called Standard (now Reference) Man was also developed which provided a consistent basis for biokinetics and dosimetry of radionuclides. The development of protective standards for internal emitters and their scientific basis has been well and fully documented by J. Newell Stannard in Chapter 16 of his monumental history of radioactivity and health (Stannard and Baalman Jr. 1988).

Perhaps the event of most significance to radiation protection guidance was the decision by the United States to place atomic energy under civilian rather than military control. Resolving the ongoing controversy over who should control what, President Harry S. Truman signed into law the McMahon Bill, the Atomic Energy Act of 1946, creating a new federal agency, the Atomic Energy Commission (AEC) to oversee all aspects of the atomic energy program. Under executive order, all personnel and properties were transferred to this new agency, and remained the property of the government. The Act was primarily concerned with weapons development, and thus the big concerns were security and national defense. Although the AEC was given broad powers under the Act, conspicuously omitted was virtually any reference to worker or public health and safety, or protection of the environment. Jones has noted that the word "safety" appears in the McMahon Bill only four times, and there is no reference whatsoever to radiation protection (Jones 2005).

It is of no small interest to note that the numerous radiation protection activities of the MED were largely carried out internally to the MED and, because of the war effort, with great rapidity, unlike the rather more leisurely efforts at radiation protection that had preceded the MED. As Hacker noted: "Ironically, the leaders of prewar standard setting played no major roles in the wartime project that posed the greatest challenge to radiation safety, the development of atomic bombs" (Hacker 1987).

One can speculate on why this was the case. Security concerns and a lack of political clout and Nobel laureates certainly played a role, as did the fact that those setting the standards pre-World War II were for the most part concerned with controlling medical exposures from X-rays. Thus, they were not involved with atomic and nuclear physics per se. But it also bears mentioning that two of the original seven health physicists appointed in the MED—John Rose and Herbert Parker—were, in fact, medical physicists and Parker went on to lead the effort at the Hanford site during the war, having been handpicked to do so.

3.7 INTERLUDE

The mid-twentieth century marks a shift in emphasis with respect to radiation protection guidance to promulgation of standards and guides. In effect, the regulatory era had begun, and would become increasingly proscriptive and detailed. The prescient Herb Parker, perhaps somewhat tongue-in-cheek, noted this in his keynote address at the 1971 Mid-Year Topical Symposium of the Health Physics Society titled "Radiation Protection Standards: Quo Vadis." In keeping with the Latin in the symposium name, which translates to "where are we going," he titled his talk "Festina Lente," which translates to "make haste slowly." Noting the rise in the number of radiation protection standards proposed by such bodies as the NCRP, the ICRP, and the ANSI to a total of 114 in 1971, as compared with only 53 five years previously, and using data points from two other prior years, Parker derived an exponential equation from which he predicted 9,662 voluntary standards in 2001, some 30 years into the future (Parker 1971), equivalent to a doubling time of about 4.7 years. Brodsky (1978), in his re-evaluation of the data used by Parker, added a fifth point, one standard in 1928, and calculated what he called a "very comforting" 2,470 voluntary standards in 2001. Whichever of the two calculations is closer to the actual number is moot, but it is clear that there was a veritable explosion in the number of voluntary standards.

The same holds true for regulatory standards, which, depending on how one chooses to count them, likely run into the thousands, considering all of the state and other governmental agencies in the United States alone with responsibilities of some kind with respect to radiation protection.

Thus, after the middle of the twentieth century, the evolution of radiation protection guidance rather neatly devolves into an era of standards and regulations and will be so treated in the remainder of this chapter. So large in number and diverse in content have been the regulations and standards developed in this period, that it would be impossible to consider the entire spectrum, and hence of necessity the emphasis will be on exposure limits and regulations. However, a number of excellent histories relating to this period have been written and those referenced in this chapter have been identified by an asterisk preceding them in the references. It should be noted that there are several more that could well be added to an already impressive list.

3.8 AN EXPLOSION OF RADIATION REGULATIONS AND STANDARDS: PART I

When the AEC officially assumed responsibility from the military for what had been the Manhattan Engineer District (MED) in January 1947, the primary concern was with military applications of nuclear energy. Atmospheric testing, with its injection of enormous amounts of radioactivity into the atmosphere, created the potential for worldwide exposure of members of the general public. In addition, there were a significant number of persons occupationally involved with the weapons program, and the potential number of persons working with the development of civilian applications of nuclear energy was growing rapidly, given great impetus by the declassification of MED documents. There was also growing concern among some radiological scientists that the tolerance dose was inadequate to provide the level of protection needed from possible adverse mutagenetic effects, as had been shown by the experimental studies of Herman J. Muller some two decades earlier (Muller 1927, 1941), and that new standards were needed to take this into account. The concern for adverse genetic effects, which showed a straight-line relationship with dose, was such that the idea of a tolerance or threshold dose was abandoned in favor of a risk basis for exposure standards.

Muller's discovery applied not only to genetic effects but to stochastically produced somatic effects such as leukemias and solid tumors and gave rise to the Linear Non-Threshold (LNT) dose response model, upon which regulations and standards became firmly based. Over the years, the LNT model has been the subject of much criticism and debate, but has evolved into a sort of paradigm and remained the underlying if unspoken basic criterion upon which exposure limits are based (Kathren and Brodsky 1996).

3.8.1 The 1950s: A Watershed

The NCRP wasted no time in preparing sophisticated recommendations, publishing no less than 18 reports during the ten-year period of the 1950s, covering a diversity of specific topics, including handling of radioactive wastes, monitoring and instrumentation, and, of course, exposure limits, which were by and large ultimately incorporated into regulations. Among the more significant of these was the NBS Handbook 52, published in 1953, whose title alone, "Maximum Permissible Amounts of Radioisotopes in the Human Body and Maximum Permissible Concentrations in Air and Water" is sufficient to speak to its content (NBS 1953). Handbook 52 drew largely on the work of Karl Z. Morgan at Oak Ridge (Morgan 1947), and was a milestone with respect to exposure standards. After only seven years, in 1959, Handbook 52 was superseded by a much-enlarged Handbook 69 that not only considered occupational exposures, but also put forth standards of exposure for members of the general public. Permissible dose recommendations reducing the permissible whole-body penetrating occupational exposure limit to 3 mSv (300 mrem) per week, or 150 mSv annually were put forth in Handbook 59, published in 1954, and Handbook 63 was devoted exclusively to protection from neutron radiations. Brodsky, Kathren, and Willis (1995) have noted that these NCRP reports provided a foundation for standards of practice with respect to radiation protection and subsequent regulations that prevail unto this day.

The ICRP also was quite busy. Recognizing that the previously proposed limit of 1 R per week was too close to the threshold for damage, it proposed in its comprehensive 1951 report a whole-body exposure limit of 0.5 R per week at the body surface, equivalent to 0.3 R (3 mSv) in free air. It also included a limit of 0.1 μg of radium in the body, and established committees, formalized in 1953, for specific topical areas, such as external radiation. Perhaps most significantly, concerned with the potential health impacts associated with atmospheric testing on nuclear explosives, the ICRP recognized the need for protection standards for the general public, setting these at one-tenth the proposed occupational limits ICRP 1951 (ICRP 1955; Clarke and Valentin 2005). Finally, at the close of the decade, the details on permissible limits for internal dose were reported by ICRP Committee II. Largely the work of Karl Z. Morgan and his committee, this report was published as the entire Vol. 3 of the journal *Health Physics*, as well as in a stand-alone version by the ICRP and remained the de facto standard for internal dose for many years (ICRP 1959b, 1960).

But perhaps the single most significant event was the Atomic Energy Act of 1954, adding the authority for licensing and regulations to the domain of the AEC (Mazuzan and Walker 1985). The AEC had begun distributing radionuclides for civilian uses in 1946, but there was no formal radiation safety program or regulatory requirements, except for informal inspections carried out by AEC staff. The Act gave the AEC preemptive authority over by-product material (as fission products were termed), nuclear reactors, and materials such as enriched uranium, plutonium, and tritium necessary to weapons production. It did not give the AEC control over X-rays, accelerators, or naturally occurring radioactivity except insofar as these were used at AEC-owned facilities. The Act led to a reorganization of the AEC, administratively separating the regulatory aspects from the promotional aspects. Congressional oversight for the AEC was the purview of the Joint Committee on Atomic Energy (JCAE) of the United States which at its annual hearing in February 1955 implored the AEC to keep licensing procedures simple and expeditious to encourage private participation. Accordingly, the AEC developed a licensing protocol for users of by-product material and, under the expanded authority of the Act, the AEC prepared a comprehensive set of regulations, based in large measure on the 1953 NCRP recommendations, that later became Title 10, Part 20 of the Code of Federal Regulations (Mazuzan and Walker 1985).

The 1954 Act prompted other safety standards and regulations. There were already NCRP recommendations with respect to X-ray safety, particle accelerators, disposal of specific wastes, and even burial of cadavers containing radioactive isotopes, as well as those already mentioned on dose limitation, but these only covered a portion of the waterfront. The NCRP had adopted a policy of discouraging the adoption of its recommendations into regulations, but in 1955 changed to a more neutral policy of neither recommending nor opposing state legislation, stating this change in the preface to its comprehensive report "Regulation of Radiation Exposure by Legislative Means" (NBS 1955). This report included a complete evaluation of the need for legislation along with a suggested state radiation protection Act that complemented and completed the AEC regulatory authority for radiation protection.

Also in the decade of the 1950s, the U.S. Public Health Service (PHS), itself not a regulatory agency but an advisory body to the states, promoted radiation protection activities by the states by providing training and other administrative support. An outstanding contribution by the PHS was a comprehensive guide for inspection of diagnostic X-ray facilities that, because of the color of its cover, became known as the "Yellow Book" (Ingraham, Terrill Jr., and Moeller 1953). The PHS jumped into the breach left by the limitations of the 1954 Act, offering training courses and other support to the states with respect to radiation safety. Perhaps the most significant contribution of the PHS has been its preparation and publication in 1954 of the "Radiological Health Handbook"), a compendium of useful information to those engaged in radiation protection activities that was the idea of Simon Kinsman, a colorful and exuberant PHS officer who developed and taught many of the PHS short courses in radiation protection. This publication has endured, with several updates, for more than 60 years and is still widely used today, but is no longer published as a government document but rather as a commercial venture.

The decade of the 1950s also saw the start of state regulatory programs, and more importantly, the establishment of the agreement state program by the AEC. In exchange for establishing a program with regulatory limits that were no more nor any less restrictive than those of the AEC, the AEC would relinquish regulatory control of by-product materials to the states who were encouraged to make their regulations comprehensive, and include sources of ionizing radiation, such as naturally occurring radioactivity and X-ray and other machine sources that did not fall within the purview of the AEC. To sweeten the pot, so to speak, the AEC provided funding and training to those states with which agreements were made. Led by California, states began to adopt, or at least considered adopting, radiation protection regulations and the startup of inspection and licensure programs for the whole spectrum of radioactive sources and ionizing radiation generating apparatus. This brought X-rays, particle accelerators, and naturally occurring radioactivity into the regulatory tent. Today, every state has its own regulatory and licensing program, carrying out regular inspections and issuing licenses patterned after those of the Nuclear Regulatory Commission (NRC) which is a successor organization of the AEC. Coordination among the states is accomplished through the Council of Radiation Program Directors (CRCPD), an active organization that meets annually to share information and new developments.

On the international scene, this extraordinarily productive decade also saw the beginning of official international governmental collaborations. As more nations acquired the capability to build and stockpile nuclear weapons, President Dwight D. Eisenhower, in his Atoms for Peace speech, urged the United Nations General Assembly to create an international agency that would establish a worldwide system of nuclear material inspection and control. This was instrumental in the formation of the IAEA by the UN in 1957, and the creation of the United Nations Scientific Committee on the Effects of Atomic Radiation. The latter dealt primarily with periodic compilations of scientific data, issuing recommendations and reports used by standards setting bodies. Also on the international front were the entries of the World Health Organization and International Labor Office into the radiation protection standards arena.

3.8.2 From Tolerance Dose to the LNT Paradigm

Certainly, a major, albeit subtle, change that took place during the decade of the 1950s was the philosophical pivot away from protection standards based on the tolerance or threshold dose concept to risk-based levels as indicated by the Linear Non-threshold Theory response to radiation exposure (LNT). The groundwork for this change had already been laid by the work of (Muller 1927, 1941) who in his studies with fruit flies (*Drosophila melanogaster*) had determined that the relationship between induced genetic mutations were linear with dose and had no threshold. Not until 1940 was Muller's work given consideration from the standpoint of protection standards. In that year, the ACXRP met to propose a reduction in the permissible exposure and concluded that a five-fold reduction from 0.1 R per day dose to 0.02 R per day was warranted based on potential genetic effects. However, one member of the committee who was unable to attend the meeting took strong exception to this change, noting that the damage from the extant daily limit 0.1 R was "...so slight that one can just as well stop there," and that a lower dose would be a hindrance to the medical applications of radium, which was then widely used for treatment of cervical, uterine, and other cancers (Hacker 1987). At the next meeting of the ACXRP, in September 1941, this view prevailed and the decision was made to postpone this action pending further development of knowledge.

The formation of the MED and the exigencies of wartime security put radiation protection standards on hold insofar as the public, at least, was concerned. There was, however, great activity with the MED itself. The basis for radiation protection exposure limits and other standards within the MED was the tolerance dose, and the underlying reasoning has been fully and clearly characterized by Simeon T. Cantril, one of the physicians who worked on the plutonium project at the Hanford site (Cantril 1951). Cantril summarized in a brief concluding section to his report the tolerance doses for radiations and radioactive materials. The whole-body tolerance dose for external X-radiation and gamma radiation exposure was given at 0.1R (~1 mSv) per day, this being considered a level at which, or below which, the body was capable of repairing any radiation-induced damage, and was hence a threshold. Cantril acknowledged non-threshold events, pointing out that these needed to be measurable or observable before they can be accepted. He also gave as an example the genetic effects noted by Muller in his *Drosophila* experiments but stated that the majority of radiation-induced effects had a threshold (Cantril 1951). He further noted in his discussion that inheritable effects had only been seen in lower forms of life, and mentioned without a specific citation that research done at the National Cancer Institute in which mice were exposed to up to 8.8 R (~88 mSv) per day showed no evidence of mutagenesis through several generations.

Cantril's report represented the prevailing opinion on the basis of radiation protection dose limiting standards at the conclusion of World War II and for a few years thereafter. Additional impetus to the conversion to the LNT was provided by a series of hearings by the Joint Committee on Atomic Energy of the U.S. Congress (JCAE) (Joint Committee on Atomic Energy (JCAE) 1960a,b). But the shift toward the LNT had already begun, and it moved very swiftly, fueled to a large degree by concern about the possible mutagenic effects of nuclear testing which was injecting large quantities of radioactivity into the atmosphere. This was not unnoticed and was incorporated into the recommendations of both the Americans and the British as well who developed the concept of a risk-based maximum permissible dose (MPD) which showed, at least for genetic effects, no threshold and a linear relationship with dose (NBS 1954; ICRP 1955). Additional impetus came from a series of hearings by the JCAE in the middle to late 1950s on fallout from weapons testing which focused attention on the genetic dose problem, and by a committee, the Biological Effects of Atomic Radiation (BEAR), appointed by the National Academy of Sciences/National Research Council (NAS) in 1955 at the behest of the AEC. The first report of the BEAR Committee was issued the following year and raised serious questions about the genetic effects of radiation, unequivocally stating that even small radiation doses could have serious adverse consequences over the lifetime of an individual (NAS/NRC 1956). Similar studies and conclusions were reached by other bodies, including the NCRP and ICRP, but may have been attributable to use of the same databases and membership overlap among the various groups (Jones 2005). Somewhat later, and perhaps with some equivocation, the United Nations Scientific Committee on the Effects of Atomic Radiation (UNSCEAR) (1958) gave credence to the LNT which was used along with a threshold model to make numerical estimates of dose effects (Kathren 1996).

Adoption of the LNT was an important step and addressed both scientific concerns and what had been a growing public concern over the inheritable genetic effects of atmospheric nuclear weapons testing. As new data regarding low level radiation effects were developed, and especially the data from (in some instances, admittedly weak and controversial) radioepidemiological studies of nuclear workers, the LNT took on the status of a paradigm among regulators and standards setters alike. It was not, however, well received by nuclear-based industries, who were concerned at the reduction of permissible doses which was restrictive to their operations and frequently required additional expenses, not only because of personnel restrictions, but also for new and expensive equipment for measurement and control. Six decades later, the LNT remains firmly entrenched as the basis for radiation protection exposure limits, although there is a body of opposition and growing scientific data and support for the concept of hormesis that suggests that very low doses of ionizing radiation may apparently be beneficial (Sanders 2010). Indeed, the shape of the dose response curve at low levels of exposure has been the subject of considerable discussion and disagreement, sometimes acrimonious, over these six-plus decades, but the LNT and its application has been and remains an important factor in reassessment and continual lowering of radiation exposure limits.

The change from the tolerance dose basis to the LNT basis led to the concept of ALARA, an acronym derived from "As Low As Reasonably Achievable." The concept was first introduced in 1954 in the NCRP Report, and later in ICRP Publication 1 where it was initially called ALAP, the acronym for As Low As Practicable (ICRP 1959b). This was a great philosophical change, for now instead of working up to the permissible exposure limits, the idea was to keep exposures to levels as far below the limits as could reasonably be done, considering economic and technical factors. Economic considerations were a key component. Ultimately, a dose of one person-rem was assigned a value of $1,000. In other words, if it cost no more than $1,000 to reduce an individual or collective dose by one person-rem by whatever means, those means should be taken. ALARA has now expanded into a full-blown program and various job are tasked with an ALARA review. Records are kept of personnel exposures and compared with previous years for evaluation in terms of ALARA, and all physical measures such as protective shielding are evaluated on the basis of ALARA.

3.9 AN EXPLOSION OF RADIATION REGULATIONS AND STANDARDS: PART II

The last half-century has seen an explosion of voluntary and regulatory radiation protection standards. Regulatory standards have been dynamic, but increasingly restrictive. Voluntary standards have led the regulatory changes and have also led to requirements for radiation measurement instrumentation and techniques. Lawsuits by persons who claimed to have suffered radiation injuries, particularly "downwinders" from weapons tests or large nuclear facilities, have encouraged the development of protection guidance, and a compensation scheme has been established for employees of the Department of Energy and its contractor employees who meet certain work criteria and who, based on their reconstructed exposure, develop cancer. The tolerance dose basis for protection standards has been replaced with a risk-based system in which the magnitude of stochastic risks such as cancer are considered to be a linear function of dose, with no threshold below which there is zero probability of an effect. Thus, the LNT is the driver or basis for establishing exposure limits for low level exposures and has led to reductions in these limits over the years.

3.9.1 Proliferation of Regulations and Standards

The decade of the 1950s was, from the standpoint of radiation protection guidance, highly productive. In addition to what might be termed technological and legislative gains, increases in public awareness and concern had been achieved, particularly with respect to genetic mutations attributable to atmospheric testing of nuclear weapons. The momentum of the decade of the 1950s did not slow, and if anything, the years that followed saw a veritable flood of radiation protection regulations, standards, and guides, along with a proliferation of organizations, both private and governmental, and national and international. Everyone, or so it seemed, wanted to get in on the act, and there was a duplication of efforts, as well as the issuance of numerous reports, voluntary standards, and regulations, and jousting for position and control. Unfortunately, as there was no overall or comprehensive or cohesive plan, or single agency with overall responsibility for radiation protection, the argument could well be made that in many instances, such standards and legislation as were proposed and even adopted were based on politics, economics, or special interests and each agency within the U.S. government was free to establish its own standards based on its own perceived needs (Palmiter and Tompkins 1965).

The lack of centralization led to a plethora of federal and state agencies concerned with some aspect of radiation protection regulation, a problem that was addressed unsuccessfully several times by proposed legislation in Congress to establish a centralized single agency with the responsibility and authority for radiation standards. This, however, was not to be, and the regulatory situation on the Federal level remains fragmented and confusing. Although dominated by three Federal agencies (none of which existed as such in 1960), there exist an additional dozen or so Federal agencies that have more circumscribed roles.

The situation with non-governmental standards bodies was similar to that in government, in that there was a proliferation of standards and standards setting bodies. In some instances, the latter were set up primarily to promote or benefit the point of view of a particular segment of those engaged in the nuclear industry. Included in this latter subgroup were standards setting bodies established by professional organizations, which did much to establish and standardize professional qualifications of various types of practitioners. It bears mentioning that, largely because of the newness of the nuclear field, other than in medical areas, there were no professional licensure or certification requirements, although this would soon change. Much like the situation with government, the process was dominated by the two old line expert standards setting bodies, whose small cadre of members were drawn from experts and who served by invitation. These two voluntary bodies, the NCRP and ICRP, were at the forefront of the establishment of standards for exposure and concentration limits, and soon expanded their efforts into the broader area of radiation protection generally. These bodies, not always consistent with each other, along with the irregularly issued comprehensive scientific literature reviews and evaluations of UNSCEAR and the NAS Committee on Biological Effects of Ionizing Radiation (BEIR), took the lead in developing the scientific bases for protection standards, and the reports and recommendations produced by these bodies were by and large those that drove the regulatory machinery. Although generally accepted and rather widely read within the radiation protection and radiation biology communities, these studies were not always without controversy, especially among individuals with special interests.

3.9.2 The ICRP

The oldest of the voluntary radiation protection standards organizations is the ICRP, established in 1928. Prior to 1960, the ICRP had published eight recommendations or commentaries on same, all in scientific or medical journals, but in 1959 it put forth its first self-published effort under its own imprimatur. To date, ICRP has self-published 131 reports. ICRP Publication 1, published in 1959, made a great impact. Among other things, it recommended lowering the whole-body permissible exposure standard, based on the dose to the gonads, the blood-forming organs, and the lens of the eye to 0.1 rem (~1 mSv) per week and 3 rem (~30 mSv) in 13 weeks. Consistent with the NCRP recommendations of the previous year, a lifetime limit of 5 (N-18) rem was proposed (where N is the worker's age in years). There were limits for other organs, and most significantly, the definition of "permissible dose" was provided, along with its biological basis. The document took a sort of middle posture between the tolerance dose and LNT bases, noting that the standards were set to provide no discernable ill effect in an individual and in populations might be discernable by statistical techniques (ICRP 1959a).

Shortly thereafter appeared the report of ICRP Committee II, already briefly discussed in Section 3.8.2. This report, an impressively large tabulation of 40 pages of text and 190 pages of tables, plus another 150 pages of references in the 1960 version published as the entire Vol. 3 of the journal *Health Physics* (ICRP 1960), was a landmark. It introduced the concept of a critical organ, provided equations and derivations of the same for calculation of organ doses based on continuous intake, established maximum permissible concentrations in air and water, and set standards for exposure of the general population. Perhaps most significantly, it served as the basis for the regulations of the AEC as put forth in Title 10 Code of Federal Regulations Part 20 (10 CFR 20), and clearly established the leadership of ICRP with respect to protection standards for radioactivity in the body.

Subsequently, in 1977, the Committee II report was replaced by ICRP Publication 30, a multivolume compilation issued piecemeal that provided detailed information regarding internal emitters and introduced the concepts of Annual Limit on Intake (ALI) to replace the Maximum Permissible Body Burden and Derived Air Concentration (DAC) to replace the Maximum Permissible Concentrations (MPC) put forth in the Committee II report (ICRP 1979). This was not done without some resistance from the radiation protection community itself, but now, nearly 40 years later, are well established and universally used in radiation protection work. Then, beginning with Publication 56 in 1989 and continuing through the 1990s, the ICRP issued a series of reports which provided tabulations of dose coefficients for individual radionuclides to workers members of the public from intakes of individual radionuclides, provided detailed biokinetic models, and the scientific and technical basis.

While ICRP contributions to internal dosimetry have been great, and are generally accepted as the standard worldwide, ICRP has also made important contributions in other areas. New concepts were introduced in Publication 26, the 1977 recommendations of the Commission, and included SI units, definition of several new dose quantities, optimization, committed dose, tissue weighting factors and equivalent dose, and consideration of both stochastic (e.g., carcinogenesis) and deterministic (e.g., skin burns and cataracts) effects. Permissible dose limits were strictly risk-based and consistent with the LNT concept which was fully embraced. Quantitative risk factors which had been developed from the experience of the Japanese survivors of the atomic bombings were used to determine recommended dose limits based on an assumed annual mortality risk of 10^{-2} per Sv (ICRP 1977, 1991; Kocher 1991). Inheritable genetic effects were no longer assumed to constitute the primary basis for setting permissible dose limits, and were assumed to constitute one-fourth of the total stochastic risk. The idea of a tolerance dose persisted in a sense only for deterministic effects for which there was a threshold limit based on somatic effects. The emphasis was now clearly based on limiting cancers and inheritable genetic effects, and rather than working up to a limit, keeping exposures as far as practicable lower than the specified limit. This was re-emphasized in the 1990 recommendations of the Commission (ICRP 1991).

ICRP Publication 26 also introduced the concept of effective dose equivalent as a means of combining both internal and external doses (or dose to various parts of the body plus the whole-body external dose) into a single value equivalent to the risk from that numerical dose of external radiation. This was accomplished by applying stochastic risk factors, termed tissue weighting factors, for various tissues and organs to equate the risk from the dose equivalent received within those organs and tissues. Thus, the effective dose equivalent was a single value that provided the total risk to an individual from all exposures, external, internal, or fractions of the body.

The new ICRP concepts and protocols represented a major change and were not immediately accepted by the radiation protection and regulatory communities. One new concept in particular that met with considerable controversy and resistance when introduced was the 50-year committed dose in which the entire dose to a specific organ or to the whole body over a 50-year period following intake is assigned to the individual in the year of intake. Although this ultimately found its way into the regulations and into general practice, this did not occur without much discussion and many hours of committee activity as well as some accommodation by the regulatory agencies. The new occupational dose limit of 20 mSv y^{-1} averaged over five years with up to 50 mSv in any one year proposed by ICRP (ICRP 1991) introduced an additional measure of stringency and was resisted, or at best grudgingly accepted, by those in nuclear installations. Among the reasons cited for the resistance were difficulties and additional costs, additional record-keeping, the apparent incongruity of assigning a 50-year committed dose to the year of intake rather than incrementally each year, and potential operational needs for additional personnel. In any risk-based system, the latter would be contradictory to the LNT, although the standards setters essentially ignored this.

3.9.3 The NCRP

The NCRP was not idle, and, if anything, even more prolific than the ICRP. As of 1975, the NCRP had published no less than 175 reports, several commentaries, lectures, and annual reports. The current organization consists of two Council level committees, plus several program area committees, each with several subcommittees, plus an advisory panel on non-ionizing radiation. Although there is considerable overlap and occasional disagreement, much of the work of the NCRP and ICRP is complementary. As a national body, the NCRP is more concerned with what is of most interest to the United States, and its emphasis is more of a broad practical or applications nature. Like the ICRP, it has made some significant contributions to radiation protection standards and also to measurements.

Among the significant publications of the NCRP have been those dealing with X-ray protection. The first NCRP X-ray protection guidance appeared in 1931 and dealt almost exclusively with medical applications, covering virtually the broad spectrum of X-ray protection measures as known at the time (NBS 1931). This report has undergone several revisions over the years and in recent years has been parceled into several reports dealing with radiation protection for specific topical areas of X-rays and other machine-produced radiations. Over the years, the NCRP recommendations have had continued wide acceptance and have been extensively incorporated into the radiation control regulations of most, if not all, states. The most recent report (NCRP 2004) deals exclusively with structural shielding design for medical X-ray imaging facilities and is the established standard for shielding design of medical, dental, and other X-ray facilities.

In 1987 and 1993, the NCRP issued new recommendations with respect to limitation of exposure to radiation (NCRP 1987, 1993). The limits proposed for doses were risk-based, and followed the ICRP to some extent in this regard, but with a number of significant differences and exceptions that allowed some measure of leniency. The annual occupational effective dose exposure limit remained at 50 mSv y^{-1} based on stochastic effects, with a lifetime limit of 10 mSv times the age of the individual (NCRP 1993). Lower limits were proposed for various special groups, such as those below 18 years of age and for the general population. For the general population, the annual limit was 1 mSv, expressed in terms of equivalent dose. For the embryo/fetus, the recommended total effective dose was specified as 5 mSv in the 1987 publication and significantly reduced by an order of magnitude in 1993 (NCRP 1987, 1993).

In its 1987 recommendations, the NCRP introduced the concept of Negligible Individual Risk Level (NIRL), defining it as a level of effective dose that could be dismissed. In other words, it was basically a threshold below which the risk was considered so small as to be for all practical purposes ignorable. The NIRL was set at 0.01 mSv y^{-1} based on a risk of 10^{-7} per Sv (NCRP 1987). In the 1993 report, the concept remained essentially the same, but the NIRL became the NIR with the word "Level" dropped. The NCRP also introduced practical guidance and higher exposure limits for special situations such as emergency situations. And, in keeping with the 1990 recommendations of the ICRP, it introduced the concept of Annual Reference Levels of Intake (ARLI) set at the same effective dose level of 20 mSv set by the ICRP Annual Limits on Intake (NCRP 1993; ICRP 1991).

3.9.4 The Federal Radiation Council

Along with growing public and scientific concerns over worldwide fallout from atmospheric testing of nuclear devices, the need for coordination among the various governmental agencies promulgating radiation protection criteria led to the establishment of the Federal Radiation Council (FRC) by Executive Order on August 14, 1959. Chaired by the Secretary of Health, Education and Welfare, the purpose of this body was not regulatory but advisory; its primary purpose was to advise the President on radiation matters affecting health and to provide guidance and assistance to Federal agencies formulating radiation standards (Chadwick 1961; Hacker 1994; Palmiter and Tompkins 1965). The hope was that the FRC would play a major, if not the dominant role, in evaluating and setting radiation protection standards for fallout, as well as coordinating with other agencies. Less than a year after its formation, the FRC issued recommendations on radiation protection guidance for federal agencies, simultaneously with its first staff report entitled, "Background Material for the Establishment of Radiation Protection Standards." Other reports quickly followed and were initially well received, at least by the radiation protection community. But a lack of authority, staffing, feuding with the AEC regarding standards for radiation from fallout, and the perception of some that the FRC was ineffective led to its demise as an independent agency a scant decade after its formation, when it was folded into the newly created Environmental Protection Agency. During its brief existence within the EPA, the role of the FRC was limited to developing Executive Orders providing standards for overall radiation dose and concentration limits for signature by the President (Brodsky, Kathren, and Willis 1995).

3.9.5 Electronics Products

Public concern about stray X-ray emissions from television sets led to the passage of the Radiation Control Health and Safety Act in the late 1960s. Television sets of the day, especially the newer color television sets, were growing larger and using higher voltages, and some of them were attaining high voltages as great as 25 keV and generating significant amounts of X-rays as an unwanted by-product of the high voltage rectifier. The Food and Drug Administration (FDA) was designated as the agency with regulatory authority to develop standards and methods to assure the radiological safety of electronic products, including not only television sets but microwave ovens, cold cathode discharge tubes, diagnostic X-ray units, and cabinet X-ray units, such as those used for screening at airports and other security locations. Accordingly, the Bureau of Radiological Health (BRH) within the FDA was assigned the role of developing product-oriented standards and instituted a program that required actual measurement of emissions before a product could be marketed. This stimulated manufacturers to devise techniques to limit stray X-ray emissions from consumer products, a program that has by and large been successful. Commendably, the FDA did not do this in a vacuum, but rather in coordination with the Occupational Safety and Health Administration (OSHA), EPA, and Consumer Product Safety Commission via a memorandum of understanding to minimize duplication and conflicting requirements (Little 1980). Subsequently, the role of the BRH was expanded and the agency name changed to Center for Devices and Radiological Health (CDRH). The CDRH now has regulatory authority over all types of devices and equipment used in medicine, as well as equipment and devices emitting radiation (Brodsky, Kathren, and Willis 1995).

3.9.6 The Environmental Protection Agency

The passage of the National Environmental Policy Act in 1969 was a major event that led to the formation of the EPA, a new federal agency that was given broad authority to collect and analyze radiation exposure from natural background, medical practice, occupational exposure, and fallout from weapons tests and reexamine the scientific bases used to estimate radiation risks and benefits associated with radiation exposure to derive appropriate balances between benefits and risks (Jones 2005). In addition, and perhaps even more significant from the standpoint of radiation protection regulation, the following year, the EPA was given additional power to regulate hazardous materials, including radioactivity in the environment, through the passage of additional legislation in the form of the Clean Air Act (CAA) and Safe Drinking Water Act (SDWA), and in 1980 through the Comprehensive Environmental Response, Compensation and Liability Act (CERCLA). Although authority for enforcement of radiation protection dose limits and concentration standards in the workplace remained the province of the AEC, the EPA assumed responsibility for radiation standards outside the boundaries of nuclear facilities, although exactly who was responsible for exactly what was not well defined (Walker 1992).

Acting under the provisions of the Uranium Mill Tailings Radiation Control Act (UMTRCA), the EPA has been instrumental in identifying and remediating sites where tailings from uranium mines and mills were disposed. Under the mistaken impression that these wastes were radiologically harmless, they were initially stored in piles, covering large tracts of land, and subsequently utilized as fill material for construction and for soil conditioning. Many homes were constructed on mill tailings and, consequently, had radon levels well in excess of the standards developed by the EPA. In accordance with the requirements of the CAA, as amended, the EPA also developed standards for the large task of remediation of sites where uranium mine and mill tailings had been stored and utilized (Jones 2005).

The EPA assumed responsibility for control of the radon decay chain, establishing standards for radon air concentrations in homes and public buildings, and licensure of contractors under the provisions of the 1988 Indoor Radon Abatement Act. Standards were also established for radon levels in water. The EPA ultimately established an agreement program whereby the states would assume control for, among other things, licensure of radon testers and mitigation contractors, and inspection requirements for sale of homes. The radon control program went through a number of phases but finally in November 2015 the EPA released the National Radon Action Plan that codified the many separate and disparate aspects of the program.

3.9.7 The AEC Fissions: The NRC and DOE

Heavy criticism of the AEC for its dual role of promotion and regulation of nuclear energy, as well as increasing concerns about meeting American energy needs, led to the passage of the Energy Reorganization Act of 1974, which abolished the AEC and created in its stead two new separate and independent agencies (Walker and Wellock 2010). The regulatory functions of the AEC were placed under the purview of the newly created Nuclear Regulatory Commission (NRC) which began operations on January 19, 1975, while the military and civilian applications were assigned to the newly created Energy Research and Development Agency (ERDA), which after a short time became the cabinet level Department of Energy (DOE) with responsibilities for all aspects of energy, including nuclear. The fission of the AEC into two new agencies created the need for two sets of protection standards and exposure limits, as the NRC did not regulate the DOE. This added to the plethora of governmental agencies at both the federal and state level that promulgated radiation protection standards and guidance. Initially these protection standards and exposure limits were effectively carbon copies of each other, but over time differences developed, especially with respect to the implementation of the 50-year committed effective dose recommendation of the ICRP, which heavily impacted the DOE.

The NRC retained regulatory control over 10 CFR 20, and the DOE published similar requirements as Title 10 Part 835. In addition to these regulations, the NRC published a number of Regulatory Guides providing often quite detailed and specific advice on how various aspects of the regulations could be met. These were not mandatory, but served as guides, compliance with which was tantamount to approval of licenses. Although these guides were standards, deviation from them was permitted if it could be shown that the alternative proposed by the licensee or applicant was equivalent or superior from a protection standpoint, a very difficult task from a practical standpoint.

Both the DOE and NRC strongly supported the preparation of voluntary consensus standards by non-governmental voluntary standards bodies, and cooperatively contributed manpower, ideas, and sometimes funding for this purpose. In addition to the NCRP, prominent among these voluntary standards organizations are the American National Standards Institute (ANSI), the American Society for Testing and Materials (ASTM), and the International Electrotechnical Commission (IEC). The emphasis of the standards produced by the latter three organization is related to instrumentation and quality assurance of measurements. The DOE had an especially vigorous program of sponsorship of voluntary standards, largely through ANSI, and continued research related to radiation biology and health physics at the national laboratories, which it retained administratively, ultimately naming the Pacific Northwest National Laboratory (PNNL) in this regard. The NRC program was limited in scope to preparing or supporting applied radiation protection standards in consonance with its regulatory mission, while the DOE had a much broader scope and larger budget. In support of voluntary standards, the two agencies worked together closely and with little friction. The DOE lead person was Edward J. Vallario who emphasized the preparation of voluntary consensus standards largely through ANSI, reinvigorating its long-established but largely dormant N13 Committee on radiation protection. He worked closely with the Health Physics Society, serving as the chair of the revitalized Standards Committee of that organization (HPSSC) for nearly a decade. The HPSSC became the Secretariat for ANSI N13, which gave it an additional capacity to carry out its mission. Under Vallario's leadership, the HPSSC committee structure was reorganized into sub-groups and greatly expanded to include many areas of radiation protection that had long been ignored, issuing new standards as well as revising existing standards. Both the quality and quantity of standards produced by N13 was high and remains so today.

As an interesting aside, both the NRC and DOE endorsed the idea of an acronym derived from ALARA, incorporating it into their regulations but by different names. The concept was introduced into the NRC regulations in 1974 and originally was known as ALAP, derived from As Low As Practicable, as it was first called in ICRP Publication 1 (ICRP 1959a). The DOE preferred the original term ALAP instead of ALARA, which the NRC had adopted, but both meant the same thing. The two agencies sparred over the terminology for a number of years. At one point the DOE offered a compromise by changing the term to ALATEP, derived from As Low As Technically and Economically Practical, but this never really gained any traction. Despite the fact that the NCRP and a number of other groups preferred ALAP, voicing concerns about the vague meaning of the word reasonably, the NRC won out and the first DOE guide, published in 1980, used the NRC preferred acronym (Kathren, Selby, and Vallario 1980).

3.9.8 Other Regulatory Agencies

Over the years, in addition to the big three—the Nuclear Regulatory Commission, the Environmental Protection Agency, and the Department of Energy—a veritable alphabet soup of Federal agencies have been given or have assumed the responsibility for some specific aspect of radiation protection and have prepared standards or regulatory requirements to that effect. Ten representative agencies, listed alphabetically, are:

1. Bureau of Mines

2. Consumer Product Safety Commission

3. Department of Defense (DOD)

4. Department of State

5. Department of Transportation (DOT)

6. Food and Drug Administration (FDA)

7. Mining Safety and Health Administration (MSHA)

8. National Institute of Standards and Technology (NIST)

9. Occupational Safety and Health Administration (OSHA)

10. U.S. Postal Service (USPS)

Although the role played by these is typically small and highly specific, for completeness they bear mentioning. For example, the USPS regulations are specific to radioactive material sent through the mail, while the Bureau of Mines is concerned with standards for respiratory protection used in mines and in various nuclear operations. In most cases, the dose limits specified are identical to those established by the three large agencies.

All the states also have important regulatory responsibilities, and to that end have regulations that apply to areas not covered by the Atomic Energy Act. These include machine sources, natural radioactivity such as radium, and, if the agreement includes it, with byproduct material.

3.10 INTO THE NEW MILLENNIUM

The intense pace and exponential growth of standards of all types that marked the last half of the twentieth century slowed as the millennium neared. The above discussion, and in particular Section 3.9 demonstrate how radiation protection standards have evolved and indicate that prior to World War II, there was relatively little done; there were no regulations and very few voluntary standards dealing with radiation protection. The emphasis was on medical applications of X-rays and radium, and there was a sort of latent euphoria with regard to safety. After the hiatus resulting from World War II, this relaxed laissez-faire attitude underwent a reversal as secrecy was stripped away and the knowledge and understanding developed during the war years were declassified and made public. The new knowledge and fears with respect to biological effects of radiation exposure, the continuation of the nuclear weapons program in the United States and its expansion across the globe, along with the promotion and swift applications of radiation-based technologies produced concerns on the part of the public, as well as the radiological science community, resulting in a veritable explosion of standards, regulations, and guides of all kinds. The latent euphoria was replaced with an underlying fear that although much benefit could be derived from the fruits of radiation applications in everyday life, there was also an associated risk, and care needed to be taken to ensure that the risk or detriment was smaller than the benefit derived. Then, too, there was the big fear of the public relating to nuclear war, and the early postwar years saw a great deal of effort devoted to civil defense and the creation of protective guidance and standards dedicated to survival in the aftermath of a nuclear attack.

As the twentieth century advanced, the public became more sophisticated and concerned about the potential for adverse radiological impacts on people and the environment, fueling to some extent the vast expansion of radiation protection guidance that was a hallmark of the latter half of the twentieth century. The proliferation of regulations and standards was the result of the efforts of governmental agencies and voluntary standards organizations to assuage these concerns, drawing on new understanding as well as the radiological expertise of countless scientists, engineers, and other professionals to establish realistic and generally accepted radiation safety criteria. Although governmental and non-governmental agencies worked together toward a common goal, the overall effort was not particularly planned or well-organized; there appeared to be no discernable master plan or guidance, and standards were frequently created on an ad hoc basis and often when fostered by special interest groups. Thus, the situation was chaotic with overlap, duplication, inconsistencies, omissions and gaps in coverage of the problem, and false starts, as well as fragmentation of efforts and control. At least in the United States, the involvement of many different governmental agencies, despite formal agreements among agencies, is inefficient, leading to confusion, disharmony, and sometimes even irreconcilable differences, with requirements for approvals and hearings from numerous agencies resulting in unavoidable and frequently unacceptable delays or demands relative to the issuance of licenses.

3.10.1 Site Remediation

As regulations and standards caught up, as it were, with the needs of the times, there was a shift in emphasis and attention to site cleanup activities and to forward-looking planning for eventual decommissioning of nuclear sites. Licensees were required to submit plans and, in some cases, establish trust funds for decommissioning and disposal of radioactive wastes at the conclusion of site operations. Such requirements were written into the regulations of a number of states as well as the NRC, and where these did not exist, were made a condition of the license.

In 1974, under the provisions of the Energy and Water Appropriations Act, Public Law No. 105-61, the Formerly Used Sites Remedial Action Program (FUSRAP) was initiated with the stated purpose of cleanup or control of sites that were part of early energy and weapons programs, but had not been cleaned up or remediated because of the more lenient standards in place at the time the sites were closed down. Responsibility for FUSRAP was initially given to the DOE. Most of the 46 sites identified were in the northeastern United States and were primarily contaminated with uranium, thorium, and radium. In addition, a number of sites were contaminated with toxic chemicals, including volatile organic compounds and heavy metals such as lead. Standards for the latter were established by the EPA under provisions of CERCLA to accomplish its cleanup goals. In 1997, FUSRAP responsibility was transferred to the Army Corps of Engineers which, after cleanup was complete, would return the sites to the DOE. FUSRAP was an early example of multi-agency cooperation; the DOE and Corps of Engineers signed a memorandum of understanding, and there was close cooperation with the EPA because of CERCLA requirements.

Also in 1997, an important step was taken with regard to cleanup standards for radioactively contaminated sites applicable to all Federal agencies. This was the establishment of a standards for cleanup of radioactively contaminated sites applicable to all Federal Agencies, prepared by the NRC, DOE, EPA and the Department of Defense, and jointly published in 1997 and revised in 2000 as a report of each of three civilian agencies under the title Multi-Agency Radiation Survey and Site Investigation Manual (MARSSIM) (NRC 2000). MARSSIM provided standards and general guidance for planning, conducting and documenting site surveys, and thus far seems to have been quite successful in achieving its goals.

3.10.2 Some Concluding Remarks

This chapter has attempted to broadly trace the evolution of radiation protection guidance through legislation and voluntary standards since the discovery of X-rays and radioactivity over the first 100 years following the twin discoveries of X-rays and radioactivity. Thus, it provides an overview of a very complex topic which in and of itself could likely fill far more pages than were available here. Initially, radiation was considered benign, even beneficial, and even as injuries appeared, protection guidance was slow to develop, not very comprehensive, and its application was purely voluntary. There was little scientific evidence available to support exposure limits and so these limits were initially set very high under the assumption that there was a safe or threshold tolerance dose from which full recovery was possible. Following the development of the atomic bomb during World War II, with its associated research into biological effects, and the recognition and promotion of numerous potentially beneficial applications involving radiation, came the realization of the necessity for more stringent radiation protection guidance in the form of regulation and voluntary standards. The tolerance dose basis for permissible dose, concentrations, and releases to the environment was abandoned in favor of the LNT, which provided a greater measure of conservatism. Taken as a whole, this led to a period of intense development of legislation and regulation, as well as the development of voluntary standards that only recently has slowed a little as the necessary standards have been developed. The growth was, to use a bit of slang, like Topsy, with many different agencies vying for primacy at least in some area related to radiation protection. The near frenetic growth that held sway in past years seems to be giving way to a more reasoned approach that will likely include a reassessment of the soundness of the basis for dose limits and the continued preparation of voluntary consensus standards in support of the commercial sector. Nuclear power seems potentially poised for a rebound and expansion, and if this turns out to be the case, it will likely provide, along with increased medical applications, a strong impetus for continued standards development.

REFERENCES

Anonymous. 1901. "A Death Associated with the Use of the Roentgen Rays." *Boston Medical and Surgical Journal*. 144 (26).

Berrington, A., S. C. Darby, H. A. Weiss, and R. Doll. 2001. "100 Years of Observation on British Radiologists: Mortality from Cancer and Other Causes 1897–1997." *The British Journal of Radiology* 74 (882):507–519.

Brecher, Ruth, and Edward M. Brecher. 1969. *The Rays: A History of Radiology in the United States and Canada*: Baltimore, MD: Williams and Wilkins.

Brenner, D. J., and E. J. Hall. 2003. "Mortality Patterns in British and Us Radiologists: What Can We Conclude?" *The British Journal of Radiology* 76 (901):1–2.

Brodsky, Allen B. 1978. *CRC Handbook of Radiation Measurement and Protection Volume 1: Physical Science and Engineering Data*: West Palm Beach, FL: CRC Press.

Brodsky, Allen, Ronald L. Kathren, and Charles A. Willis. 1995. "History of the Medical Uses of Radiation: Regulatory and Voluntary Standards of Protection." *Health Physics* 69 (5):783–823.

Brown, Percy. 1936. "American Martyrs to Science through the Roentgen Rays." *The American Journal of the Medical Sciences* 192 (2):277.

Bushong, Stewart C. 1995. "History of Standards, Certification, and Licenser in Medical Health Physics." *Health Physics* 69 (5):824–836.

Cantril, S. T. 1951. "Biological Bases for Maximum Permissible Exposures." In *Industrial Medicine on the Plutonium Project*: New York, NY: McGraw-Hill, 36–74.

Chadwick, Donald R. 1961. "V. The Federal Radiation Council." *Health Physics* 4 (3):223–227.

Clarke, Roger, and Jack Valentin. 2005. "A History of the International Commission on Radiological Protection." *Health Physics* 88 (5):407–422.

Codman, Ernest Amory. 1902. *A Study of the Cases of Accidental X-Ray Burns Hitherto Recorded*: Boston, MA: Puritan Press.

Daniel, J. 1896a. "The X-Rays." *Medical Record* 4 (1896):562.

Daniel, J. 1896b. "The X-Rays." *Science* 3:562.

Edison, T. A. 1896. "Notes." *Nature* 53:421.

Evans, Robley D. 1933. "Radium Poisoning A Review of Present Knowledge." *American Journal of Public Health and the Nation's Health* 23 (10):1017–1023.

Evans, Robley D. 1974. "Radium in Man." *Health Physics* 27 (5):497–510.

Evans, Robley D. 1981. "Inception of Standards for Internal Emitters, Radon and Radium." *Health Physics* 41 (3):437–448.

Evans, Robley D., and C. Goodman. 1940. "Determination of the Thoron Content of Air and Its Bearing on Lung Cancer Hazards in Industry." *Journal of Industrial Hygiene and Toxicology* 22:88–99.

Frei, G. A. 1896. "X-Rays Harmless with the Static Machine." *Electrical Engineering* 22:651.

Glasser, Otto. 1993. *Wilhelm Conrad Röntgen and the Early History of the Roentgen Rays*: San Francisco, CA: Norman Publishing.

Grigg, Emanuel R. N. 1965. *The Trail of the Invisible Light: From X-Strahlen to Radio(Bio) Logy*: Springfield, IL: Charles C. Thomas Publisher.

Hacker, Barton C. 1987. *The Dragon's Tail: Radiation Safety in the Manhattan Project, 1942-1946*: Berkeley, CA: University of California Press.

Hacker, Barton C. 1994. *Elements of Controversy: The Atomic Energy Commission and Radiation Safety in Nuclear Weapons Testing, 1947-1974*: Berkeley, CA: University of California Press.

Ingraham II, S. C., J. G. Terrill Jr., and D. W. Moeller 1953. *Guide for Inspection of Medical and Dental Diagnostic X-Ray Installations in the Public Health Service*: Washington, DC: U.S. Department of Health, Education and Welfare.

International Commission on Radiological Protection. 1955. "International Recommendations on Radiological Protection." *The British Journal of Radiology* Supplement 6.

International Commission on Radiological Protection. 1959a. *ICRP Publication 1: Recommendations of the ICRP*. New York, NY: Pergamon Press.

International Commission on Radiological Protection. 1959b. *ICRP Publication 2: Permissible Dose for Internal Radiation*. New York, NY: Pergamon Press.

International Commission on Radiological Protection. 1960. "Report of the ICRP Committee 2 on Permissible Dose for International Radiation (1959)." *Health Physics* 3:1–180.

International Commission on Radiological Protection. 1977. ICRP Publication 26: Recommendations of the ICRP. *Ann ICRP* 1 (3).

International Commission on Radiological Protection. 1979. Limits for intakes of radionuclides by workers. ICRP Publication 30 (and supplements). *Ann ICRP* 2 (3-4).

International Commission on Radiological Protection. 1991. ICRP Publication 60: 1990 Recommendations of the International Commission on Radiological Protection. *Ann ICRP* 21 (1-3).

International Congress of Radiology. 1929. "International Congress of Radiology Recommendations: X-Ray and Radium Protection." *Radiology* 12:519.

Joint Committee on Atomic Energy (JCAE). 1960a. *Hearings, Subcommittee on Research, Development and Radiation*: Washington, DC: Government Publications Office, Congress of the United States.

Joint Committee on Atomic Energy (JCAE). 1960b. *Selected Materials on Radiation Protection Standards: Their Basis and Use*: Washington, DC: Government Publications Office, Congress of the United States.

Jones, Cynthia G. 2005. "A Review of the History of Us Radiation Protection Regulations, Recommendations, and Standards." *Health Physics* 88 (6):697–716.

Kassabian, Mihran Krikor. 1907. *Roentgen Rays and Electro-Therapeutics: With Chapters on Radium and Phototherapy*: JB Lippincott Company.

Kathren, R. L. 1978. *Radioactivity in the Environment: Sources, Distribution and Surveillance*. New York, NY: Harwood Academic Press.

Kathren, R. L., and Allen Brodsky. 1996. "Chapter 6: Radiation Protection." In R. A. Gagliardi, P. R. Almond and N. Knight (eds), *A History of the Radiological Sciences: Radiation Physics*: Reston, VA: Radiology Centennial.

Kathren, R. L., R. W. Baalman, W. J. Bair, and H. M. Parker. 1986. *Herbert M. Parker: Publications and Other Contributions to Radiological and Health Physics*: Columbus, OH: Battelle Press.

Kathren, R. L., J. M. Selby, and E. J. Vallario. 1980. *A Guide to Reducing Radiation Exposure to as Low as Reasonably Achievable (Alara)*: US Department of Energy. DOE/EV/1B30-T5.

Kathren, R. L. 1962. "Early X-Ray Protection in the United States." *Health Physics* 8 (5):503–511.

Kathren, R. L. 1964. "William H. Rollins (1852–1929): X-Ray Protection Pioneer." *Journal of the History of Medicine and Allied Sciences* 19:287–294.

Kathren, R. L. 1996. "Pathway to a Paradigm: The Linear Nonthreshold Dose-Response Model in Historical Context: The American Academy of Health Physics 1995 Radiology Centennial Hartman Oration." *Health Physics* 70 (5):621–635.

Kathren, R. L., and Paul L. Ziemer. 1980. *The First Fifty Years of Radiation Protection—A Brief Sketch*: Pergamon Press.

Kaye, G. 1927. "W. C. Some Fundamental Aspects of Roentgen Rays and the Protection of the Roentgen-Ray Worker." *American Journal of Roentgenology* 18:401–425.

Kocher, D. C. 1991. "Perspective on the Historical Development of Radiation Standards." *Health Physics* 61 (4):519–527.

Leonard, Charles Lester. 1897. "Application of the Roentgen Rays to Medical Diagnosis." *Journal of the American Medical Association* 29:1157.

Leonard, Charles Lester. 1898. "The X-Ray 'burn': Its Productions and Prevention. Has the X-Ray Any Therapeutic Properties?" *New York Medical Journal* 68:18–20.

Little, M. S. 1980. Food and drug administration radiation protection standards and recommendations for electronic products: the development process. *Health Phys.* 38:365–378.

Matanoski, Genevieve M., Raymond Seltser, Philip E. Sartwell, Earl L. Diamond, and Elizabeth A. Elliott. 1975. "The Current Mortality Rates of Radiologists and Other Physician Specialists: Specific Causes of Death." *American Journal of Epidemiology* 101 (3):199–210.

Mazuzan, George T., and J. Samuel Walker. 1985. *Controlling the Atom: The Beginnings of Nuclear Regulation, 1946-1962*. Berkeley, CA: Univ of California Press.

Morgan, Karl E. 1947. "Tolerance Concentrations of Radioactive Substances." *The Journal of Physical Chemistry* 51 (4):984–1003.

Morton, W. J. 1896. "Notes." *Nature* (53):421.

Mould, Richard Francis. 1993. *A Century of X-Rays and Radioactivity in Medicine: With Emphasis on Photographic Records of the Early Years*: CRC Press.

Muller, Hermann Joseph. 1927. "Artificial Transmutation of the Gene." *Science* 66 (1699):84–87.

Muller, H. J. 1941. "The Role Played by Radiation Mutations in Mankind." *Science* 93:438.

Mutscheller, A. 1928. "Safety Standards of Protection against X-Ray Dangers." *Radiology* 10 (6):468–476.

Mutscheller, A. 1934. "More on X-Ray Protection Standards." *Radiology* 22 (6):739–747.

Mutscheller, A. 1925. "Physical Standards of Protection against Roentgen-Ray Dangers." *American Journal of Roentgenology* 13:65.

National Bureau of Standards. 1931. *X-Ray Protection*. NBS Handbook 15. Washington, DC: U.S. Government Printing Office.

National Bureau of Standards. 1941. *Safe Handling of Radioactive Luminous Compound*. Edited by NBS Handbook 27. Washington, DC: U.S. Government Printing Office.

National Bureau of Standards. 1953. *Maximum Permissible Amounts of Radioisotopes in the Human Body and Maximum Permissible Concentrations in Air and Water*. Edited by NBS Handbook HB 52. Washington, DC: U.S. Government Printing Office.

National Bureau of Standards. 1954. *NCRP Report 17: Permissible Dose from External Sources of Ionizing Radiation*. Edited by NBS Handbook 59: Washington, DC: U.S. Government Printing Office.

National Bureau of Standards. 1955. *Regulation of Radiation Exposure by Legislative Means*. Edited by NBS Handbook 61: Washington, DC: U.S. Government Printing Office.

National Council on Radiation Protection and Measurements. 1987. *NCRP Report 91: Recommendations on Limits for Exposure to Ionizing Radiation*: Bethesda, MD: National Council on Radiation Protection and Measurements.

National Council on Radiation Protection and Measurements. 1993. *NCRP Report 116: Limitation of Exposure to Ionizing Radiation*: Bethesda, MD: National Council on Radiation Protection and Measurements.

National Council on Radiation Protection and Measurements. 2004. *NCRP Report 147: Structural Shielding Design for Medical X-Ray Imaging Facilities*: Bethesda, MD: National Council on Radiation Protection and Measurements.

Palmiter, C. C., and P. C. Tompkins. 1965. "Guides, Standards and Regulations from the Federal Radiation Council Point of View." *Health Physics* 11:865–868.

Parker, H. M. 1948. "Health-Physics, Instrumentation, and Radiation Protection." *Advances in Biological and Medical Physics* 1:223.

Parker, H. M. 1950. "Tentative Dose Units for Mixed Radiations." *Radiology* 54 (2):257–262.

Parker, H. M. 1971. "Radiation Protection Standards, Festina Lente." Midyear Symposium of the Health Physics Society, Columbia.

Portmann, U. V. 1933. "Roentgen Therapy." In O. Glasser (ed). *The Science of Radiology*: Springfield, IL: Charles C. Thomas Publisher, 212.

Rollins, W. H. 1902a. "Non-Radiable Cases for X-Light Tubes." *Electrical Review* 40:795.

Rollins, W. H. 1898. "Diaphragms for X-Light Tubes." *Electrical Review* 33:107.

Rollins, W. H. 1899. "Supplement to Note Xi." *Electrical Review* 34:81.

Rollins, W. H. 1900. "Non-Radiable Cases for X-Light Tubes." *Electrical Review* 36:359.

Rollins, W. H. 1901a. "The Control Guinea Pigs." *The Boston Medical and Surgical Journal* 146:430.

Rollins, W. H. 1902b. "Non-Radiable Cases for X-Light Tubes." *Electrical Review* 36:359.

Rollins, W. H. 1902c. "Some Conclusions from Experiments on Guinea Pigs Which Are of Importance in the Treatment of Disease by X-Light." *The Boston Medical and Surgical Journal* 144:317.

Rollins, W. H. 1903. "A Grouping of Some of the Axioms Mentioned." *The Boston Medical and Surgical Journal* 149:387.

Rollins, W. H. 1904. *Notes on X-Light*. Privately Printed.

Rollins, W. H. 1901b. "X-Light Kills." *The Boston Medical and Surgical Journal* 144:173.

Röntgen, Wilhelm Conrad. 1895. "Über eine neue Art von Strahlen." *Sitsungsberichte der Physikalisch-medizinischen Gesellschft zu Würzburg* 9:132–141.

Röntgen, Wilhelm Conrad. 1896. "On a New Kind of Rays." *Science* 3 (59):227–231.

Rowland, R. E. 1994. *Radium in Humans: A Review of US Studies*: Washington, DC: U.S. Department of Energy.

Russ, S. 1916. "The Injurious Effects Caused by X-Rays." *American Journal of Roentgenology* 12:38.

Stannard, James Newell, and R. W. Baalman Jr. 1988. *Radioactivity and Health: A History*: Richland, WA: U.S. Department of Energy.

Taylor, L. S. 1958a. "Brief History of the National Committee on Radiation Protection and Measurements (NCRP) Covering the Period 1929–1946." *Health Physics* 1 (1):3–10.

Taylor, Lauriston S. 1979. *Organization for Radiation Protection. Operations of the ICRP and NCRP: 1928–1974*: U.S. Department of Energy.

Taylor, L. S. 1933. "Roentgen Ray Protection." In O. Glasser (ed). *The Science of Radiology*: Springfield, IL: Charles C. Thomas Publisher, 332–343.

Taylor, L. S. 1958b. "History of the International Commission on Radiological Protection." *Health Physics* 1:97–104.

Taylor, L. S. 1971. *Radiation Protection Standards*: Cleveland, OH: CRC Press.

Taylor, L. S. 1989. "80 Years of Quantities and Unites: Personal Reminiscences, Part I." *ICRU News* 1:6–14.

Tesla, Nikola. 1896. "On the Roentgen Streams." *Electrical Review New York* 29 (23):277.

Thomson, E. 1896a. "Roentgen Rays Act Strongly on the Tissues." *Electrical Engineering* 22 (25 Nov):534.

Thomson, E. 1896b. "Some Notes on Roentgen Rays." *Electrical Engineering* 22 (18 Nov):520.

Thomson, E. 1898. "Roentgen Ray Burns." *American X-Ray Journal* A3:451–453.

U.S. Nuclear Regulatory Commission (NRC). 2000. *Multi-agency radiation survey and site investigation manual*. Report NUREG 1575. Bethesda, MD: U.S. Nuclear Regulatory Commission.

Walker, J Samuel. 1992. *Containing the Atom: Nuclear Regulation in a Changing Environment, 1963-1971*: Berkeley, CA: University of California Press.

Walker, J. Samuel, and T. R. Wellock. 2010. *Short History of Nuclear Regulation, 1946-1999*: Washington, DC: Nuclear Regulatory Commission.

Weart, Spencer R., and Spencer R. Weart. 2009. *Nuclear Fear: A History of Images*: Cambridge, MA: Harvard University Press.

Webster, Edward W. 1995. "X Rays in Diagnostic Radiology." *Health Physics* 69 (5):610–635.

Williams, F. H. 1898. "X-Ray in Medicine." *Medical News* 72:609.

Williams, F. H., and F. J. Ell. 1986. *The Institute of Nuclear Medicine: The First 25 Years*: Kent, United Kingdom: C.W. Printing, Ltd.

Wintz, H., and W. Rump. 1931. *Protective Measures against Dangers Resulting from the Use of Radium, Roentgen and Ultra-Violet Rays. Report C.H. 1054*: Geneva, Switzerland: League of Nations.

CHAPTER 4

Radiation Detection and Measurement

Joseph C. McDonald

CONTENTS

4.1	Introduction	124
4.2	Area Monitoring Instruments	126
	4.2.1 Current-Mode Detectors	126
	4.2.1.1 Ionization Chambers	126
	4.2.2 Pulse-Mode Detectors	128
	4.2.2.1 Proportional Counter Detectors	129
	4.2.2.2 Geiger–Mueller Detectors	129
	4.2.2.3 Scintillation Detectors	130
	4.2.3 Neutron and Mixed Field Instruments	130
	4.2.3.1 Passive Detectors	132
	4.2.3.2 Active Detectors	132
	4.2.3.3 High-Energy Neutron Instruments	134
	4.2.3.4 Operational Considerations	134
4.3	Personal Dosimeters	135
	4.3.1 Passive Dosimeters	135
	4.3.1.1 Photographic Dosimeters	135
	4.3.1.2 Thermoluminescent Dosimeters	137
	4.3.1.3 Optically Stimulated Luminescence Dosimeters	140
	4.3.1.4 Direct-Reading Dosimeters	142
	4.3.2 Active Dosimeters	142
	4.3.2.1 Electronic Personal Dosimeters	142
	4.3.3 Dose Calculation Methodology	143
	4.3.4 Neutron and Mixed Field Dosimeters	144
	4.3.4.1 Nuclear Emulsion Dosimeters	144
	4.3.4.2 Thermoluminescent Detectors	144
	4.3.4.3 Etched-Track Detectors	146
	4.3.4.4 Superheated Drop (Bubble) Detectors	146
	4.3.4.5 Personal Neutron Accident Dosimeters (PNAD)	147

4.4   Measurement Methods and Procedures   148
      4.4.1   National Standards and Reports   149
      4.4.2   Regulatory Guidance   151
      4.4.3   Calibration and Testing   151
              4.4.3.1   Photon Calibrations   153
              4.4.3.2   Beta Particle Calibrations   154
              4.4.3.3   Neutron Calibrations   156
              4.4.3.4   Surface Contamination Monitors   158
      4.4.4   Measurement Traceability   159
      4.4.5   Statistics of Radiation Measurements   160
              4.4.5.1   Uncertainty Analysis   162
References   164

4.1   INTRODUCTION

For radiation protection purposes, it is necessary to determine the equivalent dose. Since the equivalent dose is defined, within the body, as the weighted absorbed dose in an organ or tissue in the body, this quantity cannot be measured directly. Operational quantities, such as the personal dose equivalent and ambient dose equivalent, are determined using personal dosimeters and area survey meters. Air monitors and contamination survey instruments measure activity rather than dose, but their readings can contribute information for the determination of internal dose (refer to Chapters 7 and 8). Additional quantities are also relevant to specific situations. The dose resulting from the uptake of radionuclides is further discussed in Chapters 7 and 8. When doses due to external and internal radiations have been determined, the effective dose can be computed. The detection and measurement of external ionizing radiation, along with the methods for calibration and testing, are described in the present chapter.

The term "dosimeter" normally refers to a personal dosimeter that is worn on the body and is used to determine the personal dose equivalent received by the individual wearing the device. Personal dosimeters are generally not electrically powered, although there are some electronic personal dosimeter types, and they will store a reading by means of a radiation-induced change in a detector element. The term "instrument" refers to an electrically powered device that can measure exposure, or dose equivalent, as a function of time. Instruments may be portable battery-powered devices, or installed devices that are alternating current (AC) line powered. Radiation protection instruments that make use of pulse counting techniques are generally used as detectors of radioactivity rather than dose equivalent.

Dosimeters and radiation-measuring instruments can be classified as being either active or passive. Active devices are often battery-powered and can display, store readings, or communicate readings via the internet to a data recording system. Passive devices are not powered and may not have the capability to display a reading. The most common type of passive detector is the personal dosimeter worn on the body of a worker. One example of a passive detector that can display a reading without separate circuitry is a quartz fiber (direct-reading) dosimeter that is normally worn as a supplemental dosimeter. Electronic personal dosimeters (EPDs) are active devices that contain signal-processing electronics to determine personal dose. Many EPDs have capabilities to display exposure, exposure rate, dose equivalent, or dose equivalent rate.

There are additional dosimeters and instruments for specific tasks. Extremity monitoring for workers handling radioactive sources requires the use of ring or wrist dosimeters. Job-specific supplemental dosimeters, such as those for measuring the dose to the lens of the eye, may also be needed. Emergency and accident dosimetry requires the use of a number of types of special dosimeters and instruments. Radiation areas that contain high-intensity sources, or radiation-generating devices capable of delivering high dose rates, require the use of dosimetry and instrumentation that is capable of measuring large values of dose or dose rate. The quantity absorbed dose is measured for values above about 0.1 Gy when deterministic effects can occur and the quality factor is not used (ICRP 2007).

Area monitoring is a basically different operation from individual monitoring. The purpose of area monitoring is to determine the exposure rate or dose equivalent rate at a place that may be occupied by a worker or other individual. This value is taken to represent the radiation intensity that could be received when a person is present. Its value will be posted at the entrance to an area where radiation may be present or may be produced, for example, by a radiation-generating device. The area monitored can also be much larger than a specific work area. Environmental monitoring may consist of measuring the dose equivalent rates in areas surrounding a nuclear power plant. Passive devices such as thermoluminescent dosimeters (TLDs) may be used for this purpose, along with active electronic instruments, such as high-sensitivity pressurized ionization chambers.

In order to verify the accuracy of the reading provided by a dosimeter or instrument, it is necessary to calibrate the device using a reference radiation standard that is traceable to the National Institute of Standards and Technology (NIST). A measurement standard may be either a radiation source or a radiation field produced by a radiation-generating device. Calibration involves a comparison of the indication obtained during the measurement of a quantity to the accepted, or conventionally true, value for that quantity under a set of reference conditions. After this measurement, or series of measurements, is completed, the device's indication can be adjusted so that it displays the conventionally true value of the quantity (the value accepted by agreement). However, when calibrating a passive personal dosimeter, there is no direct method of adjustment to the response of the dosimeter. For a passive type of device, a numerical coefficient can be generated and applied during the readout process in order to make the correction.

Passive dosimeters are not calibrated in the same way as active instruments. Groups of personal dosimeters may be given a known exposure to, for example, gamma rays from a ^{137}Cs source, and individual calibration factors for those dosimeters can then be computed. The method used to establish measurement quality assurance for the calibration process is to determine the proficiency of the dosimetry services and to grant them accreditation, after successful testing and on-site auditing. Two governmental programs implement the quality specifications and evaluate dosimetry service proficiency. The National Voluntary Laboratory Accreditation Program (NIST 2005) and the Department of Energy Laboratory Accreditation Program (United States Department of Energy 2011) require proficiency tests of dosimetry services and on-site audits to verify competence before accreditation can be granted. Both programs use ANSI Standard N13.11 (ANSI 2015) as the basis for the proficiency tests.

4.2 AREA MONITORING INSTRUMENTS

4.2.1 Current-Mode Detectors

Active radiation-detecting instruments are used to measure the ambient dose equivalent, $H^*(d)$, which is the dose equivalent that would be present at a depth d in a spherical tissue-equivalent phantom. The phantom is the 30-cm diameter ICRU sphere (ICRU 1985) that is intended to represent the radiological properties, including backscatter, of an individual. During calibration, the instrument's electronics and case enclosure are assumed to provide adequate backscatter. Ambient dose equivalent measurements are needed for posting information at the entrance to a radiation area, or to confirm that required values for dose equivalent rate are not exceeded in a radiation area or occupied space. Such measurements are also needed for radiological engineering, and to monitor shielding for possible changes in radiation intensity in an area. The cavity ionization chamber is a widely used current-mode detector based on principles of collecting ions that are produced in a gas in order to generate a current, or charge, that is proportional to radiation intensity and can be used to evaluate the ambient dose equivalent rate.

4.2.1.1 Ionization Chambers

An ionization chamber consists of a gas-filled cavity (often that gas is air) that is surrounded by a wall that is either formed of an electrically conductive material or has an electrically conductive coating on its inner surface. The collection of either positive or negative ions is accomplished using a potential of a few hundred volts that is applied between the conducting chamber wall and a conducting electrode mounted near the geometric center, or along the center line, of the chamber. This electrode is maintained at ground potential and is electrically connected to an electrometer. An electrometer is essentially a voltmeter with high input impedance and low leakage current, capable of measuring extremely low values of electrical current or charge. Charge, or current, is determined by measuring the potential produced across either a precision capacitor or a precision resistor, respectively (Figure 4.1).

Ionization chambers can be constructed in a number of shapes, various volumes, and using a number of materials for chamber walls. Materials having radiation absorption and interaction properties similar to either air or human tissue are normally used for chamber walls and collecting electrodes. Teflon® or polyethylene insulators are reasonably tissue or air equivalent in terms of radiation absorption and interaction properties, and therefore an ionization chamber and its constituent parts may be considered to form a homogeneous tissue-equivalent material for the purpose of dose equivalent evaluation.

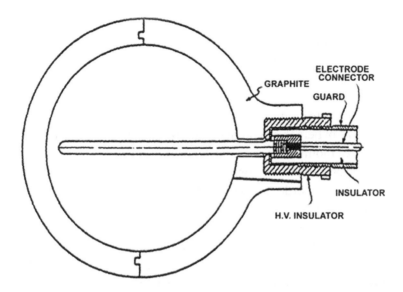

FIGURE 4.1   A cross-sectional drawing shows the principal components of an ionization chamber. A spherical chamber is a type often used in a calibration laboratory (Attix 2008).

The absorbed dose to the wall of an ionization chamber, D_w, is given by the Bragg–Gray equation (Attix 2008),

$$D_w = s_{wg} \frac{q}{m} \frac{W_g}{e}, \tag{4.1}$$

where s_{wg} is the ratio of electron mass stopping powers of the wall and gas (ICRU 1984), q is the measured charge per unit of mass of gas, m, and W_g / e is the average energy per ion pair produced in gas by radiation (ICRU 1979).

Historically, the quantity exposure has been disseminated by national measurement laboratories. Air kerma is currently used. The relationship between air kerma and exposure can be expressed in the following equation:

$$K_{air} = 2.58 \times 10^{-4} \left(\frac{W}{e} \right) \left(\frac{1}{1-g} \right), \tag{4.2}$$

where W/e is the mean energy per unit charge expended in air by electrons, and g is the fraction of the initial kinetic energy of secondary electrons dissipated in air through radiative processes. The value accepted by NIST for W/e in air is 33.97 J/C. The values of g for ^{60}Co and ^{137}Cs beams are 0.32% and 0.16%, respectively (NIST 1988).

An ionization chamber-based survey meter is pictured below. The chassis contains a battery-operated power supply for the measurement circuitry, and it has an analog meter display. The ionization chamber has a thin entrance window to permit the detection of beta particles. There is also a cover for the entrance window when photon measurements are made. The thickness of the cover is sufficient to establish charged particle equilibrium for the gamma ray sources most commonly encountered in radiation protection (Figure 4.2).

Instruments with digital displays and high-sensitivity pressurized ionization chamber type detectors are available. Such devices also rely on the use of the basic type of ionization chamber detection methodology. Large volume pressurized ionization chambers may be used as perimeter dose rate monitors since they have high sensitivity, water-resistant enclosures, and are capable of being used outdoors. Ionization chamber based monitors are also used as alarming area monitors in radiation facilities.

Ionization chambers can also be constructed using both a thin entrance window and a thin collecting electrode in order to serve as transmission monitor chambers. A thin window with a conductive coating on its inner surface serves as the high-voltage electrode. A second thin window, also with conductive coatings for a collecting electrode, and a guard ring complete the assembly. The outer surfaces of the windows are also coated to provide electrical shielding. Such chambers are normally mounted between a radiation source, such as an X-ray machine, and the item being irradiated in order to monitor the quantity of radiation delivered, without severely perturbing the beam (see Section 4.3.1).

4.2.2 Pulse-Mode Detectors

The basic construction of an ionization chamber, a gas or air-filled chamber with an electrically conducting wall and a central collecting electrode, can also be employed in other detection devices by using a different method of ion collection. Increasing the potential applied to the chamber wall will result in a transition from the measurement of current, or charge, that is constant over a fairly large region of applied potential. When the potential

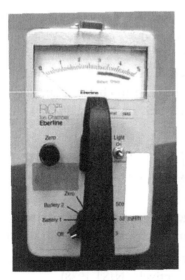

FIGURE 4.2   Example of an ionization chamber-based survey meter. The ionization chamber's thin entrance window is on the bottom side of the instrument (right-hand view). A cover has been slid out of the way for beta particle measurements (Johnson 2009).

increases to a certain value, it is found that pulses detected from ionizations in the chamber have amplitudes that are proportional to the ionization density of the secondary charged particles generated. Continuing to increase the potential results in a region of limited proportionality followed by a region where there are avalanche discharges from each photon detected, results in large, constant amplitude pulses that are simpler to detect and record than those produced by a proportional counter, but there is no information available from those pulses about the ionization density of the secondary charged particles (Knoll 2010).

4.2.2.1 Proportional Counter Detectors

A proportional counter is essentially an ionization chamber, operated at a higher potential, wherein pulses are detected rather than charge measured. Pulse heights produced in an ionization chamber are small, but when the potential is increased, there is a region of operation where pulse heights increase with applied potential. This type of operation can provide additional information that is useful for characterizing the radiation being measured (Knoll 2010).

The signal generated in a proportional counter from the deposition of radiation energy is proportional to the number of ion pairs produced within the counter. The filling gases for proportional counters usually consist of combinations of methane and argon. One such combination is known as P10. In general, proportional counters are used for surface contamination measurement or air monitoring rather than dosimetry, but they can be calibrated to display dose rate. Pulse heights produced by a proportional counter are dependent upon the ionization density of charged particles produced by the radiation incident. This property can be employed in a specifically designed proportional counter to measure the dose deposited by radiations of different LET (refer to Chapter 2, Section 2.6.4 for a discussion of LET). The distribution of dose in LET, or lineal energy, y, can be measured in order to separate, for example, the dose due to neutrons and dose due to gamma rays in a mixed field.

4.2.2.2 Geiger–Mueller Detectors

Increasing the potential between the collecting electrode and the chamber wall of an ionization chamber results in operation where pulse height is no longer proportional to the ionization density of the incoming radiation. All pulses are of the same amplitude and are large. This occurs as a result of a Townsend avalanche when gas gain is sufficiently increased. This type of operation results in a simplification of the detector's design, and the operation of the associated electronics. The Geiger–Mueller (G-M) tube may also be filled with P-10 gas, and variations of gas mixtures can be used to reduce dead-time and pulse pile-up thereby increasing counting rates. The G-M detector can be calibrated from readings taken using a reference radiation. When used as an instrument to measure dose or air kerma, an over-response is found at low photon energies. Energy response is improved by enclosing the detector in metals that selectively attenuate lower energy photons. Modified G-M detectors have also been used for measurements in mixed neutron gamma ray fields (see Section 4.2.3).

4.2.2.3 Scintillation Detectors

Organic compounds, such as anthracene dissolved in xylene or toluene, have been investigated for use as scintillators (Kallmann 1950). This early work led to the development of more advanced liquid scintillators that could be contained in large enclosures of various configurations (Knoll 2010). Many organic and inorganic compounds have been used as scintillators for the detection and measurement of radiation. Among the earliest scintillating materials used in combination with photomultipliers were organic compounds such as naphthalene ($C_{10}H_8$) and anthracene ($C_{14}H_{10}$). An inorganic compound frequently used for the detection and measurement of gamma rays and gamma ray spectra is NaI activated with 1% Tl. Other high atomic number materials have been successfully employed as inorganic scintillators including CsI (Tl), bismuth germanate (BGO), cadmium zinc telluride (CZT), and more recently $LaCl_3:Ce^{3+}$ and $LaBr_3:Ce^{3+}$ have been investigated. Lithium Iodide LiI(Eu) can also be used for gamma ray detection, but it is more often used to detect low-energy neutrons because of its ^6Li content, which has a large cross section for neutrons.

Light pulses are produced in inorganic scintillators as a result of the energy deposition by secondary electrons from gamma ray interactions in the scintillator material. Electrons in the crystal lattice of the scintillator may then be imparted with enough energy to populate lattice sites, whose energies lie between the valence band and the conduction band. Subsequent recombination of electrons and holes can give rise to visible photons. Elements such as thallium or europium, when added in low concentration to the material used to grow a scintillator crystal, create lattice sites that enable the production of visible photons and thus create one of the properties needed for an effective inorganic scintillator. It is desirable that scintillation light pulses be large and that the light intensity be proportional to the incoming photon energy. The material must also be transparent to the wavelengths of light produced so that the pulses can be detected with a photomultiplier or photodiode, and it is highly desirable to have pulses with short rise and decay times permitting high count rates without pulse pile-up.

Gamma rays can interact with matter by means of the photoelectric effect (see Chapter 2, Section 2.5.2.1), Compton scattering (see Chapter 2, Section 2.5.2.2), and pair production (see Chapter 2, Section 2.5.2.3). All of these processes lead to the deposition of energy in the form of secondary electrons that can be detected in a solid, liquid or gas via an optical or electronic signature. Only the lowest energy gamma rays usually deposit all of their energy in a single interaction, while most must undergo multiple interactions to deposit their full energy. Scintillators are efficient at detecting these energy depositions, and are therefore effective radiation detection and measuring devices. Using a scintillation detector requires the calibration of the instrument in terms of a quantity such as air kerma. A standard reference photon source such as ^{137}Cs may be used for calibration.

4.2.3 Neutron and Mixed Field Instruments

Neutrons are uncharged particles that are indirectly ionizing. In addition, neutron fields are nearly always accompanied by a photon component. Measurements in such mixed fields require specialized detectors and data reduction methods. Interactions between neutrons

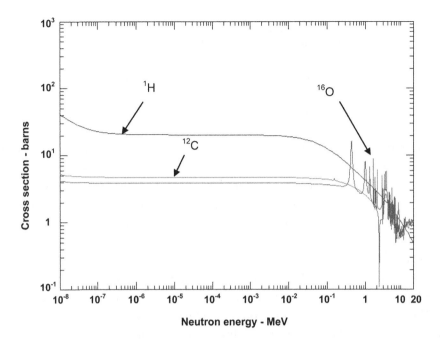

FIGURE 4.3   Neutron elastic scattering cross sections for three elements found in tissue. Above a neutron energy of about 0.1 MeV, the ^{16}O cross section becomes complex due to the presence of resonances (Korea Atomic Energy Research Institute 2015).

and matter are strongly dependent upon nuclear cross sections. One of the neutron interactions having a large cross section is the elastic scattering of a neutron and a nucleus. The data in Figure 4.3 shows that the elastic scattering cross section for hydrogen, ^{1}H, is almost a factor of ten larger than the corresponding cross section for ^{12}C over most of the energy range. In an elastic scattering event, the collision is most effective in the transfer of energy if the mass of the atom is very close to the mass of the neutron. This is the case for the hydrogen atom, whose nucleus is a proton, because the masses of the neutron and proton are nearly equal. Materials containing large amounts of hydrogen atoms, such as water, paraffin, and polyethylene, are very efficient in scattering and absorbing neutrons. For such materials, multiple collisions result in energy losses until the neutron eventually has approximately the same thermal energy as the material itself. Neutrons with energies that are equivalent to the thermal energy of a material at room temperature are referred to as thermal neutrons. The energy of thermal neutrons is approximately 0.025 eV at 20°C. The process of neutrons slowing down as a result of elastic collisions in matter is also known as moderation. Both active and passive detectors rely on the use of materials with large neutron cross sections. Three neutron interactions having large thermal neutron cross sections are shown in Figure 5.3 (ICRU 2001). Therefore, most neutron detectors incorporate compounds containing ^{6}Li, ^{10}B, or ^{3}He.

When a neutron has lost most of its energy through elastic or inelastic scattering interactions and is at thermal energy, it can be captured by a nucleus with the resulting emission of a gamma ray. This interaction can be denoted as ^{1}H(n,γ)^{2}H. The neutron, n, is incident

TABLE 4.1   Neutron Detection Interactions

Reaction	Q (MeV)	σ_{th} (b)
$^6Li(n,\alpha)^3H$	4.65	941
$^{10}B(n,\alpha)^7Li$	2.32 (93%)	3838
	2.8 (7%)	
$^3He(n,\alpha)^3H$	0.5	5530

on a hydrogen nucleus, ^1H, and is captured to form a deuterium atom, ^2H, that consists of a neutron and a proton. The process is called radiative capture, and a gamma ray with an energy of 2.223 MeV is emitted. This interaction is one reason why neutron fields are nearly always accompanied by photons. Table 4.1 lists the values for Q, the amount of energy released by the reaction, and the thermal neutron cross sections σth.

4.2.3.1 Passive Detectors

Neutron interactions in plastic films can leave permanent damage sites that can be viewed and recorded. This phenomenon forms the basis of the etched-track detector (see Section 4.3.4.3). The damage sites, or tracks, from secondary charged particles produced by neutron interactions with the hydrogenous film can be enhanced by means of chemical or electrochemical processing. Identification and counting of tracks can be carried out manually using optical microscopy. Laser-based counting techniques have also been developed. The most common type of etched-track dosimeter film used for personal dosimetry purposes makes use of the plastic polyallyl diglycol carbonate (PADC) (ICRU 2001). The neutron interactions in PADC are (n,α) and (n,f) (fission) for energies above 1.5 MeV. However, the response to low-energy neutrons can be enhanced by using a cover over the film that contains ^6Li or ^{10}B in order to produce (n,α), and (n,p) reactions. Detectors of this type have been incorporated into multi-purpose personal dosimeters (Rathbone 2010) (see Section 4.3.1.3).

4.2.3.2 Active Detectors

Neutron area monitors make use of the principle of detecting thermal neutrons resulting from the moderation of fast neutrons. Some devices employ cylindrical proportional counters filled with BF_3 or ^3He gas, or a LiI scintillation detector enriched in ^6Li. Detectors are surrounded by a cylindrical hydrogenous moderating material such as polyethylene, with a thickness that will optimize the response as a function of neutron energy. Another similar type of survey meter uses a 9-inch diameter spherical polyethylene moderator (ICRU 2001). The fluence response of the detector within the sphere, as a function of neutron energy, has a shape that approximates the fluence-to-dose equivalent conversion factor over a wide range of energies. Cadmium and other metal filters are also incorporated to alter the neutron energy response. The quantity displayed by these devices is usually the ambient dose equivalent rate, $\dot{H}^*(10)$.

Multi-sphere spectrometers can be used to determine the neutron energy distribution, from which can be calculated the ambient dose equivalent rate for neutrons (Thomas and

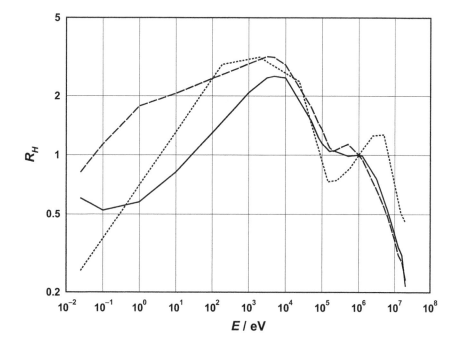

FIGURE 4.4 Neutron energy response, R_H, of the ambient dose equivalent for three types of moderator-based neutron survey instruments: Anderson-Braun (solid line), Leake design (dashed line), and the Burgkhardt design (dotted line) (ICRU 2001).

Lewis 1981). Such systems can cover the energy range from thermals to about 20 MeV. Their energy resolution is not exceptional, and the spectrum unfolding process is subject to some uncertainties. The system uses a series of polyethylene moderator spheres, of differing sizes, that can enclose a scintillator such as LiI(Eu). Gamma ray pulse heights can be subtracted from the measured pulse-height spectrum. Although the system can be difficult to set up for data collection, it is still a very useful method for determining neutron energy distributions in field conditions. The detectors are most often proportional counters filled with gases that have large neutron cross sections. The ambient dose equivalent responses of several moderator-based neutron area survey meters are shown in Figure 4.4.

A neutron instrument that relies on a different type of detection principle uses a tissue-equivalent plastic proportional counter. This device has a wall constructed of an electrically conducting tissue-equivalent plastic (Shonka, Rose, and Failla 1958), and is filled with a hydrogenous tissue-equivalent gas mixture (ICRU 1983). Pulses are generated by neutron interactions in the chamber wall, producing charged particles that are detected in the helical proportional counting region in the center of the instrument. The distribution of energy losses in the tissue-equivalent gas medium can be displayed as a function of lineal energy or computed to yield dose equivalent rate values. In addition, the doses due to photons and neutrons can be derived from the distributions in dose as a function of lineal energy (McDonald et al. 1984) (Figure 4.5).

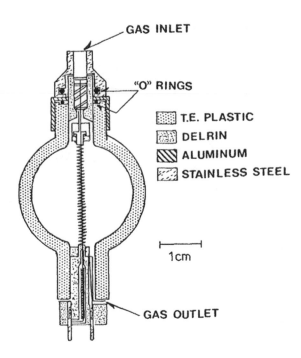

FIGURE 4.5 Cross-sectional view of a tissue-equivalent proportional counter. This model is a laboratory instrument with an A-150 plastic wall, and a cylindrical proportional counting region defined by the helical electrode shown in the cutaway view (ICRU 1983).

Tissue-equivalent ionization chambers have also been developed and have walls constructed of plastic mixtures. The plastics have atomic compositions and neutron interaction properties that are reasonably close to those for the composition of human muscle tissue. They can also be filled using a tissue-equivalent gas mixture that has a similar atomic composition, so that measurement of neutron dose using the Bragg–Gray theory can be carried out (ICRU 1983).

4.2.3.3 High-Energy Neutron Instruments

The response of moderator-based portable survey instruments, such as those mentioned above, is limited to the range of approximately 20 MeV. However, accelerator facilities may have neutron fields that extend in energies from thermal to several GeV. A modified moderator-based survey instrument has been developed using tungsten or tungsten carbide powder added to a polyethylene moderator in order to generate spallation neutrons in tungsten nuclei and enhance the high-energy response of above 8 MeV (Olsher et al. 2000). Radiation protection for high-energy accelerators is discussed in *A Guide to Radiation and Radioactivity Levels Near High Energy Particle Accelerators* (Sullivan 1992) and in ANSI/HPS Standard N43.1 (ANSI 2011).

4.2.3.4 Operational Considerations

In addition to whole-body personal dosimeters, specially constructed dosimeter types are used for particular applications. Ring and wrist dosimeters are needed for tasks such as the handling of radioactive sources. In order to assess dose to the lens of the eye, purpose-built dosimeters are needed for this task. Dosimeters having high sensitivity are also needed for environmental measurements, and dosimeters for use near high-energy accelerators have also been employed. For facilities where a criticality may occur, dosimeters such as the personal nuclear accident dosimeter are also used (see Figure 4.16).

Area survey meters with special characteristics are also needed for specific measurements. These include high-range survey meters, along with high-sensitivity meters for environmental-level measurements. Ultra-high sensitivity detectors have also been developed for Homeland Security applications. However, these particular instruments are not to be considered as radiation protection instruments, but rather detection and identification devices. Large high-sensitivity instruments are needed as portal monitors to detect possible worker contamination.

National standards, such as ANSI N323 AB (ANSI 2013) and N42.17 A,B,C (ANSI 1989a,b, 2003b) include specifications for operational tests of portable survey instruments. These include evaluation of: battery lifetime, alarm audible intensity, radiation energy and intensity response, accuracy, linearity, over-range response, and other operational characteristics.

4.3 PERSONAL DOSIMETERS

4.3.1 Passive Dosimeters

Personal dosimeters are worn on the trunk or other areas that are likely to receive the greatest portion of dose equivalent. They are issued according to relevant regulations and standards. These dosimeters are designed to measure protection quantities such as the personal dose equivalent $H_p(d)$, where d may be at a depth of 10 mm in tissue for penetrating radiation, or 0.07 mm for non-penetrating radiation. When personal dosimeters are calibrated in terms of the protection quantities, they are exposed to the relevant radiations that are usually determined in terms of quantities of air kerma or absorbed dose. The dosimeters are placed on an appropriate phantom, and the value of the protection quantity at the point just below the dosimeter's detecting element is computed using a conversion coefficient. Several types of personal dosimeters have been, and are still being, used. The following sections describe the features of personal dosimeters.

Passive dosimeters are those devices that generally do not have a power source and do not provide a displayed reading of the measured quantity. Some exceptions to this rule are described in this section. Specific devices with power sources are included. They are worn by radiation workers performing tasks that have a requirement to continuously display personal dose equivalent. Active dosimeters, for example, electronic personal dosimeters, are discussed in Section 4.3.2.

4.3.1.1 Photographic Dosimeters

Photographic film dosimeters have been used for many years, but at the moment they have been largely supplanted by thermoluminescent dosimeters and optically stimulated luminescence dosimeters for photons and beta particles. For neutrons, albedo TLDs, etched-track detectors, and superheated emulsion (bubble) detectors are currently used. Nevertheless, there are a few NVLAP-accredited dosimetry services offering film dosimeters.

Film dosimeters have advantages and disadvantages. A developed film dosimeter offers a virtually permanent record of an exposure that can be re-read in the future. Film dosimeters, when used in pairs, have sufficient sensitivity to cover a wide range of expected exposures, and they are relatively inexpensive. Disadvantages include the difficulties associated with chemical processing and the related waste disposal problems. Film dosimeters are not reusable, and they are not as resistant to environmental conditions such as temperature and humidity when compared to thermoluminescent dosimeters.

Photographic films consist of emulsions applied to bases made from cellulose acetate or polyester. The gelatin emulsion coating is compounded using silver bromide grains. Ionizing radiation incident on the emulsion causes some silver ions to lose their charge and become neutral silver atoms. A chemical developing process fixes the silver and makes the exposed area of the film opaque. The silver-containing emulsion in a film dosimeter is sensitive to both visible light and ionizing radiation, so films are stored in light-tight envelopes or other opaque enclosures. After exposure to radiation, the degree of film darkening is measured using an optical densitometer. The amount of light absorbed by the developed film varies with the amount of ionizing radiation to which the film was exposed. The quantity measured by an optical densitometer had been referred to as optical density, but this term is no longer used. The current terminology is absorbance, as given by:

$$A = \log_{10}\left(\frac{I_0}{I}\right), \tag{4.3}$$

where I_0 is the intensity of light with no film present, and I is the intensity of light measured when the film is in place. The absorbance of an exposed film is non-linear. Since the absorbance of irradiated film dosimeters has a limited range of linearity, it is difficult to measure exposure or dose over a wide range of intensity. Therefore, two films, or a film having emulsions on both sides, may be used to cover a wider range of exposure. Photographic film over-responds to low-energy photons. This is due to the increasing mass attenuation coefficients of silver for energies below a few hundred keV. For energies of a few tens of keV, attenuation of the opaque film envelope significantly reduces the sensitivity of the emulsion, resulting in a peak in sensitivity at approximately 40 keV.

For an emulsion that responds equally well to both photons and beta particles, one method of discriminating between the two radiations is to place material that will absorb beta particles but transmit photons with minimal absorption. A reading is then taken to provide a measurement of the photon dose. A second reading, with the material removed, can be taken to provide a measurement of photon dose plus beta dose. The difference of the two readings will yield the beta dose. This method is often used when measurements are taken with an ionization-chamber-based portable survey instrument. Cadmium has a large thermal neutron cross section and is used as a filter to discriminate between thermal and fast neutron dose.

4.3.1.2 Thermoluminescent Dosimeters

Thermoluminescent dosimeters (TLDs) are widely used for radiation protection dosimetry, and their operation makes use of the physical properties of crystalline compounds. Such compounds include LiF infused with small quantities of Mg and Ti that are referred to as "dopants." These materials are added during the crystal-growing process to produce lattice sites that can be occupied by electrons. The notation for the doped LiF dosimeter material is written as LiF: Mg, Ti. Other commonly employed TLD materials include: Al_2O_3:C, LiF: Mg, Cu, P, CaF_2: Mn, $Li_2B_4O_7$:Mn, $CaSO_4$:Mn. These materials and others are available in various forms from commercial suppliers of dosimeters and dosimetry systems. In recent years, LiF: Mg,Cu,P phosphors have been employed because of their improved dosimetric characteristics. This phosphor has improved sensitivity and is more nearly tissue-equivalent than other commonly used TL materials. For mixed-field measurements, this material has a lower response to neutrons than LiF: Mg, Ti. But it does exhibit a complex glow curve that can be affected by the heating technique.

The process of thermoluminescence as applied to radiation detection and measurement relies on the behavior of electrons in crystalline solids. Electrons are fermions that have non-integer spins, obey Fermi-Dirac statistics, and the Pauli Exclusion Principle restricts their occupation of certain atomic orbits. The result is the constraint of electrons to atomic orbitals that have been given the identifying letters s, p, d, and f, that correspond to the angular momentum quantum numbers 0, 1, 2, and 3. When atoms form a crystalline lattice, the density of the structure is such that many molecular orbitals are present, resulting in regions, or bands, in energy. Bands are separated by an energy gap, where no electrons are normally present. Metals do not have an energy gap and are therefore good electrical conductors. Insulators have large energy gaps of several eV, making the transition of electrons to the conduction band very difficult. Semiconductors have smaller energy gaps of about 1 eV or less (Kittel 1991). LiF is an insulator with a wide energy gap of 13.6 eV, but in the case of LiF: Mg, Ti, the added dopants of magnesium and titanium provide lattice sites within the energy gap that can be occupied by electrons.

When a TLD is irradiated, the deposited energy can provide electrons that are in the valence band with enough energy to transition to the conduction band. This transition leaves behind a positive charge in the valence band known as a hole. The electron transitioning to the conduction band may lose some energy and drop into an energy level in the band gap that is present because of the added dopants of Mg and Ti. This energy level is known as an electron trap. A corresponding energy level near the valence band is known as a hole trap. Either electron traps or hole traps may be luminescence centers. When an electron or hole recombines at such a center, luminescence is produced. Heating of the TLD provides energy to trapped electrons causing them to enter the conduction band. As the temperature of the TLD is increased, electrons or holes recombine at a luminescence center, visible light is emitted as a function of time, and the resulting signal, detected by a photomultiplier, is known as a glow curve. It can be shown that the peak height, or the area under the glow curve, is proportional to dose. The application of heat for emptying traps is a relatively inefficient process. The probability of an electron migrating from a trap can be given as a function of time by an equation that is similar to the Arrhenius equation governing chemical reactions (Galwey and Brown 2002):

$$p = ae^{-E/kT}, \tag{4.4}$$

where p is the probability of trap escape as a function of time, α is a constant, E is the energy per molecule, k is the Boltzmann constant, and T is the temperature. This type of assumption forms the starting point for the Randall–Wilkins theory of thermoluminescence (Randall and Wilkins 1945). Additional detailed theories have been developed more recently (Horowitz et al. 1998; Nail et al. 2002).

When LiF: Mg, Ti crystals are irradiated and subsequently heated quickly to a few hundred °C, a blue luminescence is observed having a wavelength of approximately 400 nm. The energy of a transition resulting in a wavelength 400 nm would be ~3 eV. Therefore, electron transitions leading to luminescence would have taken place well within the band gap. Electron traps are normally associated with the Mg centers, and the hole traps with the Ti centers (Nunn et al. 2008).

After irradiation, TLDs are read out by heating and recording the light emitted as a function of temperature. A plot of light intensity as a function of temperature is referred to as a glow curve. Glow curves may be simple, with one or two peaks, or more complex, with several peaks corresponding to the emptying of a number of groups of trapping sites. The light emission produced by heating is detected using a photomultiplier with appropriate optical filters to ensure detection of thermoluminescence while rejecting spurious light from heating of the surroundings or the substrate. Signal-processing electronics record glow curves and perform data analysis to produce computations of the dose delivered to the TLD.

Heating of the TLD can be accomplished using a resistance heater upon which the TLD is placed. The temperature of the heater is measured, usually with a thermocouple bonded to the metal heater pan. Since TLDs are heated to several hundred °C, they are surrounded by an oxygen-free gas, such as nitrogen, in order to reduce spurious light and damage to the TL crystals. Nitrogen gas can also be used as a source of heating. Optical filters are interposed between the heated TLD and the photomultiplier tube in order to filter out infra-red radiation. An example of a glow curve produced by an irradiated LiF TLD is shown in Figure 4.6 below.

A determination of dose from the glow curve shown in Figure 4.6 can be carried out by initially assuming the light intensity of the glow curve arose from a single trapping site. The area under this curve is proportional to dose. Additional information can be gained by determining the contributions to the glow curve from various trapping sites. First-order, second-order and general-order kinetics analysis yielded a deconvolution into seven sub-peaks arising from different trapping sites. The sub-peaks may have different dosimetric characteristics, so this analysis can potentially account for phenomena that affect the accuracy of the measurement (Chung et al. 2005).

All dosimeters show some degree of non-linearity of response as a function of exposure or dose. LiF: Mg, Ti shows a linear response up to approximately 10^3 R, whereas CaF$_2$: Mn

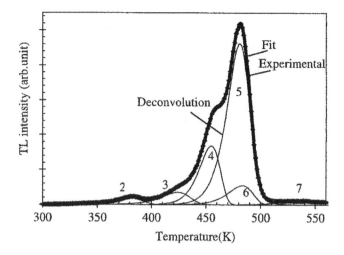

FIGURE 4.6 Example of LiF TL-100 glow curve and analysis by deconvolution methods (Chung et al. 2005).

responds linearly with exposure up to nearly 10^5 R. CaF_2:Mn is useful for high exposures; however, it is not as tissue-equivalent as LiF: Mg, Ti, due to its large effective atomic number of 16.3.

The deviation from linear response as a function of exposure (or absorbed dose) is known as supralinearity. The effect is most pronounced for the lithium borate dosimeter ($Li_2B_4O_7$: Mn). This behavior is due primarily to the increase in the sensitivity of high temperature traps in thermoluminescent materials and the intrinsic efficiency of the complex thermoluminescence process in that detector.

The responses as a function of photon energy for several common thermoluminescent materials show differences that are due to the physical properties of those materials. The over-response in the region of approximately 30 keV is partially due to the dependence of photon absorption upon the effective atomic number Z_{eff}, and the intrinsic sensitivity of the recombination centers in the crystal structure. The effective Z of LiF, is 8.2. Tissue has an effective atomic number of 7.4. The Z_{eff} of $Li_2B_4O_7$ is 7.3, and there is an under response in that energy region. In order to make the response of LiF: Ti, Mg closer to being closer to unity as a function of energy, metal filters of Cu or Sn are placed over the detectors in many TLD-based personal dosimeters. Calibration of TLD-based personal dosimeters is discussed in Section 4.4.3 (Figure 4.7).

The personal dosimeter shown below in Figure 4.8 is worn by radiation workers who are permitted to enter areas that may contain sources of neutrons as well as gamma rays. Supplemental dosimeters may also be required. Personal neutron dosimetry methods are discussed in Section 4.3.4 on Neutron and Mixed Field Dosimeters.

Extremity dosimeters are worn when there is an indication that the largest portion of dose equivalent will be received in a particular area such as the fingers or other extremities. Many extremity dosimeters use thermoluminescent detectors that are mounted as part of a plastic ring or wrist band. The dosimetric characteristics of extremity dosimeters are similar to those of the thermoluminescent detectors described above.

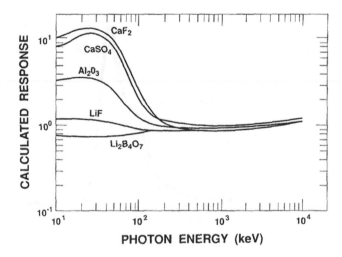

FIGURE 4.7 Relative response as a function of photon energy for a number of thermoluminescent detector materials (ICRU 1992).

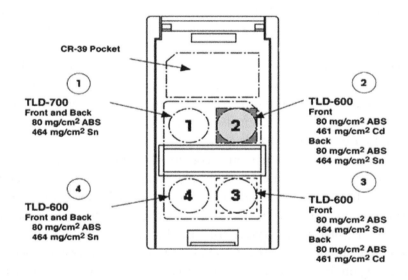

FIGURE 4.8 Detector placement in a Hanford personal dosimeter containing LiF dosimeters covered with various metal filters. Since this is a multi-purpose dosimeter, there are also provisions for neutron detectors such as CR-39 plastic etched-track detector and TLD-600 detectors that are enriched with ^{6}Li which has a large cross section for neutrons (Rathbone 2010).

4.3.1.3 Optically Stimulated Luminescence Dosimeters

Optically stimulated luminescence (OSL) dosimeters have many characteristics in common with TLDs. Both detector types are composed of doped crystalline materials, and both produce visible light whose intensity is proportional to the quantity of radiation incident. One of the principal materials used as an OSL detector is aluminum oxide doped with carbon, Al_2O_3: C. This material is a sensitive radiation detector, but it is also sensitive to visible light. Al_2O_3: C dosimeters must be placed in light-tight enclosures until they are read out.

Whereas TLDs are heated to a few hundred °C, to produce luminescence, OSL dosimeters are stimulated using light provided by a laser or light-emitting-diode (LED). The stimulating light source can be provided by a frequency-doubled-YAG laser generating green light with a wavelength of 532 nm, or a single or multiple LED array at about 535 nm (Miller et al. 2008). The energy deposited in the OSL dosimeter by the laser, or LED, stimulates trapped electrons that can then combine with trapped holes at recombination centers, known as F-centers or color centers, that are lattice sites within the Al_2O_3: C crystal. The wavelengths of the emitted light range from about 370–420 nm, centering at 420 nm, or deep blue. Since the stimulating light is close in wavelength to the emitted light, band-pass optical filters are used to filter out the green excitation light and transmit the blue OSL emission to the photomultiplier that detects the emitted OSL in much the same way as the thermoluminescent light is detected.

Al_2O_3: C OSL dosimeters respond to photons in the range from 5 keV to 40 MeV. The beta particle response ranges from 150 keV to about 10 MeV (Agyingil, Mobit, and Sandison 2006). Both OSL dosimeters and LiF: Mg, Ti, TLD over-respond to low-energy photons. This is primarily due to the photoelectric cross sections of the materials for low-energy photons, which is dependent on atomic number. Al_2O_3 and LiF have effective atomic numbers of 10.2 and 8.2, respectively. Therefore, the use of metal filters helps to reduce this over-response, and they are incorporated into the dosimeter holder. This holder also needs to be light-tight, since Al_2O_3: C to reduce the possible effect of ambient visible light (Figure 4.9).

OSL dosimeters can detect beta particles, and commercially available personal dosimeters combine an OSL detector with a CR-39 etched-track detector to provide a detection capability for neutrons.

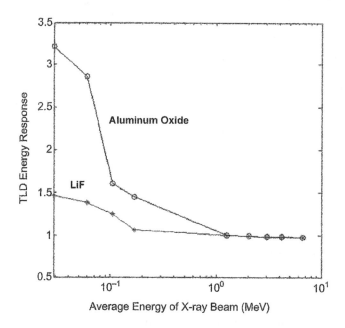

FIGURE 4.9 Response of Al_2O_3:C OSL dosimeters as a function of average photon energy (upper curve). The response for LiF: Mg, Ti TLD is shown in the lower curve (Agyingil, Mobit, and Sandison 2006).

4.3.1.4 Direct-Reading Dosimeters

In addition to the personal dosimeter worn by a radiation worker, another passive dosimeter may be required for entry into a specific radiation area. This device is known as a supplemental dosimeter and may consist of one or more passive dosimeters, depending upon the requirements for entry into the radiation area. A simple, passive device is the direct-reading dosimeter that is basically a quartz fiber electroscope with an optical readout. This dosimeter is worn on the body in the same area as the personal dosimeter. These devices are small, self-powered, and relatively sensitive. However, the range of readout is limited. Therefore, a dosimeter having the appropriate range of detection must be issued to the worker.

Within the direct-reading dosimeter tube, there is a closed container, basically an ionization chamber, within which radiation-induced ions neutralize the charges on the quartz fiber that has been charged before entry into the radiation area. In one type of direct-reading dosimeter, the degree of discharge is measured by viewing the shadow of the quartz fiber on an optical scale calibrated in terms of a radiation quantity such as milliroentgens. Other types such as the one shown in Figure 4.10 are read out using a separate instrument. Two requirements must be kept in mind. First, the anticipated exposure should be known, and a dosimeter of appropriate sensitivity used. The range of direct-reading dosimeters is not large, and devices are provided with a range of sensitivities. Second, the user must be sure that the dosimeter has been charged, and an initial reading of the device should be taken and recorded. The initial reading may be zero, but this is not always the case. A small reading (a few percent of full scale) is acceptable. In addition to beta-gamma sensitive direct-reading dosimeters, some manufacturers offer neutron-sensitive dosimeters that contain hydrogenous (plastic) walls. Neutron interactions with hydrogen atoms produce charged particles and gamma rays that are detected by the ionization chamber. Calibration is performed using a secondary standard neutron source. The performance requirements for direct-reading dosimeters can be found in ANSI Standard N322 (ANSI 1997).

4.3.2 Active Dosimeters

4.3.2.1 Electronic Personal Dosimeters

A number of electronic personal dosimeters (EPD) have been developed, and they have many features in common with both survey meters and personal dosimeters. Electronic

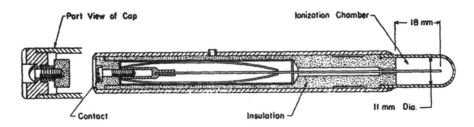

FIGURE 4.10 Schematic drawing of an ionization chamber dosimeter. Initially, the moveable fiber is given an electrostatic charge to zero the instrument. Ionizing radiation then discharges and moves the fiber in proportion to its intensity. (Reprinted courtesy of the NIST, U.S. Department of Commerce. Not copyrightable in the United States, 1988).

personal dosimeters should be used only for the purpose of recording the dose equivalent received by the worker wearing the device, and not as area survey meters. Performance requirements for these devices are given in ANSI Standard N42.20 (ANSI 2003a), and some dosimetry services using EPDs have been accredited by NVLAP. The features of EPDs that make them useful personal dosimeters are their ability to provide an immediate reading and to alarm on approach to a high-dose rate area. These devices can also, like TLD and OSL dosimeters, store accumulated values of personal dose that can be downloaded via the internet for recording.

4.3.3 Dose Calculation Methodology

Area monitoring instruments display readings of air kerma rate or ambient dose equivalent rate. Passive personal dosimeters require the use of calculations to determine personal dose equivalent. Dose calculation algorithms have been developed to process hundreds, or in some cases thousands, of readings obtained from personal dosimeters at facilities and laboratories dealing with radiation and radioactive materials. These calculations are one part of the process to determine personal dose equivalent.

Most passive personal dosimeters contain several thermoluminescent, or optically stimulated luminescence, elements. The response of these elements is determined by means of calibration exposures. The adjusted element readings are generated and used in the dose calculation process. Different dosimeter elements have different density thicknesses, for example: 1000 mg cm^{-2}, 300 mg·cm^{-2}, or 7 mg·cm^{-2}. These correspond, respectively, to whole-body tissue, the lens of the eye, and skin. The dose is determined for each case. Although the materials that comprise dosimeter elements are relatively tissue-equivalent, corrections are necessary in some cases. For neutron exposures, the energy distribution and radiation weighting factor must be known for the particular location, so that personal dose equivalent calculations can be made. Personal neutron dosimeters have some capability to discriminate between fast and thermal neutrons. This is usually accomplished by having two neutron-sensitive dosimeter elements, one of which is open and the other covered by cadmium. Cadmium has a large cross section for thermal neutrons of approximately 4000 b, up to about 0.4 eV. Thermal neutrons are those in equilibrium at standard room temperature and having a mean energy of 0.025 eV. Cadmium will absorb these neutrons, but its cross section decreases rapidly above about 1 eV. Therefore, the cadmium covered dosimeter element will not record thermal neutrons. The radiation weighting factor for thermal neutrons is 5, whereas the radiation weighting factor for higher energy neutrons varies up to a maximum of 20 for energies between 100 keV to 2 MeV (ICRP 2007). Therefore, discriminating between the responses due to neutrons with different radiation weighting factors is important. Evaluating personal dose in mixed low LET fields (gamma rays and beta particles, for example) also makes use of similar techniques. Filters placed above dosimeter elements absorb beta particles, and a non-filtered dosimeter element will record beta plus gamma dose. Determination of the photon dose fraction of a mixed neutron gamma ray field can make use of detectors with very different sensitivities to those radiations, such as ^{7}LiF: Mg, Ti and ^{6}LiF: Mg, Ti, or instruments that can discriminate between neutron and gamma ray signals (see Figure 4.5).

Multi-element personal dosimeters contain a number of detectors with various responses as a function of energy. These dosimeters can provide values for a number of different types of radiation with a single measurement. Analysis of the readings from these detectors involves mathematical corrections for their response characteristics. One method to solve this complex problem relies on the use of an iterative approach involving repetitive computations and corrections that simulate a learning process likened to the action of neural networks in the brain (Cassata et al. 2002).

Calculation of personal dose equivalent based on readings obtained from detector elements within a personal dosimeter is complex and requires evaluating the energy and type of radiation giving rise to the signals. In addition, the detector elements must have a large enough sensitivity range to enable accurate determination of personal dose equivalent over a wide range of energies and intensities.

4.3.4 Neutron and Mixed Field Dosimeters

4.3.4.1 Nuclear Emulsion Dosimeters

Nuclear emulsions had been used for many years, but production of the NTA® type emulsion ceased and it is no longer available. However, similar emulsions are still produced. Neutron interactions with the silver grains in the emulsion produce charged particle tracks that can be developed and counted. The gelatin of the emulsion is rich in hydrogen, so elastic scattering results in proton recoils that will create tracks.

The gelatin also contains nitrogen and neutrons can interact via the $^{14}N(n, p)^{14}C$ reaction. Tracks, or stars, can be produced by spallation reactions such as $^{12}C(n, n')3\alpha$. Thermal neutrons can also be produced by the (n, γ) reaction when using a cadmium converter. The recognition of a proton track becomes more difficult at energies greater than about 10 MeV, because of the wide separation of grains along the path of the sparsely ionizing proton. However, the spallation cross section increases with neutron energy and the response of the detector becomes determined by the tracks of spallation products. Also, stars are formed due to spallation reactions with the emulsion constituents.

Nuclear emulsions must be enclosed in a light-tight covering, and care must be taken to reduce the effects of excessive temperature and humidity. The energy dependence of a nuclear track film is shown in Figure 4.11.

4.3.4.2 Thermoluminescent Detectors

The basic operating principles for thermoluminescent dosimeters have been discussed in Chapter 5, Section 5.3.1.1. TLDs for use in neutron and mixed neutron-photon fields incorporate materials that have large neutron cross sections, such as ^{6}Li. The natural lithium used in certain variations of the dosimetric compound LiF: Mg, Ti contains the isotopes ^{6}Li and ^{7}Li. A version of the TL material manufactured using ^{7}Li, which has a smaller neutron cross section is more sensitive to photons. The detector material manufactured with LiF that has been enriched in ^{6}Li has a higher sensitivity to neutrons, due to the large cross-section $^{6}Li(n, \alpha)^{3}H$, ($\sigma_{th} = 940$ b).

The albedo neutron dosimeter is shown in Figure 4.12. This dosimeter contains both ^{6}Li and ^{7}Li enriched LiF: Mg, Ti TLDs enclosed in a boron-loaded holder. A thinner portion

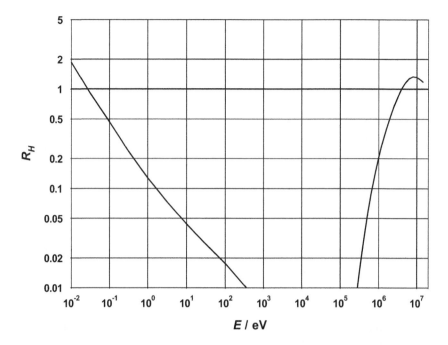

FIGURE 4.11   The personal dose equivalent response, R_H, as a function of neutron energy, E, normalized to unity for Am-Be neutrons (ICRU 2001).

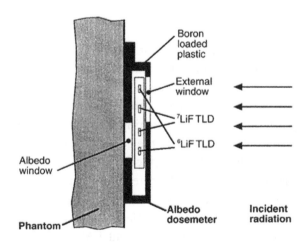

FIGURE 4.12   Cross section of an albedo neutron dosimeter showing TLDs and components of the dosimeter holder (ICRU 2001).

of the holder is labeled as the albedo window. Albedo neutrons are those scattered from the underlying hydrogenous phantom. When worn on the body, albedo neutrons will also be detected by the dosimeter. In addition, direct neutron interactions are detected. From analysis of the ratios of individual element readings, information about the neutron energy content in the incident field can be deduced (ICRU 2001).

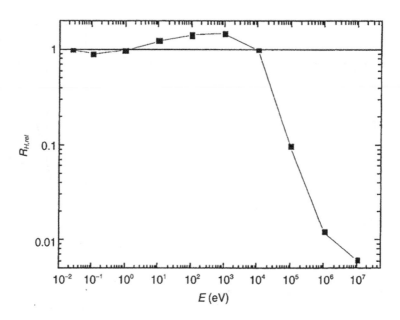

FIGURE 4.13 Personal dose equivalent, H_p (10), response as a function of neutron energy for an albedo dosimeter (ICRU 2001).

The TLD-albedo neutron dosimeter response with respect to personal dose equivalent $H_p(10)$, as a function of neutron energy, is shown in the figure below (Figure 4.13).

4.3.4.3 Etched-Track Detectors

The charged particles produced by neutron interactions in plastic can result in damage to the plastic that can be made visible. The damage is generally permanent, but may be partly restored, or may be modified over time, and that process is influenced by temperature, humidity, and the local presence of oxygen or other gases. The particle tracks may be viewed with an optical microscope after etching with a suitable solvent. The process of electrochemical etching increases the size of the etched pits such that they are visible and can be counted automatically by low-power optical systems. In electrochemical etching, an alternating electric field is applied across the detector foil during chemical etching

Alpha particles and fission fragments can produce damage tracks in polycarbonate, and are sensitive to recoil nuclei from neutron scattering. The energy threshold of sensitivity to neutrons by this detection mechanism is about 1 MeV. The energy dependence can be improved by the use of converter materials covering different sections of the polyallyl diglycol carbonate (PADC) detector (Figure 4.14).

4.3.4.4 Superheated Drop (Bubble) Detectors

The operational principle of bubble detectors is based on the use of small droplets of a liquid that is at a temperature above its normal boiling point (superheated) and suspended in a viscous medium. The droplets remain in the liquid phase until a neutron interacts with an atom in one of the droplets. Secondary charged particles from neutron interactions

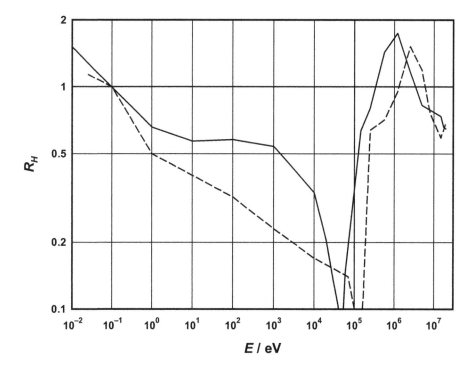

FIGURE 4.14 Energy dependence of the personal dose equivalent response for two operational etched-track dosimetry systems using PADC and appropriate converters, one chemical etch (dashed line), and the other electrochemical etch (solid line). Curves are normalized to 1 mSv⁻¹ for neutrons from an ²⁴¹Am-Be source (ICRU 2001).

transfer energy to the droplet and cause localized evaporation. As is the case with many dosimeters that rely on the production of a visible change, temperature sensitivity is a problem. Bubble detectors can be reused. A piston cap on top of the detector enclosure can be screwed down to increase pressure and cause the bubbles to condense. Another variant of the superheated drop detector uses an aqueous gel that allows bubbles to rise to the surface and create a sound that is detected electronically. Noise and vibration can interfere with acoustical bubble detection (Figure 4.15).

4.3.4.5 Personal Neutron Accident Dosimeters (PNAD)
Personal nuclear accident dosimeters are issued to those working in areas where a criticality event is possible. The radiation field that might be produced by a nuclear criticality is complex, and PNADs must have basic capabilities for not only measuring personal dose, but also providing information relative to the neutron spectrum that was produced. The PNAD shown below contains a TLD and activation foils that can be read out after an incident to provide spectral and intensity information (Rathbone 2010). Each of the foils responds to different parts of the incident neutron spectrum, and with calculations, the device can serve as a crude spectrometer. Information obtained from this dosimeter can be used to determine neutron dose from the criticality (Figure 4.16).

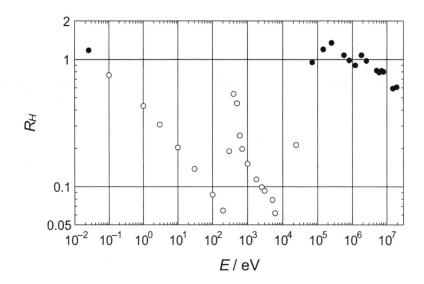

FIGURE 4.15   Ambient dose equivalent response of a superheated emulsion detector, as a function of neutron energy. The values are normalized to unity for neutrons from an Am-Be source. The full circles are experimentally determined response values, the open circles depict results from a Monte Carlo simulation (ICRU 2001).

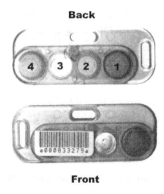

FIGURE 4.16   Hanford PNAD with activation foils and TLD chip. The activation foils contained in this PNAD are: Position 1: Indium (Cd cover), Position 2: Sulfur (bare), Position 3: Indium (bare), Position 4: Copper (Cd cover), and a TLD-700 chip.

4.4 MEASUREMENT METHODS AND PROCEDURES

Many measurements of quantities of ionizing radiation are performed on a daily basis in the workplace. These measurements are difficult to carry out for a number of reasons. The dose rates are low, and even if a measurement using a personal dosimeter is carried out over weeks or months, the indication of the dosimeter may be close to the level of noise. The direction of the radiation may not be known, and that direction is likely to be variable. It is not possible to measure the dose at points within the body, and so dosimeters worn on the surface of the body only provide an estimate of the dose at depths greater than one centimeter. Additional information can be provided by investigating the properties of the radiation environment in the workplace, and this information can help to improve the determination of the equivalent dose.

Workers that are issued personal dosimeters, or use area survey meters, are instructed in their use. However, it is not always possible to guarantee that everything will be perfect in the workplace. Rather than being clipped to clothing on the front of the chest, personal dosimeters may be attached to a necklace along with identifications and other items. During work, the necklace is free to swing away from the body. Therefore, the geometry established in the calibration laboratory, where the dosimeter is mounted on a flat plastic phantom, is not replicated. This adds to the uncertainty of the workplace measurement.

Survey meters may be used in conditions that can also very different from those in the calibration laboratory. Instruments may be transported to a workplace in a non-air conditioned vehicle and used in extreme environmental conditions. Such situations add to the uncertainties in measured values, and accurate corrections to those values are difficult to make. During use of the instruments, it may be difficult to maintain the distance between a possibly contaminated surface and the surface of a detector so that short-ranged particulate radiation can be accurately measured.

Air monitoring instruments may also be subject to conditions whereby the uncertainty of measurements is increased. During the course of measurements, there may be cross-talk from atmospheric radon. Detectors make use of filters and signal-processing techniques to minimize the ambiguity due to this source of uncertainty. Efforts must be made to minimize the sources of uncertainty, and estimates of the measurement uncertainty should be performed and reported with the recorded worker dose.

4.4.1 National Standards and Reports

The American National Standards Institute (ANSI) develops consensus standards dealing with many topics, including radiation protection dosimetry. Two important sets of ANSI standards deal with performance specifications for portable radiation protection instrumentation, along with test and calibration procedures. They are N42.17 parts A, B, C (ANSI 1989a,b; 2003b) and N323 AB (ANSI 2013) The ANSI N42.17A standard provides minimum performance criteria for portable instruments used in normal environmental conditions. N42.17B specifies minimum performance criteria for occupational airborne radioactivity monitoring instrumentation. N42.17C provides performance specifications for portable instrumentation for use in extreme environmental conditions. N323AB establishes test and calibration requirements for portable radiation protection survey instrumentation.

The ANSI standards include specifications for standard test conditions in a calibration facility, such as temperature, pressure, and humidity. The quantities and their units that radiation protection instruments are expected to measure and display are described. Requirements for the basic operability of instruments are given. Radiation response is tested for accuracy using sources and certifications that are traceable to NIST or an equivalent national metrology institute. Specifications and tests are also provided for the angular dependence of response. Many of the tests and specifications are consistent with recommendations from the National Council on Radiation Protection and Measurements, the International Commission on Radiation Units and Measurements (ICRU), the International Electrotechnical Commission, and ISO, the International Organization for Standardization.

ANSI N323 AB (ANSI 2013) provides calibration and calibration-related requirements for portable radiation protection instruments. The standard includes acceptance testing requirements that are evaluated before instruments are put into use. The activities related to the commissioning of equipment used for radiation protection are outlined. Included in these activities are maintenance and repair. Instrument calibration is described, and part of the evaluation of instruments includes adjustment to meet requirements. The accuracy and uncertainty of measurements are specified, and requirements for those quantities are provided. Daily response checks for instruments are described, as is the connection to re-calibration. Calibration frequency is an important, but complex concept that is explained. This process is dependent upon the extent of use, the operational environment, the construction of the particular instrument type, and accumulated data about the device type. Procedures similar to those described in the ICRU's *Report 76* on measurement quality assurance (2006) are included in N323 AB (ANSI 2013). Requirements for the instrument calibration facility are provided, and it is specified that calibration standards be traceable to NIST, or an equivalent national metrology institute.

ANSI N13.11 standard (ANSI 2015) describes the proficiency testing program for personal dosimeters (NIST 2005). This standard includes six test categories that evaluate the performance of personal dosimeters exposed to radiations that are representative of a wide range of workplace fields. The broad categories of photons, beta particles, and neutrons require testing to a variety of radiation energies and intensities, including mixtures of energies. The accident dosimetry category extends to possible exposures of 5 Gy that are randomly chosen. Neutron categories are intended to reflect realistic workplace situations where various mixtures of neutrons and photons are encountered. Testing for the angular response of dosimeters is included for specific angles.

The N13.11 standard defines a dosimeter performance quantity known as the bias B, or mean value of the performance quotient, p. This performance quotient is defined as the difference between the dose value reported by the participant in the test and the dose delivered, divided by the dose delivered by the irradiating laboratory. The standard deviation of the bias is also evaluated. The bias is defined as:

$$B = \bar{P} = \frac{1}{n}\sum_{i=1}^{n} P_i \tag{4.5}$$

Unlike survey instruments, most personal dosimeters store a signal that can only be read out once. Therefore, numbers of dosimeters are exposed to a single quantity of the test radiation and the mean value of their readings is recorded for the N13.11-specified test. The performance criterion for the test is defined by the requirement,

$$|B| + S \leq L, \tag{4.6}$$

where B is the bias, S is the standard deviation of the values of P, and L is a tolerance level corresponding to the required level of performance. The value of L is, for most tests, 0.4, but for the accident-dose category a value of 0.3 is required. Both 10 CFR 20 (United States Nuclear Regulatory Commission 2011) and 10 CFR 835 (United States Department of Energy 1998) require dosimetry services to be accredited by the National Voluntary Laboratory Accreditation Program (NVLAP), which requires successful performance with respect to the tolerance level set in the N13.11 standard. In addition, performance tests are specified in ANSI/HPS N13.32, *Performance Testing of Extremity Dosimeters* for extremity dosimeters (ANSI 2008). After a successful onsite assessment, accreditation can be granted, and proficiency is tested every two years thereafter.

An important element of the N13.11 testing is the evaluation of measurement uncertainty. As mentioned earlier, the measurement of a quantity is not complete without a statement of the measurement uncertainties. *NCRP Report 158: Uncertainties in the Measurement and Dosimetry of External Radiation* (National Council on Radiation Protection and Measurements 2007), provides useful information regarding the measurement uncertainties for dosimetric methods and materials. In addition, a large amount of technical discussion regarding personal dosimeters and area survey instruments is contained in this report.

4.4.2 Regulatory Guidance

The Code of Federal Regulations documents, 10 CFR 20 (United States Nuclear Regulatory Commission 2011) and 10 CFR 835 (United States Department of Energy 1998), specify detailed requirements for radiation protection dosimetry. 10 CFR Part 20 contains procedures for protection against ionizing radiation that results from activities conducted under licenses issued by the Nuclear Regulatory Commission. The regulations control the receipt, possession, use, transfer, and disposal of licensed material in order to place limits on the total dose to an individual and ensure it does not exceed the standards for radiation protection given in the regulations.

NRC licensees are required to develop a radiation protection program appropriate to the scope of licensed activities. Occupational dose limits for adults are given, and specifications for the determination of external dose from airborne radioactive material are included. Requirements for the determination of internal exposure are also included.

Regulatory Guides and NUREG documents are available online from the NRC Office of Standards Development, and include technical information relating to a variety of subjects. DOE Guides are also available online from the DOE Occupational Health Office (Division 8).

4.4.3 Calibration and Testing

Calibration consists of performing a set of operations to establish, under specified conditions, the relationship between values indicated by a dosimetric device and the known, or conventionally true, value of the quantity to be measured. A "conventionally true value" is a value accepted by convention or agreement. The results of measuring a quantity must be accompanied by an estimate of the measurement uncertainty. This uncertainty provides an indication of the reliability of the measurement and also facilitates comparisons of measurements to reference values for the quantity (ISO and International Electrotechnical Commission 2004; National Council on Radiation Protection and Measurements 2007).

Two properties of a calibration measurement are accuracy and precision. Accuracy is the closeness of agreement between the result of a measurement and a true, or conventionally true, value of the quantity being measured. Precision is the closeness of agreement among a group of values obtained by replicate measurements of a quantity, under specified conditions (ISO and International Electrotechnical Commission 2004). Measurement precision is usually expressed numerically by computing the standard deviation, variance, or coefficient of variation of the distribution of measured values.

For external-radiation protection purposes, there are four basic types of radiation that are employed in a calibration laboratory: photons, beta particles, alpha particles, and neutrons. However, activity calibrations for surface-deposited alpha and beta sources are not normally performed for external dosimetry purposes. These radiations all have distinctly different physical properties, and they are characterized by different radiation quantities and units. The primary physical quantities for photons, beta particles, and neutrons are: air kerma, absorbed dose, and fluence, respectively. These quantities can be measured or calculated and specified at a reference point in the radiation field.

The methods for generating the reference radiations are also different. X-ray photons are generated using radiation generators to produce filtered and collimated beams. Spectra may be wide or narrow, as with fluorescence beams. Gamma ray sources are used to produce narrow spectra and may be used with or without collimation. For dosimeter calibration purposes, beta-particle sources are mounted in special configurations (Böhm 1986) and are used at short distances without collimation. Neutron sources are operated in large irradiation facilities in order to minimize the scattering effects of the shielded room. Each of these situations requires calculations that include the correction coefficients required to produce the desired radiation protection quantity at the calibration position.

The protection quantities for survey instruments and personal dosimeters are ambient dose equivalent and personal dose equivalent, respectively. Ambient dose equivalent calibrations of survey instruments are intended to be receptor-free with no phantom present. The definition of personal dose equivalent assumes the presence of a phantom. Radiation protection area survey instruments are assumed to provide sufficient backscatter resulting from the presence of their electronics and enclosures. Personal dosimeters are designed to measure personal dose equivalent at a point in the body just below the dosimeter. Calibrations of personal dosimeters are, therefore, performed on phantoms. The properties of these phantoms are described in the applicable national and international standards, disseminated by ANSI and ISO, which have been discussed earlier. For personal dosimeters worn on the body, the phantom is a 30 × 30 × 30-cm rectangular solid of polymethylmethacrylate (PMMA). Phantoms for extremity dosimeters such as wrist/ankle dosimeters require the use of a cylindrical PMMA phantom with dimensions of 7.3 cm diameter and

30 cm in length. Finger ring dosimeters are calibrated on PMMA rods with dimensions of 1.9 cm and 30 cm length. Specialized dosimeters, such as those designed to measure dose to the lens of the eye make use of holders that are worn on the head. These dosimeters are also calibrated using rod phantoms.

The International Commission on Radiological Protection (ICRP 2007) has recommended that the accuracy for effective dose, the summation of equivalent doses in tissues or organs that are multiplied by the appropriate tissue weighting factor, when it is at the level of 20 mSv per year, averaged over five years, should be within a factor of 1.5. Expressed as a percentage, it would imply ±50%. For other situations encountered in radiation protection, such as accident dosimetry, where the quantity measured is absorbed dose and may range from 100 mGy to 10 Gy, the IAEA recommends that accuracy should be approximately ±25% (IAEA 1996). ANSI Standard N323 (ANSI 2013) recommends that accuracy for photon measurements using area survey instruments should be 5% for most intensities, and 10% for rates less than $1.0\ \mu Gy \cdot h^{-1}$; the accuracy for neutrons: 10%; and the accuracy for beta particles: 10%, but 20% for dose rates less than $1.0\ \mu Gy \cdot h^{-1}$. The range of recommended accuracies for measurements is dependent upon the level of hazard, that is, a greater hazard requires a tighter specification of measurement accuracy for the determination of dose. The specifications of accuracy also reflect the practical limitations of the current instrumentation.

4.4.3.1 Photon Calibrations

The primary dosimetric quantity to be determined is air kerma rate. The air kerma rate at the point of test in the calibration laboratory is determined using a secondary standard ionization chamber. Once the air kerma rate is known at the point of test in the radiation field, conversion coefficients, $\bar{c}_{K,d,a}$, can be applied to compute the personal dose equivalent rate as a function of depth and angle. The time duration of the irradiation then determines the delivered personal dose equivalent. The radiation protection quantities obtained are the personal dose equivalents, $H_p(0.07)$ and $H_p(10)$. An outline of the procedure for calibration follows:

- The effects of scattered radiation present at the point of test (reference point) from supports and the walls of the facility should be known and not exceed 5%.

- An appropriate reference standard instrument is used, such as an air-equivalent ionization chamber calibrated for a range of energies. More than one instrument may be needed for different source intensities. The calibrations for the ionization chambers should be traceable to national standards.

- Conversion coefficients specified in ANSI/HPS N13.11 (ANSI 2015) and ISO 4037-2 (ISO 1997) are applied to yield reference radiation protection quantities.

- A 30 × 30×15-cm rectangular phantom, constructed of polymethylmethacrylate (PMMA) is used for whole-body dosimeter irradiations.

- The uniformity of irradiation as a function of position on the face of the phantom is determined.

- The phantom may need to be rotated to perform irradiations as a function of the angle of radiation incidence.

- Irradiations can be carried out using either collimated beams or an uncollimated radiation source.

- A monitor ionization chamber should be used to correct for possible fluctuations in the output of an X-ray source and to determine irradiation durations.

- Requirements for extremity dosimeter irradiations can be found in ANSI/HPS N13.32 (ANSI 2008) and ISO 4037-2 (ISO 1997) (Figure 4.17).

4.4.3.2 Beta Particle Calibrations

The absorbed-dose rates to tissue due to beta particles emanating from any of the three reference beta-particle sources specified in ISO Standard 6980-1 (ISO 2006a) are measured using an extrapolation chamber. Because of the limited range of these beta-particle sources, Bragg–Gray conditions cannot be established in an ionization chamber with a

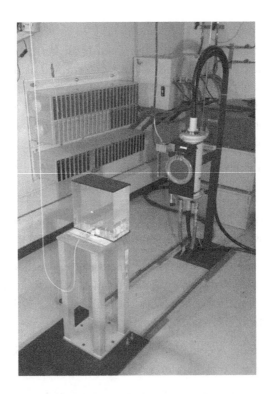

FIGURE 4.17 X-ray calibration range showing 30 × 30 × 15-cm PMMA phantom (left), with X-ray tube and transmission monitor ionization chamber at the right. The cable for a quality control ionization chamber, mounted in the phantom, can be seen at the left.

fixed electrode separation. In order to overcome this difficulty, the extrapolation method is used. A diagram showing the internal construction of the extrapolation ionization chamber is shown in Figure 4.18.

Beta particles have finite and relatively short ranges in tissue and similar materials, such as plastics and water. An ionization chamber that is designed to measure beta dose must, therefore, have a thin entrance window of the type shown in Figure 4.18. The window, indicated in the figure as w, is a thin plastic coated on its interior side with electrically-conductive graphite. The collecting electrode, indicated as a, also comprises a graphite coating on a plastic block, p. Electrical contact with the collecting electrode is made using a wire marked $\mathbf{c}$. Ions formed by interactions between incoming beta particles and the air molecules within the chamber are indicated as dots labeled as ℓ. The area labeled as g also has a graphite coating and serves as a guard ring. A high voltage is applied between the entrance window and the collecting electrode a (Figure 4.19).

This distance, and consequently the air volume inside the chamber, should be sufficiently small to not disturb the beta particle flux in order to satisfy the Bragg–Gray conditions. The ionization current produced is measured as a function of the distance between the electrodes, and by extrapolating this function to the origin it is possible to determine, with appropriate corrections, the absorbed dose rate to tissue. The basic elements of the procedure for beta-source calibration are as follows.

- Nearly all primary calibration laboratories and accredited secondary calibration laboratories make use of the same primary standard ionization chamber for determining absorbed dose rate to tissue for beta radiation (Böhm 1986).

- The irradiator used with the ionization chamber has a specific geometry and is used with any one of three beta-emitting sources: ^{204}Tl, ^{85}Kr, and ^{90}Sr/^{90}Y.

- Thin plastic filters are used to flatten the field at the irradiation position.

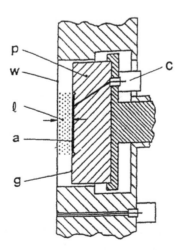

FIGURE 4.18 Cross-sectional view of thin window extrapolation ionization chamber for the measurement of beta-particle dose (ISO 2004b).

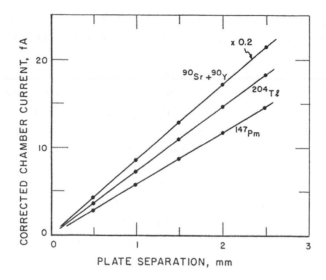

FIGURE 4.19 Current measurements as a function of ionization chamber electrode separation, referred to as extrapolation curves for the beta-particle sources specified in ISO 6980-2 (ISO 2004a).

- There is an electrically controlled shutter for timing the duration of the irradiation. The use of a transmission monitor is not practical due to the short range of the beta particles.

- Measurements of the beta absorbed dose rate are performed using the extrapolation chamber described above.

- Instruments and dosimeters can be placed at a reference position determined using a supplied metal rod of precise length.

- Dosimeters may be placed on PMMA phantoms of rectangular or cylindrical design.

- The quantity determined using the extrapolation chamber is absorbed dose to tissue (Figure 4.20).

4.4.3.3 Neutron Calibrations

The primary quantities needed for the calibration of neutron dosimeters or area monitors are fluence and fluence rate. Dose equivalent quantities are computed using fluence-to-dose equivalent conversion coefficients. This calculation is shown in the following equation:

$$H = \int h_f\,(E) f_{\;E} dE \qquad (4.7)$$

where H can be either the ambient or personal dose equivalent. The quantity $h_\Phi(E)$ is the energy-dependent fluence-to-dose equivalent conversion coefficient, and Φ_E is the energy distribution of the neutron fluence. The neutron fluence-to-personal dose equivalent conversion coefficient also depends on the angle of radiation incidence, energy, and direction distributions of the fluence. This is denoted as $\Phi_{E,\Omega}$ (ICRU 2001).

FIGURE 4.20 The beta source and shutter mechanism is shown at the right, and a circular plastic beam-flattening filter is mounted between the source enclosure and the cylindrical PMMA extremity phantom. A thin-window quality control ionization chamber is at the lower left.

Unlike photons or beta particles, there is no standard instrument used by the primary calibration laboratory. In the case of neutrons, two sources serve as the U.S. national standards: a bare ^{252}Cf source, and that same source mounted at the center of a D_2O-filled stainless-steel sphere covered with cadmium (ISO 2001). The free-field fluence rate, φ, is determined from:

$$\varphi = \frac{B_\Omega}{\ell^2},$$ (4.8)

where ℓ is the distance from the center of the source to the point of test. The neutron angular source strength, B_Ω, is defined in ISO 8529-1 (ISO 2001). Calculations of the free-field dose equivalent rate at the reference position allow calibrations of neutron survey instruments and neutron personal dosimeters to be performed. However, the effects of neutron scattering from the irradiation facility's walls and floor, and the air in the room, need to be taken into account. Procedures for evaluating these effects are described in ISO 8529-2 (ISO 2000). The basic principles that are taken into account for neutron calibrations are as follows:

- ^{252}Cf neutron source emission rates are determined by NIST using the activation of $MnSO_4$ which serves as the national standard method.

- The effect of neutron scatter in the irradiation room must be determined to correct the effect upon the free-field fluence rate at the irradiation position.

- The distance from source to detector is set at 50 cm, and the effect of air attenuation is taken into account.

- Dosimeters are mounted on a 40 × 40 × 15-cm PMMA phantom which may be rotated for determining response as a function of the angle of incidence.

- Irradiation duration is controlled by timing, and a quality control ionization chamber placed near the irradiation position to confirm the duration of the exposure.

- Periodic calculations of the neutron emission rate for the source are performed to account for the relatively short 2.64 y half-life and ingrowth of ^{250}Cf (Figure 4.21).

4.4.3.4 Surface Contamination Monitors

Surface contamination may be specified in terms of activity, the number of disintegrations (spontaneous nuclear transformations) per unit time, whereas the response of monitoring instruments is displayed as counts per unit time. In general, there is no simple, known relationship between surface emission rate and activity. Thus, there is a need for calibration sources that are specified primarily in terms of surface emission rate as well as activity. Traceability of calibration sources to international or national standards can be established by a system of reference transfer instruments. ISO Standard 8769 (ISO 2016) deals with

FIGURE 4.21 A D_2O-filled sphere surrounds a ^{252}Cf source that will be brought into place by the pneumatic transfer system (vertical tube above). Dosimeters are mounted on the surface of a 40 × 40 × 15-cm PMMA phantom. A spherical-tissue-equivalent plastic quality control ionization chamber is shown just behind the D_2O-filled sphere.

alpha emitters, beta emitters, and photon emitters having maximum photon energies less than 1.5 MeV. This standard also specifies reference radiations for the calibration of surface contamination monitors, using large-area sources, in terms of surface emission rates. Calibration of these sources should be traceable to national standards. NUREG-1507 (United States Nuclear Regulatory Commission 1997), *Minimal Detectable Concentrations with Typical Survey Instruments for Various Contaminants and Field Conditions*, provides guidance for calibrations of contamination monitors.

4.4.4 Measurement Traceability

National standards documents including ANSI N13.11 (ANSI 2015), N323 (ANSI 2013), N42.17ABC (ANSI 1989a,b, and 2003b), and N42.20 (ANSI 2003a) recommend that the calibration of radiation measurement devices, such as personal dosimeters and area survey instruments, be traceable to national reference standards. Measurement traceability requires the establishment of an unbroken chain of calibrations to specified references. The implications of these words are that documentation, such as a calibration certificate, must be maintained in order to verify that a local secondary standard instrument, such as an ionization chamber, had been calibrated at NIST or at an accredited dosimetry calibration laboratory. The documentation explains what is meant by an "unbroken chain."

Calibration certificates must include: an identification of the standard specification used or a description of the method of calibration, a description of the conditions under which the calibration was made (such as ambient temperature, pressure, humidity), a statement of the evaluated uncertainty of the measurement, and an indication that the calibration laboratory participates in a measurement quality assurance program operated by NIST.

Accredited dosimetry calibration laboratories must conform to the specifications given by ISO Standard 17025 *General requirements for the competence of testing and calibration laboratories* (ISO and International Electrotechnical Commission 2005). Measurement quality assurance also helps to establish and confirm traceability of calibrations that have been performed at an accredited calibration laboratory. Transfer-standard ionization chambers are first calibrated at NIST and sent to the participating laboratory for a set of measurements. The transfer standards are returned to NIST and re-calibrated. A report is issued by NIST comparing the sets of measurements.

NIST Special Publication 812 (NIST 1991) and NIST/NVLAP Handbook 150-2D (NIST 2004) form the basis for the requirements of the MQA interaction and the operation of an accredited calibration laboratory. *ICRU Report 76: Measurement Quality Assurance for Ionizing Radiation Dosimetry* (ICRU 2006), discusses the elements of quality assurance and their application to radiation measurements and calibrations.

It should be noted that there are some areas of radiation protection dosimetry for which traceability to national standards is not possible to establish. Since there is no national standard for high-energy radiations, such as those found in the vicinity of research accelerators, it is not possible to demonstrate traceability for dosimetric measurements of those radiations. No national standards exist for radiation fields that would be present in a criticality event, and there are also no national standards for internal dose. For these situations, comparison measurements have been performed that at least demonstrate the degree of uniformity in the computation of dose in such situations (Sims 1989; Hui, Loesch, and McDonald 1997; and Stewart et al. 2012).

4.4.5 Statistics of Radiation Measurements

Many measurements of ionizing radiation quantities are repeated a number of times. It is often necessary to perform a series of measurements in order to compute a statistical average value for a quantity and to provide an estimate of the uncertainty associated with that value. As mentioned earlier, a report of a value for a measured quantity must include an estimate of the uncertainty associated with that value. ANSI Standards N13.11 (ANSI 2015), N323 (ANSI 2013), N42.17A,B,C (ANSI 1989a,b, and 2003b), N42.20 (ANSI 2003a), and International Standards: ISO 17025 (ISO and International Electrotechnical Commission 2005), 4037-1,2,3 (ISO 1996, 1997, 1999), 6980-1,2,3 (ISO 2004a, 2006a,b), 8529-1,2,3 (ISO 2001, 2004a, 1998), along with NIST Technical Note 1297 (NIST 1994), and the ISO Guide to the Expression of Uncertainty in Measurement (ISO 1995) all provide information about evaluating the uncertainty of measurement results.

Radiation measurements performed using personal dosimeters and area survey meters may represent a single data point. The readout of a personal TLD represents a single number due to the fact that the characteristic of the device is a destructive readout. An exception is the OSL-based dosimeter that has the capability of multiple readouts. Measurements using area survey meters are normally single point measurements, or perhaps represent a few repeat measurements. In many of these cases, statistical analysis of the few readings is not carried out. During calibrations or proficiency tests of dosimeters or instruments, repeat measurements of the magnitude of a reference radiation are always performed, including repeat measurements using a group of passive personal dosimeters. Proficiency tests of dosimeters like ANSI N13.11 (ANSI 2015) specify how many repeat measurements are required, and the performance requirement includes a simple assessment of uncertainty with the inclusion of the standard deviation.

The test quantities that are defined in the ANSI N13.11 (ANSI 2015) standard include the following:

The performance quotient, P_i defined as:

$$P_i = \frac{H_R(d)_i - H_p(d)_i}{H_p(d)_i}, \tag{4.9}$$

where $H_p(d)_i$ is the conventional quantity value (true value by convention or agreement) of personal dose equivalent assigned by the irradiating laboratory to a dosimeter, and $H_R(d)_i$ is the corresponding personal dose equivalent reported by the test participant. For the accident category, absorbed dose replaces the personal dose equivalent.

The bias B is defined as:

$$B \equiv \bar{P} = \frac{1}{n}\sum_{i=1}^{n} P_i, \tag{4.10}$$

where the summation is taken over all n values of P_i for a particular test in a radiation category and a phantom depth.

A performance criterion:

$$B^2 + S^2 \leq L^2, \qquad (4.11)$$

where S is the standard deviation of the distribution of individual dosimeter readings and L is the tolerance level, or limit of acceptable agreement between given and recorded dose. The tolerance level ranges from 0.24 to 0.30 depending on the radiation category.

The standard deviation of the values of the performance quotient, P_i is given as:

$$S = \sqrt{\frac{\sum_{i=1}^{n}(P_i - \bar{P})^2}{n=1}}, \qquad (4.12)$$

where the sum is extended over all n values of P_i for a particular test depth and category, and $\bar{P}$:

$$\bar{P} = \frac{1}{n}\sum_{i=1}^{n} P_i, \qquad (4.13)$$

The performance testing standards for area monitoring instruments, such as ANSI N323AB (ANSI 2013) and N42.17A,B,C (ANSI 1989a,b, 2003b) refer to statistical evaluation of quantities as given in the *ISO Guide to the Expression of Uncertainty in Measurement* (ISO 1995), including:

Relative Error, ε_{REL} : The difference between instrument's reading, M, and the conventionally true value, CTV, of the quantity being measured divided by the conventionally true value multiplied by 100%.

$$\varepsilon_{REL} = \frac{(M - \text{CTV})}{\text{CTV}} \times 100\% \qquad (4.14)$$

Variance, σ^2 : The expected value of the sum of differences between measured values and their mean, raised to the power of 2, then divided by one less than the number of measurements.

$$\sigma^2 = \frac{1}{n-1}\sum_{i=1}^{n}(x_i - \bar{x})^2, \qquad (4.15)$$

Standard deviation, s_N, is a statistical measure of the scattering of a set of data, and the generalized formula, as distinct from the formula used the proficiency tests and defined above, is:

$$s_N = \sqrt{\frac{1}{N}\sum_{i=1}^{N}(x_i - \bar{x})^2}, \qquad (4.16)$$

Standard error of the mean, σ_M, is an indication of the variation among sample means of the same population. It provides an estimate of the variation between samples. The standard deviation indicates the variation in one specific sample.

$$\sigma_M = \frac{\sigma}{\sqrt{N}} \tag{4.17}$$

σ_M, the standard error of the mean is the standard deviation σ of the large distribution divided by the $\sqrt{N}$, where N is the sample size.

The lower limit of detection, L_D is the lowest quantity of a value that can be distinguished from the absence of that value within a stated confidence limit. An estimate of the lower limit of detection for a dosimetric measurement can be given by evaluating the mean dose equivalent values from unirradiated and irradiated dosimeters, denoted as H_0 and H_1, respectively. Standard deviations for these quantities S_0 and S_1 are calculated and the lower limit of detection is computed using:

$$L_D = 2 \frac{\left[t_p S_0 + \left(\frac{t_p S_1}{H_1} \right)^2 H_0' \right]}{\left[1 - \left(\frac{t_p S_1}{H_1} \right)^2 \right]} \tag{4.18}$$

where t is the t distribution for $n-1$ degrees of freedom and p value of 0.95. H_0' is the average of unirradiated dosimeter readings without subtracting background (United States Department of Energy 1987).

4.4.5.1 Uncertainty Analysis

The determination of measurement uncertainty is an essential part of measurement quality assurance. The types of uncertainty components include: *Type A*, or uncertainties that can be evaluated using statistical means, and *Type B*, referring those uncertainties evaluated by other means. Previously, these two types were also referred to as random, and systematic, respectively. The usual approach to the evaluation of *Type A* uncertainties is to calculate the standard deviation of the mean, σ_{mean}, of a series of independent measurements. It is often assumed that the original distribution is a normal distribution. The larger the sample size, the smaller the standard error of the mean.

The *Type A* and *Type B* uncertainties, when the distribution is not known, can be estimated by assuming a rectangular probability distribution with a constant value between a lower and an upper limit. The estimate is usually taken as the expectation value x_i of the distribution,

$$x_i = \frac{(a_+ + a_-)}{2}, \tag{4.19}$$

where a_+ is the upper limit, and a_- is the lower limit of the rectangular distribution. The standard uncertainty associated with x_i is given by,

$$\frac{a}{\sqrt{3}} \tag{4.20}$$

The combined standard uncertainty of a measurement result is taken as the estimated standard deviation of the result. A coverage factor, k, can be applied to produce an expanded uncertainty. A conventional coverage factor is taken as $k = 2$. A coverage factor larger than 2 may be required for certain regulatory applications. Therefore, an expanded uncertainty may be needed to define an interval that assumed to contain the measurement result. This can be written as:

$$U = ku_c(y), \tag{4.21}$$

where U is the expanded uncertainty, k is the coverage factor, and $u_c(y)$ is the combined uncertainty of the measurement result y (ISO 1995).

When personal dosimeters and area survey meters are calibrated before being used in the workplace, there are certain sources of uncertainty for which corrections can be applied. Such sources are referred to as influence quantities. Examples of influence quantities may include: ambient temperature and pressure, humidity, AC line or battery voltage, and background radiation. Extensive discussions of sources of uncertainty for personal dosimeters and area survey meters can be found in *NCRP Report 158* (NCRP 2007).

Since personal dosimeters are worn on the body, they are subject to environmental effects such as the possibility for high temperatures, water damage, and dirt infiltration into the dosimeter holder. These effects are not controlled as they are in the calibration laboratory (see Section 4.3). The magnitude of these sources of uncertainty can only be estimated. On the other hand, certain operational checks can be performed with survey instruments before they are used. These include exposure to a small, stable, radioactive source as a check on response before use, and most meters have a capability to check the battery voltage. Electronic personal dosimeters may also have such capabilities, but passive personal dosimeters do not. Personal dosimeters are also "single-shot" devices. A set of repeat measurements in the field is not possible using a single passive personal dosimeter. A set of repeat measurements using a survey meter could be performed to give an estimate of the statistical uncertainty associated with the value of the reading. The statistical uncertainty associated with the reading of a survey meter is also dependent on the nature of circuitry and the display. Analog meter displays may show fluctuations in their indication. Digital-display instruments may indicate fluctuations in numerals displayed. Estimates of statistical uncertainties based on observations in both cases are difficult, but those estimates could be included as part of the random, *Type A*, uncertainty.

It is necessary to evaluate the uncertainty associated with the quantity effective dose, and the methods outlined above will yield that estimate. The discussion of effective dose, equivalent dose, and tissue weighting factors is explored in detail in Chapter 2.

REFERENCES

Agyingil, E. O., P. N. Mobit, and G. A. Sandison. 2006. "Energy Response of an Aluminium Oxide Detector in Kilovoltage and Megavoltage Photon Beams." *Radiation Protection Dosimetry* 18 (1):28–31.

American National Standards Institute. 1989a. "Occupational Airborne Radioactivity Monitoring Instrumentation". In *American National Standard N42.17B*. New York, NY: Institute of Electrical and Electronic Engineers.

American National Standards Institute. 1989b. "Performance Specifications for Health Physics Instrumentation—Portable Instrumentation for Use in Extreme Environmental Conditions". In *American National Standard N41.17C*. New York, NY: Institute of Electrical and Electronic Engineers.

American National Standards Institute. 1997. "Inspection, Test, Construction, and Performance Requirements for Direct Reading Electrostatic/Electroscope Type Dosimeters". In *American National Standard N322*. New York, NY: Institute of Electrical and Electronic Engineers.

American National Standards Institute. 2003a. "American National Standard Performance Criteria for Active Personnel Radiation Monitors". In *American National Standard N42.20*. New York, NY: Institute of Electrical and Electronic Engineers.

American National Standards Institute. 2003b. "American National Standard Performance Specifications for Health Physics Instrumentation-Portable Instrumentation for Use in Normal Environmental Conditions". In *N42.17A*. New York, NY: Institute of Electrical and Electronic Engineers.

American National Standards Institute. 2008. "American National Standard: Performance Testing of Extremity Dosimeters". In *American National Standard ANSI/HPS N13.32*. McLean, VA: Health Physics Society.

American National Standards Institute. 2011. "American National Standard: Radiation Safety for the Design and Operation of Particle Accelerators". In *ANSI/HPS N43.1*. McLean, VA: Health Physics Society.

American National Standards Institute. 2013. "American National Standard: Radiation Protection Instrumentation Test and Calibration, Portable Survey Instruments". In *ANSI N323AB*. New York, NY: Institute of Electrical and Electronic Engineers.

American National Standards Institute. 2015. "American National Standard: Personnel Dosimetry Performance—Criteria for Testing". In *ANSI N13.11*. McLean, VA: Health Physics Society.

Attix, Frank Herbert. 2008. *Introduction to Radiological Physics and Radiation Dosimetry*. New York, NY: John Wiley & Sons.

Böhm, J. 1986. *The National Primary Standard of the PTB for Realizing the Unit of the Absorbed Dose Rate to Tissue for Beta Radiation*. Braunschweig, Germany: PTB-Bericht.

Cassata R., J. M. Moscovitch, J. E. Rotunda, and K. J. Velbeck. 2002. "A New Paradigm in Personal Dosimetry Using Lif: Mg, Cu, P." *Radiation Protection Dosimetry* 101 (1–4):27–42.

Chung, K. S., H. S. Choe, J. I. Lee, J. L. Kim, and S. Y. Chang. 2005. "A Computer Program for the Deconvolution of Thermoluminescence Glow Curves." *Radiation Protection Dosimetry* 115 (1–4):343–349.

Galwey, Andrew K., and Michael E. Brown. 2002. "Application of the Arrhenius Equation to Solid State Kinetics: Can This Be Justified?" *Thermochimica Acta* 386 (1):91–98.

Horowitz, Y. S., S. Mahajna, L. Oster, Y. Weizman, D. Satinger, and D. Yossian. 1998. "The Unified Interaction Model Applied to the Gamma-Induced Supralinearity and Sensitisation of Peaks 4 and 5 in Lif: Mg, Ti (Tld-100)." *Radiation Protection Dosimetry* 78 (3):169–194.

Hui, T. E., R. M. Loesch, and J. C. McDonald. 1997. "The Second Internal Dosimetry Intercomparison Study of the Us Department of Energy." *Radiation Protection Dosimetry* 72 (2):131–138.

International Atomic Energy Agency. 1996. *International Basic Safety Standards for Protecting against Ionizing Radiation and for the Safety of Radiation Sources.* Vienna: International Atomic Energy Agency.

International Commission on Radiation Units and Measurements. 1979. *ICRU Report 31: Average Energy Required to Produce an Ion Pair.* Washington, DC: International Commission on Radiation Units and Measurements.

International Commission on Radiation Units and Measurements. 1983. *ICRU Report 36: Microdosimetry.* Bethesda, MD: International Commission on Radiation Units and Measurements.

International Commission on Radiation Units and Measurements. 1984. *ICRU Report 37: Stopping Powers for Electrons and Positrons.* Bethesda, MD: International Commission on Radiation Units and Measurements.

International Commission on Radiation Units and Measurements. 1985. *ICRU Report 39: Determination of Dose Equivalents Resulting from External Radiation Sources.* Bethesda, MD: International Commission on Radiation Units and Measurements.

International Commission on Radiation Units and Measurements. 1992. *ICRU Report 47: Measurement of Dose Equivalents from External Photon and Electron Radiations.* Bethesda, MD: International Commission on Radiation Units and Measurements.

International Commission on Radiation Units and Measurements. 2001. *ICRU Report 66: Determination of Operational Dose Equivalent Quantities for Neutrons.* Ashford, Kent, U.K.: Nuclear Technology Publishing.

International Commission on Radiation Units and Measurements. 2006. "ICRU Report 76: Measurement Quality Assurance for Ionizing Radiation Dosimetry." *Journal of the ICRU* 6 (2):1–50.

International Commission on Radiological Protection. 2007. ICRP Publication 103: The 2007 Recommendations of the International Commission on Radiological Protection. *Ann ICRP* 37:(2–4).

International Organization for Standardization. 1995. *Guide to the Expression of Uncertainty in Measurement.* Geneva: International Organization for Standardization.

International Organization for Standardization. 1996. "X and Gamma Reference Radiation for Calibrating Dosemeters and Doserate Meters and for Determining Their Response as a Function of Photon Energy—Part 1: Radiation Characteristics and Production Methods". In *ISO 4037 Part 1.* Geneva: International Organization for Standardization.

International Organization for Standardization. 1997. "X and Gamma Reference Radiation for Calibrating Dosemeters and Doserate Meters and for Determining Their Response as a Function of Photon Energy—Part 2. Dosimetry for Radiation Protection over the Energy Ranges from 8 keV to 1,3 MeV and 4 MeV to 9 MeV". In *ISO 4037 Part 2.* Geneva: International Organization for Standardization.

International Organization for Standardization. 1998. "Reference Neutron Radiations - Part 3: Calibration of Area and Personal Dosimeters and Determination of Response as a Function of Energy and Angle of Incidence". In *ISO 8529 Part 3.* Geneva: International Organization for Standardization.

International Organization for Standardization. 1999. "X and Gamma Reference Radiation for Calibrating Dosemeters and Doserate Meters and for Determining Their Response as a Function of Photon Energy—Part 3: Calibration of Area and Personal Dosemeters and the Measurement of Their Response as a Function of Energy and Angle of Incidence". In *ISO 4037 Part 3.* Geneva: International Organization for Standardization.

International Organization for Standardization. 2000. "Reference Neutron Radiations—Part 2: Calibration Fundamentals of Radiation Protection Devices Related to the Basic Quantities Characterizing the Radiation Field". In *ISO 8529 Part 2.* Geneva: International Organization for Standardization.

International Organization for Standardization. 2001. "Reference Neutron Radiations - Part 1: Characteristics and Methods of Production". In *ISO 8529 Part 1*. Geneva: International Organization for Standardization.

International Organization for Standardization. 2004a. "Reference Beta-Particle Radiation - Part 2 : Calibration Fundamentals Related to Basic Quantities Characterizing the Radiation Field". In *ISO 6980 Part 2*. Geneva: International Organization for Standardization.

International Organization for Standardization. 2004b. "X and Gamma Reference Radiation for Calibrating Dosemeters and Doserate Meters and for Determining Their Response as a Function of Photon Energy - Part 4: Calibration of Area and Personal Dosemeters in Low Energy X Reference Radiation Fields". In *ISO 4037 Part 4*. Geneva: International Organization for Standardization.

International Organization for Standardization. 2006a. "Reference Beta-Particle Radiation - Part 1: Methods of Production". In *ISO 6980 Part 1*. Geneva: International Organization for Standardization.

International Organization for Standardization. 2006b. "Reference Beta-Particle Radiation - Part 3: Calibration of Area and Personal Dosemeters and the Determination of Their Response as a Function of Beta Radiation Energy and Angle of Incidence". In *ISO 6980 Part 3*. Geneva: International Organization for Standardization.

International Organization for Standardization. 2016. "Calibration of Surface Contamination Monitors - Alpha-, Beta- and Photon Emitters". Geneva: International Organization for Standardization.

International Organization for Standardization, and International Electrotechnical Commission. 2004. "International Vocabulary of Basic and General Terms in Metrology (Vim)". In *ISO VIM (DGuide) 99999*. Geneva: International Organization for Standardization.

International Organization for Standardization, and International Electrotechnical Commission. 2005. "General Requirements for the Competence of Testing and Calibration Laboratories". In *ISO/IEC 17025*. Geneva: International Organization for Standardization.

Johnson, Michelle Lynn. 2009. *Radiation Protection Instrument Manual PNL-Ma-562*. Hanford, WA: Pacific Northwest National Laboratory.

Kallmann, Hartmut. 1950. "Scintillation Counting with Solutions." *Physical Review* 78 (5):621.

Kittel, C. 1991. *Introduction to Solid State Physics*. 8th edn. New York, NY: John Wiley and Sons.

Knoll, Glenn F. 2010. *Radiation Detection and Measurement*. New York, NY: John Wiley & Sons.

Korea Atomic Energy Research Institute. 2015. Nuclear Data Center, Cross Section Plotter—KAERI Nuclear Data Evaluation Laboratory (KAERI/NDEL) Nuclear Data Evaluation Laboratory PO Box 105 Yusong, Daejon 305–600, Republic of Korea.

McDonald, J. C., R. V. Griffith, P. A. Plato, and J. A. Miklos. 1984. "Measurement of Gamma Ray Dose from a Moderated 252cf Source." *Radiation Protection Dosimetry* 9 (2):113–118.

Miller, S. D., M. K. Murphy, K. L. Simmons, C. Yahnke, C. Yoder, B. Markey, M. Salasky, and A. M. Crabtree. 2008. "A High-Precision, Tissue-Equivalent Dosimeter for Nuclear Accident and Radiation Oncology Applications Based on Optically Stimulated Luminescence (Osl) in Al 2 O 3: C." *Radiation Measurements* 43 (2):875–878.

Nail, I., Y. S. Horowitz, L. Oster, and S. Biderman. 2002. "The Unified Interaction Model Applied to Lif: Mg, Ti (Tld-100): Properties of the Luminescent and Competitive Centers During Sensitisation." *Radiation Protection Dosimetry* 102 (4):295–304.

National Council on Radiation Protection and Measurements. 2007. *NCRP Report 158: Uncertainties in the Measurement and Dosimetry of External Radiation*. Bethesda, MD: National Council on Radiation Protection and Measurements.

National Institute of Standards and Technology. 1988. *Calibration of X-Ray and Gamma-Ray Measuring Instruments*. Gaithersburg, MD: National Institute of Standards and Technology.

National Institute of Standards and Technology. 1991. *Criteria for the Operation of Federally-Owned Secondary Calibration Laboratories*. Gaithersburg, MD: National Institute of Standards and Technology.

National Institute of Standards and Technology. 1994. *NIST Technical Note 1297: Guidelines for Evaluating and Expressing the Uncertainty of NIST Measurement Results*. Gaithersburg, MD: National Institute of Standards and Technology

National Institute of Standards and Technology. 2004. *Calibration Laboratories Technical Guide for Ionizing Radiation Measurements*. Gaithersburg, MD: National Institute of Standards and Technology.

National Institute of Standards and Technology. 2005. *NIST Handbook - 150 2005 Edition, National Voluntary Laboratory Accreditation Program, Procedures and General Requirements*. Gaithersburg, MD: National Institute of Standards and Technology.

Nunn, A. A., S. D. Davis, J. A. Micka, and L. A. DeWerd. 2008. "LiF: Mg, Ti TLD Response as a Function of Photon Energy for Moderately Filtered X-Ray Spectra in the Range of 20–250 kVp Relative to C60o." *Medical Physics* 35 (5):1859–1869.

Randall, J. T., and M. H. F. Wilkins. 1945. "Phosphorescence and Electron Traps I." *Proceedings of the Royal Society Series A* 184 (999):365–389.

Rathbone, B. A. 2010. *PNL-Ma-842 Hanford External Dosimetry Technical Basis Manual*. Richland, WA: Pacific Northwest National Laboratory.

Shonka, Francis R., John E. Rose, and G. Failla. 1958. "Conducting Plastic Equivalent to Tissue, Air, and Polystyrene". In *Second United Nations Conference on Peaceful Uses of Atomic Energy* 21. New York, NY: United Nations, 184.

Sims, C. S. 1989. "Nuclear Accident Dosimetry Intercomparison Studies." *Health Physics* 57 (3):439–448.

Stewart, F. A., A. V. Akleyev, M. Hauer-Jensen, J. H. Hendry, N. J. Kleiman, T. J. Macvittie, B. M. Aleman, A. B. Edgar, K. Mabuchi, and C. R. Muirhead. 2012. "ICRP Publication 118: ICRP Statement on Tissue Reactions and Early and Late Effects of Radiation in Normal Tissues and Organs–Threshold Doses for Tissue Reactions in a Radiation Protection Context." *Ann ICRP* 41 (1):1–322.

Sullivan, Anthony H. 1992. *A Guide to Radiation and Radioactivity Levels near High Energy Particle Accelerators*: Ashford, Kent, U.K.: Nuclear Technology Publishing.

Thomas, D. J., and V. E. Lewis. 1981. Standardisation of neutron fields produced by the H-3(d,n)He-4 reaction. *Nucl. Instrum. Meth.* 179, 397–404.

United States Department of Energy. 1987. *Department of Energy Standard for the Performance Testing of Personnel Dosimetry Systems*. Washington, DC: U.S. Department of Energy.

United States Department of Energy. 1998. *Occupational Radiation Protection*. Washington, DC: U.S. Department of Energy.

United States Department of Energy. 2011. *Laboratory Accreditation Program for External Dosimetry*. Washington, DC: U.S. Department of Energy.

United States Nuclear Regulatory Commission. 1997. *Nureg-1507: Minimum Detectable Concentrations with Typical Radiation Survey Instruments for Various Contaminants and Field Conditions*. Rockville, MD: Nuclear Regulatory Commission.

United States Nuclear Regulatory Commission. 2011. *10 CFR Part 20: Standards for Protection against Radiation*. Rockville, MD: Nuclear Regulatory Commission.

CHAPTER 5

Reference Individuals Defined for External and Internal Radiation Dosimetry

Wesley E. Bolch

CONTENTS

5.1 Definition and Purpose 170
    5.1.1 Technical Basis for the ICRP Reference Individual 170
    5.1.2 Historical Development 171
    5.1.3 Forms of Dose Assessment and the Appropriate Role of the Reference Individual 174
5.2 Anatomical Aspects of the ICRP Reference Individual 176
    5.2.1 Total-Body Measurements 176
    5.2.2 Individual Organ Systems 178
    5.2.3 Elemental Tissue Compositions and Mass Densities 184
    5.2.4 The Embryo, Fetus, and Pregnant Female 188
5.3 Computational Realizations of the ICRP Reference Individual Anatomy 191
    5.3.1 Stylized Computational Phantoms 192
    5.3.2 Voxel Computational Phantoms 192
    5.3.3 Hybrid Computational Phantoms 194
5.4 Physiological Aspects of the ICRP Reference Individual 196
    5.4.1 Daily Water Balance 196
    5.4.2 Respiratory Volumes and Capacities 196
    5.4.3 Time Budgets and Ventilation Rates 196
    5.4.4 Transit Times of Luminal Content in the Alimentary Tract 197
    5.4.5 Urinary and Fecal Excretion Rates 199
    5.4.6 Bone Remodeling Rates 200
    5.4.7 Physiology of the Developing Fetus and Mother 201
5.5 Comparison of ICRP Reference Data with That of Asian Populations 201

5.6   Use of the ICRP Reference Individual in External and Internal Dosimetry   204
    5.6.1   Dose Coefficients for Internal Occupational and Environmental
        Exposures   205
        OIR Part 2   206
        OIR Part 3   206
        OIR Part 4   206
        OIR Part 5   206
    5.6.2   Dose Coefficients for External Occupational Exposures   207
    5.6.3   Dose Coefficients for External Environmental Exposures   207
    5.6.4   Dose Coefficients for Medical Exposures   207
References   209

5.1 DEFINITION AND PURPOSE

5.1.1 Technical Basis for the ICRP Reference Individual

The computation of absorbed dose to organs and tissues in the human body following exposure to ionizing radiation requires a variety of data on the exposed individual. These data can be broadly categorized as either anatomic or physiologic in nature. Anatomic data include items such as the height and weight of the exposed individual, the masses of individual internal organs, and the elemental composition of these organ tissues (Bolch et al. 2016). Additional information will also need to be determined, such as the shape, depth, and relative position of individual organs, and, for organs where the target region for dose assessment is a radiosensitive cell layer, the location of that layer must also be identified. These items of information are key to the use of radiation transport models of various forms of ionizing radiation—photons, electrons, neutrons, and heavy charged particles—for the purpose of organ dose assessment from external sources (occupational, environmental, and medical exposures). In situations where the exposure is internal to the body—following inhalation, ingestion, or wound intake of radionuclides—additional data on the physiological characteristics of the exposed individual must also be known. These include an array of data including breathing rates, chemical dissolution and mechanical transfer rates in the respiratory tract airways, transfer rates of materials through differing segments of the alimentary tract, blood flow rates and percentage volume distribution in various organs and tissues, transfer rates from blood to organ tissues and from organ tissues back to blood, bone remodeling rates, and urinary and fecal excretion rates. These physiologic data allow the construction of intake and systemic biokinetic models for individual radionuclides (and their chemical forms) for the assessment of internal radiation dose in occupational and environmental exposure scenarios. For medical applications, many of these same data are required for organ dosimetry following the administration of radiopharmaceuticals.

In order to establish a consistent, reliable, and reproducible framework for radiological protection guidance worldwide, the International Commission on Radiological Protection (ICRP) has issued a number of reports following comprehensive reviews of relevant datasets on human anatomy and physiology. The end result of these reviews is the establishment of "reference values" for all anatomic and physiologic parameters needed for prospective dose assessment following internal or external radiation exposure. Anatomic reference values are used to construct 3D computational models of the 12 member family of ICRP Reference Individuals to include the male and female newborn, 1-year-old, 5-year-old, 10-year-old, 15-year-old, and the adult (nominally 35 years of age) (ICRP 1975, 2002). These models—which have taken differing computational forms, as discussed in Section 5.3 and Chapter 6—are used to transport radiation particles either externally incident upon the body or emitted internally within the body following radionuclide intake. Physiologic reference values are in turn used to construct reference intake and systemic biokinetic models of radionuclide distribution, retention, recirculation, and excretion (see Chapter 6). These ICRP reference anatomic, intake, and systemic biokinetic models are used in combination, along with radiation decay and emission data, to then establish ICRP reference external and internal dose coefficients—defined as the organ or effective dose per unit external exposure (fluence or air kerma), or the organ or effective dose per unit radionuclide activity intake, respectively (see Chapter 8). The entire ICRP System of Radiological Protection is thus built upon first defining Reference Parameter Values—both anatomic and physiologic—for its age and gender-dependent reference individuals.

5.1.2 Historical Development

Immediately following World War II, there was a tremendous expansion in the development of radioactive materials, radiation devices, and radiological systems for industry, commercial, and medical use. Consequently, the ICRP recognized the urgent need to establish permissible levels of radiation exposure and internal body burden to workers in these employment sectors. In so doing, there needed to be a consistent computational framework to make calculations of internal and external radiation dose. In September of 1949, representatives from the United States, the United Kingdom, and Canada held the Chalk River Conference on Permissible Dose at which consensus on the anatomic and physiological definition of Standard Man was initially established. First, data on organ masses were taken from the work of Cook (1948) and Lisco (1949). Second, values of the chemical composition of the total body and specific organ tissues were taken from the work of Hawk, Oser, and Summerson (1947), although it was agreed that more data collection was warranted, particularly with respect to tissue concentrations of trace elements. Third, patterns of food intake and urinary/fecal excretion, as well as the duration of occupation exposures, had to be established. The participants of the conference decided that these values would be based on average values for normal levels of exertion in temperate geographical locations. Other data were established regarding water balance, respiration rates, and retention of particulate matter in the lungs. These initial reference parameter values for Standard Man were subsequently modified at the 6th International Congress of Radiology (ICRP 1951), the Tripartite Conference on Permissible Dose (Tripartite Conference on Permissible Dose 1953), and the Seventh International Congress of Radiology (ICRP 1953). In 1955 and 1960, the ICRP issued its first two reports on permissible dose from internal radiation to workers in which the Standard Man model was applied (ICRP 1955, 1960).

In 1963, Committee 2 of the ICRP requested that the Commission establish a new Task Group for the revision of the Standard Man concept (ICRP 1975). In its establishment, the terminology was changed from Standard Man to Reference Man. The new Task Group was assigned three major tasks: (1) to revise and extend the definition of Reference Man which represented a typical radiation worker, (2) to report on the extent to which individual anatomic and physiologic parameters might differ from established ICRP reference parameter values, and (3) to expand the Reference Man concept to include variations of age and gender so that radiological protection guidance could be established for members of the general public. Once formed, the Task Group focused on the following tasks and objectives:

- The Task Group would limit its attention to those anatomic and physiologic data that are known or thought to be significant to internal and external dose assessment.

- The Task Group agreed that it was neither feasible nor necessary to specify Reference Man as representative of any given well-defined population group. Due to the limited nature of data sources worldwide, the Task Group defined Reference Man to be between 20 and 30 years, to weigh 70 kg, to be 170 cm in standing height, and to live in a climate with an average temperature of between 10^0 and 20^0 C. He is Caucasian and is Western European or North American in habitat and custom.

- The Task Group agreed that it was not feasible to define Reference Man as an "average" or "median" individual of any specified population group and that it was not necessary that he be defined in any such precise statistical sense. However, in many cases, reference parameter values were taken as average or median values from reported studies. Still, the emphasis of their final selections was on standardization and not statistical accuracy. In many cases, data were rounded to only two significant figures, and expert opinion and subjective interpretation of the data were applied.

- The Task Group attempted to provide the reported range of parameter values in the studied population from which the Reference Parameter Value was selected or established.

- The Task Group attempted to clearly distinguish its adoption of a Reference Parameter Value from other data sources in the literature. The intent was to provide the reader with an understanding of the basis for Reference Parameter Value selection, of the variability in primary and secondary data sources, and in some cases, how little relevant data were available to the Task Group for it to make its selection.

The final report of the Task Group on Reference Man appeared in 1975 as ICRP Publication 23 (ICRP 1975). This report—totaling 480 pages in length—was divided into three major sections and two appendices. Section I defined anatomic values of Reference Man, and itself was divided into 13 chapters: the total body, the integumentary system, the skeletal system, hematopoietic and lymphatic systems, the skeletal muscles, the cardiovascular system, the digestive system, the respiratory system, the urogenital system, the endocrine system, the central nervous system, special sense organs, and pregnancy.

Section II then presented gross and elemental content of tissues and organs of Reference Man. Finally, Section III provided physiologic data for Reference Man, including the daily balance of some 51 naturally occurring elements. Appendix I provided the first systematic tabulation of photon-specific absorbed fractions (defined as the amount of photon energy per unit mass deposited in a target organ as a fraction of the photon energy emitted in a source organ). These specific absorbed fraction (SAF) values were derived from work of the Medical Internal Radiation Dose (MIRD) Committee of the Society of Nuclear Medicine in its early version of a computational stylized phantom representing the ICRP Reference Man (Snyder et al. 1969, 1974). The Chair of the ICRP Task on Reference Man was Dr Walter S. Snyder of Oak Ridge National Laboratory, who was also Chair of the MIRD Committee at this time. Appendix II of ICRP Publication 23 defined the various symbols and acronyms used throughout the report. Supplemental to ICRP Publication 23 were early intake models of inhalation and ingestion which were to be later formally published in ICRP Publication 30 (ICRP 1980). The bases for these early models were published in articles appearing in the February 1966 issue of the journal *Health Physics* (Eve 1966; Dolphin and Eve 1966; Bates et al. 1966). While ICRP Publication 23 was written to support the standardization of external and internal dosimetry for radiological protection, the document was monumental in its scope and detail and, since its publication, has been used widely by many other scientific professionals—medicine, toxicology, and industrial hygiene, to name just a few. In 1984, however, ICRP Committee 2 launched an effort to revise and update ICRP Publication 23 for two main reasons. First, additional information had been published regarding radionuclide biokinetics and dosimetry since the initial review by the Task Group on Reference Man in the early 1970s. Second, regulatory bodies were placing an increased emphasis on radiological protection standards for members of the general public, and thus more information was needed to better characterize—through reference parameter values—the non-adult members of the ICRP Reference Individual series.

While the original intent of the new Task Group on Reference Man was to fully update ICRP Publication 23 (ICRP 2002), funding and manpower issues led to an alternative approach in which key revisions would appear in different ICRP Publications supplemented by peer-reviewed journal articles by task group members. The final effort to revise Publication 23 led to the following series of papers (Leggett and Williams 1991, 1995) and ICRP publications (ICRP 1994a, 1995b, 2001, 2002, 2006):

Leggett, R.W. and Williams, L.R. 1991. "Suggested reference values for regional blood volumes in humans." *Health Phys.* 60 (2): 139–154 (1991).

Leggett, R.W. and Williams, L.R. 1995. "A proposed blood circulation model for reference man." *Health Phys.* 69 (2): 187–201.

ICRP Publication 66 "Human Respiratory Tract Model for Radiological Protection" (1994)

ICRP Publication 70 "Basic Anatomical and Physiological Data for Use in Radiological Protection: The Skeleton" (1995)

ICRP Publication 88 "Doses to the Embryo and Fetus from Intakes of Radionuclides by the Mother" (2001)

ICRP Publication 89 "Basic Anatomical and Physiological Data for Use in Radiological Protection: Reference Values" (2002)

ICRP Publication 100 "Human Alimentary Tract Model for Radiological Protection" (2006)

The organization of ICRP Publication 89 is as follows. Following an opening Chapter 1 defining the basis of ICRP reference parameter values, Chapter 2 provides a convenient summary table of both prenatal and postnatal reference parameter values, each divided into sections on anatomic and then physiologic data. Chapter 3 is devoted to the embryo and fetus, while Chapter 4 focuses on total-body data including data on anatomy, body composition, and physiology. Chapter 5 to 11 follow a layout similar to the organization of the original Publication 23 report, where anatomic and physiologic data are given according to individual organ systems. Chapter 12 provides data for the adult pregnant female, while elemental composition of the body tissues is provided in Chapter 13. It should be noted that many researchers developing computational anatomic phantoms of the ICRP Reference Individuals have also made use of International Commission on Radiation Units and Measurements (ICRU) Report 46 (ICRU 1992) which also provided organ-specific and, when available, age-specific elemental compositions of internal organs and tissues.

Another important issue regarding ICRP Reference Parameter Values is that they are defined by the Commission for the purpose of standardization of dosimetric and biokinetic models used to compute organ and effective dose coefficients for prospective radiological protection. As a result, they are fixed and not subject to uncertainty. This issue is specifically noted in the Commission's most recent recommendation document—ICRP Publication 103—in para.166 (ICRP 2007):

> (166) The Commission is aware of the uncertainty or lack of precision in radiation dose models and efforts are undertaken to critically evaluate and to reduce them wherever possible. For regulatory purposes, the dosimetric models and parameter values that the Commission recommends are reference values. These are fixed by convention and are therefore not subject to uncertainty. Equally the Commission considers that the biokinetic and dosimetric models which are needed for the purpose of dose assessment are defined as reference data and, therefore, are also fixed and not applied with an uncertainty. These models and values are re-evaluated periodically and may be changed by ICRP on the basis of such evaluations when new scientific data and information are available.

5.1.3 Forms of Dose Assessment and the Appropriate Role of the Reference Individual

The assessment of organ or effective dose from either external or internal radiation exposure may be broadly placed into four categories. These are defined based upon (1) whether the dose assessment is to be made to a "specific individual" or to an "unspecified individual," and (2) whether the dose assessment is to be made "retrospectively" or "prospectively." Table 5.1 provides specific examples and characteristics of each dose assessment category.

The ICRP Reference Individual finds its greatest application for either retrospective or prospective dose assessment to unspecified individuals. These would include hypothetical individuals exposed either in the past or in the future, respectively, where organ and effective doses are to be compared to dose constraints, diagnostic reference levels, and/or dose limits for monitoring adherence to radiological protection guidance and/or regulations. To facilitate these dose estimates, ICRP provides reference dose coefficients—both for external radiation fields and for internal intakes of radionuclides—which are based on ICRP-defined reference anatomic models, reference intake models, and reference systemic biokinetic models. They may apply to either males or females, and one of the six postnatal reference ages as defined previously.

TABLE 5.1   Characteristics and Examples of Retrospective and Prospective Dose Assessments for Specific or Unspecified Individuals

Dose Assessment	Specific Individual	Unspecified Individual
Retrospective	Radionuclide intake/external exposure has already occurred Known, real individual exposed or contaminated Either personal, physical, or physiological information available (e.g., bioassay data or dosimeter badge reading) Possible exposure setting: • Worker with positive urine bioassay data • Worker with a high dosimetry badge reading • Real individual, member of the public, cohort member in an epidemiological study (if detailed and individualized data are available)	Radionuclide intake/external exposure has already occurred Individual from a reference category: • Individual represents any person in the specified category • No personal information available Possible exposure setting: • Child in a city exposed to last year's releases from a nuclear facility—external exposure or radionuclide intake • Hypothetical adult member of the population exposed to past releases • Cohort members in epidemiological studies where individual data are limited
Prospective	Radionuclide intake or external exposure is expected to occur Known, real individual to be exposed Either personal, physical, or physiological information available (e.g., bioassay data or dosimeter badge reading) Possible exposure setting: • Treatment planning for a real patient • Planned exposure for a given worker	Radionuclide intake or external exposure may occur in the future Individual from a reference category: • Individual represents any person in the specified category • No personal information available Possible exposure setting: • Hypothetical farmer near a future nuclear facility • Hypothetical male worker at a future nuclear facility • Hypothetical pregnant woman exposed to a unit intake

Source: Table 3.1, *NCRP Report 164* (National Council on Radiation Protection and Measurements 2009).

When the intended object of the dose assessment, however, is a specific individual, the dose assessment may permit the incorporation of unique information on this real exposed individual. If the exposure (real or predicted) leads to significantly high organ doses thought to exceed tissue reaction thresholds, medical interventions might be warranted or should be planned. To make these decisions, medical personnel will want dose estimates, not as approximated to the ICRP Reference Individual, but as they occurred, or will occur, in the specific individual. In so doing, dose estimates may employ non-reference computational phantoms with body heights and weights that differ from ICRP Reference Individual. These may include voxel phantoms constructed directly from individual-specific radiological images. For internal exposures, parameter values in the various ICRP reference intake and systemic biokinetic models may be altered as individual-specific knowledge of their values in the real exposed individual is known or discovered.

If, however, the prospective or retrospective dose assessment is made to the specific individual for the purpose of verifying compliance with radiological protection guidance and dose limits on effective dose, then these individualized adjustments to ICRP reference models and data are much more restricted. In fact, they can only be altered to better assess estimates of the particle fluence/air kerma for external exposures, or the radionuclide intake for internal exposures. Once these unique exposures metrics are established for the real individual, computation of effective dose should only be made using ICRP reference dose coefficients. This restriction is based upon the understanding that the effective dose—for retrospective or prospective radiological protection—is a dose quantity belonging only to the ICRP Reference Individual, not the real exposed individual. In the case of an inhalation exposure of ^{137}Cs, for example, air concentration measurements in the real environment, and values of the breathing rate for the real exposed individual, could be used to determine a best estimate of the worker's inhalation intake (in Bq). The next step in the dose assessment, however, is to assume the Reference Individual, not the real worker, inhaled that same amount of radioactivity. ICRP reference dose coefficients (effective dose per unit intake) are then used to compute the effective dose to the Reference Individual for comparison to regulatory guidance or dose limits.

5.2 ANATOMICAL ASPECTS OF THE ICRP REFERENCE INDIVIDUAL

In this section, we summarize the major elements of the anatomic aspects of the ICRP Reference Individual as defined in ICRP Publication 89 (ICRP 2002). Data on total-body measurements, individual organ systems, and elemental tissue compositions are given here, with the focus on identifying the major literature sources upon which the Task Group on Reference Man made its final decisions on reference parameter values. Data on the embryo and fetus are also reviewed briefly.

5.2.1 Total-Body Measurements

Reference parameter values formally defining the total-body characteristics of ICRP Reference Individuals include: standing height, total-body mass, body surface area, lean body mass, and mass of body fat. Table 5.2 summarizes the body heights, masses, and surface areas for each of the 12 ICRP Reference Individuals. Reference values for height

TABLE 5.2 Reference Values for Height, Mass, and Surface Area of the Total Body

Age	Height (cm)		Mass (kg)		Surface Area (m²)	
	Male	Female	Male	Female	Male	Female
Newborn	51	51	3.5	3.5	0.24	0.24
1 year	76	76	10	10	0.48	0.48
5 years	109	109	19	19	0.78	0.78
10 years	138	138	32	32	1.12	1.12
15 years	167	161	56	53	1.62	1.55
Adult	176	163	73	60	1.90	1.66

Source: Table 2.9, "ICRP Publication 89" (ICRP 2002).

were selected as central estimates for European populations addressed in Eveleth and Tanner (1976, 1990). Values for adult males and females are based on data for 18-year-old males and females, respectively, noting that longitudinal studies indicate that maximal heights are attained by age 18 years. Reference values of body mass are also based on European data on body growth, together with consideration of increasing in total-body mass after maximum heights are attained. The reference value for body mass of the ICRP Reference Adult Male (73 kg) is 10% greater than central estimates of body mass for 18-year-old European males. The reference value for body mass of the ICRP Reference Adult Female (60 kg) is 10% greater than central estimates of body mass for 16-year-old European females. Reference values for body surface area (in m²) are given by the following expression:

$$SA = \alpha_0 H^{\alpha_1} M^{\alpha_2} \tag{5.1}$$

where $\alpha_0 = 0.0235$, $\alpha_1 = 0.42246$, and $\alpha_2 = 0.15456$, with height in cm, and mass in kg. The values were derived by Gehan and George (1970) from measurements on 401 subjects with surface areas ranging from 0.11 to 2 m².

Total-body mass can be divided into two compartments—lean body mass and fat. Lean body mass (LBM) includes stroma of adipose tissue and structural lipids in cells such as membrane lipids. Since neutral fat does not bind water or electrolytes, all of the body's water and electrolytes are usually assigned to the LBM compartment. Components of the total body that are almost entirely contained in the LBM, such as water and potassium, are often expressed as a fraction of LBM rather than of total-body mass. The ICRP reference value for the water content of LBM in adult males and females was set at 73% (published range was 71% to 74%). Body fat and thus total-body mass are much more variable than LBM.

The LBM of the ICRP Reference Individuals can thus be determined as the difference between reference total-body mass and reference of body fat, which are summarized below in Table 5.3. Body fat corresponds to two essential entities—"essential" fat and "nonessential" fat, where the former are the constituents of cells and are roughly 2% of LBM. Non-essential fat is composed of closely packed fat cells in a loose connective tissue called adipose tissue and is found mainly in the subcutaneous tissue layer.

TABLE 5.3   Reference Values for the Mass of Body Fat[a]

Age	Mass (g)	
	Male	**Female**
Newborn	370	370
1 year	2300	2300
5 years	3600	3600
10 years	6000	6000
15 years	9000	14 000
Adult	14 600	18 000

Source: "ICRP Publication 89," p.76 (ICRP 2002).
[a] Excludes essential body fat. Includes interstitial fat and yellow bone marrow.

5.2.2 Individual Organ Systems

A major task of ICRP Publication 89 (ICRP 2002) needed for both the construction of anatomic computational models of the ICRP Reference Individuals and the development of radionuclide-specific biokinetic models was the establishment of reference values for organ mass. These are summarized as a function of age and gender in Table 5.4. Separate data columns are provided for the ICRP Reference Adult Male and Adult Female, as well as the ICRP Reference 15-year-old Male and 15-year-old Female. With the exception of the reference masses for the sex-specific organs (testes, ovaries, uterus, and prostate), and of the urethra, and the 5-year-old and 10-year-old's brains, all values at ages below 15 years are the same for both the reference male and reference female.

Values of reference organ mass are taken primarily from collective review of autopsy studies in both ICRP Publications 23 (ICRP 1975) and 89 (ICRP 2002). As such, there is always the question of the degree to which the blood content of the measured organ samples was still present when mass data were collected. This issue is significant in the correct interpretation of ICRP reference organ masses—are they to be considered exclusive or inclusive of blood, where the former would represent their in vivo mass and the latter only the organ parenchyma? The authors of ICRP Publication 89 (ICRP 2002) considered the issue of blood inclusion in their report. The published reference organ masses (given here in Table 5.4) are the organ masses exclusive of blood. The issue of the proper interpretation of ICRP reference organ masses has complicated the construction of computational phantoms of the ICRP Reference Individuals, whereas the opposite (and incorrect) interpretation had been taken by the phantom modelers (see related discussion in Section 5.3).

In some cases, ICRP reference masses are assigned based upon assumptions of organ mass as a percentage of total-body mass. For example, reference masses of the tongue are based upon the assumption that this tissue represents 0.1% of total-body mass.

In comparing ICRP reference organ masses between those assigned in ICRP Publication 23 (ICRP 1975) and those assigned in ICRP Publication 89 (ICRP 2002), it is noted that some values have remained unchanged, whereas other reference values have been revised. For example, the references values given in ICRP Publication 89 for the adult male and

TABLE 5.4   Reference Values for Masses of Organs and Tissues for the ICRP Reference Individuals

Organ/Tissue	Newborn	1 Year	5 Years	10 Years	15 Years M	15 Years F	Adult M	Adult F
Adipose tissue[a]	930	3800	5500	8600	12 000	18 700	18 200	22 500
Separable adipose tissue, excluding yellow marrow	890	3600	5000	7500	9500	16 000	14 500	19 000
Adrenals (2)	6	4	5	7	10	9	14	13
Alimentary system								
Tongue	3.5	10	19	32	56	53	73	60
Salivary glands	6	24	34	44	68	65	85	70
Esophagus								
Wall	2	5	10	18	30	30	40	35
Stomach								
Wall	7	20	50	85	120	120	150	140
Contents	40	67	83	117	200	200	250	230
Small intestine								
Wall	30	85	220	370	520	520	650	600
Contents	56	93	117	163	280	280	350	280
Large intestine								
Right colon								
Wall	7	20	49	85	122	122	150	145
Contents	24	40	50	70	120	120	150	160
Left colon								
Wall	7	20	49	85	122	122	150	145
Contents	12	20	25	35	60	60	75	80
Rectosigmoid								
Wall	3	10	22	40	56	56	70	70
Contents	12	20	25	35	60	60	75	80
Liver	130	330	570	830	1300	1300	1800	1400
Gallbladder								
Wall	0.5	1.4	2.6	4.4	7.7	7.3	10	8
Contents	2.8	8	15	26	45	42	58	48
Pancreas	6	20	35	60	110	100	140	120
Brain	380	950	1310/1180	1400/1220	1420	1300	1450	1300
Breasts	–	–	–	–	15	250	25	500
Circulatory system								
Heart—with blood[a]	46	98	220	370	660	540	840	620
Heart—tissue only	20	50	85	140	230	220	330	250
Blood	290	530	1500	2500	4800	3500	5600	4100
Eyes	6	7	11	12	13	13	15	15
Fat (storage fat)[a]	370	2300	3600	6000	9000	14 000	14 600	18 000
Integumentary system								
Skin	175	350	570	820	2000	1700	3300	2300
Muscle, skeletal	800	1900	5600	11 000	24 000	17 000	29 000	17 500
Pituitary gland	0.1	0.15	0.25	0.35	0.5	0.5	0.6	0.6

(Continued)

TABLE 5.4 (CONTINUED)   Reference Values for Masses of Organs and Tissues for the ICRP Reference Individuals

Organ/Tissue	Newborn	1 Year	5 Years	10 Years	15 Years M	15 Years F	Adult M	Adult F
Respiratory system								
Larynx	1.3	4	7	12	22	15	28	19
Trachea	0.5	1.5	2.5	4.5	7.5	6	10	8
Lung—with blood[a]	60	150	300	500	900	750	1200	950
Lung—tissue only	30	80	125	210	330	290	500	420
Skeletal system								
Total skeleton[a]	370	1170	2430	4500	7950	7180	10 500	7800
Bone, cortical	135	470	1010	1840	3240	2960	4400	3200
Bone, trabecular	35	120	250	460	810	740	1100	800
Bone, total[a]	170	590	1260	2300	4050	3700	5500	4000
Marrow, active	50	150	340	630	1080	1000	1170	900
Marrow, inactive	0	20	160	630	1480	1380	2480	1800
Cartilage	130	360	600	820	1140	920	1100	900
Teeth	0.7	5	15	30	45	35	50	40
Miscellaneous	20	45	55	90	155	145	200	160
Spleen	9.5	29	50	80	130	130	150	130
Thymus	13	30	30	40/35	35	30	25	20
Thyroid	1.3	1.8	3.4	7.9	12	12	20	17
Tonsils (2 palatine)	0.1	0.5	2	3	3	3	3	3
Urogenital system								
Kidneys (2)	25	70	110	180	250	240	310	275
Ureters (2)	0.77	2.2	4.2	7.0	12	12	16	15
Urinary bladder	4	9	16	25	40	35	50	40
Urethra	0.48/0.14[b]	1.4/0.42	2.6/0.78	4.4/1.3	7.7	2.3	10	3
Testes (2)	0.85	1.5	1.7	2	16	–	35	–
Epididymes (2)	0.25	0.35	0.45	0.60	1.6	–	4	–
Prostate	0.8	1.0	1.2	1.6	4.3	–	17	–
Ovaries (2)	0.3	0.8	2.0	3.5	–	6	–	11
Fallopian tubes (2)	0.25	0.25	0.35	0.50	–	1.1	–	2.1
Uterus	4.0	1.5	3	4	–	30	–	80
Total body (kg)[c]	3.5	10	19	32	56	53	73	60

Source: Table 2.8, "ICRP Publication 89" (ICRP 2002).

[a] This entry duplicates other mass information in this table and should not be included in the whole-body sum of reference values for tissue masses.

[b] Male (M)/female (F) values.

[c] The body components listed above represent 96% of the total-body mass. Separable connective tissues and certain lymphatic tissues account for most of the remaining 4% of body mass.

female liver (1800 g and 1400 g, respectively), the adult male and female thyroid (20 g and 17 g, respectively), and the adult male and female kidneys (310 g and 275 g, respectively) remained unchanged from their previous values given in ICRP Publication 23. It is noted that pediatric reference organ masses were not explicitly reported in Publication 23.

Examples of changes in reference organ masses between these two documents include those of the adult brain, lungs, and breasts. Brain masses increased from 1400 g (adult male) and 1200 g (adult female) in Publication 23 to values of 1450 g (adult male) and 1300 g (adult female) in Publication 89. Lung masses (exclusive of its blood content) increased from 470 g (adult male) and 385 g (adult female) in Publication 23 to values of 500 g (adult male) and 420 g (adult female). Additionally, breast masses decreased slightly from 26 g to 25 g for the Reference Adult Male between the two documents, but increased from 360 g to 500 g for the Reference Adult Female.

One other important change in the definition of the ICRP Reference Adult Male and Adult Female between Publications 23 and 89 is the assignment of reference masses for bone marrow. In ICRP Publication 23, reference masses for total bone marrow were assigned as 3000 g in the adult male and 2600 g in the adult female. Each of these masses was further partitioned 50/50 into red bone marrow (now called active marrow) and yellow bone marrow (now called inactive marrow). Thus, there were 1500 g and 1300 g of active marrow assigned to the Reference Adult Male and Reference Adult Female, respectively. In Publication 89, references masses of active and inactive bone marrow for the Reference Adult Male are assigned as 1170 g and 2480 g, respectively (3650 g in total), and for the Reference Adult Female are assigned as 900 g and 1800 g, respectively (2700 g in total). These changes reflect, in part, a better understanding of the ratio of active to total bone marrow (marrow cellularity or marrow cellularity factor, CF) in different bones of the skeleton. Again, ICRP Publication 89 provides explicit reference values for marrow masses as a function of age and gender, whereas only adult values were reported in ICRP Publication 23.

A good example of the importance of considering the explicitly defined characteristics of the ICRP Reference Adult in the assignment of reference organ masses is shown below in Figure 5.1 for the liver. As noted previously, ICRP defines its Reference Individuals as Western European or North American and between 20 and 30 years of age. This definition thus conforms with the assignment of 1800 and 1400 g for the reference adult male and female liver, respectively (upper curves below and at the age of peak liver mass). The adult age-dependence and population-specific variations are thus not accounted for in these assignments of single adult male and adult females values of liver mass.

One of the major advances given in Publication 89 over Publication 23 was the more definitive assignment of the distribution of total blood volume within the various tissues and organs of the ICRP reference adults as shown in Table 5.5. When developing biokinetic models for different radionuclides, especially those with radiological half-lives comparable to or shorter than radionuclide residence times in the body, the circulating blood should be considered as a unique and explicit source tissue. These reference values are assigned by ICRP based upon the extensive reviews and analysis of the literature by Williams and Leggett (1989), and by Leggett and Williams (1991, 1995). It is noted that pediatric values of blood distribution were not reported in ICRP Publication 89, but are—at the time of this writing—being considered by ICRP Committee 2. (Wayson et al. 2018).

Table 5.6 reports reference values of the lengths of the alimentary tract organs. These values reported in ICRP Publication 89 were later adopted within ICRP Publication 100 defining the new ICRP Human Alimentary Tract Model (HATM) (ICRP 2006). In

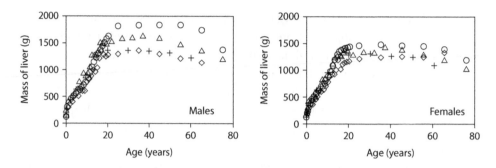

FIGURE 5.1 Mass of the liver: (O) Western data, (Δ) Japanese data, (+) Chinese data, and (◇) Indian data.

TABLE 5.5 Reference Values for Regional Blood Volumes and Blood Flow Rates in Adults

Organ or Tissue	Blood Content (% Total Blood Volume)		Blood Flow Rate (% Cardiac Output)	
	Male	Female	Male	Female
Fat	5.0	8.5	5.0	8.5
Brain	1.2	1.2	12	12
Stomach and esophagus	1.0	1.0	1.0	1.0
Small intestine	3.8	3.8	10	11
Large intestine	2.2	2.2	4.0	5.0
Right heart	4.5	4.5	–	–
Left heart	4.5	4.5	–	–
Coronary tissue	1.0	1.0	4.0	5.0
Kidneys	2.0	2.0	19	17
Liver	10	10	6.5 (arterial)	6.5 (arterial)
			25.5 (total)	27.0 (total)
Pulmonary	10.5	10.5	–	–
Bronchial tissue	2.0	2.0	2.5	2.5
Skeletal muscle	14	10.5	17	12
Pancreas	0.6	0.6	1.0	1.0
Skeleton	7.0	7.0	5.0	5.0
Red marrow	4.0	4.0	3.0	3.0
Trabecular bone	1.2	1.2	0.9	0.9
Cortical bone	0.8	0.8	0.6	0.6
Other skeleton	1.0	1.0	0.5	0.5
Skin	3.0	3.0	5.0	5.0
Spleen	1.4	1.4	3.0	3.0
Thyroid	0.06	0.06	1.5	1.5
Lymph nodes	0.2	0.2	1.7	1.7
Gonads	0.04	0.02	0.05	0.02
Adrenals	0.06	0.06	0.3	0.3
Urinary bladder	0.02	0.02	0.06	0.06
All other tissues	1.92	1.92	1.39	1.92
Aorta and large arteries	6.0	6.0	–	–
Large veins	18	18	–	–

Source: "ICRP Publication 89," p. 21 (ICRP 2002).

TABLE 5.6   Reference Values for the Lengths of Alimentary Tract Segments

					15 Years		Adult	
	Newborn	1 Year	5 Years	10 Years	Male	Female	Male	Female
Esophagus	10	13	18	23	27	26	28	26
Small intestine	80	120	170	220	270	260	280	260
Large intestine								
Right colon	14	18	23	28	35	30	34	30
Left colon	16	21	26	31	35	35	38	35
Rectosigmoid	15	21	26	31	35	35	38	35
Large intestine Total	45	60	75	90	100	100	110	100

Source: Table 2.11, "ICRP Publication 89" (ICRP 2002).

previous ICRP publications, the large intestine was subdivided into four sections—ascending colon, transverse colon, descending colon, and sigmoid colon. In both Publications 89 and 100, these divisions were revised, primarily to better conform with data on material transit times, into the right colon (ascending and half of the transverse colon), left colon (other half of the transverse colon and descending colon), and rectosigmoid colon (sigmoid colon with the inclusion of the rectum). In earlier stylized models of the ICRP Reference Individuals, crude geometric models of the colon were constructed which did not fully conform to these reported reference lengths. More importantly, no attempt was made in the era of stylized phantoms to model the small intestines, where in lieu of an anatomically realistic model, a simplistic tissue-filled cuboid was placed to represent both small intestine wall and lumen contents. Newer models of human anatomy—both voxel and hybrid (as discussed later in Section 5.3)—make explicit use of these reference lengths in constructing anatomical 3D models of these reference individuals.

Reference parameter values pertaining to the skeletal tissues are summarized in Tables 5.7 through 5.10. It is of particular note that in Tables 5.7 and 5.8—which give the ratio of bone mineral into its cortical and trabecular divisions, and values of surface-to-volume ratios, respectively—are reported only for the reference adults. The age dependence of these values in the pediatric members of the ICRP Reference Individuals still need to be fully defined and reported. In the absence of more data, ICRP typically applies these adult reference parameters when working with its pediatric reference individuals. Table 5.9 reports the age dependence of the mass density of cortical bone. Computational phantom modelers typically assume that these densities equally apply to the bone trabecular of skeletal spongiosa. It is noted that the age-dependent densities given in Table 5.9 from ICRP Publication 23 differ slightly from those reported in ICRU Report 46 as shown below. A summary of all reference skeletal tissue masses by both age and gender is shown in Table 5.10.

Other tissues that are defined in ICRP Publication 89, and are of great importance to constructing 3D anatomical phantoms of these reference individuals, particularly when considering external radiation exposures, are the skin and eye lens. Table 5.11 reports reference values for the mass of the total skin, as well as its division into epidermis and

TABLE 5.7   Reference Values for the Division of Bone Mass in the Adult Male or Female

Compact bone	80%
Trabecular bone	20%

Source: Table 2.18, "ICRP Publication 89" (ICRP 2002).

TABLE 5.8   Reference Values for the Volume and Surface Area of Bone in the Adult Male

Volume of bone tissues (i.e., inside the periosteal envelope and outside the endosteal envelope)	
All bone tissue	2710 cm³
Cortical bone	2130 cm³
Trabecular bone	580 cm³
Surface:volume ratio	
Cortical bone	3 mm²/mm³ (30 cm²/cm³)
Trabecular bone	18 mm²/mm³ (180 cm²/cm³)
Total surface area	
All bone	17 m²
Cortical bone	6.5 m²
Trabecular bone	10.5 m²

Source: Table 2.19, "ICRP Publication 89" (ICRP 2002).

dermis. Reference values for the mass of the skin are based on estimates of the skin thickness and body surface area. The derived values are reasonably consistent with central estimates based on reported values that exclude the hypodermis.

These values, along with reported values of reference skin thickness (Lee et al. 2010), have been used to define anatomic models of the skin in the ICRP reference phantoms.

Another important tissue to be modeled regarding external exposures is the eye lens. Recent radiation epidemiological data suggest a significantly lower dose threshold for the induction of radiogenic cataracts (ICRP 2012a). Consequently, detailed models of the eye have been constructed for the development of dose coefficients for eye lens dose. Many of these eye models are based on the ocular tissue dimensions reported by Charles and Brown (1975). The reference data shown below for eye lens depth and size are taken from this study and are consistent with what was reported previously in ICRP Publication 23 (Table 5.12).

5.2.3 Elemental Tissue Compositions and Mass Densities

While reference masses for internal organs are needed to construct volumetric anatomic computational models of the ICRP Reference Individuals, equally important is knowledge of the elemental composition of these tissues when these computational phantoms are used for radiation transport simulation. ICRP and ICRU have established reference elemental compositions for all tissues of the body through a two-step

TABLE 5.9   Reference Values for the Density of Skeletal Components

	Densities of Bone (g/cm³)	
	ICRP Publication 89	**ICRU Report 46**
Whole skeleton, adults	1.3	–
Dry, mineralized collagenous bone matrix, adults	2.3	–
Hydrated cortical bone		
Newborn	1.65	1.72
1 year	1.66	1.71
5 years	1.70	1.75
10 years	1.75	1.79
15 years	1.80	1.83
Adult	1.90	1.92

Source: Table 2.20, "ICRP Publication 89" (ICRP 2002).

TABLE 5.10   Reference Values for the Masses of Skeletal Tissues and Skeletal Calcium (g)

	Newborn	1 Year	5 Years	10 Years	15 Years Male	15 Years Female	Adult Male	Adult Female
Total skeleton	370	1170	2340	4500	7950	7180	10 500	7800
Bone	170	590	1260	2300	4050	3700	5500	4000
Active marrow	50	150	340	630	1080	1000	1170	900
Inactive marrow	0	20	160	630	1480	1380	2480	1800
Cartilage	130	360	600	820	1140	920	1100	900
Miscellaneous	20	50	70	120	200	180	250	200
Skeletal calcium	28	100	240	460	830	760	1180	860

Source: Table 2.24, "ICRP Publication 89" (ICRP 2002).

[a] As defined here, the skeleton does not include periarticular tissue or blood, but does include teeth, periosteum, and blood vessels (masses included in "miscellaneous").

TABLE 5.11   Reference Values for the Mass of Epidermis, Dermis, and Total Skin (g)

	Males			Females		
Age	**Epidermis**	**Dermis**	**Total Skin**	**Epidermis**	**Dermis**	**Total Skin**
Newborn	12	163	175	12	163	175
1 year	24	326	350	24	326	350
5 years	39	531	570	39	531	570
10 years	56	764	820	56	764	820
15 years	100	1900	2000	80	1620	1700
Adult	120	3180	3300	85	2215	2300

Source: Table 2.27, "ICRP Publication 89" (ICRP 2002).

process. The first step is to define reference elemental compositions for the elements H, C, N, O, P, and S in the major body tissue constituents—water, fat, protein, carbohydrates, and bone ash, where the latter additionally includes its Ca content as shown in Table 5.13. When next considering individual body organs and tissues, the percentage mass distribution of these five tissue constituents are assigned via literature review of

TABLE 5.12   Reference Values for Eye Lens Depth and Size in Adult Males and Females

	Lens Depth and Size (cm)
Anterior aspect of lens to anterior pole of cornea	0.3–0.4
Anterior aspect of lens to anterior aspect of closed lid	0.8
Equator of lens to anterior of corneal border	0.3
Equatorial diameter of lens	0.9
Axial thickness of lens	0.4

Source: Table 2.28, "ICRP Publication 89" (ICRP 2002).

TABLE 5.13   Elemental Composition of Body Tissue Constituents

Component	H	C	N	O	P	S	Ca
Water	11.2	–	–	88.8	–	–	–
Fat	11.8	77.3	–	10.9	–	–	–
Protein	6.6	53.4	17.0	22.0	–	1.0	–
Carbohydrate	6.2	44.5	–	49.3	–	–	–
Bone ash	0.2	–	–	41.4	18.5	–	39.9

Source: Table 13.1, "ICRP Publication 89" (ICRP 2002).

biochemical analyses of autopsy specimens. For example, ICRP Publication 23 (ICRP 1975) defines the normal reference adult liver as composed of 71% water, 18% protein, 7% lipid, and 2% glycogen (carbohydrate). Some 1.3% of the adult liver includes other trace elements such as Cl and K. These values are normalized to 100% and then the data of Table 5.13 are used to compute elemental mass percentages for the adult liver.

Elemental compositions for the soft tissues and gender-specific body tissues are reported in Tables 5.14 and 5.15 for the ICRP Reference Individuals, with the exclusion of the reference newborn. Unique elemental tissue compositions for the reference newborn are given in separate Tables 5.16 and 5.17. Elemental data for the soft tissues in ICRP Publication 89 (ICRP 2002) is largely adopted from ICRU Report 46 (ICRU 1992). It is noted that a greater degree of age dependence of some tissue compositions are given in that ICRU report, whereas an age distinction is made in ICRP Publication 89 only for the reference newborn. Age-dependent elemental compositions of the skeletal tissues are given in Table 5.18. These data are inclusive of the reference newborn and are taken primarily from ICRU Publication 70.

One limitation with the elemental composition data reported in Chapter 13 of ICRP Publication 89 is its lack of documentation of reference values of mass density in the ICRP Reference Individuals. In constructing 3D anatomic models of the reference adults and children, "reference volumes" must be derived as the ratio of a reference mass and a reference density. In Publication 89, there are limited summaries of specific gravity (ratio of tissue density to that of water density). For example, ICRP Publication 89 reports that the specific gravity of the newborn and adult kidneys is 1.035 and 1.05, respectively. No similar data are given for the specific gravity of the liver. In the elemental composition tables

TABLE 5.14   Composition of Soft Tissues for Children and Adults[a]

Organ/Tissue	Elemental Composition (% by mass)									
	H	C	N	O	Na	P	S	Cl	K	Other
Adrenals[b]	10.5	25.6	2.7	60.2	0.1	0.2	0.3	0.2	0.2	–
Alimentary tract										
Tongue[c]	10.2	14.3	3.4	71.0	0.1	0.2	0.3	0.1	0.4	–
Esophagus	10.5	25.6	2.7	60.2	0.1	0.2	0.3	0.2	0.2	–
Stomach	10.6	11.5	2.2	75.1	0.1	0.1	0.1	0.2	0.1	–
Small intestine	10.6	11.5	2.2	75.1	0.1	0.1	0.1	0.2	0.1	–
Large intestine	10.6	11.5	2.2	75.1	0.1	0.1	0.1	0.2	0.1	–
Liver	10.3	18.6	2.8	67.1	0.2	0.2	0.3	0.2	0.3	–
Gallbladder[b]	10.5	25.6	2.7	60.2	0.1	0.2	0.3	0.2	0.2	–
Pancreas	10.6	16.9	2.2	69.4	0.2	0.2	0.1	0.2	0.2	–
Blood	10.2	11.0	3.3	74.5	0.1	0.1	0.2	0.3	0.2	0.1 Fe
Brain	10.7	14.5	2.2	71.2	0.2	0.4	0.2	0.3	0.3	–
Heart	10.4	13.9	2.9	71.8	0.1	0.2	0.2	0.2	0.3	–
Eyes	9.6	19.5	5.7	64.6	0.1	0.1	0.3	0.1	–	–
Fat	11.4	59.8	0.7	27.8	0.1	0.1	0.1	–	–	–
Skin	10.0	20.4	4.2	64.5	0.2	0.1	0.2	0.3	0.1	–
Muscle	10.2	14.3	3.4	71.0	0.1	0.2	0.3	0.1	0.4	–
Pituitary gland[b]	10.5	25.6	2.7	60.2	0.1	0.2	0.3	0.2	0.2	–
Respiratory tract										
Trachea[b]	10.5	25.6	2.7	60.2	0.1	0.2	0.3	0.2	0.2	–
Larynx	9.6	9.9	2.2	74.4	0.5	2.2	0.9	0.3	–	–
Lung	10.3	10.5	3.1	74.9	0.2	0.2	0.3	0.3	0.2	–
Spleen	10.3	11.3	3.2	74.1	0.1	0.3	0.2	0.2	0.3	–
Thymus[b]	10.5	25.6	2.7	60.2	0.1	0.2	0.3	0.2	0.2	–
Thyroid	10.4	11.9	2.4	74.5	0.2	0.1	0.1	0.2	0.1	0.1 I
Tonsils[b]	10.5	25.6	2.7	60.2	0.1	0.2	0.3	0.2	0.2	–
Urogenital system										
Kidneys	10.3	13.2	3.0	72.4	0.2	0.2	0.2	0.2	0.2	0.1 Ca
Ureters[b]	10.5	25.6	2.7	60.2	0.1	0.2	0.3	0.2	0.2	–
Urinary bladder	10.5	9.6	2.6	76.1	0.2	0.2	0.2	0.3	0.3	–
Urethra[b]	10.5	25.6	2.7	60.2	0.1	0.2	0.3	0.2	0.2	–
Epididymes[b]	10.5	25.6	2.7	60.2	0.1	0.2	0.3	0.2	0.2	–

Source: Table 13.2, "ICRP Publication 89" (ICRP 2002).
[a]   Based on composition data tabulated in *ICRU Report No. 46* (ICRU 1992).
[b]   Composition assigned to bulk soft tissue.
[c]   Composition assigned to muscle.

of ICRU Report 46, reference values of mass density are provided alongside the data on elemental tissue compositions.

Both ICRP Publication 89 and ICRU Report 46 (ICRU 1992) clearly note that many factors—such as age, gender, diet, and health status—can significantly impact the composition of the body tissues. For example, cirrhosis of the liver due to chronic alcoholism may cause the lipid content of the liver to change from ~5% to 19% by mass with its water

TABLE 5.15 Composition of Gender-Specific Tissues for Children and Adults[a]

Organ/Tissue	Elemental Composition (% by mass)								
	H	C	N	O	Na	P	S	Cl	K
Male									
Breast	11.4	59.8	0.7	27.8	0.1	–	0.1	0.1	–
Testes	10.6	9.9	2.0	76.6	0.2	0.1	0.2	0.2	0.2
Prostate	10.5	25.6	2.7	60.2	0.1	0.2	0.3	0.2	0.2
Female									
Breast	11.6	51.9	–	36.5	–	–	–	–	–
Ovaries	10.5	9.3	2.4	76.8	0.2	0.2	0.2	0.2	0.2
Fallopian tubes[b]	10.6	31.5	2.4	54.7	0.1	0.2	0.2	0.1	0.2
Uterus[b]	10.6	31.5	2.4	54.7	0.1	0.2	0.2	0.1	0.2

Source: Table 13.3, "ICRP Publication 89" (ICRP 2002).
[a] Based on composition data tabulated in *ICRU Report No. 46* (ICRU 1992).
[b] Composition assigned to bulk soft tissue.

content decreasing ~75% to 64% by mass. Values of elemental composition of body tissues given in ICRP Publication 89 are thus for normal, healthy tissues.

5.2.4 The Embryo, Fetus, and Pregnant Female

Due to the need for radiological protection guidance on radiation exposure to the pregnant female—either as a radiation worker or a member of the general public—an additional member of the ICRP series of reference individuals includes the embryo and developing fetus. Anatomic reference values are thus needed for both the total-body mass of the fetus, and for the individual fetal organs. For the embryo, ICRP typically assumes that the radiation dose to the uterine wall will suffice as a surrogate tissue for the embryo in exposures very early in pregnancy.

Reference values of fetal total-body mass, as a function of fetal age from 8 weeks to 38 weeks, are given in Table 5.19. At the time of the development of ICRP Publication 89 (ICRP 2002), the most recent set of fetal growth curves were given in an extensive compilation of biometric data by Guihard-Costa et al. (1995). Their study was designed to establish a set of normalized data for clinical characterization of fetal growth and development. This study involved nearly 5000 fetuses, some studied post-mortem and others in vivo via ultrasound. It is noted that ICRP defines fetal age in weeks as the post-conception (PC) age, and not the time since last menstrual period (LMP), which is another common convention in reporting fetal age.

Reference values for individual fetal organs as a function of fetal age (weeks PC) are given in Table 5.20. These values are derived from a variety of data sources. For example, reference masses of the fetal brain are taken from a best fit of measurements taken from the study by Guihard-Costa et al. (1995) on 291 samples of fetal brain prior to fixation. Reference masses of the fetal thyroid are taken from a composite age-dependent fit of the data from Aboul-Khair et al. (1966), Evans et al. (1967), Ares et al. (1995), and Costa et al. (1986). Masses

TABLE 5.16   Composition of Soft Tissues for Newborns[a]

Organ/Tissue	Elemental Composition (% by mass)									
	H	C	N	O	Na	P	S	Cl	K	Other
Adrenals[b]	10.6	16.3	2.0	71.0	–	–	0.1	–	–	–
Alimentary tract										
Tongue[c]	10.4	10.3	2.4	76.2	0.1	0.1	0.1	0.2	0.2	–
Esophagus	10.6	16.3	2.0	71.0	–	–	0.1	–		
Stomach	10.6	11.5	2.2	75.1	0.1	0.1	0.1	0.2	0.1	–
Small intestine	10.6	11.5	2.2	75.1	0.1	0.1	0.1	0.2	0.1	–
Large intestine	10.6	11.5	2.2	75.1	0.1	0.1	0.1	0.2	0.1	–
Liver	10.6	12.6	2.7	73.3	0.1	0.3	0.2	0.2	0.3	–
Gallbladder[b]	10.6	16.3	2.0	71.0	–	–	0.1	–	–	–
Pancreas	10.6	16.9	2.2	69.4	0.2	0.2	0.1	0.2	0.2	–
Blood	10.0	13.1	4.0	72.0	0.1	0.1	0.2	0.2	0.2	0.1 Fe
Brain	10.8	5.5	1.1	81.6	0.2	0.3	0.1	0.2	0.2	–
Heart	10.6	7.5	1.8	79.3	0.2	0.1	0.1	0.2	0.2	–
Eyes	9.6	19.5	5.7	64.4	0.1	0.1	0.3	0.1	–	–
Fat	11.1	29.7	0.9	58.0	0.1	0.1	0.1	–	–	–
Skin	10.4	10.4	2.8	75.5	0.2	0.1	0.2	0.3	0.1	–
Muscle	10.4	10.3	2.4	76.2	0.1	0.1	0.1	0.2	0.2	–
Pituitary gland[b]	10.6	16.3	2.0	71.0	–	–	0.1	–	–	–
Respiratory tract										
Trachea[b]	10.6	16.3	2.0	71.0	–	–	0.1	–	–	–
Larynx	9.6	9.9	2.2	74.4	0.5	2.2	0.9	0.3	–	–
Lung	10.6	7.6	1.8	79.2	0.2	0.2	0.1	0.2	0.1	–
Spleen	10.5	8.6	2.4	77.6	0.2	0.2	0.1	0.2	0.2	–
Thymus[b]	10.6	16.3	2.0	71.0	–	–	0.1	–	–	–
Thyroid	10.4	11.9	2.4	74.5	0.2	0.1	0.1	0.2	0.1	0.1 I
Tonsils[b]	10.6	16.3	2.0	71.0	–	–	0.1	–	–	–
Urogenital system										
Kidneys	10.7	6.4	1.6	80.4	0.2	0.2	0.1	0.2	0.2	–
Ureters[b]	10.6	16.3	2.0	71.0	–	–	0.1	–	–	–
Urinary bladder	10.5	9.6	2.6	76.1	0.2	0.2	0.2	0.3	0.3	–
Urethra[b]	10.6	16.3	2.0	71.0	–	–	0.1	–	–	–
Epididymes[b]	10.6	16.3	2.0	71.0	–	–	0.1	–	–	–

Source: Table 13.5, "ICRP Publication 89" (ICRP 2002).

[a]   Based on composition data tabulated in *ICRU Report No. 46* (ICRU 1992).

[b]   Composition assigned to bulk soft tissue (Ziegler et al. 1976).

[c]   Composition assigned to muscle.

of fetal active bone marrow are taken from reported values by Hudson (1965) and power-function fits to this data by Luecke, Wosilait, and Young (1995).

ICRP Publication 89 additionally provides anatomic data on the adult pregnant female as a function of fetal age as needed to construct both anatomic pregnant female computational phantoms and associated biokinetic models of radionuclide circulation, placental

TABLE 5.17   Composition of Gender-Specific Tissues for Newborns[a]

Organ/Tissue	Elemental Composition (% by mass)								
	H	C	N	O	Na	P	S	Cl	K
Male									
Breast[b]	11.1	29.7	0.9	58.0	0.1	–	0.1	0.1	–
Testes[c]	10.6	16.3	2.0	71.0	–	–	0.1	–	–
Prostate[c]	10.6	16.3	2.0	71.0	–	–	0.1	–	–
Female									
Breast[b]	11.1	29.7	0.9	58.0		–	0.1	0.1	–
Ovaries	10.6	16.3	2.0	71.0	0.1	–	0.1		–
Fallopian tubes[c]	10.6	16.3	2.0	71.0	–	–	0.1	–	–
Uterus[c]	10.6	16.3	2.0	71.0	–	–	0.1	–	–

Source: Table 13.6, "ICRP Publication 89" (ICRP 2002).
[a] Based on composition data tabulated in *ICRU Report No. 46* (ICRU 1992).
[b] Adipose tissue composition.
[c] Composition assigned to bulk soft tissue (Ziegler et al. 1976).

TABLE 5.18   Age-Dependent Element Composition of the Skeleton[a]

Organ/Tissue	Elemental Composition (% By Mass)										
	H	C	N	O	Na	Mg	P	S	Cl	Ca	Fe
Active marrow	10.5	41.4	3.4	43.9	0.1	0.2	0.2	0.2	–	–	0.1
Inactive marrow	11.5	64.4	0.7	23.1	0.1	–	0.1	0.1	–	–	–
Cartilage	9.6	9.9	2.2	74.4	0.5	–	2.2	0.9	0.3	–	–
Teeth	2.2	9.5	2.9	42.1	–	0.7	13.7	–	–	28.9	–
Bone mineral[a]											
Newborn	4.2	16	4.5	50.2	–	0.3	8.0	0.3	–	16.5	
1 year	4.1	16	4.5	49.3	–	0.3	8.5	0.3	–	17	
5 years	4.0	16	4.5	46.9	0.1	0.2	9.0	0.3	–	19	
10 years	3.9	16	4.4	45.6	0.1	0.2	9.5	0.3	–	20	
15 years	3.8	16	4.3	45.2	0.2	0.2	9.5	0.3	–	20.5	
Adult	3.5	16	4.2	44.5	0.3	0.2	9.5	0.3	–	21.5	–

Source: Table 13.4, "ICRP Publication 89"
[a] Based on data summarized in "ICRP Publication 70" (ICRP 1995b).

TABLE 5.19   Reference Values for Body Mass of the Fetus

Fetal Age (Weeks)	Mass (g)
8	4.7
10	21
15	160
20	480
25	990
30	1700
35	2700
38	3500

Source: Table 2.1, "ICRP Publication 89" (ICRP 2002).

TABLE 5.20 Reference Values for Organ Mass in the Developing Fetus

Organ/Tissue	Fetal Age (Weeks)							
	8	10	15	20	25	30	35	38
Brain	3.9	6.7	23	62	120	200	300	370
Thyroid	0.011	0.022	0.077	0.18	0.36	0.63	1.0	1.3
Heart	0.038	0.15	1.1	3.0	6.0	9.9	15	20
Adrenals (2)	0.016	0.06	0.38	0.98	1.9	3.0	4.6	6.0
Marrow, active	0.070	0.30	2.4	6.9	14	24	38	50
Kidneys (2)	0.024	0.13	1.3	3.8	7.6	13	20	25
Liver	0.21	0.87	6.5	19	38	63	100	130
Lungs	0.096	0.63	5.8	15	26	38	51	60
Pancreas	0.39	0.69	1.5	2.3	3.1	3.8	4.5	5.0
Spleen	0.00049	0.0035	0.069	0.36	1.1	2.7	5.8	9.5
Thymus	0.011	0.022	0.45	1.5	3.2	5.8	9.7	13

Source: Table 2.4, "ICRP Publication 89" (ICRP 2002).

TABLE 5.21 Analysis of Weight Gain During Pregnancy

Tissues and Fluids	Increase in Mass (g) up to:			
	10 Weeks	20 Weeks	30 Weeks	38 Weeks
Fetus	5	300	1500	3400
Placenta	20	170	430	650
Amniotic fluid	30	350	750	800
Uterus	140	320	600	970
Breasts	45	180	360	405
Blood	100	600	1300	1450
Extracellular, extravascular fluid[a]	0	30	80	1480
Unaccounted maternal stores	310	2050	3480	3340
Total	650	4000	8500	12 500
Extracellular, extravascular fluid	0	500	1530	4700
Total	650	4500	10 000	14 500

Source: "ICRP Publication 89," p. 231 (ICRP 2002).

[a] No edema. The following values are representative of generalized edema.

transfer, and fetal tissue uptake. An example of this information is shown in Table 5.21, given the absolute increase in reference tissue mass of the mother as fetal age progresses from 10 weeks post-conception to full term at 38 weeks post-conception. These values are taken from data from Hytten (1980) and Goldberg et al. (1993), as reviewed by Munro and Eckerman (1998).

5.3 COMPUTATIONAL REALIZATIONS OF THE ICRP REFERENCE INDIVIDUAL ANATOMY

In order to compute values of absorbed fraction and specific absorbed fractions needed for internal dose assessment, as well as dose coefficients for external exposures, radiation transport methods must be applied to the emission, scatter, absorption, and movement of radiation particles within the tissues of the ICRP Reference Individual. Various forms of computational anatomic models have been applied to generate a virtual three-dimensional representation of the anatomy of the ICRP Reference Individuals. They are summarized briefly in this section of the chapter.

5.3.1 Stylized Computational Phantoms

The earliest computational phantoms for modeling the ICRP Reference Individuals were developed at Oak Ridge National Laboratory in the late 1960s to late 1970s. An historical review of these earlier phantoms is given by Poston, Bolch, and Bouchet (2002). In the early 1980s, the ORNL series was updated to represent not only the ICRP reference adults, but the entire postnatal ICRP age series (Cristy 1980). The series included an adult male, a newborn, and individuals of 1, 5, 10 and 15 y of age developed from anthropological data (legs, trunk, and head) and from age-specific organ masses published in ICRP Publication 23 (ICRP 1975). Although some of the organ shapes and centroids were still obtained using the similitude rule from the Snyder–Fisher adult model (Snyder, Ford, and Warner 1978), these phantoms represented a great improvement for pediatric dosimetry over similitude pediatric phantoms (those created via uniform volumetric reductions in the adult phantom). These phantoms also included new regions and improvements, such as a stylized breast-tissue region for all ages, the inclusion of the model of the heart developed by Coffey (Coffey, Cristy, and Warner 1981), and a model of the thyroid. While the ORNL pediatric model series was initially published in 1980 (Cristy 1980), these models were not readily utilized until 1987 following the publication of ORNL *Specific Absorbed Fractions of Energy at Various Ages from Internal Photon Sources* (Cristy and Eckerman 1987). In this Report, the only major change in the phantom series was that the 15-year-old model was assumed also to represent the average adult female. For this purpose, the breast, uterus, and ovaries were modified according to published reference average values (ICRP 1975). The phantoms were used with a Monte Carlo photon transport code to calculate SAFs of energy in all five pediatric phantoms, as well as in the adult male, for 12 photon energies (0.01 to 4 MeV). Electron transport was not considered in these simulations and the electron energy was assumed to be locally deposited. The ORNL stylized phantoms were revised in 1996 to include new models of the esophagus and neck regions. An extensive revision was additionally made by Han, Bolch, and Eckerman (2006) to incorporate into the ORNL phantom series:

- updated anatomic models of the kidneys (Bouchet et al. 2003), GI tract (Mardirossian et al. 1999), and the head and brain (Bouchet and Bolch 1999); and

- adjusted organ masses to match the age-dependent reference masses given in ICRP Publication 89 (ICRP 1975)

The revised ORNL series is presently available via the MCNP Medical Physics Geometry code package (http://mcnp.lanl.gov).

5.3.2 Voxel Computational Phantoms

Stylized mathematical models of human anatomy provide a relatively efficient geometry for use with Monte Carlo radiation transport codes. Nevertheless, these models are only approximations of the true anatomical features of individuals for which dose estimates are required. An alternative class of anatomic models is based on three-dimensional imaging techniques, such as magnetic resonance imaging (MRI) or computed tomography (CT). These voxel models of human anatomy represent large arrays of image voxels that are individually assigned both a tissue type (e.g., soft tissue, bone, air, etc.) and an organ identity (heart wall, femur shaft, tracheal airway, etc.). Thus, image segmentation is needed to process the original image into a format acceptable for radiation transport using codes.

Two general anatomic sources exist for imaging to construct voxel models of human anatomy: cadavers and live subjects (typically medical patients). Each data source has its distinct advantages and disadvantages. Cadaver imaging generally offers a substantially improved opportunity for full anatomic coverage, including the extremities. With CT image acquisition, higher-resolution scans can be performed, as radiation dose considerations and patient motion are not of concern. For these same reasons, multiple scans on the same cadaver can be performed using different technique factors (kilovolt peak, milliampere, filtration, etc). Problems associated with cadaver imaging include tissue shrinkage, body fluid pooling, air introduction to the GI tract, collapse of the lungs, and general organ settling (Bolch et al. 2010). Perhaps the greatest disadvantage, however, is that cadaver imaging will most likely not involve the use of CT tissue contrast agents needed for soft tissue image segmentation. CT is generally the imaging modality of choice for construction of full-body tomographic computational models (Zaidi and Xu 2007). Skeletal tissues are more readily defined under CT imaging, whereas image distortions of skeletal tissues are problematic under MRI. With live patient imaging, CT contrast agents provide acceptable definitions of soft tissue borders needed for image segmentation, and thus MRI does not offer a distinct advantage over CT for this data source. In cadaver imaging, the lack of contrast agents hinders border definitions for soft tissue organs. To compensate, an MRI of the cadaver might be considered. However, the frozen state of the cadaver tissues can alter both spin-lattice and spin-spin relaxation times (T1 and T2, respectively), complicating and distorting MRIs of the subject. Thus, CT imaging is recommended for both live and deceased subjects upon whom tomographic models are to be constructed. With continuing advances in multi-detector, multi-slice CT imaging, scan times are minimized, offering improved opportunities for live-subject imaging as potential data sources. CT-MRI fusion is another possibility for input data to image segmentation. However, patient motion artifacts within MRIs, and their generally lower-image resolutions, currently restrict opportunities for CT-MRI co-registration to the head region where patient motion is more easily minimized.

Following image acquisition, the next step in model construction is image segmentation. In cases where the CT image provides strong tissue contrast, automated pixel-growing methods of image segmentation can be applied to rapidly delineate organ boundaries. In these methods, a central pixel is tagged and the algorithm groups all neighboring pixels within a defined gray-level interval (e.g., skeletal regions, air lumen of nasal passages, certain soft tissue organs under CT contrast). For those organs with poor tissue contrast, organ boundaries must be segmented manually. Final review of the tomographic model by medical personnel trained in radiographic anatomy is highly recommended. Standardized software packages can offer all necessary tools for construction of tomographic computational models.

A comprehensive review of voxel phantoms available for dosimetric evaluation are given in Zaidi and Xu (2007) and Xu and Eckerman (2009). Since most voxel phantoms were developed from cadaver, patient, or volunteer subjects, many are individual-specific with regard to organ and body morphometries and thus do not exactly conform to ICRP reference characteristics. Exceptions include the MAX and NORMAN phantoms of the adult male, and the FAX and NAOMI phantoms of the adult female, which were adjusted to conform to specifications given in ICRP Publication 89 (Jones 1997; Dimbylow 2005; Kramer et al. 2006). Furthermore, the University of Florida series of pediatric voxel phantoms were also adjusted through voxel modification to conform to the anatomical characteristics of the ICRP reference children at their age-interpolated values of height, weight, and organ mass (Lee et al. 2006). In 2009; ICRP published a pair of adult voxel phantoms representing the ICRP Publication 89 Adult Reference Male and Adult Reference Female for use in implementation of the 2007 recommendations and a systematic update of dose coefficients for both internal and external radiation exposures (Schlattl, Zankl, and Petoussi-Henss 2007; ICRP 2009). The creation of the ICRP Publication 110 reference phantoms thus represented the very first time ICRP had published its own anatomic models of the ICRP Reference Individuals, in this case, the adult (ICRP 2009). The vast majority of previous ICRP publications had been based upon dosimetry computations using the ORNL stylized phantom series. The ICRP Publication 110 phantoms are shown in Figure 5.2.

5.3.3 Hybrid Computational Phantoms

As both existing phantom types have their own distinct drawbacks regarding both anatomical realism and flexibility in morphometry alteration, investigators have sought new methods for anatomical modeling that provide and preserve both of these important features. Phantoms developed under this new approach can thus be termed "hybrid phantoms," as they retain both the anatomic realism of voxel phantoms and the flexibility of stylized phantoms. In one approach to hybrid phantom construction, the three-dimensional surface equations used to define organ boundaries within existing stylized phantoms are replaced with NURBS (non-uniform rational basis-spline) surfaces (Piegl 1991). NURBS is a mathematical modeling technique widely used in three-dimensional computer graphics and film animation. NURBS surfaces can precisely represent not only standard analytic shapes (as needed to model organs in low-contrast images), but they can additionally define complex free-form surfaces required for certain intricately shaped internal organs and organ systems.

NURBS surfaces were adopted by Segars (2001) in the development of the NCAT Phantom, replacing the previous stylized Mathematical Cardiac Torso Phantom constructed from the MIRD-5 stylized model (Segars, Lalush, and Tsui 2001). The NCAT phantom is based upon the anatomy of the National Library of Medicine's Visible Human project image set, and has been widely adopted in studies of torso anatomy for cardiac SPECT images (He et al. 2005; Lalush, Jatko, and Segars 2005; Lomsky et al. 2005; Mair, Gilland, and Sun 2006; Tobon-Gomez et al. 2008). Organ shapes in the NCAT phantom are much more realistic than those of Mathematical Cardiac Torso Phantom, while they maintain the flexibility to consider anatomical variations as well as both cardiac and respiratory motion. While

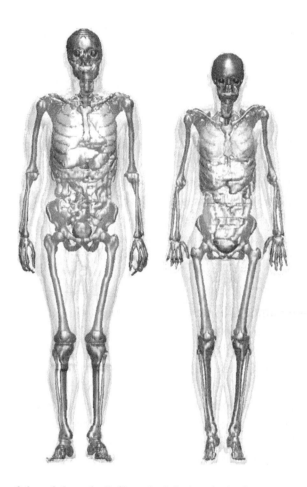

FIGURE 5.2 Images of the adult male (left) and adult female (right) computational phantoms of the ICRP (ICRP 2009). The following organs can be identified by different surface colors: breast, bones, colon, eyes, lungs, liver, pancreas, small intestine, stomach, teeth, thyroid, and urinary bladder. Muscle and adipose tissue are semi-transparent. For illustration purposes, the voxelized surfaces have been smoothed.

in theory, the NCAT phantom can be rescaled to pediatric body dimensions, this approach becomes more and more uncertain at younger ages due to age- and gender-specific changes in organ shape, size, and position. However, the NURBS approach can be readily applied to pediatric CT images with subsequent adjustments to construction of either pediatric phantoms of ICRP reference dimensions, or of patient-specific dimensions as required in medical dose reconstruction studies. Hybrid phantoms of the ICRP Reference Individuals have been developed by researchers at the University of Florida (adult, pediatric, and pregnant female) (Lee et al. 2010; Maynard et al. 2011; Maynard et al. 2014), Vanderbilt University (adult and pediatric) (Stabin et al. 2012), and RPI (adults and pregnant females) (Zhang et al. 2009; Xu et al. 2007). More recently, the ICRP has established a task group to systematically convert the ICRP Publication 110 (ICRP 2009) reference voxel phantoms into a polygon mesh format, thus allowing the incorporation of finer tissue structures within a single anatomic framework to include the radiosensitive cell layers of the skin, eyes, respiratory tract airways, and alimentary tract walls (Nguyen et al. 2015; Yeom et al. 2016a; Yeom et al. 2016b; Kim et al. 2017).

5.4 PHYSIOLOGICAL ASPECTS OF THE ICRP REFERENCE INDIVIDUAL

In addition to anatomical definitions and associated reference values for the anatomy of the ICRP Reference Individuals, standardized data are equally needed to establish reference physiology as a function of age and gender in the development of reference intake and systemic biokinetic models for radionuclides. These include a variety of parameters, such as daily water balance, time budgets of exposure, respiratory volumes and capacities, ventilation rates, transit times in the segments of the alimentary tract, urinary and fecal excretion rates (needed for bioassay interpretation), and bone remodeling rates (for bone-seeking radionuclides). Reference physiology of the development fetus is also needed for radiological protection guidance during pregnancy.

5.4.1 Daily Water Balance

To maintain a constant amount of water, the body must eliminate an amount equivalent to that ingested in fluids and food, plus that produced by metabolism. The regulation of total-body water is based primarily on a sensing system that responds to plasma osmolality, with Na^+, Cl^-, and HCO_3^- being the major solutes in plasma. Water is absorbed in the upper portions of the small intestine and distributed by way of the lymphatic fluids and blood into the tissues and cells of the body. It is eventually excreted through the kidneys/urinary bladder, skin, lungs, and intestines. Reference values for the daily water balance in adults are summarized in Table 5.22, which are based upon Task Group review of a variety of sources in Section 4.4.2 of ICRP Publication 89 (ICRP 2002).

5.4.2 Respiratory Volumes and Capacities

Table 5.23 summarizes various volumes and capacities of the respiratory tract as a function of both age and gender, as taken from previous reference values given in ICRP Publication 66 (ICRP 1994a). These values are used to define key parameters regarding particle deposition in the airways of differing regions of the respiratory tract. Values of TLC, FRC, and VC are graphically defined in Figure 5.3.

5.4.3 Time Budgets and Ventilation Rates

During dose assessment from the inhalation of radionuclides to unspecified individuals—either workers or members of the general public—assumptions must be made regarding ventilation rates which of course vary with the level of exertion, and change with subject age and associated size and development of the respiratory tract airways. Under these conditions, reference values of daily time budgets and breathing rates must be assumed as defined by ICRP reference parameter values. Tables 5.24 and 5.25 give these ICRP reference values for members of the general public and for adult radiation workers, respectively. These data are assembled from prior analyses documented in ICRP Publications 66 and 71 (ICRP 1994a, 1995c). Changes from reference breathing rates from ICRP Publication 23 (ICRP 1975) to those given more recently in ICRP Publication 89 (ICRP 2002) are shown in Table 5.25.

TABLE 5.22   Reference Values for Water Balance in Adults

	Male	Female
Water intake in food and fluids (mL/day)	2600	1960
Oxidation of food (mL/day)	300	225
Losses (mL/day)		
Urine	1600	1200
Insensible loss[a]	690	515
Sweat	500	375
Feces	110	95

Source: Table 2.30, "ICRP Publication 89" (ICRP 2002).
[a] Assumed to be divided equally between the lungs and skin.

TABLE 5.23   Reference Values for Respiratory Volumes/Capacities[a]

					15 Years		Adult	
	3 Months	1 Year	5 Years	10 Years	Male	Female	Male	Female
TLC (l)	0.28	0.55	1.6	2.9	5.4	4.5	7.0	5.0
FRC (l)	0.15	0.24	0.77	1.5	2.7	2.3	3.3	2.7
VC (l)	0.20	0.38	1.0	2.3	4.0	3.3	5.0	3.6
V_D (l)[b]	0.014	0.020	0.046	0.078	0.13	0.11	0.15	0.12

Source: Table 2.31, "ICRP Publication 89" (ICRP 2002).
[a] Rounded values from Table 7 in "ICRP Publication 66" (ICRP 1994a). See Annex B of "ICRP Publication 66" for published data on various populations.
[b] These are secondary values calculated by scaling the airway dimensions for body height.
TLC, total lung capacity; FRC, functional residual capacity; VC, vital capacity; V_D, dead space.

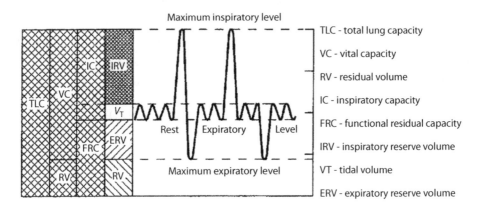

FIGURE 5.3   Respiration quantities.

5.4.4 Transit Times of Luminal Content in the Alimentary Tract

For the development of intake models of radionuclide ingestion, the ICRP revised its model of the GI tract originally given in ICRP Publication 30 (ICRP 1980) to include a more physiological realistic model of the full alimentary tract (transit through the mouth and esophagus were included), with kinetics that differentiated solid and liquid intakes,

TABLE 5.24 Reference Values for Daily Time Budgets and Ventilation Parameters at Each Level of Exertion for Members of the General Public

Exercise Level	3 Months			1 Year			5 Years		
	h	m³/h	m³	h	m³/h	m³	h	m³/h	m³
Sleep	17	0.09	1.5	14	0.15	2.1	12	0.24	2.9
Sitting				3.3	0.22	0.73	4.0	0.32	1.3
Light exercise	7.0	0.19	1.3	6.7	0.35	2.3	8.0	0.57	4.6
Heavy exercise									
Total			2.8			5.1			8.8

Exercise Level	10 Years			15 Years (male)			15 Years (female)		
	h	m3/h	m3	h	m3/h	m3	h	m3/h	m3
Sleep	10	0.31	3.1	10	0.42	4.2	10	0.35	3.5
Sitting	4.7	0.38	1.8	5.5	0.48	2.6	7.0	0.40	2.8
Light exercise	9.3	1.1	10.3	7.5	1.38	10.4	6.8	1.3	8.8
Heavy exercise				1.0	2.92	2.9	0.25	2.6	0.65
Total			15.2			20.1			15.8

Exercise Level	Adult (male)			Adult (female)		
	h	m3/h	m3	h	m3/h	m3
Sleep	8.0	0.45	3.6	8.5	0.32	2.7
Sitting	6.0	0.54	3.2	5.4	0.39	2.1
Light exercise	9.8	1.5	14.7	9.9	1.3	12.9
Heavy exercise	0.25	3.0	0.75	0.19	2.7	0.52
Total			22.2			18.2

Source: Table 2.34, "ICRP Publication 89" (ICRP 2002).

TABLE 5.25 Reference Values for Daily Ventilation Rates for Adult Workers

	Air Breathed (m³/day)		
	Sedentary Worker		Heavy Worker
Activity	Male	Female	Male
Sleeping (8 h)	**3.6 (3.6)**[a]	**2.6 (2.9)**	**3.6 (3.6)**
Occupational (8 h)			
1/3 sitting	**9.6 (9.6)**[b]	*7.9 (9.1)*	
2/3 light exercise			
7/8 light exercise			**13.5** *(9.6)*
1/8 heavy exercise			
Non-occupational (8 h)			
4/8 sitting			
3/8 light exercise	**9.7** *(9.6)*	**8.0** *(9.1)*	**9.7** *(9.6)*
1/8 heavy exercise			
Total air breathed (m³)	**22.9** *(22.8)*	**18.5** *(21.1)*	**26.8** *(22.8)*

Source: Table 2.35, "ICRP Publication 89" (ICRP 2002).
[a] From Table B.17 in "ICRP Publication 66" (ICRP 1994a).
[b] Values in parentheses from "ICRP Publication 23" (ICRP 1975).

TABLE 5.26   Reference Values for Transit Times of Luminal Contents through Major Segments of the Alimentary Tract

| | Age Group | | | | |
| | | | | Adult | |
	Newborn	1 Year	5–15 Years	Males	Females
Mouth					
Solids	–	15 s	15 s	15 s	15 s
Liquids	2 s	2 s	2 s	2 s	2 s
Total diet	2 s	12 s	12 s	12 s	12 s
Esophagus—fast (90%)					
Solids	–	8 s	8 s	8 s	8 s
Liquids	4 s	5 s	5 s	5 s	5 s
Total diet	4 s	7 s	7 s	7 s	7 s
Esophagus—fast (90%)					
Solids	–	45 s	45 s	45 s	45 s
Liquids	30 s	30 s	30 s	30 s	30 s
Total diet	30 s	40 s	40 s	40 s	40 s
Stomach					
Solids	–	75 mins	75 mins	75 mins	105 mins
Liquids—caloric	75 mins	45 mins	45 mins	45 mins	60 mins
Liquids—non-caloric	10 mins	30 mins	30 mins	30 mins	30 mins
Total diet	75 mins	70 mins	70 mins	70 mins	95 mins
Small intestine[a]	4 h	4 h	4 h	4 h	4 h
Right colon[a]	8 h	10 h	11 h	12 h	16 h
Left colon[a]	8 h	10 h	11 h	12 h	16 h
Rectosigmoid[a]	12 h	12 h	12 hs	12 h	16 h

Source: Table 2.37, "ICRP Publication 89" (ICRP 2002).
[a] Intestinal transit times apply to all material.

some age dependence of the transfer rates, and a differentiation of reference transit rates in the adult male and adult female. Table 5.26 summarizes these reference parameter values that were used later in the development of the Human Alimentary Tract Model (HATM) of ICRP Publication 100 (ICRP 2006).

5.4.5 Urinary and Fecal Excretion Rates

The volume of urine excreted daily varies with age, gender, diet, exercise, and other factors. In adults, the 24-hour urine volume is typically 1200 to 2000 mL. A reasonable central estimate a 73-kg adult male may be about 1600 mL/day, or about 22 mL/kg/day. A similar per body mass value was estimated in for children ages 6 to 11 years. With excessive water intake, urine output can be as high as 10% of the kidney's glomerular filtration rate or approximately 250 mL/kg/day. During prolonged periods of high water loss or low water intake, urine output may decrease to as little as 6–7 mL/kg/day. Reference values for urinary excretion rates for the ICPR reference individuals are given in Table 5.27.

The mass of feces excreted per day varies substantially from one person to another, and from one population to another, due largely to differences in dietary fiber intake. In a study

TABLE 5.27   Reference Values for Daily Urinary Excretion

Age	Excretion	
	Male	Female
Newborn	300	300
1 year	400	400
5 years	500	500
10 years	700	700
15 years	1200	1200
Adult	1600	1200

Source: Table 2.41, "ICRP Publication 89" (ICRP 2002).

TABLE 5.28   Reference Values for the Mass of Feces Excreted per Day

Age	Mass (g/day)	
	Male	Female
Newborn	24	24
1 year	40	40
5 years	50	50
10 years	70	70
15 years	120	120
Adult	150	120

Source: Table 2.38, "ICRP Publication 89" (ICRP 2002).

of six healthy adult subjects, increasing dietary fiber intake from 17 to 45 g/day for 3 weeks increased fecal mass from 79 ± 6.6 g/day to 228 ± 29.9 g/day (mean ± SD) (Cummings et al. 1976). Fecal mass averaged 51 g/day in six healthy male subjects on a low-fiber diet and 157 g/day when the same subjects were placed on a high-fiber diet (Beyer and Flynn 1978). ICRP reference values for daily fecal excretion rate are given in Table 6.29. These values are based on reported estimates, supplemented with data on relative intakes of food and fluids as a function of age and gender. As far as practical, fecal excretion data associated with unusually high or low intakes of dietary fiber were excluded (Table 5.28).

5.4.6 Bone Remodeling Rates

Bone remodeling refers to the process of bone mineral turnover that replaces existing bone but changes the shape and total amount of bone very slowly or not at all. Bone modeling refers to local influences that alter the size and shape of the growing bones. Bone growth refers to the process that increases the volume of bones.

The remodeling process is carried out by certain cells on the bone surfaces, namely osteoclasts, osteoblasts, and their precursors. Each surface is always in one of three functional states—forming, resorbing, or quiescent. Bone-resorbing surfaces are scalloped by Howship's lacunae containing osteoblasts and poorly characterized mono-nuclear cells. Bone-forming surfaces are covered by osteoid seams and osteoblasts.

Table 5.29 gives ICRP reference values for bone remodeling for both cortical and trabecular bone as a function of age. A value of 3% per year is estimated as the average rate of

TABLE 5.29   Reference Values for Bone Remodeling Rates

	Remodeling Rate (%/year)	
Age	Cortical Bone	Trabecular Bone
Newborn	300	300
1 year	105	105
5 years	56	66
10 years	33	48
15 years	19	35
Adult	3	18

Source: Table 2.43, "ICRP Publication 89" (ICRP 2002).

remodeling of cortical bone in adult humans, based on histological data and estimates of the rate of turnover of radionuclides in adult humans. A value of 18% per year is estimated as the average rate of remodeling of trabecular bone in the adult human, based on histological data, estimates of the ration of turnover of radionuclides in adult humans, and the presumption that the amount of remodeling per unit area of bone surface is the same in trabecular bone as it is in cortical bone. Remodeling rates vary substantially with age during adulthood, and thus the values given here are estimated averages.

5.4.7 Physiology of the Developing Fetus and Mother

Chapter 12 of ICRP Publication 89 (ICRP 2006) provides some information on physiological changes during pregnancy needed to support biokinetic models of intake and radiation exposure of the fetus. These include descriptive summaries of data on basal metabolic rate, fluid intake, respiratory function, gastrointestinal function, liver function, renal function, and cardiovascular function. Table 5.30 compares reference values of blood flow in the pregnant and non-pregnant adult female.

5.5 COMPARISON OF ICRP REFERENCE DATA WITH THAT OF ASIAN POPULATIONS

A considerable amount of information has been published during the past decade on characteristics of several Asian populations. Included are reports on Japanese populations (Tanaka, Kawamura, and Nakahara 1979; Tanaka 1992), Chinese populations (Wang, Chen, and Zhu 1999), and Indian populations (Jain et al. 1995). The most extensive effort has been the five-year effort conducted under the auspices of the IAEA. In this effort, characteristics of populations in Bangladesh, China, India, Japan, Republic of Korea, Pakistan, Philippines, and Vietnam were examined and compared (IAEA 1998). This IAEA report presented comparative information on height, weight, other anthropomorphic measurements, organ masses, daily dietary intake, pulmonary function, and water balance. Also included in the IAEA report were the results of a model prepared by Tanaka giving suggested reference values for Asian male and female subjects at these six ages: newborn; 1, 5, 10, and 15 years; and adult (Tanaka and Kawamura 1996).

The authors of this report noted major questions that arose relating to the adequate and appropriate characterization of reference values for Asian populations. These uncertainties

TABLE 5.30 Reference Values for Blood Flow to Organs of the Non-Pregnant and Pregnant Adult Female Near Term

Organ/Tissue	Blood Flow Rate (% Cardiac Output)	
	Non-Pregnant	Pregnant
Fat	8.5	7.8
Brain	12.0	8.8
Gastrointestinal tract	17.0	12.5
Heart	5.0	3.7
Kidneys	17.0	16.6
Liver	27.0[a]	20.0[a]
Arterial	(6.5)	(4.8)
Portal	(20.5)	(15.2)
Lungs	2.5	1.8
Muscle	12.0	8.8
Pancreas	1.0	0.7
Skeleton	5.0	3.7
Skin	5.0	8.7
Spleen	3.0	2.2
Thyroid	1.5	1.1
Uterus	0.4	12.0
Breast	0.4	3.5
Other	3.2	3.3
Cardiac output (1/min)	5.9	7.3

Source: Table 2.44, "ICRP Publication 89" (ICRP 2002).
[a]Total of value in parentheses.

included: (1) significant variations between, and even within, national populations; and (2) secular trends within a given population as a result of changes in food distribution and dietary habits. The problem of population variations on a regional or even national scale is analogous to the difficulty in defining a worldwide "Reference Man" given differences in major ethnic populations.

As an example of a secular change, data on the height of 17-year-old Japanese males were compared for the years 1977 and 1991. The mean height had increased from 169.1 to 170.6 cm and this shift was evident throughout the distribution of heights. The authors also noted that there was no acceleration of growth in Western European and North American countries and secular trends could be ignored. In contrast, the acceleration of growth in body height and mass should be considered for developing countries.

In addition to secular trends, the authors had to deal with variations within national populations from different geographical locations, as well as different ethnic and income distributions. Limited funding also restricted the measurement program, and national data obtained for other purposes were sometimes included in the analyses.

Tables 5.31 and 5.32 display values of standing heights and total-body masses, respectively, among the various countries of the IAEA report, in comparison to the ICRP reference values of ICRP Publication 89 (ICRP 2006). In all cases, mean values of height and

TABLE 5.31   Height in Adult Asian Males and Females Compared
with ICRP Reference Values[a]

	Males (cm)[b,c]	Females (cm)[b]
Pakistan	171 ± 6.4	158 ± 6.7
China	169 ± 5.8	158 ± 5.4
Japan	168 ± 5.7	155 ± 5.2
Republic of Korea	167 ± 5.5	155 ± 4.9
Bangladesh	164 ± 12.8	155 ± 5.6
Vietnam	164 ± 5.2	154 ± 4.5
Philippines	163 ± 13.8	151 ± 5.4
India	163 ± 7.5	151 ± 6.5
ICRP reference values	176	163

Source: "ICRP Publication 89," p.84 (ICRP 2002).
[a] Modified from IAEA report (IAEA 1998).
[b] Mean ± SD.
[c] Country entries sorted according to male values.

TABLE 5.32   Total-Body Mass in Adult Asian Males and
Females Compared with ICRP Reference Values

	Males (kg)[b,c]	Females (kg)[b]
Pakistan	63.9 ± 8.1	52.6 ± 8.5
Republic of Korea	63.8 ± 7.7	54.5 ± 6.5
Japan	63.6 ± 8.8	52.3 ± 7.4
China	58.3 ± 6.4	51.1 ± 6.4
Bangladesh	57.8 ± 9.0	47.9 ± 7.9
Philippines	56.6 ± 8.3	49.2 ± 8.7
Vietnam	51.8 ± 5.4	46.8 ± 5.3
India	51.5 ± 8.5	44.2 ± 8.0
Tanaka model	60	51
ICRP reference values	73	60

Source: "ICRP Publication 89," p.85 (ICRP 2002).
[a] Modified from IAEA report (IAEA 1998).
[b] Mean ± SD.
[c] Country entries sorted according to male values.

weight of adults in these Asian countries are smaller in magnitude than those given by
ICRP reference values. This result has led to efforts in several of these countries—China,
Korea, and Japan in particular—to launch national efforts to devise their own national
definitions of reference individuals, including associated computational anatomic models.
Comparisons of country-specific total-body masses of males and of females, in compari-
son to ICRP reference values, are next shown in Tables 5.33 and 5.34, respectively.

Finally, as an example of organ-specific deviations from ICRP reference values, Table 5.35
gives age-specific total lung masses (inclusive of pulmonary blood) for three of the Asian
countries—Japan, China, and India—in comparison to the Tanaka Reference Asian model
and the ICRP Reference model. At 10 years of age, the two reference models generally agree

TABLE 5.33   Total-Body Mass (Mean Values) in Asian Males as a Function of Age Compared with ICRP Reference Values

	Newborn	1 Year	5 Years	10 Years	15 Years	Adult[b]
Pakistan	3.2	–[c]	20.3	34.2	51.6	63.9
Republic of Korea	–	–	–	30.7	53.2	63.8
Japan	3.2	9.6	19.0	32.5	57.2	63.6
China	3.2	9.1	16.3	27.0	48.6	58.3
Bangladesh	2.4	8.1	16.4	27.2	43.9	57.8
Philippines	–	9.3	15.2	24.3	43.1	56.6
Vietnam	3.0	7.6	14.8	23.5	40.9	51.8
India	2.9	8.5	14.6	22.9	38.3	51.5
Tanaka model	–	11	19	30	54	60
ICRP reference values	3.5	10	19	32	56	73

Source: "ICRP Publication 89," p.85 (ICRP 2002).
[a] Modified from IAEA report (IAEA 1998).
[b] Entries ranked according to adult values.
[c] Data not available.

TABLE 5.34   Total-Body Mass (Mean Values) in Asian Females as a Function of Age Compared with ICRP Reference Values

	Newborn	1 Year	5 Years	10 Years	15 Years	Adult[b]
Pakistan	–[c]	–	–	30.6	49.3	54.5
Republic of Korea	3.3	–	15.7	19.1	46.9	52.6
Japan	3.2	9.1	18.6	32.8	51.6	52.3
China	3.1	8.5	15.8	27.1	46.3	51.1
Bangladesh	2.5	7.0	16.4	26.7	42.5	49.9
Philippines	–	9.0	15.2	25.7	43.3	49.2
Vietnam	2.9	7.8	14.5	27.0	40.5	46.8
India	2.8	8.1	14.2	22.9	38.7	44.2
Tanaka model	–	11	19	31	49	51
ICRP reference values	3.5	10	19	32	53	60

Source: "ICRP Publication 89," p.85 (ICRP 2002).
[a] Modified from IAEA report (IAEA 1998).
[b] Entries ranked according to adult values.
[c] Data not available.

regarding reference lung masses. At earlier ages, however, the Tanaka Reference model lung masses exceed those of the ICRP, while at older ages, the Tanaka Reference model generally reported lower lung masses than given in the ICRP model.

5.6 USE OF THE ICRP REFERENCE INDIVIDUAL IN EXTERNAL AND INTERNAL DOSIMETRY

In this closing section, we briefly summarize the current status of ICRP publications of reference dose coefficients for use in external and internal dose assessment. The three foundation documents upon which the most recent dose coefficients are computed include (1) ICRP Publication 103 defining the current values for radiation and tissue weighting factors

TABLE 5.35   Mass of Lungs (Inclusive of Blood) as a Function of Age in Asian Populations Compared with
ICRP Reference Values[a]

		Mass (g)					
		Newborn	1 Year	5 Years	10 Years	15 Years	Adult[b]
Males	Japan	90	190	320	550	910	1170
	China	61	210	360	560	940	1060
	India	63	120	250	460	650	840
	Tanaka model	–	190	320	520	930	1200
	ICRP reference values[b]	60	150	300	500	900	1200
Females	Japan	90	190	260	450	640	910
	China	57	190	350	470	770	840
	India	63	98	210	410	600	670
	Tanaka model	–	190	310	540	710	910
	ICRP reference values[b]	60	150	300	500	750	950

Source: "ICRP Publication 89," p. 85 (ICRP 2002).
[a] Asian values taken from IAEA report (IAEA 1998).
[b] From this report.

of the effective dose (ICRP 2007), (2) ICRP Publication 107 on reference radionuclide
decay data (ICRP 2008b), and (3) ICRP Publication 110 on Adult Reference Computational
Phantoms (ICRP 2009). ICRP Committee 2 is presently developing publications outlining
reference pediatric phantoms and reference pregnant female phantoms to extend these
efforts to members of the general public.

5.6.1 Dose Coefficients for Internal Occupational and Environmental Exposures

The earliest values of dose coefficients for use in occupational internal radionuclide exposure assessment (both inhalation and ingestion intake pathways) were issued as a multivolume series in ICRP Publication 30 (ICRP 1980). These dose coefficients used decay
scheme data from ICRP Publication 38 (ICRP 1983) and computational values of specific
absorbed fractions given in ICRP Publication 23 (ICRP 1975) using the MIRD stylized
adult phantoms. Biokinetic models were rather simplistic in form (sums of exponential
terms) in which radionuclide uptake from blood to the various source organs was assumed
to be instantaneous, and radionuclide elimination from each source organ was assumed
to go directly to excreted activity (no recirculation back to blood). In the subsequent years
since that publication series, ICRP dose coefficients were significantly enhanced to include
values for members of the general public (using the age-dependent ORNL stylized phantoms) and the creation of physiologically realistic systemic biokinetic models. Inhalation
modeling was substantially improved via new models for particle deposition, clearance,
and dosimetry as documented in ICRP Publication 66 (ICRP 1994a). In 1994, a major revision to the ICRP Publication 30 (ICRP 1980) adult reference dose coefficients were given
in ICRP Publication 68 (ICRP 1994b). Age-dependent dose coefficients for both inhalation
and ingestion were issued in a five-part series as ICRP Publications 56, 67, 69, 71, and 72
(ICRP 1990, 1993, 1995a,c, 1996). In 2012, a comprehensive compendium of ICRP dose

coefficients, based upon ICRP Publication 60 (ICRP 1991) radiation and tissue weighting factors, was issued as ICRP Publication 119 (ICRP 2012b).

Following the release of ICRP Publication 103 (new radiation and tissue weighting factors) (ICRP 2007), ICRP Publication 107 (new radionuclide decay data) (ICRP 2008b), and 110 (new reference voxel phantoms) (ICRP 2009), ICRP Committee 2 embarked on a major update of the ICRP 30 and ICRP 56–72 publication series for internal dose coefficients for the reference adult male and reference adult female. The series is entitled Occupational Intakes of Radionuclides (OIR). OIR Part 1 provides an introduction to the entire OIR series, including detailed chapters on: (1) monitoring and assessment of internal occupational exposures to radionuclides; (2) biokinetic and dosimetric models; (3) methods of individual and workplace monitoring; (4) monitoring programs; (5) general aspects of retrospective dose assessment and retrospective dose verification; and (6) data provided for elements and radioisotopes in the various chapters of the subsequent OIR document series (ICRP 2015b). Two important annexes are given in OIR Part 1. Annex A provides revisions to the Human Respiratory Tract Model from that published in ICRP Publication 66. Annex B gives a summary of the evolution of ICRP's system biokinetic models. Values of specific absorbed fraction for source/target organ pairs in the ICRP Reference Adult Male and Adult Female have been computed and are available in ICRP Publication 133 (ICRP 2016a).

At the time of this writing, the subsequent OIR document series will cover the following elements, providing adult reference values for dose coefficients for both inhalation and ingestion, as well as new dose coefficients reporting the effective dose per unit bioassay content (ICRP 2016b). Furthermore, the reports will provide dose coefficients for external exposure to radioactive noble gases in occupational settings (Veinot et al. 2017a).

OIR Part 2

Hydrogen (H), Carbon (C), Phosphorus (P), Sulphur (S), Calcium (Ca), Iron (Fe), Cobalt (Co), Zinc (Zn), Strontium (Sr), Yttrium (Y), Zirconium (Zr), Niobium (Nb), Molybdenum (Mo), and Technetium (Tc).

OIR Part 3

Ruthenium (Ru), Antimony (Sb), Tellurium (Te), Iodine (I), Caesium (Cs), Barium (Ba), Iridium (Ir), Lead (Pb), Bismuth (Bi), Polonium (Po), Radon (Rn), Radium (Ra), Thorium (Th), and Uranium (U).

OIR Part 4

Lanthanides series, actinium (Ac), protactinium (Pa) and transuranic elements

OIR Part 5

Fluorine (F), Sodium (Na), Magnesium (Mg), Potassium (K), Manganese (Mn), Nickel (Ni), Selenium (Se), Molybdenum (Mo), Technetium (Tc), and Silver (Ag).

In a parallel effort, ICRP Committee 2 has plans to update the age-dependent dose coefficients of the ICRP Publication 56–72 series for use in environmental radiological protection. The new series of documents will be issued as the EIR series—Environmental Intakes of Radionuclides. In the EIR series, the revised physiologically realistic biokinetic compartmental models of the OIR series will be adopted and extended to include, when data are available, age-dependent transfer coefficients. In addition, new values of specific absorbed fraction will be used based upon the ICRP reference pediatric voxel phantoms.

5.6.2 Dose Coefficients for External Occupational Exposures

The first document published by the ICRP on dose coefficients for occupational external exposures to reference adults was ICRP Publication 51 (ICRP 1988a). This document was subsequently updated in 1996 as ICRP Publication 74 (ICRP 1997), which were jointly issued as ICRU Report 57 (ICRU 1998). In these documents, dose coefficients were given for broad beams of monoenergetic radiations in standardized orientations (e.g., AP for Anterior-Posterior incident) including photons, electrons, neutrons, and various other charged particles. The energy ranges were limited, and most of the data were obtained using the ORNL series of stylized adult phantoms. In 2010, a major update to its reference values for external dose coefficients was issued in ICRP Publication 116 using the newly developed ICRP 110 reference adult voxel phantoms (ICRP 2010). This document provided both male and female organ-specific equivalent dose per particle fluence and the gender-averaged effective dose per particle fluence. Members of the task group that compiled this work also contributed substantially to external dose coefficients of relevance to both cosmic ray exposures to air-crew in ICRU Report 84 (ICRU 2010), and to space radiation exposures to astronauts in ICRP Publication 123 (ICRP 2013).

5.6.3 Dose Coefficients for External Environmental Exposures

Values of organ and effective dose from external exposure to radionuclides concentrated in air, water, and soil have been published both by the GSF laboratory in Munich, Germany (Petoussi et al. 1991; Petoussi-Henss et al. 2012) and the Oak Ridge National Laboratory (Eckerman and Ryman 1993; Veinot et al. 2017b; Hiller et al. 2017; Bellamy et al. 2017). However, the ICRP itself has never officially provided reference values for environmental external exposure. Following the 2011 Fukushima nuclear power accident in Japan, the ICRP established a new task group to compile age-dependent external dose coefficients for use in environmental dose assessment. This work includes the use of the ICRP Publication 110 (ICRP 2009) in reference to adult voxel phantoms and updated versions of the UF/NCI pediatric voxel phantoms. These dose coefficients and the resulting ICRP publication are anticipated to be available in 2018.

5.6.4 Dose Coefficients for Medical Exposures

ICRP literature on dose coefficients for medical exposures has been limited to date to those supporting dose assessments in diagnostic nuclear medicine. Task Group 36, a joint task group of ICRP Committee 2 and Committee 3, has issued three reports on organ and effective dose per administered activity (Sv/Bq) for various commonly used radiopharmaceuticals. These include ICRP Publication 53 (ICRP 1988b), ICRP Publication 80 (ICRP 1998), and ICRP Publication 106 (ICRP 2008a). In 2015, an extensive compendium of radiopharmaceutical dose coefficients from these three prior publications was issued as ICRP Publication 128 (ICRP 2015a). The biokinetic models typically used by Task Group 36 are of the retention equation format of ICRP Publication 30 (ICRP 1980), in which values of fractional uptake and biological half-life are assigned to each identified source organ. Almost without exception, these biokinetic models are based upon observed radio-pharmaceutical uptake, retention, and excretion data obtained from nuclear medicine imaging studies in adult patients. Age-dependent dose coefficients are developed using adult biokinetic models, but with age-dependent values of specific absorbed fractions from the ORNL stylized model series. Future updates to ICRP Publication 128 dose coefficients will utilized pediatric phantom SAFs currently in development.

The two other broad categories of diagnostic imaging for which dose coefficients could be developed include diagnostic fluoroscopy and computed tomography. To date, ICRP has not embarked on developing reference dose coefficients for these two imaging modalities. For diagnostic fluoroscopy, this can be accomplished in one of two ways. One approach would be to provide organ and effective doses for the entire procedure which would entail estimates of beam orientation, beam location, field size, irradiation time, X-ray intensity, and X-ray beam energy spectra for each segment of the diagnostic fluoroscopic examination. The other approach would be to provide reference dose coefficients for both organs and the effective dose for individual fluoroscopic and radiographic fields ranging across an array of clinical parameters, and then let the user assemble them into estimates of exam-specific cumulative organ and effective dose for each reference individual.

For computed tomography, the same approach could be applied. Various professional organizations now have established reference protocols for various CT imaging examinations, giving the anatomical landmarks defining the CT scan length, as well as values of body-size specific technique factors (pitch, tube voltage, mAs, and beam collimation). These reference imaging protocols could be applied to various computational phantoms representing the ICRP series of reference children and adults. Newer dose reduction technologies could also be accommodated, such as tube current modulation and iterative reconstruction whereby lower tube currents can be applied and yet yield acceptable diagnostic quality images (Stepusin et al. 2017).

While these efforts may be pursued, these dose coefficients would only apply to the small array of ICRP Reference Individuals—newborn, 1-year-old, 5-year-old, 10-year-old, 15-year-old, and adults, both males and females. They would only suffice for intercomparison of imaging protocols and changes in technique factors. They would not be applicable to dose reconstructions to individual patients, many of which would not conform to the body morphometries and ages of the 12-member ICRP series of reference individuals. For patient-specific medical dosimetry to include nuclear medicine in addition to fluoroscopy and computed tomography, the use of an extended phantom library covering a broad range of patient body morphometries would be a far more efficient approach to patient organ dosimetry as needed for dose tracking, radiation epidemiology, and benefit/risk optimization (Bolch et al. 2010).

REFERENCES

Aboul-Khair, S. A., T. J. Buchanan, J. Crooks, and A. C. Turnbull. 1966. "Structural and Functional Development of the Human Foetal Thyroid." *Clin. Sci.* 31 (3):415–24.

Ares, S., I. Pastor, J. Quero, and G. Morreale de Escobar. 1995. "Thyroid Gland Volume as Measured by Ultrasonography in Preterm Infants." *Acta. Paediatr.* 84 (1):58–62.

Bates, D. V., B. R. Fish, T. F. Hatch, T. T. Mercer, and P. E. Morrow. 1966. "Deposition and Retention Models for Internal Dosimetry of the Human Respiratory Tract. Task Group on Lung Dynamics." *Health. Phys.* 12 (2):173–207.

Bellamy, M. B., K. G. Veinot, M. M. Hiller, S. A. Dewji, K. F. Eckerman, C. E. Easterly, N. E. Hertel, and R. W. Leggett. 2017. "Effective Dose Rate Coefficients for Immersions in Radioactive Air and Water." *Radiat. Prot. Dosimetry* 174 (2):275–86. doi: 10.1093/rpd/ncw103.

Beyer, P. L., and M. A. Flynn. 1978. "Effects of High- and Low-Fiber Diets on Human Feces." *J. Am. Diet. Assoc.* 72 (3):271–7.

Bolch, W., C. Lee, M. Wayson, and P. Johnson. 2010. "Hybrid Computational Phantoms for Medical Dose Reconstruction." *Radiat. Environ. Biophys.* 49 (2):155–68. doi: 10.1007/s00411-009-0260-x.

Bolch, W. E., N. Petoussi-Henss, F. Paquet, and J. Harrison. 2016. "ICRP Dose Coefficients: Computational Development and Current Status." *Ann. ICRP* 45 (1 Suppl):156–77. doi: 10.1177/0146645316636010.

Bouchet, L. G., and W. E. Bolch. 1999. "Five Pediatric Head and Brain Mathematical Models for Use in Internal Dosimetry." *J. Nucl. Med.* 40 (8):1327–36.

Bouchet, L. G., W. E. Bolch, P. Blanco, B. Wessels, J. A. Siegel, D. A. Rajon, I. Clairand, and G. Sgouros. 2003. "MIRD Pamphlet No. 19: Absorbed Fractions and Radionuclide S Values for Six Age-Dependent Multi-Region Models of the Kidney." *J. Nucl. Med.* 44 (7):1113–47.

Charles, M. W., and N. Brown. 1975. "Dimensions of the Human Eye Relevant to Radiation Protection." *Phys. Med. Biol.* 20 (2):202–18.

Coffey, J. L., M. Cristy, and G. G. Warner. 1981. "MIRD Pamphlet No. 13: Specific Absorbed Fractions for Photon Sources Uniformly Distributed in the Heart Chambers and Heart Wall of a Heterogeneous Phantom." *J. Nucl. Med.* 22 (1):65–71.

Cook, M. J. 1948. *A Survey Report of the Characteristics of the Standard Man.* Oak Ridge, TN: Oak Ridge National Laboratory.

Costa, A., V. De Filippis, M. Panizzo, G. Giraudi, E. Bertino, R. Arisio, M. Mostert, G. Trapani, and C. Fabris. 1986. "Development of Thyroid Function between Vi-Ix Month of Fetal Life in Humans." *J. Endocrinol. Invest.* 9 (4):273–80. doi: 10.1007/BF03346925.

Cristy, M. 1980. *Mathematical Phantoms Representing Children of Various Ages for Use in Estimates of Internal Dose.* Oak Ridge, TN: Oak Ridge National Laboratory.

Cristy, M., and K. F. Eckerman. 1987. *Specific Absorbed Fractions of Energy at Various Ages from Internal Photon Sources.* Oak Ridge, TN: Oak Ridge National Laboratory.

Cummings, J. H., M. J. Hill, D. J. Jenkins, J. R. Pearson, and H. S. Wiggins. 1976. "Changes in Fecal Composition and Colonic Function Due to Cereal Fiber." *Am. J. Clin. Nutr.* 29 (12):1468–73.

Dimbylow, P. 2005. "Development of the Female Voxel Phantom, NAOMI, and Its Application to Calculations of Induced Current Densities and Electric Fields from Applied Low Frequency Magnetic and Electric Fields." *Phys. Med. Biol.* 50 (6):1047–70.

Dolphin, G. W., and I. S. Eve. 1966. "Dosimetry of the Gastrointestinal Tract." *Health Phys.* 12 (2):163–72.

Eckerman, K. F., and J. C. Ryman. 1993. *Federal Guidance Report No. 12: External Exposure to Radionuclides in Air, Water, and Soil.* Washington, DC: U.S. Environmental Protection Agency.

Evans, T. C., R. M. Kretzschmar, R. E. Hodges, and C. W. Song. 1967. "Radioiodine Uptake Studies of the Human Fetal Thyroid." *J. Nucl. Med.* 8 (3):157–65.

Eve, I. S. 1966. "A Review of the Physiology of the Gastrointestinal Tract in Relation to Radiation Doses from Radioactive Materials." *Health Phys.* 12 (2):131–61.

Eveleth, P. B., and J. M. Tanner. 1976. *Worldwide Variation in Human Growth*. Cambridge, U.K.: Cambridge University Press.

Eveleth, P. B., and J. M. Tanner. 1990. *Worldwide Variation in Human Growth*, 2nd edn. Cambridge, U.K.: Cambridge University Press.

Gehan, E. A., and S. L. George. 1970. "Estimation of Human Body Surface Area from Height and Weight." *Cancer Chemother. Rep.* 54 (4):225–35.

Goldberg, G. R., A. M. Prentice, W. A. Coward, H. L. Davies, P. R. Murgatroyd, C. Wensing, A. E. Black, M. Harding, and M. Sawyer. 1993. "Longitudinal Assessment of Energy Expenditure in Pregnancy by the Doubly Labeled Water Method." *Am. J. Clin. Nutr.* 57 (4):494–505.

Guihard-Costa, A. M., J. C. Larroche, P. Droulle, and F. Narcy. 1995. "Fetal Biometry: Growth Charts for Practical Use in Fetopathology and Antenatal Ultrasonography." *Fetal. Diagn. Ther.* 10 (4):215–78.

Han, E. Y., W. E. Bolch, and K. F. Eckerman. 2006. "Revisions to the ORNL Series of Adult and Pediatric Computational Phantoms for Use with the MIRD Schema." *Health Phys.* 90 (4):337–56. doi: 10.1097/01.HP.0000192318.13190.c400004032-200604000-00004 [pii].

Hawk, P. B., B. L. Oser, and W. H. Summerson. 1947. *Practical Physiological Chemistry*, 12th edn. Philadelphia, PA: The Blakeston Company.

He, B., Y. Du, X. Song, W. P. Segars, and E. C. Frey. 2005. "A Monte Carlo and Physical Phantom Evaluation of Quantitative in-111 Spect." *Phys. Med. Biol.* 50 (17):4169–85.

Hiller, M. M., K. Veinot, C. E. Easterly, N. E. Hertel, K. F. Eckerman, and M. B. Bellamy. 2017. "Reducing Statistical Uncertainties in Simulated Organ Doses of Phantoms Immersed in Water." *Radiat. Prot. Dosim.* 174 (4):439–48. doi: 10.1093/rpd/ncw240.

Hudson, G. 1965. "Bone Marrow Volume in the Human Foetus and Newborn." *Br. J. Haematol.* 11:446–52.

Hytten, F. E. 1980. "Nutrition." In F. Hytten and G. Chamberlain (eds), *Clinical Physiology in Obstetrics*. Oxford, U.K.: Blackwell Scientific Publications.

International Atomic Energy Agency. 1998. *Compilation of Anatomical, Physiological, and Metabolic Characteristics for Reference Asian Man*. Vienna, Austria: International Atomic Energy Agency.

International Commission on Radiation Units and Measurements. 1992. *ICRU Report No. 46: Photon, Electron, Proton and Neutron Interaction Data for Body Tissues*. Bethesda, MD: International Commission on Radiation Units and Measurements.

International Commission on Radiation Units and Measurements. 1998. *ICRU Report No. 57: Conversion Coefficients for Use in Radiological Protection against External Radiation*. Bethesda, MD: International Commission on Radiation Units and Measurements.

International Commission on Radiation Units and Measurements. 2010. "ICRU Report 84: Reference Data for the Validation of Doses from Cosmic-Radiation Exposure of Aircraft Crew." *J. ICRU* 10 (2):1–35.

International Commission on Radiological Protection. 1951. "International Recommendations on Radiological Protection." *Br. J. Radiol.* 24 (277):46–53.

International Commission on Radiological Protection. 1953. *Seventh International Congress of Radiology and Associated Conferences*. Copenhagen, Denmark, July 22–29, 1953.

International Commission on Radiological Protection. 1955. Recommendations of the ICRP - Report on the International Subcommittee 11 on Permissible Dose for Internal Emitters. *Br. J. Radiol.* (Suppl. 6).

International Commission on Radiological Protection. 1960. *ICRP Publication 2: Recommendations of the ICRP, Report of Committee II on Permissible Dose for Internal Radiation*. Oxford, U.K.: Pergamon Press.

International Commission on Radiological Protection. 1975. "ICRP Publication 23: Report on the Task Group on Reference Man." In *Annals of the ICRP*. Oxford, U.K.: Pergamon Press.

International Commission on Radiological Protection. 1980. *ICRP Publication 30: Limits for Intakes of Radionuclides by Workers*. Oxford, U.K.: Pergamon Press.

International Commission on Radiological Protection. 1983. *ICRP Publication 38: Radionuclide Transformations - Energy and Intensity of Emissions*. Ann ICRP3.

International Commission on Radiological Protection. 1988a. *ICRP Publication 51: Data for Use in Protection from External Radiation*. New York, NY: International Commission on Radiological Protection.

International Commission on Radiological Protection. 1988b. "ICRP Publication 53: Radiation Dose to Patients from Radiopharmaceuticals." *Ann. ICRP* 18 (1–4):1–137.

International Commission on Radiological Protection. 1990. "ICRP Publication 56: Age-Dependent Doses to Members of the Public from Intake of Radionuclides - Part 1." *Ann. ICRP* 20 (2):1–22.

International Commission on Radiological Protection. 1991. *ICRP Publication 60: 1990 Recommendations of the International Commission on Radiological Protection*. Ann ICRP.

International Commission on Radiological Protection. 1993. "ICRP Publication 67: Age-Dependent Doses to Members of the Public from Intake of Radionuclides: Part 2 - Ingestion Dose Coefficients." *Ann. ICRP* 23 (3–4):1–167.

International Commission on Radiological Protection. 1994a. "ICRP Publication 66: Human Respiratory Tract Model for Radiological Protection." *Ann. ICRP* 24 (1–3):1–482.

International Commission on Radiological Protection. 1994b. *ICRP Publication 68: Dose Coefficients for Intakes of Radionuclides by Workers*. Oxford, U.K.: International Commission on Radiological Protection.

International Commission on Radiological Protection. 1995a. *ICRP Publication 69: Age-Dependent Doses to Members of the Public from Intake of Radionuclides: Part 3 - Ingestion Dose Coefficients*. Elmsford, NY: International Commission on Radiological Protection.

International Commission on Radiological Protection. 1995b. "ICRP Publication 70: Basic Anatomical and Physiological Data for Use in Radiological Protection: The Skeleton." *Ann. ICRP* 25 (2):1–180.

International Commission on Radiological Protection. 1995c. "ICRP Publication 71: Age-Dependent Doses to Members of the Public from Intake of Radionuclides - Part 4 Inhalation Dose Coefficients." *Ann. ICRP* 25 (3–4):1–405.

International Commission on Radiological Protection. 1996. "ICRP Publication 72: Age-Dependent Doses to the Members of the Public from Intake of Radionuclides - Part 5 Compilation of Ingestion and Inhalation Coefficients." *Ann. ICRP* 26 (1):1–91.

International Commission on Radiological Protection. 1997. *ICRP Publication 74: Conversion Coefficients for Use in Radiological Protection against External Radiation*. New York, NY: International Commission on Radiological Protection.

International Commission on Radiological Protection. 1998. "ICRP Publication 80: Radiation Dose to Patients from Radiopharmaceuticals - Addendum 2 to ICRP Publication 53." *Ann. ICRP* 28 (3):1–126.

International Commission on Radiological Protection. 2001. *ICRP Publication 88: Doses to the Embryo and Fetus from Intakes of Radionuclides by the Mother*. Elmsford, NY: International Commission on Radiological Protection.

International Commission on Radiological Protection. 2002. "ICRP Publication 89: Basic Anatomical and Physiological Data for Use in Radiological Protection - Reference Values." *Ann. ICRP* 32 (3–4):1–277.

International Commission on Radiological Protection. 2006. "ICRP Publication 100: Human Alimentary Tract Model for Radiological Protection." *Ann. ICRP* 36 (1–2):1–366.

International Commission on Radiological Protection. 2007. "ICRP Publication 103: Recommendations of the International Commission on Radiological Protection." *Ann. ICRP* 37 (2–4):1–332.

International Commission on Radiological Protection. 2008a. "ICRP Publication 106: Radiation Dose to Patients from Radiopharmaceuticals—Addendum 3 to ICRP Publication 53." *Ann. ICRP* 38 (1–2):1–197.

International Commission on Radiological Protection. 2008b. "ICRP Publication 107: Nuclear Decay Data for Dosimetric Calculations." *Ann. ICRP* 38 (3):1–26.

International Commission on Radiological Protection. 2009. "ICRP Publication 110: Adult Reference Computational Phantoms." *Ann. ICRP* 39 (2):1–165.

International Commission on Radiological Protection. 2010. "ICRP Publication 116: Conversion Coefficients for Radiological Protection Quantities for External Radiation Exposures." *Ann. ICRP* 40 (2–5):1–257.

International Commission on Radiological Protection. 2012a. "ICRP Publication 118: ICRP Statement on Tissue Reactions and Early and Late Effects of Radiation in Normal Tissues and Organs: Threshold Doses for Tissue Reactions in a Radiation Protection Context." *Ann. ICRP* 41 (1–2):1–322.

International Commission on Radiological Protection. 2012b. "ICRP Publication 119 - Compendium of Dose Coefficients Based on ICRP Publication 60." *Ann. ICRP* 41 (Suppl.):1–130.

International Commission on Radiological Protection. 2013. "ICRP Publication 123: Assessment of Radiation Exposure of Astronauts in Space." *Ann. ICRP* 42 (4):1–339.

International Commission on Radiological Protection. 2015a. "ICRP Publication 128: Radiation Dose to Patients from Radiopharmaceuticals - a Compendium of Current Information Related to Frequently Used Substances." *Ann. ICRP* 44 (2S):1–321.

International Commission on Radiological Protection. 2015b. "ICRP Publication 130 - Occupational Intakes of Radionuclides: Part 1." *Ann. ICRP* 44 (2):1–188.

International Commission on Radiological Protection. 2016a. "ICRP Publication 133 - the ICRP Computational Framework for Internal Dose Assessment for Reference Adults: Specific Absorbed Fractions." *Ann. ICRP* 45 (2):1–74.

International Commission on Radiological Protection. 2016b. "Occupational Intakes of Radionuclides: Part 2." *Ann. ICRP* 45 (3/4):1–352.

Jain, S. C., S. C. Metha, B. Kumar, A. R. Reddy, and A. Nagaratnam. 1995. "Formulation of the Reference Indian Adult: Anatomic and Physiologic Data." *Health Phys.* 68 (4):509–22.

Jones, D. G. 1997. "A Realistic Anthropomorphic Phantom for Calculating Organ Doses Arising from External Photon Irradiation." *Radiat. Prot. Dosim.* 72 (1):21–29.

Kim, H. S., Y. S. Yeom, T. T. Nguyen, C. Choi, M. C. Han, J. K. Lee, C. H. Kim et al. 2017. "Inclusion of Thin Target and Source Regions in Alimentary and Respiratory Tract Systems of Mesh-Type ICRP Adult Reference Phantoms." *Phys. Med. Biol.* 62 (6):2132–52. doi: 10.1088/1361-6560/aa5b72.

Kramer, R., H. J. Khoury, J. W. Vieira, and V. J. Lima. 2006. "Max06 and Fax06: Update of Two Adult Human Phantoms for Radiation Protection Dosimetry." *Phys. Med. Biol.* 51 (14):3331–46.

Lalush, D. S., M. K. Jatko, and W. P. Segars. 2005. "An Observer Study Methodology for Evaluating Detection of Motion Abnormalities in Gated Myocardial Perfusion Spect." *IEEE Trans. Biomed. Eng.* 52 (3):480–5.

Lee, C., C. Lee, J. L. Williams, and W. E. Bolch. 2006. "Whole-Body Voxel Phantoms of Paediatric Patients - UF Series B." *Phys. Med. Biol.* 51 (17):4649–61.

Lee, C., D. Lodwick, J. Hurtado, D. Pafundi, J. L. Williams, and W. E. Bolch. 2010. "The UF Family of Reference Hybrid Phantoms for Computational Radiation Dosimetry." *Phys. Med. Biol.* 55 (2):339–63. doi: S0031-9155(10)33216-7 [pii] 10.1088/0031-9155/55/2/002.

Leggett, R. W., and L. R. Williams. 1991. "Suggested Reference Values for Regional Blood Volumes in Humans." *Health Phys.* 60 (2):139–154.

Leggett, R. W., and L. R. Williams. 1995. "A Proposed Blood Circulation Model for Reference Man." *Health Phys.* 69 (2):187–201.

Lisco, H. 1949. *Biological and Medical Divisions Quarterly Progress Report (November 1948 to February 1949).* Chicago, IL: Argonne National Laboratory.

Lomsky, M., J. Richter, L. Johansson, H. El-Ali, K. Astrom, M. Ljungberg, and L. Edenbrandt. 2005. "A New Automated Method for Analysis of Gated-Spect Images Based on a Three-Dimensional Heart Shaped Model." *Clin. Physiol. Funct. Imaging* 25 (4):234–40. doi: 10.1111/j.1475-097X.2005.00619.x.

Luecke, R. H., W. D. Wosilait, and J. F. Young. 1995. "Mathematical Representation of Organ Growth in the Human Embryo/Fetus." *Int. J. Biomed. Comput.* 39 (3):337–47.

Mair, B. A., D. R. Gilland, and J. Sun. 2006. "Estimation of Images and Nonrigid Deformations in Gated Emission Ct." *IEEE Trans. Med. Imaging* 25 (9):1130–44.

Mardirossian, G., M. Tagesson, P. Blanco, L. G. Bouchet, M. Stabin, H. Yoriyaz, S. Baza, M. Ljungberg, S. E. Strand, and A. B. Brill. 1999. "A New Rectal Model for Dosimetry Applications." *J. Nucl. Med.* 40 (9):1524–31.

Maynard, M. R., J. W. Geyer, J. P. Aris, R. Y. Shifrin, and W. Bolch. 2011. "The UF Family of Hybrid Phantoms of the Developing Human Fetus for Computational Radiation Dosimetry." *Phys. Med. Biol.* 56 (15):4839–79. doi: S0031-9155(11)83784-X [pii] 10.1088/0031-9155/56/15/014.

Maynard, M. R., N. S. Long, N. S. Moawad, R. Y. Shifrin, A. M. Geyer, G. Fong, and W. E. Bolch. 2014. "The UF Family of Hybrid Phantoms of the Pregnant Female for Computational Radiation Dosimetry." *Phys. Med. Biol.* 59 (15):4325–43. doi: 10.1088/0031-9155/59/15/4325.

Munro, N. B., and K. F. Eckerman. 1998. "Impacts of Physiological Changes During Pregnancy on Maternal Biokinetic Modelling." *Radiat. Protect. Dosimetry* 79 (1–4):327–33.

National Council on Radiation Protection and Measurements. 2009. *NCRP Report No. 164: Uncertainties in Internal Radiation Dose Assessment.* Bethesda, MD: National Council on Radiation Protection and Measurements.

Nguyen, T. T., Y. S. Yeom, H. S. Kim, Z. J. Wang, M. C. Han, C. H. Kim, J. K. Lee et al. 2015. "Incorporation of Detailed Eye Model into Polygon-Mesh Versions of ICRP-110 Reference Phantoms." *Phys. Med. Biol.* 60 (22):8695–707. doi: 10.1088/0031-9155/60/22/8695.

Petoussi, N., P. Jacob, M. Zankl, and K. Saito. 1991. "Organ Doses for Foetuses, Babies, Children, and Adults from Environmental Gamma Rays." *Radiat. Prot. Dosim.* 37 (1):31–41.

Petoussi-Henss, N., H. Schlattl, M. Zankl, A. Endo, and K. Saito. 2012. "Organ Doses from Environmental Exposures Calculated Using Voxel Phantoms of Adults and Children." *Phys. Med. Biol.* 57 (18):5679–713. doi: 10.1088/0031-9155/57/18/5679.

Piegl, L. 1991. "On NURBS: A Survey." *IEEE Comp Graph Appl.* 11:55–71.

Poston, J. W., W. E. Bolch, and L. G. Bouchet. 2002. "Mathematical Models of Human Anatomy." In H. Zaidi and G. Sgouros (eds), *Monte Carlo Calculations in Nuclear Medicine: Therapeutic Applications.* Bristol, U.K.: Institute of Physics Publishing, 108–132.

Schlattl, H., M. Zankl, and N. Petoussi-Henss. 2007. "Organ Dose Conversion Coefficients for Voxel Models of the Reference Male and Female from Idealized Photon Exposures." *Phys. Med. Biol.* 52 (8):2123–45. doi: S0031-9155(07)37425-3 [pii] 10.1088/0031-9155/52/8/006.

Segars, W. P. 2001. *Development and Application of the New Dynamic NURBS-Based Cardiac-Torso (Ncat) Phantom.* Ph.D. dissertation, Biomedical Engineering, University of North Carolina.

Segars, W. P., D. S. Lalush, and B. M. Tsui. 2001. "Modeling Respiratory Mechanics in the Mcat and Spline-Based MCAT Phantom." *IEEE Trans. Nucl. Sci.* 48 (1):89–97.

Snyder, W. S., M. R. Ford, and G. G. Warner. 1978. *MIRD Pamphlet No. 5, Revised: Estimates of Specific Absorbed Fractions for Photon Sources Uniformly Distributed in Various Organs of a Heterogeneous Phantom.* New York: Society of Nuclear Medicine.

Snyder, W. S., M. R. Ford, G. G. Warner, and H. L. Fisher. 1969. *MIRD Pamphlet No. 5: Estimates of Absorbed Fractions for Monoenergetic Photon Sources Uniformly Distributed in Various Organs of a Heterogeneous Phantom.* New York: Society of Nuclear Medicine.

Snyder, W. S., M. R. Ford, G. G. Warner, and S. B. Watson. 1974. *A Tabulation of Dose Equivalent Per Microcurie-Day for Source and Target Organs of an Adult for Various Radionuclides.* Oak Ridge, TN: Oak Ridge National Laboratory.

Stabin, M. G., X. G. Xu, M. A. Emmons, W. P. Segars, C. Shi, and M. J. Fernald. 2012. "RADAR Reference Adult, Pediatric, and Pregnant Female Phantom Series for Internal and External Dosimetry." *J. Nucl. Med.* 53 (11):1807–13. doi: 10.2967/jnumed.112.106138.

Stepusin, E. J., D. J. Long, K. Ficarrotta, D. E. Hintenlang, and W. E. Bolch. 2017. "Physical Validation of a Monte Carlo-Based Phantom-Derived Approach to Computed Tomography Organ Dosimetry under Tube Current Modulation." *Med. Phys.* 44 (10):5423–5432. doi: 10.1002/mp.12461.

Tanaka, G. 1992. *Reference Japanese. Vol 1: Anatomical Data.* Chiba, Japan: National Institute of Radiological Sciences.

Tanaka, C., and H. Kawamura. 1996. *Anatomical and Physiological Characteristics for Asian Reference Man.* Hitachinaka, Japan: National Institute of Radiological Sciences.

Tanaka, G. I., H. Kawamura, and Y. Nakahara. 1979. "Reference Japanese Man: Mass of Organs and Other Characteristics of Normal Japanese." *Health Phys.* 36 (3):333–46.

Tobon-Gomez, C., C. Butakoff, S. Aguade, F. Sukno, G. Moragas, and A. F. Frangi. 2008. "Automatic Construction of 3d-ASM Intensity Models by Simulating Image Acquisition: Application to Myocardial Gated Spect Studies." *IEEE Trans. Med. Imaging* 27 (11):1655–67.

Tripartite Conference on Permissible Dose. 1953. *Tripartite Conference on Permissible Dose.* Harriman, NY, March 30–April 1, 1953.

Veinot, K. G., S. A. Dewji, M. M. Hiller, K. F. Eckerman, and C. E. Easterly. 2017a. "Organ and Effective Dose Rate Coefficients for Submersion Exposure in Occupational Settings." *Radiat. Environ. Biophys.* 56 (4):453–62. doi: 10.1007/s00411-017-0705-6.

Veinot, K. G., K. F. Eckerman, M. B. Bellamy, M. M. Hiller, S. A. Dewji, C. E. Easterly, N. E. Hertel, and R. Manger. 2017b. "Effective Dose Rate Coefficients for Exposure to Contaminated Soil." *Radiat. Environ. Biophys.* 56 (3):255–67. doi: 10.1007/s00411-017-0692-7.

Wang, J., R. Chen, and H. Zhu. 1999. *Data of Anatomical, Physiological, and Metabolic Characteristics of Chinese Reference Man.* Beijing, China: Atomic Energy Press.

Williams, L. R., and R. W. Leggett. 1989. "Reference Values for Resting Blood Flow to Organs of Man." *Clin. Phys. Physiol. Meas.* 10 (3):187–217.

Xu, X. G., and K. F. Eckerman. 2009. *Handbook of Anatomical Models for Radiation Dosimetry, Series in Medical Physics and Biomedical Engineering.* London, U.K.: Taylor & Francis.

Xu, X. G., V. Taranenko, J. Zhang, and C. Shi. 2007. "A Boundary-Representation Method for Designing Whole-Body Radiation Dosimetry Models: Pregnant Females at the Ends of Three Gestational Periods: RPI-P3, -P6 and -P9." *Phys. Med. Biol.* 52 (23):7023–44. doi: S0031-9155(07)59960-4 [pii] 10.1088/0031-9155/52/23/017.

Yeom, Y. S., H. S. Kim, T. T. Nguyen, C. Choi, M. C. Han, C. H. Kim, J. K. Lee et al. 2016a. "New Small-Intestine Modeling Method for Surface-Based Computational Human Phantoms." *J. Radiol. Prot.* 36 (2):230–45. doi: 10.1088/0952-4746/36/2/230.

Yeom, Y. S., Z. J. Wang, T. T. Nguyen, H. S. Kim, C. Choi, M. C. Han, C. H. Kim et al. 2016b. "Development of Skeletal System for Mesh-Type ICRP Reference Adult Phantoms." *Phys. Med. Biol.* 61 (19):7054–73. doi: 10.1088/0031-9155/61/19/7054.

Zaidi, H., and X. G. Xu. 2007. "Computational Anthropomorphic Models of the Human Anatomy: The Path to Realistic Monte Carlo Modeling in Radiological Sciences." *Annu. Rev. Biomed. Eng.* 9:471–500.

Zhang, J., Y. H. Na, P. F. Caracappa, and X. G. Xu. 2009. "RPI-Am and RPI-Af, a Pair of Mesh-Based, Size-Adjustable Adult Male and Female Computational Phantoms Using ICRP-89 Parameters and Their Calculations for Organ Doses from Monoenergetic Photon Beams." *Phys. Med. Biol.* 54 (19):5885–908. doi: S0031-9155(09)22503-6 [pii] 10.1088/0031-9155/54/19/015.

Ziegler, E. E., A. M. O'Donnell, S. E. Nelson, and S. J. Fomon. 1976. "Body Composition of the Reference Fetus." *Growth* 40 (4):329–41.

CHAPTER 6

Biokinetic Models

Rich Leggett

CONTENTS

6.1   Introduction   216
    6.1.1   Purposes of Biokinetic Models   216
    6.1.2   Types of Information Typically Used to Develop Biokinetic Models   217
6.2   The ICRP's Respiratory Models   218
    6.2.1   ICRP Publication 2   218
    6.2.2   ICRP Publication 30 (Task Group Lung Model)   220
    6.2.3   ICRP Publication 66 (Human Respiratory Tract Model)   221
    6.2.4   Revision of the Human Respiratory Tract Model   224
6.3   The ICRP's Gastrointestinal (GI) Models   227
    6.3.1   ICRP Publication 2   227
    6.3.2   ICRP Publication 30   227
    6.3.3   ICRP Publication 100 (Human Alimentary Tract Model)   228
6.4   The ICRP's Systemic Biokinetic Models   235
    6.4.1   The Need for Element-Specific Structures for Systemic Biokinetic Models   235
    6.4.2   Evolution of the ICRP's Structures for Systemic Biokinetic Models   237
       6.4.2.1   ICRP Publication 2   237
       6.4.2.2   ICRP Publication 30   238
       6.4.2.3   ICRP Publication 68 and the Publication 72 Series   239
    6.4.3   Case Studies of Systemic Biokinetic Models and Underlying Data   243
       6.4.3.1   Strontium   243
       6.4.3.2   Iodine   254
       6.4.3.3   Cesium   271
       6.4.3.4   Plutonium   283
    6.4.4   Biokinetic Models for Radionuclides Produced in Vivo by Decay of Parent Nuclides   295
       6.4.4.1   General Considerations   295
       6.4.4.2   Case Studies of Treatment of Radioactive Progeny Produced in Vivo   298
References   303

6.1 INTRODUCTION

6.1.1 Purposes of Biokinetic Models

A biokinetic model is a set of mathematical functions that describe the time-dependent behavior of material that enters the body via a number of possible routes of intake. Biokinetic models are used to predict the time-dependent distribution of materials in the body and their rates of elimination along specific excretion pathways. For the case of an internally deposited radionuclide, such predictions are needed to:

- derive organ and effective dose coefficients (dose per unit intake), and

- estimate actual intake of the radionuclide based on measurements of activity in urine, feces, or other biological samples, or external measurements of activity in the total body or specific regions of the body. These results may then be used to calculate organ and effective doses resulting from the intake.

The biokinetic models illustrated in this chapter are from published or upcoming reports of the International Commission on Radiological Protection (ICRP). The ICRP's biokinetic models are generally first-order compartment models, with movement between compartments defined by transfer coefficients. A transfer coefficient from Compartment A to Compartment B of a biokinetic model represents fractional transfer of the contents of A to B per unit time.

For radionuclides with reasonably well-established behavior in the human body, the ICRP's models are intended to yield central dose estimates for healthy members of the population or a subgroup of the population, such as a given age group. For radionuclides with poorly established biokinetics, the models are designed so that dose estimates are more likely to overestimate than underestimate the central values for the population or subgroup.

Radioactive material may enter the body by three main routes: inhalation, ingestion, or through the skin. Entry through the skin may occur through intravenous or intramuscular injection, absorption through intact skin directly into the bloodstream, or absorption through an opening in the skin (a wound).

ICRP provides two models for the analysis of the behavior of the material after intake but before entry into the bloodstream, namely a respiratory tract model for inhaled material and an alimentary tract model for ingested material or material transferring from the respiratory tract to upper alimentary tract via different mechanisms, for example, escalation up the tracheobronchial tree, and swallowing. The respiratory and alimentary tract models then feed into "systemic" models, that is, models describing the behavior of the material once it enters the systemic circulation (bloodstream). A separate model for the intravenous injection case is not provided by the ICRP; this case can be addressed by assigning material to a blood compartment of the appropriate ICRP biokinetic model at time zero. The ICRP also does not provide wound models, because wounds are not part of what is considered normal operations, but represent special operational situations. Several wound models are available in the technical literature, for example, in NCRP Publication 156 (NCRP 2007).

Thus, in the ICRP's dose computation scheme, information on the behavior of radionuclides in the body is condensed into three main types of biokinetic models:

- A largely generic respiratory tract model is used to describe the deposition and retention of inhaled activity in the respiratory tract and its subsequent absorption to blood or clearance to the gastrointestinal tract, e.g., by escalation from the lungs, and swallowing. A respiratory tract model generally is not entirely generic, because element-specific or compound-specific data may be used to tailor the model to specific properties of elements or compounds following their inhalation and deposition in the respiratory tract. For example, for application of the model to radioisotopes of a given element, generic parameter values describing the dissolution of material in the respiratory tract may be modified to reflect reported dissolution data for that element.

- A generic gastrointestinal tract model is used to describe the movement of swallowed and secreted activity and, together with element-specific gastrointestinal absorption fractions, to describe the rate and extent of absorption of radionuclides from the gut into blood.

- Typically, element-specific systemic biokinetic models are used to describe the time-dependent distribution and excretion of radionuclides after their absorption or injection into blood. There is a trend in ICRP documents toward use of generic model structures, together with element-specific parameter values or a mixture of generic and element-specific parameter values, for groups of elements (e.g., chemical families) that show qualitatively similar behavior in the body.

6.1.2 Types of Information Typically Used to Develop Biokinetic Models

The extent to which modelers can characterize the typical biokinetics of an inhaled or ingested element in a given population (e.g., radiation workers, adult males, adult females, infants, children, adolescents) depends on the type(s) of information available and the quality and quantity of each type of information.

As a rule, element-specific features of a biokinetic model are based on some combination of observations of the behavior of the element in human subjects (H1), the element in other mammalian species (A1), chemically similar elements in human subjects (H2), and chemically similar elements in other mammalian species (A2). Depending on the degree of biological realism in the model structure, the four primary types of information might be supplemented with considerations of mass balance and basic physiological data (P). Generic features of a biokinetic model are typically based on broadly common patterns of behavior indicated by collective data for the material of interest, for example, collective data on the behavior of inhaled elements in the respiratory tract.

In general, greater confidence can be placed in a biokinetic model based on H1 data than a model based on H2, A1, and/or A2 data of equal quality and completeness. For most elements, however, H1 data alone are not sufficient to develop a meaningful model, due either to the sparsity of such observations or to limitations in the data, such as the atypical nature of the human study groups, uncertainty in the level and pattern of intake of the element, or inaccuracy in the measurements. For such reasons, H1 data must be supplemented or replaced in many cases by surrogate data and/or physiological considerations.

Use of A1 data (interspecies extrapolation) is sometimes supported by interspecies comparisons, but more often relies on the concept of a general biological regularity across mammalian species with regard to cell and organ structure and function, biochemistry, and body temperature regulation. However, the qualitative similarities among mammalian species often do not translate into quantitatively similar behavior of radionuclides, whether or not the data are scaled to account for differences in body size or metabolic rates. Moreover, there are many examples of qualitative irregularities among mammalian species with regard to organ structure and function as well as biochemistry. For such reasons, the confidence that can be placed in a model component based on A1 data depends not only on the quality and completeness of the data for individual species but also the consistency of scaled or unscaled data for different species, and the availability of data for those species judged to be reasonably human-like with regard to pertinent physiological processes.

Use of H2 data (chemical analogy) is based on evidence that chemically similar element pairs often exhibit close physiological similarities (e.g., Sr-Ca, Ra-Ba, Cm-Am, K-Rb). However, there are counterexamples to the premise that chemical analogues are also physiological analogues (e.g., Na-K). Also, element pairs that follow virtually identical paths of movement in the body often exhibit much different kinetics (e.g., K-Cs). Thus, the confidence that can be placed in a model component based on H2 data depends not only on the quality and completeness of the data for individual chemical analogues of the element of interest, but also on the extent to which quantitative biokinetic relations between the element pairs have been established. Use of A2 data is even more problematic because both interspecies and inter-element extrapolations are involved.

In practice, one often considers some combination of the various data types, H1, H2, A1, A2, and P, when building a model.

The following discussions of each of these general types of biokinetic models summarize the evolution of the ICRP's models and describe in more detail the ICRP's most recent respiratory and gastrointestinal models and selected systemic biokinetic models.

6.2 THE ICRP'S RESPIRATORY MODELS

6.2.1 ICRP Publication 2

ICRP Publication 2 (ICRP 1960) provided the first comprehensive set of ICRP-recommended biokinetic models, including a generic respiratory tract model. The respiratory model was based mainly on results of controlled studies of the fate of inhaled radionuclides in laboratory animals, supplemented by follow-up of workers who had inhaled measurable quantities of radionuclides. The available information provided a broad picture of the different behaviors of relatively soluble and relatively insoluble forms of inhaled radionuclides in the respiratory tract.

The respiratory model of Publication 2 consists of the following assumptions:

- 25% of inhaled activity is exhaled immediately, 50% deposits in the upper respiratory tract, and 25% deposits in the lower respiratory tract.

- Activity deposited in the upper respiratory tract is swallowed immediately, i.e., assigned to the stomach.

- Any soluble material deposited in the lower respiratory tract is transferred immediately to blood.

- Half of any insoluble material deposited in the lower respiratory tract is cleared from the tract and swallowed immediately, and half clears to the environment with a biological half-time of 120 d (except for plutonium and thorium, which are assigned half-times of 1 and 4 years, respectively).

- An element-specific fraction f_1 of swallowed activity is absorbed from the gastrointestinal (GI) tract to blood and the fraction $1-f_1$ is excreted in feces.

Thus, if the inhaled material is soluble, the fraction of inhaled activity that reaches blood is $0.25 + 0.5f_1$. If the inhaled material is insoluble, the fraction reaching blood is $0.625f_1$, where 0.625 is the sum of the deposition in the upper respiratory (0.5) and half the deposition in the lower respiratory tract, that is, half of 0.25.

The respiratory model of Publication 2 was modified in ICRP Publication 10 (ICRP 1968) by changing the destination of the slowly removed (insoluble) portion of activity in the lower tract. Specifically, it was assumed in Publication 10 that the slowly removed portion is absorbed into the blood, rather than being cleared to the environment.

A schematic of the respiratory tract model of Publication 2 is shown in Figure 6.1. The modification later made in ICRP Publication 10 (ICRP 1968) is also shown.

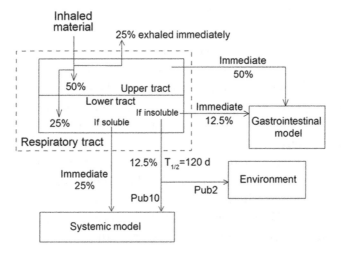

FIGURE 6.1   Respiratory model of ICRP Publication 2, also showing a modification made in ICRP Publication 10 (ICRP 1968). All percentages refer to the inhaled amount. Longer removal half-times of insoluble material from the lower respiratory tract are assigned to plutonium (1 y) and thorium (4 y).

6.2.2 ICRP Publication 30 (Task Group Lung Model)

The respiratory model of ICRP Publication 2 (ICRP 1960), and its modest modification in ICRP Publication 10 (ICRP 1968), had important limitations for purposes of radiation protection. For example, the model does not account for the fact that the total and regional depositions in the respiratory tract depend strongly on the size of inhaled particles; it addresses only two undefined categories of aerosols, called soluble and insoluble; and relatively fast clearance from the respiratory tract is not described kinetically, but is assumed to be instantaneous.

A more sophisticated respiratory tract model was developed by an ICRP task group in the mid-1960s. By that time a relatively large body of information on respiratory deposition and clearance of inhaled material had been developed from such sources as:

- Measurements of total and regional respiratory deposition of particles tagged with radionuclides in human subjects.

- Measurement of clearance of extremely insoluble particles tagged with radionuclides from different lung regions in human subjects.

- Predictions of total and regional respiratory deposition based on idealized models of lung anatomy and airflow patterns.

- Follow-up of increasing numbers of radiation workers who accidentally inhaled relatively large quantities of externally measurable radionuclides.

- A growing literature describing results of invasive and non-invasive studies of deposition and clearance of inhaled material in laboratory animals (mainly dogs, rats, mice, guinea pigs, and rabbits, with dogs being the preferred species among these animals for biokinetic modeling purposes).

The updated respiratory model, called the Task Group Lung Model (TGLM) was published in 1966 (Bates et al. 1966) but was not used by the ICRP as a basis for calculating exposure limits until the late 1970s, when a slightly modified version was adopted for use in ICRP Publication 30 (ICRP 1979). The TGLM is considered a major scientific accomplishment, in that it consolidated essentially the total relevant database into a soundly based predictive model on which to base recommendations for limits on airborne radionuclides in the workplace. The publication of the TGLM stimulated extensive research on the deposition, particle clearance, and absorption of inhaled radionuclides and led to many reports comparing new observations with predictions of the TGLM.

The structure of the TGLM as applied in ICRP Publication 30 is shown in Figure 6.2. The TGLM model includes four main regions, all of which are anatomically identifiable:

- nasal-pharynx (NP)

- tracheobronchial (TB)

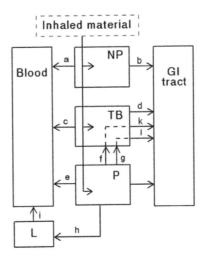

FIGURE 6.2   Structure of the Task Group Lung Model (TGLM) used in ICRP Publication 30 (ICRP 1979).

- pulmonary (P)

- lymphatic (L)

Inhaled material is deposited in regions NP, TB, and P, with regional deposition depending on the size of the inspired particles. Material is cleared from NP and TB to the blood and the GI tract, and from P to the blood, GI tract, and lymphatic region (L). A given chemical or physical form of a radionuclide is assigned to one of three clearance classes: D (days), W (weeks), or Y (years). These correspond to rapid, intermediate, or slow clearance, respectively, of material deposited in the respiratory passages. Removal half-times are assumed to be independent of particle size.

Biological half-times for the TGLM are given in Table 6.1. Deposition fractions depend on the particle size. For the default particle size of 1 μm AMAD (activity median aerodynamic diameter) assigned in ICRP Publication 30, the deposition fractional assigned to the total respiratory tract is 0.63, of which 0.30 is assigned to NP, 0.08 to TB, and 0.25 to P.

6.2.3 ICRP Publication 66 (Human Respiratory Tract Model)

In 1984, the ICRP formed a task group to review information on respiratory kinetics developed since the completion of the TGLM and to revise or replace the TGLM if warranted. In 1994, the ICRP adopted a new, age- and gender-specific respiratory model called the Human Respiratory Tract Model (HRTM), developed by that task group (ICRP 1994a). The types of information used to develop the HRTM were generally the same as those underlying the TGLM, but much more extensive information of all types had been published by the time of completion of the HRTM. In particular, the developers of the HRTM had access to much more extensive data from controlled studies of the deposition and clearance of inhaled material in healthy human subjects and in laboratory animals and from follow-up of occupationally exposed subjects. Also, more sophisticated anatomical models

TABLE 6.1   Biological Removal Half-Times (d) from Compartments of the TGLM

Region	Path	Class		
		D	**W**	**Y**
NP	A	0.01	0.01	0.01
	B	0.01	0.4	0.4
TB	C	0.01	0.01	0.01
	D	0.2	0.2	0.2
P	E	0.5	50	500
	F	NA	1.0	1.0
	G	NA	50	500
	H	0.5	50	500
L	I	0.5	50	1000

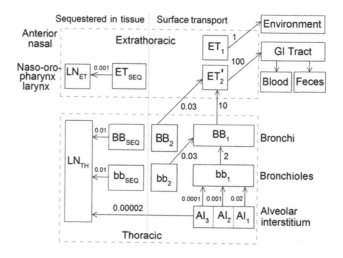

FIGURE 6.3   Structure of the original HRTM. The numbers shown are particle transport rates (d⁻¹) along the indicated paths. ET = extrathoracic, LN = lymph nodes, SEQ = sequestered.

of the lungs had become available for the purposes of predicting the initial distribution of inhaled particles in the lungs for essentially any particle size. For the most part, these anatomical lung models were developed for purposes other than radiation protection, such as for the assessment of lung deposition of chemical toxins in the workplace.

The structure of the original HRTM (ICRP 1994a) is shown in Figure 6.3 (a recently modified version is described later). The model divides the respiratory system into extrathoracic (ET) and thoracic regions. The airways of the ET region are further divided into two categories: the anterior nasal passages, in which deposits are removed by extrinsic means such as nose blowing, and the posterior nasal passages including the nasopharynx, oropharynx, and the larynx, from which deposits are swallowed. The airways of the thorax include the bronchi (compartments labeled BB_i), the bronchioles (compartments labeled bb_i), and the alveolar region (compartments labeled AI_i). Material deposited in the thoracic airways may be cleared into blood by absorption, to the GI tract by mechanical processes (that is, transported upward and swallowed), and to the regional lymph nodes via lymphatic channels.

The number of compartments in each region was chosen to allow duplication of the different kinetic phases observed in humans or laboratory animals. Particle transport rates shown beside the arrows in Figure 6.3 are reference values in units of d^{-1}. For example, particle transport from bb_1 to BB_1 is assumed to occur at a fractional rate of 2 d^{-1}, and particle transport from ET_2' to the gastrointestinal tract is assumed to occur at a fractional rate of 100 d^{-1}. These are reference values determined as best estimates of the central value in the population.

Removal by the mechanical clearances of particles indicated in Figure 6.3 is in addition to absorption to blood. The rate of absorption of a radionuclide from the respiratory tract to blood depends on the chemical and physical form of the inhaled element. Dissolved activity generally is assumed to be immediately absorbed to blood, although the HRTM allows for binding of dissolved activity to tissues of the respiratory tract and gradual absorption of bound activity to blood when indicated by specific information. Absorption is assumed to occur at the same rate in all regions of the respiratory tract except ET_1. The ICRP's default parameter values for relatively soluble, moderately soluble, and relatively insoluble aerosols imply that the absorption rate decreases with time.

The simplest form of the dissolution model within the HRTM is shown in Figure 6.4. This form applies to inhaled material with monotonically decreasing dissolution rates, which is a common situation. A more complex model is required in the less common case in which the dissolution rate does not decrease monotonically with time. For the relatively simple dissolution model shown in Figure 6.4, it is assumed that a fraction f_r of deposited material dissolves at the relatively fast rate s_r, and the remaining fraction $1-f_r$ dissolves more slowly at the rate s_s. The relatively soluble and less soluble fractions are assigned to separate compartments upon deposition.

In most applications of the HRTM, inhaled particulate material is assigned to one of three generic absorption types: Type F, representing fast dissolution and a high level of absorption to blood; Type M, representing a moderate rate of dissolution and an intermediate level of absorption to blood; and Type S, representing slow dissolution and a low level of absorption to blood. The default values (rounded central values) for material found to

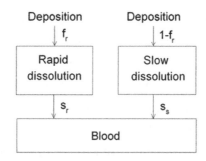

FIGURE 6.4   Model of time-dependent absorption within the HRTM generally applied when the dissolution rate of the material decreases with time (as with Type F, M, or S material). Fractions f_r and $1-f_r$ of deposited material have different dissolution rates (s_r and s_s, respectively).

TABLE 6.2   In the Original HRTM, Default Absorption Parameter Values for Type F (Soluble), M (Moderately Soluble), and S (Relatively Insoluble) Materials

Parameter	Symbol	F (fast)	M (moderate)	S (slow)
Fraction dissolved rapidly	f_r	1	0.1	0.001
Rapid dissolution rate (d⁻¹)	s_r	100	100	100
Slow dissolution rate (d⁻¹)	s_s	–	0.005	0.0001

be relatively soluble (Type F), moderately soluble (Type M), or relatively insoluble (Type S) in in vivo studies on human subjects or laboratory, or in simulated human lung fluid, are listed in Table 6.2.

The data and models used to determine age-specific features of the HRTM include:

- Measurements of ventilation rates in human subjects of all ages.

- An age-specific model of total and regional deposition of inhaled particles in the respiratory tract based on established physical principles and supported by measurements on adult human subjects and limited data for children.

- Age-specific lung clearance data for laboratory animals, primarily dogs.

The developers of the HRTM concluded that the available information supported the incorporation of age- and gender-specific deposition fractions into the HRTM, but that the data were insufficient to develop age-specific rates of particle clearance and absorption to blood. Thus, the particle clearance and absorption rates developed mainly from data for adult males are applied to both sexes and all age groups.

6.2.4 Revision of the Human Respiratory Tract Model

A revised version of the HRTM was introduced in ICRP Publication 130 (ICRP 2015), the first part of an ICRP report series on occupational intake of radionuclides. The revision was motivated by new information on deposition and clearance of material in the upper respiratory tract, intermediate-term clearance from the bronchial region, and long-term retention in the deep lungs. The new information came primarily from:

- Controlled human studies of the nasal clearance of radiolabeled insoluble particles inhaled through the nose while resting or exercising lightly.

- Controlled human studies in which relatively large tagged particles were inhaled slowly and removal from the bronchiolar region was observed.

- A 15-y follow-up of a group of workers who had a simultaneous brief inhalation exposure to particles containing cobalt-60.

- A controlled study of lung retention of activity in volunteers over ~3 y following inhalation of insoluble particles tagged with gold-195.

- A 30-y follow-up of lung retention in workers who inhaled plutonium oxide during a fire.

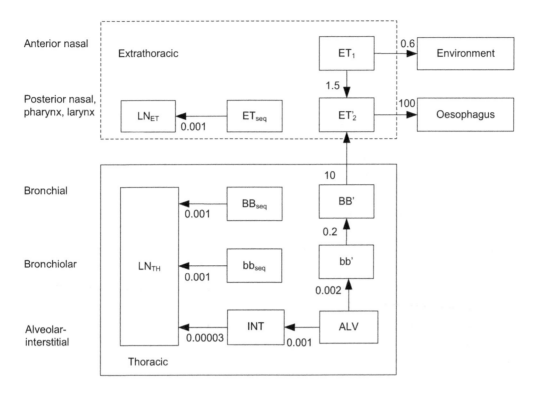

FIGURE 6.5 Structure of the revised HRTM, developed for use in updated ICRP documents on occupational intake of radionuclides. The numbers shown are particle transport rates (d⁻¹) along the indicated paths. ET = extrathoracic; BB = bronchi; bb = bronchioles; LN = lymph nodes; ALV = alveolar; INT = interstitial; TH = thoracic; seq = sequestered.

The structure of the revised HRTM is shown in Figure 6.5. The main changes to the original model structure are as follows:

- The region ET_2, formerly called ET_2', is described in the revised model by two compartments called ET_{seq} and ET_2', where the subscript "sec" indicates sequestered material. The oral passage, formerly contained in ET_2', is no longer included in this region of the respiratory tract. Compartment ET_2' is redefined as consisting of the posterior nasal passage, pharynx, and larynx.

- In each of the bronchial (BB) and bronchiolar (bb) regions, there is now one, instead of two, phases of clearance toward the throat. Compartments BB_1 and BB_2 in Figure 6.3 are replaced by compartment BB' in Figure 6.5, and compartments bb_1 and bb_2 in Figure 6.3 are replaced by compartment bb' in Figure 6.5.

- In the alveolar-interstitial (AI) region, the three AI compartments of the original HRTM have been replaced by the alveolar (ALV) and interstitial (INT) compartments. Particles are cleared from the ALV compartment either to the ciliated airways or to the INT compartment. Particles clear very slowly from INT to the lymph nodes.

TABLE 6.3 In the Revised HRTM, Default Absorption Parameter Values for Type F (soluble), M (Moderately Soluble), and S (Relatively Insoluble) Materials

Parameter	Symbol	F (fast)	M (moderate)	S (slow)
Fraction dissolved rapidly	f_r	1	0.2	0.01
Rapid dissolution rate (d^{-1})	s_r	30	3	3
Slow dissolution rate (d^{-1})	s_s	–	0.005	0.0001

The default parameter values for Types F, M, and S material (essentially, typical values of the parameters f_r, s_r, and s_s shown in Figure 6.4) were modified from the values used in the original HRTM, based on an expanded set of experimental data derived from in vivo and in vitro studies. The updated default parameter values for Types F, M, and S are listed in Table 6.3. The default parameter values describing absorption from HRTM compartments to blood (f_r, s_r, and s_s) may be replaced by element-specific values when information is sufficient to derive such values.

The parameter values of the HRTM (both the original and revised versions) are reference values, intended as typical values for healthy, non-smokers in the general population. For example, reference parameter values are used in the HRTM to describe breathing rates, sizes of airways in the respiratory tract, particle size, regional deposition fractions in the HRTM, rates of mechanical transport of deposited particles, and rates of dissolution of inhaled material.

There are circumstances in which it is appropriate to replace reference parameter values of the HRTM with exposure-specific values. The parameter values that should and should not be changed depend to some extent on the intended use of the model.

If the purpose is to calculate the "dose of record" for a worker for determination of compliance with exposure guidelines, it should be considered that in this context, the dose applies to a reference person, rather than an actual individual. As defined in ICRP Publication 103 (ICRP 2007), the dose of record is

> The effective dose of a worker assessed by the sum of the measured personal dose equivalent … and the committed effective dose retrospectively determined for the Reference Person using results of individual monitoring of the worker and ICRP reference biokinetic and dosimetric computational models. Dose of record may be assessed with site-specific parameters of exposure … but the parameters of the Reference Person shall be fixed as defined by the Commission.

Examples of parameters of the HRTM that describe the Reference Person include breathing rates and particle transport parameters (which are assumed not to depend on the type of particle deposited in the respiratory tract). Examples of material-specific parameters of the HRTM include solubility (which determines the lung-to-blood absorption parameters) and particle size.

If the HRTM is to be used to derive best estimates of tissue doses to the exposed person rather than to demonstrate compliance with regulatory limits, then it is appropriate to change not only the material-specific parameter values of the HRTM but also those

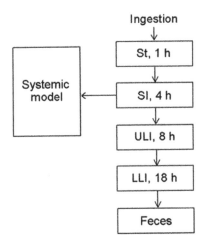

FIGURE 6.6 Gastrointestinal tract model used in ICRP Publication 2. The time given for a segment of the tract is the residence time of material in the contents of that segment. Flow of material through a segment is treated as "slug flow" rather than as first-order removal.

describing characteristics of the exposed person, insofar as information is sufficient to justify the latter changes. For example, information on the level of activity of an exposed person might be used to estimate the actual breathing rate during the exposure. On the other hand, exposure-specific information would rarely, if ever, be sufficient to justify a change in the reference particle transport rates in the HRTM.

6.3 THE ICRP'S GASTROINTESTINAL (GI) MODELS

6.3.1 ICRP Publication 2

Figure 6.6 shows the GI transit model used in ICRP Publication 2 (ICRP 1960). The GI tract is represented as a series of four segments: stomach (St), small intestine (SI), upper large intestine (ULI), and lower large intestine (LLI). This model depicts "slug flow" of swallowed material through the tract, that is, abrupt removal from a given segment of the GI tract following a reference residence time in that segment. Residence times in different segments of the tract presumably came from studies of transit of ingested non-absorbable markers such as barium sulfate, charcoal, or radiolabeled material in human subjects. The residence times of material in these segments are 1, 4, 8, and 18 h, respectively. For example, material entering St contents remains in the St contents for exactly one hour and then is abruptly removed to SI contents. Absorption to the systemic circulation is assumed to occur in SI.

6.3.2 ICRP Publication 30

Growing information on the movement of material through the GI tract indicated that the assumption of slug flow does not accurately characterize the movement of material through much of the tract and that the assumption of first-order kinetics appears to provide a workable approximation of the complicated kinetics of material in different segments of the tract. An updated GI model with first-order transfer between segments of

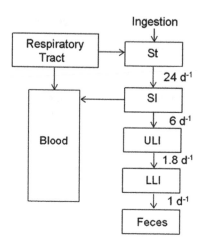

FIGURE 6.7 The GI tract model adopted in ICRP Publication 30 (Part 1) (ICRP 1979) and used by the ICRP until 2006. The transit rates through the lumen of the tract are independent of age.

the GI tract (Figure 6.7) developed by I.S. Eve (Eve 1966) was adopted in ICRP Publication 30 (ICRP 1979). Transfer rates between compartments were based on a relatively large set of reported transfer rates of non-absorbable markers in healthy human subjects, and post-mortem measurements of material in different segments of the tract in subjects who were healthy up to the time of death from a traumatic event.

As in the model of Publication 2, the Publication 30 model divides the GI tract into four segments: stomach (St), small intestine (SI), upper large intestine (ULI), and lower large intestine (LLI). Reference removal rates (d^{-1}) from the various segments are shown in Figure 6.7. Absorption of ingested activity to blood is assumed to occur in the small intestine (SI) and is described by an element-specific f_1 value representing fractional absorption of the stable element to blood. If $f_1 = 1$, the element is assumed to transfer directly from the stomach to blood or, equivalently, to pass to blood instantaneously upon entering the contents of the small intestine.

Although developed specifically for calculation of doses to workers, the GI model of Publication 30 was later used by the ICRP to estimate doses to members of the public (ICRP 1990, 1993, 1995a,b). In the applications to members of the public, the ICRP accounted for changes with age in the mass and dimensions of the GI tract and elevated absorption of some radionuclides in infants and children, but the rate of transit of activity through segments of the tract was assumed to be invariant with age.

6.3.3 ICRP Publication 100 (Human Alimentary Tract Model)

In the late 1990s, an ICRP task group was appointed to develop an age- and gender-specific biokinetic model of the full alimentary tract, including the oral cavity and esophagus, to reflect the large body of data on the transit of material through the alimentary tract that had evolved since the completion of the Publication 30 model in the mid-1960s. The new

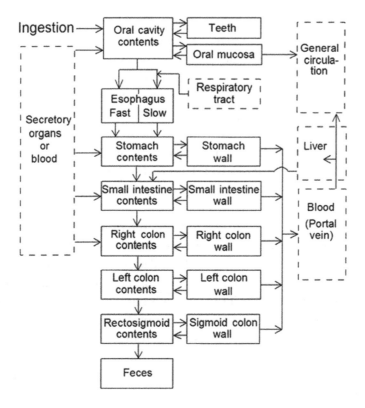

FIGURE 6.8 Structure of the HATM. The dashed boxes are not part of the HATM but are included in the schematic to show connections with the respiratory and systemic models.

model, called the Human Alimentary Tract Model (HATM), was completed in 2006 and published in ICRP Publication 100 (ICRP 2006).

The structure of the HATM is shown in Figure 6.8. The compartments and paths of movement represent the following processes:

- entry of a radionuclide into the oral cavity by ingestion or into the esophagus after mechanical clearance from the respiratory tract;

- sequential transfer through the lumen of the oral cavity, esophagus, stomach, small intestine, and segments of the colon, followed by emptying in feces;

- radionuclide deposition and retention on or between the teeth and return to the oral cavity;

- deposition and retention in the oral mucosa or walls of the stomach or intestines;

- transfer from the oral mucosa or walls of the stomach or intestines back into the lumenal contents or into blood (absorption);

- transfer from secretory organs or blood into the contents of segments of the tract.

Entry into the alimentary tract by ingestion or transfer from the respiratory tract and sequential transfer through the lumen of the tract are regarded as generic processes, in that the rates are assumed to be independent of the radionuclide or its form. The other processes addressed by the HATM occur at rates that are assumed to depend on the element and, in some cases, on the form of the element taken into the body. For example, element-specific parameter values are required to define the extent of uptake and retention on the teeth or in the walls of the tract, or transfer through the walls to blood. An element-specific process is addressed in HATM applications only if information is available to assign a non-zero transfer rate to that process. For most elements, specific information on the behavior of an element in the alimentary tract is limited to total absorption to blood.

First-order kinetics is assumed in the HATM. The residence times of material in the lumen of segments of the alimentary tract were initially estimated in terms of the mean transit time because this is the form in which data on GI tract motility generally are reported. The transit time of an atom in a region of the tract is the length of time that it resides in that region, and the transit time of a substance in a region (also called the mean transit time) is the mean of the distribution of transit times of its atoms. The first-order transfer rate or "emptying rate" used to represent a transit time T hours in a segment of the alimentary tract is 1/T per hour, and the corresponding biological half-time in the segment is (ln2) × T hours. Transit times of lumenal contents are regarded as primary parameter values of the HATM, and the first-order transfer rates derived from those transit times are regarded as secondary values.

Separate transit times were developed for transfer of ingested solids, liquids, and total diet through the mouth and esophagus, and for transit of non-caloric liquids, caloric liquids, solids, and total diet through the stomach. The material-specific values were developed for application to special cases. It is anticipated that transit values for total diet will be used as default values.

The types of information used to develop transit times in each segment of the HATM are illustrated below.

Oral cavity: A radionuclide enters the oral cavity in ingested material or by secretion of absorbed activity in saliva. As illustrated in Figure 6.9, the residence time of ingested material in the mouth is highly variable, depending on the composition and texture of food, the level of hunger, age, personal habits, customs, and other factors. Liquids typically are removed from the mouth in a single swallow in which a posterior movement of the tongue forces the liquid into the oropharynx. Solids typically are chewed for a sufficient time to reduce particles to a few cubic millimeters. For conversion of reported data to transit times, the assumptions were made that the transit time of a liquid is the time from intake to first swallow and the transit time of a solid is three-quarters of the time from intake to final swallow. The residence time of secreted saliva was assumed to be the same as that of food. Transit times in the oral cavity were assumed to be independent of age after infancy, because differences with age in measured swallowing times for specific foods may be largely offset by changes in diet. Baseline transit times for material in the oral cavity are given in Table 6.4.

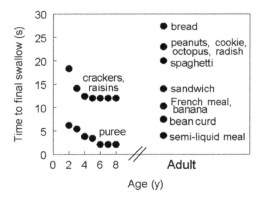

FIGURE 6.9   Illustration of data used to derive reference transit times for ingested material in the oral cavity.

TABLE 6.4   HATM Baseline Transit Times for the Oral Cavity

	Transit Time (s)	
Ingested Material	**Infant**	**Ages > 1 y**
Solids	–	15
Liquids	2	2
Total diet	2	12

Esophagus: When material is swallowed, a coordinated and sequential set of peristaltic contractions produces a zone of pressure that moves down the esophagus with the bolus in front of it. The time required for the wave to travel from the pharynx to the stomach typically is 4–12 sec. The esophagus may not be totally emptied by the original peristaltic contraction initiated by the swallow. Several secondary contractions often are required to remove the remaining material, and some material can remain for several minutes or even hours in the esophagus. In the HATM, esophageal transit is represented by two components: a fast component representing movement in front of the initial peristaltic contraction initiated by the swallow, and a slow component representing transfer of residual swallowed material. Reported mean transit times for the fast component are summarized in Figure 6.10. Baseline transit times used in the HATM for the fast and slow components of transfer of material through the esophagus are listed in Table 6.5.

Stomach: The kinetics of gastric emptying is affected by many factors, including composition of the ingested material, gender, and age. Emptying times generally increase in the order: non-caloric liquids < caloric liquids < solids. Emptying of liquids usually begins within 1–3 min of their arrival in the stomach and can be described reasonably well by a mono-exponential function, although a lag-phase of several minutes has been reported for liquids of high caloric density. Removal of the solid component typically consists of an initial lag-phase of several minutes in which there is relatively slow emptying, followed by an extended phase of nearly linear emptying. For healthy adult subjects, reported central values for observed gastric half-emptying times range from 40 to 160 min for solids, 8 to 107 min for caloric and unspecified liquids, and 15 to 35 min for liquids clearly identified

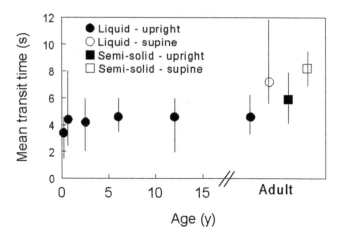

FIGURE 6.10 Differences with age, food type, and body position in transit times through the lumen of the esophagus (fast component). Symbols represent means and vertical lines represent ranges of individual observations for children and reported central values for adults.

TABLE 6.5 HATM Baseline Transit Times for the Esophagus

	Transit Time (s) for Infants		Transit Time (s) for Ages > 1 y	
Ingested Material	**Fast (90%)**	**Residual (10%)**	**Fast (90%)**	**Residual (10%)**
Solids	–	–	8	45
Liquids	4	30	5	30
Total diet	4	30	7	40

as non-caloric. The means of collected central values are approximately 90 min for solids with coefficient of variation (CV) ~30%, 35 min (CV ~60%) for caloric and unspecified liquids, and 25 min (CV ~30%) for non-caloric liquids. Reported emptying times of either solids or caloric liquids are greater on average in women than in men (Figure 6.11). Based on various measures of central tendency including the median, mean, weighted mean, and trimmed weighted mean, a typical or central half-emptying time for solids is about 75–80 min in adult males and 100–110 min in adult females; for caloric liquids, a typical half-emptying time is 30–35 min in males and 40–45 min in females; for non-caloric liquids, a typical half-emptying time for either gender is about 20–25 min. The transit time in the stomach changes from infancy to early childhood, but it is not evident that there is much change with age thereafter. Baseline transit times for material in the stomach are given in Table 6.6.

Small intestine: The motility patterns of the small intestine are organized to optimize its primary functions of digestion and absorption of nutrients and absorption of fluids and electrolytes. Movement of digested material through the small intestine after a meal is a nearly linear process, but subsequent motility complexes that clear undigested residue are spread unevenly over time. Reported transit times through the small intestine based on reproducible techniques are in the range 1.8–8 h, with most values near 3–4 h. Limited age-specific data suggest that the transit time through the small intestine is similar in children,

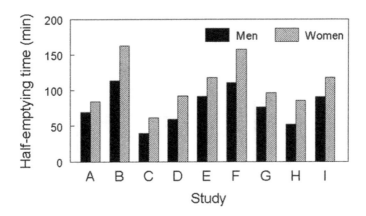

FIGURE 6.11   Comparison of gastric half-emptying times of solids in adult male and female subjects in nine studies.

TABLE 6.6   HATM Baseline Transit Times for the Stomach

Ingested Material	Transit Time (min)			
	Infant	Age 1–15 Years	Adults	
			Males	Females
Solids	–	75	75	105
Liquids				
Caloric	75	45	45	60
Non-caloric	10	30	30	30

young adults, and elderly persons. In the HATM, the baseline transit time through the small intestine is 4 h for both sexes, all ages, and all material.

Colon: The colon absorbs water and electrolytes that enter from the small intestine or in secretions and stores fecal matter until it can be expelled. Flow of material in the colon is slow and highly variable. Periods of contraction between longer periods of quiescence result in mass movements of colonic material a few times during the day. Most of the movements of the proximal colon are weak peristaltic contractions that serve to mix contents back and forth, exposing them to absorptive surfaces. Typically, 1–3 times a day, peristaltic contractions move significant amounts of material from one region of the colon to another. One mass movement may transport contents from the transverse to the sigmoid colon or rectum. The rectum serves mainly as a conduit but can also serve as a storage organ when the mass received from the sigmoid colon is too small to evoke the recto-anal inhibitory reflex that signals the need to defecate, or when this reflex is neglected. The HATM divides the colon into the right colon, left colon, and rectosigmoid, a division often used for diagnostic and experimental examinations of colonic transit. This division was chosen to make the best use of experimental data and is expected to allow best available estimates of the time-dependent distribution of activity in the colon. Central estimates of the colonic transit time by different investigators vary by about a factor of 4 (17–68 h), but most reported values are in the range 24–48 h. Collective data (Figure 6.12), as well as data

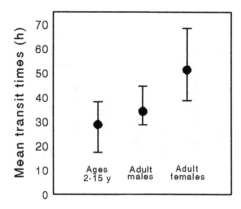

FIGURE 6.12   Ranges (vertical bars) and overall means (circles) of collected central colonic transit times for groups of normal children, adult males, or adult females.

TABLE 6.7   Baseline Transit Times for Segments of the Colon (All Material)

	Transit Time (h)				
Segment	**Infant**	**1 y**	**5–15 Years**	**Adult Male**	**Adult Female**
Right colon	8	10	11	12	16
Left colon	8	10	11	12	16
Rectosigmoid	12	12	12	12	16

from individual studies involving both genders, indicate that transit through the colon is substantially slower on average in women than in men. Mean transit times appear from collective data to be shorter on average in children than adults (Figure 6.12). Age-specific data on the time to the first appearance of ingested markers in feces that are used to diagnose bowel function suggest an increase with age in transit times from infancy to adulthood. Baseline transit times through segments of the colon are age- and gender-specific (Table 6.7). Baseline values are independent of the material entering the colon.

First-order kinetics is generally assumed in the biokinetic models used in radiation protection, in part for computational convenience, and in part because this assumption is expected to yield a reasonably good approximation to the actual behavior of radionuclides in the body in most cases. In the development of first-order rate constants for the HATM, the removal half-times of luminal contents from segments of the tract are set to produce the average residence times of stable atoms implied by the reference transit times summarized earlier. The intent is to produce reasonable central estimates of the cumulative activity of radionuclides in the contents of the segments using relatively simple kinetics.

For relatively short-lived radionuclides, first-order kinetics could overestimate decays in the lower regions of the tract, because it implies an immediate appearance of some ingested atoms in all regions of the tract. For example, an ingested radionuclide with half-life 20 min is likely to decay almost entirely between the mouth and colon because more than 10 radiological half-lives may elapse before the first appearance of the ingested material in the right colon. The HATM predicts on the basis of first-order kinetics that about 3% of the total decays in the alimentary tract would occur in the colon after ingestion of a radionuclide with half-life 20 min.

As a first-order model, the HATM depicts continuous fecal excretion of activity starting immediately after ingestion, resulting in an overestimate of early fecal excretion. For example, for an adult male, the model predicts that fecal excretion during the first half day after intake is about 3% of the ingested amount, in the absence of radiological decay or absorption to blood. By contrast, studies indicate that the first appearance of ingested markers in feces of healthy adults is usually more than 12 h.

HATM predictions of cumulative fecal excretion over periods of 1 d or longer appear to be reasonable central estimates for the population. When using the HATM or any other gastrointestinal model for interpretation of bioassay data, however, it should be kept in mind that the pattern of fecal excretion of ingested material is highly variable and difficult to predict in individual cases. This is illustrated in Figures 6.13 and 6.14, which compare HATM predictions with observed patterns of fecal excretion after ingestion of ^{85}Sr (nine subjects) and ^{26}Al (two subjects), respectively.

6.4 THE ICRP'S SYSTEMIC BIOKINETIC MODELS

6.4.1 The Need for Element-Specific Structures for Systemic Biokinetic Models

Radionuclides entering blood may distribute nearly uniformly throughout the body (e.g., ^{3}H), they may selectively deposit in a particular organ (e.g., ^{131}I in the thyroid gland), or they may show elevated uptake in a few different organs (e.g., ^{239}Pu in bone and liver). If a radionuclide that enters blood is an isotope of an essential element, that is, an isotope of an element required by the human body for good health and normal growth (e.g., ^{45}Ca or ^{55}Fe), it follows the normal metabolic pathways for that element. If it is chemically similar to an essential element (e.g., ^{137}Cs as a chemical analogue of potassium, and ^{90}Sr as a chemical analogue of calcium), it may follow the movement of the essential element in a qualitative manner but may show different rates of transfer across membranes, due to the membrane's ability to discriminate between elements on the basis of only moderately

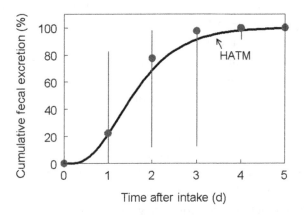

FIGURE 6.13   HATM predictions of cumulative fecal excretion (relative to five-day fecal excretion) compared with observations for ingested ^{85}Sr. Circles and vertical lines represent medians and ranges, respectively, of values determined for nine young adult males.

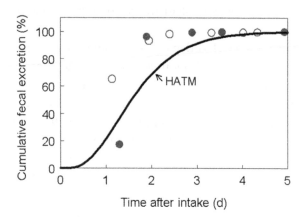

FIGURE 6.14   HATM predictions of cumulative fecal excretion (relative to five-day fecal excretion) compared with observations for two adult male subjects who ingested [26]Al in drinking water.

different physical or chemical properties. The behavior of a radioisotope of a non-essential element after its uptake to blood may also depend on such factors as the extent to which it can be sequestered by the reticuloendothelial system (the body's scavenger cells), its affinity for specific biological ligands, its filterability by the kidneys, and the ability of the body to eliminate it in liver bile or other secretions into the gastrointestinal tract. In some cases, the biokinetics of an isotope of a non-essential element may resemble that of an essential element to some extent due to common affinities for some but not all components of tissues and fluids. For example, the behavior of plutonium in blood and liver is related to that of iron, due to an affinity of plutonium for certain proteins that transport or store iron, but, as a whole, the biokinetics of plutonium in the body differs greatly from that of iron.

The ICRP's alimentary tract model and, for the most part, the ICRP's respiratory tract model, are generic models in the sense that the transfer coefficients between alimentary or respiratory tract compartments are assumed to be independent of the element. That is, the transfer rates depend on the material carrying the element rather than on properties of the element itself. By contrast, the ICRP's systemic biokinetic models usually are element-specific models, that is, the structure and parameter values were developed specifically for the element. The use of element-specific parameter values to describe the systemic behavior of radionuclides is important because: (1) a radionuclide usually must be separated from its carrier (e.g., dust particles deposited in the respiratory tract, or ingested material in the alimentary tract) before it can be absorbed into blood, and (2) chemically different elements entering the systemic circulation often show substantially different biokinetics. In fact, a generic model structure that depicts all potentially important systemic repositories and paths of transfer of all elements of interest in radiation protection would be too complex to be of much practical use. However, generic model structures have been used in ICRP documents to address the systemic biokinetics of some groups of elements, typically chemical families, known or expected to have qualitatively similar behavior in the body. For example, ICRP Publication 20 (ICRP 1973) introduced a generic model formulation for the alkaline earth elements calcium, strontium, barium, and radium, but provided element-specific values for most model parameters. In ICRP Publication 30 (ICRP 1979, 1981) a model developed for plutonium, including parameter values as well as model structure, was applied to most actinide elements. In the ICRP's series of reports on age-specific doses to members of the public from intake of environmental radionuclides (ICRP 1990, 1993, 1995a,b,c), a generic model structure was applied to selected elements that behave similarly to plutonium in the body, and a different generic structure was applied to selected elements that behave similarly to calcium in the skeleton.

6.4.2 Evolution of the ICRP's Structures for Systemic Biokinetic Models

6.4.2.1 ICRP Publication 2

ICRP Publication 2 (ICRP 1960) provided the ICRP's first reasonably comprehensive set of systemic biokinetic models for radionuclides. The systemic models adopted in Publication 2 were not designed to depict realistic paths of movement of radionuclides in the body, but were intended to approximate the cumulative activity of radionuclides in their most important systemic repositories, presumably representing the most important sites of radiation damage after absorption to blood. An absorbed radionuclide was assumed to move instantly to a few "organs of reference," with presumably elevated concentrations (activity per unit mass) of the absorbed activity. In effect, the systemic models of Publication 2 were designed to help identify the critical organ, defined as the organ whose damage by the radiation results in the greatest damage to the body; this included the dose to the lung and segments of the gastrointestinal tract, as well as doses to systemic tissues. At the time, the ICRP's radiation protection system was based on limiting the dose to the critical organ. In practice, the critical organ for an internally deposited radionuclide was generally identified in Publication 2 simply as the organ receiving the highest estimated dose.

In Publication 2, the list of organs of reference for a radionuclide generally includes a hypothetical entity called total body. Total body represents the total activity absorbed to blood, assumed for purposes of calculating total body dose to be uniformly distributed in the body. Publication 2 states that total body is listed as an organ of reference "primarily as aid in computing MPC values [maximum permissible concentrations] for mixtures, and as a check on the oversimplified model used" (ICRP 1960, p.11). It is not clear how the total-body concept fulfilled either purpose.

The systemic modeling format used in Publication 2 is illustrated in Figure 6.15, which shows the systemic model for phosphorus used in that report. The organs of reference for phosphorus are bone, liver, brain, and total body. The bone, liver, and brain collectively receive 27.7% of the absorbed phosphorus. By definition, total body receives 100% of the absorbed amount. For such cases in which the organs of reference other than total body receive far less than 100% of the absorbed amount, it is conceivable that total body could receive the highest dose, and thus be considered the critical organ.

An obvious difficulty with the total-body concept of Publication 2 as a radiation protection quantity is that it does not reflect the collective doses to all radiosensitive tissues of the body, as it is derived only from the fraction of inhaled or ingested activity estimated

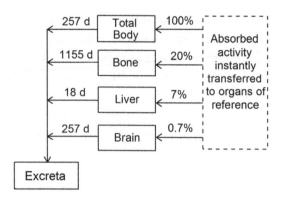

FIGURE 6.15   Systemic model for phosphorus used in ICRP Publication 2 (ICRP 1960). The times to the left of the organs of reference are biological half-times.

to be absorbed to blood. In many cases of internal deposition of radionuclides, the dose to the lungs and gastrointestinal tract from unabsorbed activity represents a major portion of the dose to tissues, and presumably a major portion of the radiogenic risk to the exposed individual.

6.4.2.2 ICRP Publication 30

ICRP Publication 30 (ICRP 1979, 1980, 1981, 1988) built on the modeling approach applied in Publication 2. The systemic biokinetic models used in Publication 30 are generally in the form of retention functions (e.g., sums of exponential terms) that may be interpreted as first-order compartmental models with one-directional flow of activity. As is the case for the systemic models of Publication 2, the systemic models of Publication 30 were designed mainly to estimate the cumulative activities of each radionuclide in its main repositories in the body. The models do not depict realistic paths of movement of radionuclides in the body, but describe only the initial distribution of elements after uptake to blood and the net biological half-times of elements in source organs, that is, the compartments in which the absorbed activity is assumed to distribute. Activity absorbed from the gastrointestinal or respiratory tract or through wounds is assumed to enter a transfer compartment, from which it moves to source regions with a specified half-time, typically 0.25 d or longer. Retention in a source organ usually is described in terms of 1–3 first-order retention components, with multiple biological half-times representing retention in multiple hypothetical compartments within a source organ. Feedback of activity from tissues to blood is not treated explicitly in Publication 30, with the exception of the model for iodine. It is generally assumed that activity leaving an organ moves directly to a collective excretion compartment, that is, radioactive decay along actual routes of excretion is not assessed. Relatively short-lived radionuclides (half-lives up to 15 d) depositing in bone generally are assigned to bone surface and longer-lived radionuclides are assigned either to bone surface or bone volume, depending on their main sites of retention in bone as indicated by available data.

The systemic biokinetic model for phosphorus recommended in Part 1 of ICRP Publication 30 (ICRP 1979) is shown in Figure 6.16. Absorbed phosphorus enters a compartment called the "Transfer compartment" (essentially, blood). Phosphorus leaves this compartment with a half-time of 0.5 d and is distributed as follows: 15% is removed from the body; 30% goes to mineral bone, and 55% is uniformly distributed in remaining tissue ("Other"). Other is divided into two compartments, one receiving 15% of activity leaving blood and having a removal half-time of 2 d, and the second receiving 40% and having a half-time of 19 d. Phosphorus is assumed to be permanently retained in bone. Based on a default assumption used in Publication 30 for bone-seeking radionuclides, a phosphorus isotope with half-life less than 15 d is assigned to bone surface, and a phosphorus isotope with longer half-life is assigned to bone volume. Thus, ^{32}P ($T_{1/2}$ = 14.26 d) that enters bone is assumed to decay on bone surface, and ^{33}P ($T_{1/2}$ = 25.34 d) that enters bone is assumed to decay in bone volume.

The systemic biokinetic models of Publication 30 were intended primarily for calculation of dose per unit intake for planning purposes rather than for retrospective evaluation of doses. For some elements, these systemic biokinetic models were developed separately from the ICRP's concurrent bioassay models. For example, urinary and fecal excretion models for plutonium, americium, and curium recommended in ICRP Publication 54 (ICRP 1989), "Individual Monitoring for Intakes of Radionuclides by Workers: Design and Interpretation," were derived independently of the concurrent systemic models for these elements used by the ICRP to derive dose coefficients. In such cases, the interpretation of bioassay might be based on inconsistent bioassay and dosimetric models. For example, to estimate dose from intake of ^{239}Pu based on urinary ^{239}Pu measurements, the intake would be estimated using the plutonium urinary excretion model recommended in ICRP Publication 54, and the tissue doses from that intake would be calculated using the plutonium systemic model from Publication 30, which predicts a faster loss of plutonium from the body than represented by the urinary and fecal excretion models for plutonium used in Publication 54. Such mismatches of the ICRP's dosimetric and bioassay models have been eliminated in recent years, as systemic models have been developed to serve both as bioassay and dosimetric models.

6.4.2.3 ICRP Publication 68 and the Publication 72 Series

During the period 1989–1996, the ICRP updated its guidance on occupational intake of radionuclides (ICRP 1994b) and issued a series of reports (referred to here as the Publication 72 series) on age-specific doses to members of the public from intake of radionuclides (ICRP 1990, 1993, 1995a,b,c). Most of the systemic models used in Publication 68 and the Publication 72 series followed the modeling scheme used in Publication 30 and illustrated in Figure 6.16, except that biological removal along explicit excretion pathways was depicted in reports completed after the appearance of ICRP Publication 60 (ICRP 1991a). The excretion pathways were included in the systemic models mainly to address doses to the urinary bladder and colon, both of which were assigned tissue weighting factors in Publication 60. The inclusion of explicit excretion pathways also had the benefit that

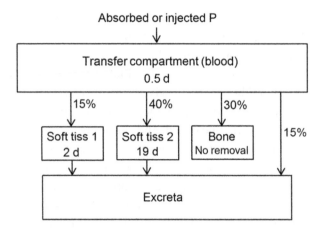

FIGURE 6.16 Systemic biokinetic model for phosphorus used in ICRP Publication 30, Part 1 (ICRP 1979), illustrating the one-directional flow of systemic activity depicted in models of Publication 30 and some later ICRP documents.

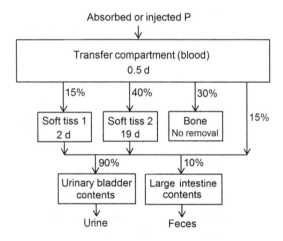

FIGURE 6.17 Systemic biokinetic model for phosphorus applied in ICRP Publication 68 (ICRP 1994b). This is a modification of the phosphorus model used in ICRP Publication 30, with biologically removed phosphorus assigned specific excretion pathways.

a single model could be used both as a bioassay model and a dosimetric model. This benefit was limited by the fact that the models used in Publication 68 and in the Publication 72 series were often just modest modifications of models originally used in Publication 30, and in some cases were designed to yield cautiously high dose estimates, rather than to depict realistic rates of biological removal of radionuclides from the body.

Figure 6.17 illustrates the systemic modeling format used for most elements in ICRP Publication 68 and the Publication 72 series. The systemic model for phosphorus shown in Figure 6.17 is modified from the phosphorus model used in Publication 30 (Figure 6.16) by channeling biologically removed phosphorus through specific excretion pathways, namely, the urinary bladder (90%) and the large intestine (10%). Comparison of model predictions with observed excretion rates for ^{32}P indicated that this resulted in a reasonably good bioassay model for phosphorus, at least with regard to interpretation of urinary excretion data.

A different modeling scheme involving more realistic paths of movement of systemic radionuclides was applied in Publication 68 and in the Publication 72 series to selected radionuclides, including iron and several so-called bone-seeking elements: calcium, strontium, barium, lead, radium, thorium, uranium, neptunium, plutonium, americium, and curium. The model structures applied to these elements depict feedback of material from organs to blood and, where feasible, physiological processes that determine the biokinetics of radionuclides. Examples of such physiological processes are bone remodeling, which results in removal of plutonium or americium from bone surface, and phagocytosis of aging erythrocytes by reticuloendothelial (scavenger) cells, which results in transfer of iron from blood to iron storage sites.

The physiologically based modeling scheme applied in Publication 68 and in the Publication 72 series to selected elements is illustrated in Figure 6.18, which shows the generic model structure used for the actinide elements thorium, neptunium, plutonium, americium, and curium. In updated ICRP documents on occupational or environmental intakes of radionuclides, this model structure is applied to a larger set of elements that exhibit generally similar behavior in the body, including additional actinide elements and all lanthanide elements.

In the generic model structure shown in Figure 6.18, the systemic tissues and fluids are divided into five main components: blood, skeleton, liver, kidneys, and other soft tissues. Blood is treated as a uniformly mixed pool. Each of the other main components is further divided into a minimal number of compartments needed to explain available biokinetic data on these five elements or, more generally, "bone-surface-seeking" elements, meaning elements that tend to bind to bone surfaces and remain there until gradually removed by bone restructuring processes. The liver is divided into compartments representing short and long-term retention. Activity entering the liver is assigned to the short-term compartment (Liver 1), from which it may transfer back to blood, to the intestines via biliary secretion, or to the long-term compartment from which activity slowly returns to blood. The kidneys are divided into two compartments, one that loses activity to urine over a period of hours or days (Urinary path), and another that slowly returns activity to blood (Other kidney tissue). The remaining soft tissue, other than bone marrow, is divided into compartments ST0, ST1, and ST2 representing rapid, intermediate, and slow return of activity to blood, respectively. ST0 is used to account for a rapid buildup of activity in soft tissues and rapid feedback to blood after acute input of activity to blood and is regarded as part of the circulating activity. The skeleton is divided into cortical and trabecular fractions, and each of these fractions is subdivided into bone surface, bone volume, and bone marrow. Activity entering the skeleton is assigned to bone surface, from which it is transferred gradually to bone marrow and bone volume by bone remodeling processes. Activity in bone volume is transferred gradually to bone marrow by bone remodeling. Activity is lost from bone

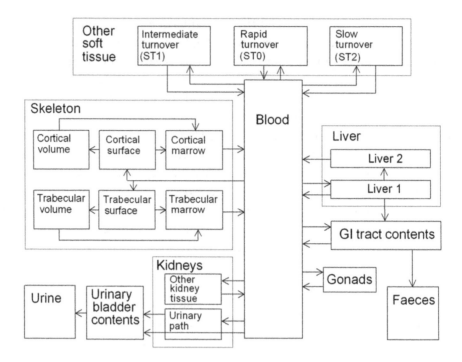

FIGURE 6.18   Model structure applied in ICRP Publication 68 and the Publication 72 series to the bone-surface seekers thorium, neptunium, plutonium, americium, and curium.

marrow to blood over a period of months and is subsequently redistributed in the same pattern as the original input to blood. The rates of transfer from cortical and trabecular bone compartments to all destinations are functions of the turnover rate of cortical and trabecular bone, assumed to be 3% and 18% per year, respectively in adults. Other parameter values in the model are element-specific.

A variation of the model structure shown in Figure 6.18 is applied in Publication 68 and in the Publication 72 series to calcium, strontium, barium, radium, lead, and uranium (Figure 6.19). These elements behave differently from the bone-surface seekers addressed above in that they diffuse throughout bone volume within hours or days after depositing in bone. In updated reports by the ICRP on occupational or environmental intake of radionuclides, the model structures shown in Figures 6.18 and 6.19 are applied to a much wider range of bone-volume-seeking elements. For example, the model structure shown in Figure 6.19 is applied to phosphorus in view of the similar behavior of phosphorus and calcium in bone (ICRP 2016). The new recycling model for phosphorus has replaced the one-directional model used in ICRP Publication 68 (Figure 6.17).

The compartments in Figure 6.18 representing bone marrow and gonads are omitted from the model structure for bone-volume seekers shown in Figure 6.19 because marrow and gonads generally are not sites of elevated accumulation of bone-volume seekers. Also, if a particular compartment or pathway shown in Figure 6.19 is not an important repository for a given bone-volume seeker, it is not considered as a separate pool in the model for that element. For example, in the models for calcium and strontium, blood is treated as

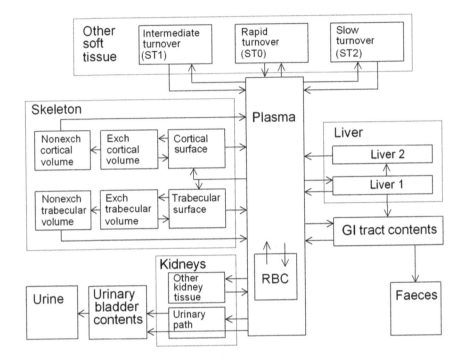

FIGURE 6.19 Model structure applied in the Publication 72 series to calcium, strontium, barium, lead, radium, and uranium. Exch = exchangeable, Non-Exch = Non-Exchangeable, RBC = red blood cells.

a single well-mixed pool (hence, the RBC compartment is removed from the model structure), and the liver and kidneys are assumed to be part of "Other soft tissues" (hence, the liver and kidney compartments are removed from the model structure).

6.4.3 Case Studies of Systemic Biokinetic Models and Underlying Data

6.4.3.1 Strontium

Except where otherwise indicated, the discussion in this section is based on the following reviews, all of which provide extensive bibliographies: ICRP Publication 20 (ICRP 1973), Leggett (Leggett 1992), ICRP Publication 67 (ICRP 1993), and ICRP Publication 71 (ICRP 1995b).

6.4.3.1.1 Summary of the Database

Strontium is a member of the alkaline earth family (Group IIA of the periodic table) and has been shown in human and animal studies to be a physiological analogue of the alkaline earths calcium, barium, and radium. The systemic biokinetics of strontium differs to some extent from that of these other three alkaline earths, due to discrimination between these elements by biological membranes and hydroxyapatite crystals of bone. For example, strontium is less effectively absorbed from the intestines and more effectively excreted by the kidney than calcium, and is lost from bone at a higher rate than calcium over the first few months after uptake to blood. On the other hand, strontium appears to be more effectively absorbed from the intestines and lost from bone at a lower rate than barium or radium. Overall, the systemic behavior of strontium is closer to that of calcium than to that of the heavier alkaline earths, barium, and radium. Nevertheless, collective biokinetic data for all four of these elements help to fill gaps in information for the individual elements, including strontium.

The biokinetics of strontium has been studied extensively in human subjects and laboratory animals. A large database related to the transfer of ^{90}Sr from food and milk to the human skeleton was developed in the 1950s and 1960s, when ^{90}Sr was accumulating in the environment as a result of nuclear weapons tests. Those data indicate that much higher transfer of ^{90}Sr from the environment to the skeleton occurs in growing children than in adults, with highest transfer occurring in infants and toddlers and a second phase of elevated uptake occurring during adolescence (Figure 6.20).

Another large study of the accumulation of environmental ^{90}Sr in humans involved a population living along the Techa River in Russia. Around 1950, a plutonium production facility released large amounts of ^{90}Sr into the river, and it was carried for long distances down the river and accumulated in fish and in gardens near the river. External measurements of the whole-body content of ^{90}Sr were obtained over a 24-year period for thousands of persons living near the river and eating the contaminated fish and vegetables (Shagina et al. 2003). Findings concerning the whole-body accumulation of environmental ^{90}Sr in the Techa River residents as a function of age are generally consistent with the autopsy measurements of fallout ^{90}Sr made in the United States and other countries in the 1950s and 1960s.

Interpretation of data on the accumulation of environmental ^{90}Sr in human populations is complicated by the facts that measured skeletal burdens were accumulated over an extended period and reflect variation with age in gastrointestinal absorption of strontium, as well as its uptake and retention by bone. More easily interpreted data are available from controlled studies of the behavior of radio-strontium in adult human subjects, as illustrated

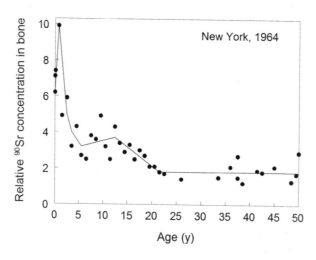

FIGURE 6.20   Results of an autopsy study of the concentration of ^{90}Sr in human bone as a function of age at death.

by the curve fits to human data shown in Figure 6.21. The results of these controlled studies indicate some age dependence in strontium biokinetics, even during adulthood. In particular, total-body retention of intravenously injected strontium is substantially greater in young adults (persons in their early twenties at the time of injection) than in mature adults (at least 25 years old at the time of injection). This is presumably related to a higher rate of addition of calcium—and hence its physiological analogues strontium, barium, and radium—to the skeleton during young adulthood, when the skeleton is still growing.

Some radio-strontium injection data are available for human subjects injected at pre-adult ages, but are for unhealthy subjects. These data do not yield a clear picture of total-body retention of radio-strontium at pre-adult ages because of their high inter-subject variability. However, the data suggest that there is elevated uptake of strontium by the skeleton at pre-adult ages, followed by faster turnover of skeletal strontium than seen in adults.

Somewhat less variable data are available from injection studies involving radio-calcium (e.g., Figure 6.22). These data indicate that uptake of calcium by the skeleton increases in the order: mature adult < young adult < pre-adult, but do not indicate whether there is faster turnover of radio-calcium at younger ages due to relatively short observation periods.

A large amount of age-specific data on the behavior of strontium in laboratory animals, particularly dogs, indicate that skeletal uptake of strontium tends to decrease with age from early life until the skeleton has fully matured (roughly age 2 y in dogs and 25 y in humans) and to a lesser extent after full maturity (Figure 6.23). Studies that include very young dogs (e.g., 1 month old at injection) indicate that after a few months the retention curves for dogs injected at young ages tend to fall below retention curves for animals injected at higher ages, presumably due to faster loss from the young skeleton as a result of faster bone turnover at younger ages.

When extrapolating biokinetic data for calcium, barium, or radium to strontium, it is necessary to account for known differences in the systemic biokinetics of these elements. For example, human studies have established that the urinary excretion rate is higher for strontium than for calcium at early times after uptake to blood, while the two elements have similar rates of loss in feces. This is illustrated in Table 6.8, which shows typical excretion rates of calcium, strontium, barium, and radium in urine and feces during the first three days after introduction of these elements to blood. Systemic radium and barium are lost from the body at a much higher rate than calcium or strontium over the first few days, due to a much higher rate of loss of barium and radium in feces.

A second established difference in the systemic behavior of the alkaline earth elements is that retention of alkaline earth elements in bone tends to decrease in the order calcium > strontium > barium ≥ radium. This difference appears to be due to discrimination among these elements by bone crystal.

Kinetic analysis of plasma disappearance curves for healthy human subjects indicates that calcium, strontium, barium, and radium initially leave plasma at a rate of several hundred plasma volumes per day. The plasma content equilibrates rapidly with an extravascular pool, presumably consisting largely of interstitial fluids, that is roughly three times the size of the plasma pool. After the initial mixing with the extravascular pool, the plasma

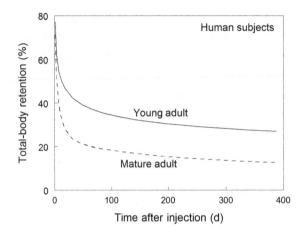

FIGURE 6.21   Central estimates of total-body retention of radio-strontium following intravenous administration to young (~21 y) or mature (≥ 25 y) adult human subjects.

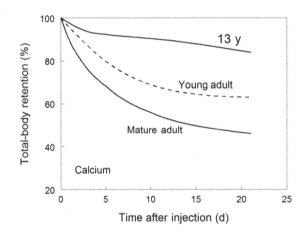

FIGURE 6.22   Central estimates of total-body retention of radio-calcium following intravenous administration to human subjects of different ages.

disappearance curve becomes much less steep but continues to decline over an extended period. The plasma disappearance curve for strontium is similar to that of calcium. The heavier alkaline earth elements barium and radium show similar kinetics to one another in plasma but leave plasma at a higher rate than calcium or strontium, due at least in part to faster loss of barium and radium to excretion pathways.

The systemic distributions of calcium, strontium, and radium as a function of time after intake has been determined from autopsy measurements on human subjects with reasonably well-established time of intake of radioisotopes of these elements. Data for early to intermediate times after intake come mainly from autopsy studies of persons who died from terminal illnesses between a few hours and a several months after administration. Data on the long-term distribution of the alkaline earth elements in humans come mainly from autopsy studies of the stable elements, although some long-term distribution data for

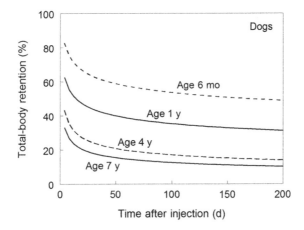

FIGURE 6.23   Total-body retention of intravenously injected ⁹⁰Sr in dogs as a function of age at administration.

TABLE 6.8   Typical Cumulative Excretion of Radioisotopes of Calcium, Strontium, Barium, and Radium in Urine and Feces over the First 3 Days after Intravenous Administration to Healthy Adult Human Subjects (Age ≥ 25 y)

	Cumulative Excretion (% of Injected Amount)		
Element	Urine	Feces	Total
Calcium	14	7	21
Strontium	31	7	38
Barium	7	50	57
Radium	2	55	57

radioactive or stable isotopes of these elements, particularly strontium and radium, are available for human subjects. The autopsy data indicate, for example, that soft tissues contain only about 1% of natural strontium in the total body of the adult human.

Following intravenous injection of ⁸⁵Sr to seriously ill human subjects, soft tissues initially contained about as much strontium as bone, but the soft tissue content fell off sharply after a few weeks while the bone content declined only slowly over the first few months (Figure 6.24). Similar results were observed for ⁴⁵Ca in these subjects, with the main exception that ⁴⁵Ca generally showed slower removal from bone than did ⁸⁵Sr.

Biokinetic studies of the alkaline earth elements in laboratory animals and human subjects reveal fast, intermediate, and slow phases of removal of these elements from bone. The fast phase occurs over the first few days after deposition and represents mainly return of deposited activity from bone surface to plasma. The intermediate phase represents mainly loss of a portion of the activity that has entered bone crystals but remains relatively exchangeable with calcium in bone fluids. This pool is referred to as "exchangeable bone volume." The slow phase represents activity that is firmly fixed in bone crystal. This pool is referred to as "non-exchangeable bone volume." The rate of removal of an alkaline

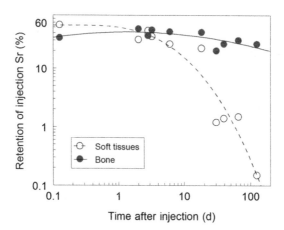

FIGURE 6.24 Retention of [85]Sr in bone and soft tissues as a function of time after intravenous injection into terminally ill human subjects. The symbols represent measured values and the curves are fits to those values.

earth element from the exchangeable bone volume pool of a given bone type (cortical or trabecular) depends on the element. The rate of removal from the non-exchangeable pool presumably depends only on the bone turnover rate, and hence is assumed to be independent of the element.

Data for laboratory animals indicate that fractional deposition on bone surfaces, as judged by the skeletal content in the first few hours after injection, is similar for calcium, strontium, barium, and radium. The initial distribution between different bones of the skeleton and between cortical and trabecular bone also appears from animal studies to be similar for all four elements.

Similar fractional deposition of these elements on bone surface is also suggested by collective data on the early behavior of alkaline earth tracers in human subjects.

It appears that the portion of injected activity released from bone over the intermediate term, that is, from the exchangeable bone volume pool, is roughly the same for all four elements when adjusted for differences in excretion rates and represents about 15%–20% if activity leaving bone surfaces. However, the removal half-time from exchangeable bone volume appears to increase in the order radium ≤ barium < strontium < calcium.

Removal from non-exchangeable bone volume occurs over a period of years and is faster for trabecular bone than for cortical bone, presumably due to the higher rate of turnover of trabecular bone than cortical bone. Reference age-specific bone turnover rates are tabulated in ICRP Publication 89 (ICRP 2002). These values are based on histomorphometric measurements on human subjects and studies of retention of certain bone-seeking radionuclides in human subjects. Most histomorphometric measurements are on ribs and the iliac crest, but some measurements are available for various long bones. Based on the available information, the turnover rates are assumed to be the same for cortical and trabecular bone early in life, but about five times greater for trabecular bone than for cortical bone in the mature adult.

6.4.3.1.2 Systemic Biokinetic Models

6.4.3.1.2.1 *ICRP Publication 2*

Only a small portion of the current biokinetic database for strontium had been developed at the time of completion of ICRP Publication 2 (ICRP 1960), which provided the ICRP's first comprehensive set of biokinetic models for radionuclides. The model for strontium applied in Publication 2 was based mainly on biokinetic studies on small laboratory animals and limited measurements of the accumulation of environmental strontium ^{90}Sr in human bone. The model identifies two "organs of reference": total body, representing strontium that reached the systemic pool, and bone. The deposition fraction for bone is 0.3, that is, it is assumed that 30% of absorbed strontium deposits in bone. The assigned biological half-time for total body is 13,000 d, and the assigned half-time for bone is 18,000 d.

6.4.3.1.2.2 *ICRP Publication 20*

A number of studies of the behavior of strontium and the physiologically related alkaline earth elements calcium, barium, and radium in human subjects were published between the late 1950s and early 1970s. ICRP Publication 20 (ICRP 1973) reviewed these data and introduced a generic model format for these four elements. Some parameter values of the models are generic (i.e., the same for all four elements), and some are element specific. Whole-body retention R(t) of an alkaline earth element at time t days after injection is described by the equation:

$$R(t) = (1-p)e^{-mt} + pE^b (t+E)^{-b} \left[(1-B)e^{-srLt} \right]$$

where

L is the rate of turnover (resorption) of compact bone

s is the ratio of turnover rates of trabecular and compact bone

B is the fraction of bone volume activity deposited in compact bone

r is an element-specific factor that corrects for redeposition of activity in new bone at sites of resorption long after injection

b is an empirically determined element-specific rate related to diffusion of activity from bone

E is an empirically determined, relatively short time period related to the turnover of an initial pool in bone

m is an empirically determined rate constant of a small early exponential loss

p is the fraction of R not in the early exponential loss

Parameter values for the alkaline earth elements were based primarily on reported rates of bone turnover in man, controlled studies of calcium, barium, strontium, and radium metabolism in human subjects, and data developed in the 1950s and 1960s on the accumulation of ^{90}Sr in food and human bone.

6.4.3.1.2.3 ICRP Publication 30

The biokinetic model for strontium applied in ICRP Publication 30 was based on the strontium model of ICRP Publication 20 but was formulated as a sum of exponential terms for computational convenience. These exponential terms were derived as curve fits to time-dependent retention values for bone and soft tissue pools generated by the strontium model of ICRP Publication 20, for the case of acute input to blood.

6.4.3.1.2.4 ICRP Publication 67

As discussed in an earlier section, ICRP Publication 67 (ICRP 1993) introduced a generic model structure for the alkaline earth elements and other bone-volume-seeking elements that explicitly depicts the recycling of activity between tissues and blood, as well as excretion of activity along explicit pathways (Figure 6.19). As in the model of ICRP Publication 20, the long-term behavior of calcium-like elements in bone is assumed to be determined by the turnover rate of bone and hence to be independent of the element, while the short to intermediate behavior in bone is assumed to depend on the element. The generic structure includes separate compartments for blood plasma and red blood cells, and separate compartments representing liver, kidneys, and remaining soft tissues ("other"). The compartments representing red blood cells, liver, and kidneys were not applied to all bone-volume-seeking elements but were included to address the relatively high accumulation of a few calcium-like elements such as lead and uranium at these sites.

In the strontium model of ICRP Publication 67, blood is treated as a uniformly mixed pool (i.e., the RBC compartment in Figure 6.19 is not used in the model) that exchanges activity with soft tissues and bone surfaces. Soft tissues are divided into three compartments corresponding to fast, intermediate, and slow return of activity to blood (compartments ST0, ST1, and ST2, respectively). The liver and kidneys are not addressed separately in the model for strontium but are included implicitly in the three soft tissue compartments, ST0, ST1, and ST2. Bone is divided into cortical and trabecular bone, and each of these bone types is further divided into bone surface and bone volume. Bone volume is viewed as consisting of two pools, one that exchanges with activity in bone surface for a period of weeks or months, and a second, non-exchangeable pool from which activity can be removed only by bone restructuring processes. Activity depositing in the skeleton is assigned to cortical and trabecular bone surface. Over a period of days, a portion of the activity on bone surface moves to exchangeable bone volume (remaining within the same bone type), and the rest returns to blood. Activity leaves exchangeable bone volume over a period of months, with part of the activity moving to bone surface and the rest to non-exchangeable bone volume. The rate of removal from non-exchangeable bone volume is assumed to be the rate of bone turnover, with different turnover rates applying to cortical and trabecular bone.

For interpretation of environmental data for ^{90}Sr (for purposes of biokinetic model development), the following gastrointestinal absorption fractions were applied on the basis of human and animal data: 0.6 for infants (ages 0–100 d), 0.4 for ages 1–15 y, and 0.3 at ages 25+ y. Linear interpolation between ages was used to estimate absorption between 100 d and 1 y and between 15 and 25 y. In the development of an outflow rate of strontium from blood, it was assumed that blood kinetics is the same for strontium and calcium at all ages. An outflow rate of 15 d^{-1} was derived.

For the adult, the rates of transfer of strontium between blood and the soft tissue compartments are set as follows in the model of ICRP Publication 67. Compartment ST0 represents the fast-exchange soft tissue compartment indicated in human studies to contain ~3 times as much calcium or strontium as blood. Considering the initially rapid buildup of strontium in soft tissues following injection, it is assumed that half of strontium leaving blood enters ST0, giving a transfer coefficient from blood to ST0 of 0.5 × 15 d^{-1} = 7.5 d^{-1}, and that transfer from ST0 back to blood is one-third this rate, or 2.5 d^{-1} (to force ST0 to contain three times as much strontium as blood). Soft tissue compartment ST1 together with ST0 is used to reproduce the early to intermediate term soft tissue content of strontium observed in human studies (Figure 6.25). This is achieved by assigning a fractional transfer from blood to ST1 of 0.1 (i.e., 10% of outflow from blood), corresponding to a transfer coefficient of 0.1 × 15 d^{-1} = 1.5 d^{-1}, and assigning a removal half-time from ST1 to blood of 6 d. Fractional deposition in a relatively non-exchangeable soft tissue pool, ST2, is set at 0.0002 (transfer coefficient = 0.0002 × 15 d^{-1} = 0.003 d^{-1}), with a removal half-time back to blood of 5 y (transfer coefficient of 0.00038 d^{-1}).

These parameter values for ST2 were set after the parameter values for ST0 and ST1 were set, and were designed to yield a total-soft tissue content of stable strontium of 1% of the total-body content.

It is assumed on the basis of data for laboratory animals and human subjects that fractional deposition on bone surfaces is the same for calcium, strontium, barium, and radium. From the collective data for these four elements, it is estimated that one-fourth of outflow from blood deposits on bone surfaces. For strontium, this yields a transfer coefficient from blood to total bone surface (cortical plus trabecular) of 0.25 × 15 d^{-1} = 3.75 d^{-1}.

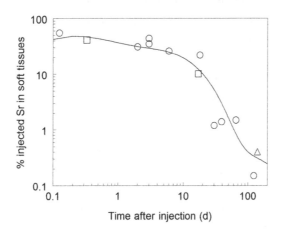

FIGURE 6.25 Comparison of predictions of the systemic model for strontium recommended in ICRP Publication 67 (ICRP 1993) with soft tissue retention data reported for adult human subjects administered ^{85}Sr intravenously. The circles, squares, and triangle represent results of three different studies.

Relative deposition of strontium or other alkaline earth elements on cortical and trabecular bone surfaces is based on the estimated calcium turnover rate of each bone type. As an average over adult ages, deposition on trabecular bone is estimated to be 1.25 times that on cortical bone. The transfer rate from blood to trabecular bone surface is $(1.25/2.25) \times 3.75$ d^{-1} = 2.08 d^{-1} and from blood to cortical bone surface is $(3.75-2.08)$ d^{-1} = 1.67 d^{-1}.

The residence time on human bone surfaces has not been determined with much precision for any of the alkaline earth elements. A removal half-time of 1 d is estimated from collective data for all four elements. The most direct information underlying this estimate consists of autoradiographic measurements of surface activity in human and canine bone samples taken at times ranging from few hours to a few days after intravenous injection of ^{45}Ca. Indirect information includes observations of the early rate of decline in whole-body retention of intravenously injected radioactive calcium, strontium, barium, and/or radium in human subjects, coupled with measurements of soft tissue retention (representing total-body retention minus bone retention).

Parameter values for exchangeable bone volume are estimated from whole-body measurements for human subjects, using data for times after bone surfaces and soft tissues have largely cleared of activity but before loss from bone resorption becomes an important consideration. Based on analysis of whole-body retention data for human subjects injected with radioisotopes of calcium, strontium, barium, or radium, the fraction of activity that moves from bone surfaces back to blood is assumed to be the same for all four elements. Specifically, five-sixths of activity leaving bone surfaces is assumed to return to blood and one-sixth is assumed to transfer to exchangeable bone volume. The transfer rate from trabecular or cortical bone surface to the corresponding exchangeable bone volume compartment is $(1/6) \times \ln(2)/1$ d = 0.116 d^{-1}, and the transfer rate from trabecular or cortical bone surface to blood is $(5/6) \times \ln(2)/1$ d = 0.578 d^{-1}.

Element-specific removal half-times from the exchangeable bone volume compartments are based in part on fits to the intermediate-term retention data from human injection studies. It is also considered that the assigned half-times should increase roughly in proportion to the likelihood of the element entering non-exchangeable sites in bone mineral, as suggested by data from in vitro experiments with hydroxyapatite crystals and whole-body retention patterns for alkaline earth elements in human subjects. A removal half-time of 80 d is assigned to strontium, compared with 100 d for calcium, 50 d for barium, and 30 d for radium. Because the data do not allow the derivation of removal half-times as a function of bone type, the same half-time is applied to cortical and trabecular exchangeable bone volume compartments.

Discrimination between alkaline earth elements by bone is accounted for by fractional transfer of activity from exchangeable to non-exchangeable bone volume. It is assumed that calcium, strontium, barium, and radium are all equally likely to become temporarily incorporated in bone mineral after injection into blood, but that the likelihood of reaching a non-exchangeable site in bone crystal decreases in the order: calcium > strontium > barium > radium. Fractional transfers of calcium, strontium, barium, and radium from exchangeable to non-exchangeable bone volume are set at 0.6, 0.5, 0.3, and 0.2, respectively, for consistency with whole-body and skeletal retention data on these elements, as well as results of in vitro measurements on hydroxyapatite crystals. The derived rate of transfer of strontium from exchangeable trabecular or cortical bone volume to the corresponding non-exchangeable bone volume compartment is $0.5 \times \ln(2)/80$ d = 0.0043 d^{-1}, and to the corresponding bone surface compartment is $0.5 \times \ln(2)/80$ d = 0.0043 d^{-1}.

Biological removal from the non-exchangeable bone volume compartments of cortical and trabecular bone is assumed to result from bone turnover. The average bone turnover rates during adulthood are estimated as 3% y^{-1} and 18% y^{-1} for cortical and trabecular bone, respectively (ICRP 2002). The corresponding transfer rates from the non-exchangeable bone volume compartments of cortical and trabecular bone to blood are 0.0000821 d^{-1} and 0.000493 d^{-1}, respectively.

Transfer coefficients describing clearance of strontium from blood to urine and feces are based on results of several studies of the early retention and excretion of radio-strontium by healthy human subjects. It is assumed on the basis of these studies that 11.5% (as a central estimate) of strontium leaving blood is transferred to the contents of the urinary bladder contents and subsequently to urine and 3.5% (also a central estimate) is transferred to the contents of the right colon contents, and subsequently to feces. Therefore, the transfer rate from blood to the urinary bladder contents is 0.115×15 d^{-1} = 1.73 d^{-1}, and from blood to the contents of the right colon contents is 0.035×15 d^{-1} = 0.525 d^{-1}.

The ICRP Publication 67 model for strontium in adults is extended to younger age groups by assuming elevated uptake by bone (i.e., elevated transfer from blood to bone surface) at younger ages, and elevated rates of removal from bone volume to blood at younger ages due to elevated bone turnover rates. Differences with age in uptake by bone are based on observations of the age-specific behavior of radioisotopes of the alkaline earth elements, including strontium in human subjects and laboratory animals. The assumed differences with age in cortical and trabecular bone turnover rates are taken from a paper by Leggett, Eckerman, and Williams (1982); these same values were later adopted as reference bone turnover rates in ICRP Publication 89 (ICRP 2002).

The following specific assumptions were used to extend the parameter values for strontium in the adult to younger ages (intake ages 100 d, 1 y, 5 y, 10 y, and 15 y).

- Fractional deposition on trabecular or cortical bone surface as a function of age is proportional to the age-specific calcium addition rate as estimated by Leggett, Eckerman, and Williams (1982).

- For a given bone type (trabecular or cortical), the rate of transfer from non-exchangeable bone volume to blood is equal to the age-specific rate of bone turnover (Leggett, Eckerman, and Williams 1982).

- For a given bone type, the rates of loss from bone surface and exchangeable bone volume (i.e., the total transfer rate from exchangeable bone volume to bone surface plus non-exchangeable bone volume) are independent of age. However, on the basis of animal studies, fractional transfer of strontium from bone surface to exchangeable bone volume is assumed to be smaller in children than in adults.

- At all ages, deposition in soft tissues and excretion pathways is proportional to the corresponding fractions for mature adults. For children, the fraction of outflow from blood left over after subtraction of bone deposition is divided into deposition fractions for soft tissues and excretion pathways in proportion to the corresponding deposition fractions for mature adults. For example, if the deposition fraction for bone surface is 0.25 in mature adult and 0.5 at age X, then deposition fractions for soft tissue compartments and excretion pathways at age X are $(1.0-0.5)/(1.0-0.25) = 2/3$ times the corresponding deposition fractions for adults. This approach is reasonably consistent with age-specific data for alkaline earth elements in dogs.

- Removal rates from soft tissue compartments are the same in children as in adults, with the exception of the rapid-turnover soft tissue compartment named ST0, for which the removal rate to blood at a given age is set so that the activity in ST0 is three times that in blood at equilibrium.

Transfer coefficients in the ICRP Publication 67 model for strontium for pre-adult ages (ages 100 d, 1 y, 5 y, 10 y, 15 y at intake) based on these assumptions are listed in columns 2–6 of Table 6.9. Parameter values for ages intermediate to those listed in the table are determined by linear interpolation between parameter values for the two bounding ages.

As illustrated in Figure 6.26, the parameter values were designed for reasonable consistency with central age-specific biokinetic data for strontium in the human body. Where definitive age-specific data for strontium in humans are lacking, the parameter values were designed to reproduce patterns of change with age in the biokinetics of strontium, as suggested by studies involving other alkaline earth elements, particularly calcium, in humans, and studies of strontium metabolism in dogs.

6.4.3.2 Iodine

6.4.3.2.1 Summary of Data Related to Iodine Kinetics

6.4.3.2.1.1 Iodine Requirements in Humans

Iodine is an essential component of the thyroid hormones thyroxine (T_4) and triiodothyronine (T_3), which regulate metabolic processes and are critical to growth and development. Several tens of micrograms of inorganic iodide are trapped daily by the adult human thyroid and used for synthesis of T_4 and T_3. T_4 is produced only in the thyroid and represents > 90% of the hormonal iodine secreted by the thyroid. About 20% of the circulating T_3 is produced in the thyroid and the rest is produced from T_4 in extrathyroidal tissues through a process involving removal of a single iodine atom from T_4. T_3 is more active than T_4 and exerts most of the effects of the thyroid hormones in the body.

Iodine is largely recycled by the body after use of T_4 and T_3 by tissues, but the body's supply must be supplemented with dietary iodine, due to obligatory losses in excreta. The World Health Organization (World Health Organization 2001) recommends daily intake of 90 μg of iodine at ages 0–59 mo, 120 μg at ages 6–12 y, 150 μg at ages greater than 12 y, and 200 μg for pregnant or lactating women, to ensure adequate production of thyroid hormones and prevention of goiter and hypothyroidism.

TABLE 6.9   Age-Specific Transfer Coefficients (d⁻¹) for the Strontium Model of ICRP Publication 67
(ICRP 1993)

Pathª	Age at Intake					
	100 d	1 y	5 y	10 y	15 y	Adult
Blood to UB Contents	0.5770	1.27	1.38	1.02	0.60	1.73
Blood to ULI	0.175	0.385	0.420	0.308	0.182	0.525
Blood to Trab Surf	2.250	1.350	1.330	2.120	3.100	2.080
Blood to Cort Surf	9.000	5.400	4.670	6.280	8.000	1.670
Blood to ST0	2.50	5.50	6.00	4.40	2.60	7.50
Blood to ST1	0.50	1.10	1.20	0.88	0.52	1.50
Blood to ST2	0.0010	0.0022	0.0024	0.0018	0.0010	0.0030
Trab Surf to Blood	0.6010	0.6010	0.6010	0.6010	0.6010	0.5780
Trab Surf to Trab Exch Vol	0.0924	0.0924	0.0924	0.0924	0.0924	0.1160
Cort Surf to Blood	0.601	0.601	0.601	0.601	0.601	0.578
Cort Surf to Cort Exch Vol	0.0924	0.0924	0.0924	0.0924	0.0924	0.1160
ST0 to Blood	0.833	1.830	2.000	1.470	0.867	2.500
ST1 to Blood	0.116	0.116	0.116	0.116	0.116	0.116
ST2 to Blood	0.00038	0.00038	0.00038	0.00038	0.00038	0.00038
Trab Exch Vol to Trab Surf	0.0043	0.0043	0.0043	0.0043	0.0043	0.0043
Trab Exch Vol to Trab Non-Ech Vol	0.0043	0.0043	0.0043	0.0043	0.0043	0.0043
Cort Exch Vol to Cort Surf	0.0043	0.0043	0.0043	0.0043	0.0043	0.0043
Cort Exch Vol to Cort Non-Exch Vol	0.0043	0.0043	0.0043	0.0043	0.0043	0.0043
Cort Non-Exch Vol to Blood	0.00822	0.00288	0.00153	0.000904	0.000521	0.0000821
Trab Non-Exch Vol to Blood	0.00822	0.00288	0.00181	0.001320	0.000959	0.0004930

ª UB = Urinary Bladder, ULI = Upper Large Intestine, Trab = Trabecular, Cort = Cortical, Surf = Surface, Vol =
Volume, Exch = Exchangeable, Non-Exch = Non-Exchangeable. ST0, ST1, and ST2 are soft tissue compartments with fast, intermediate, and slow turnover, respectively.

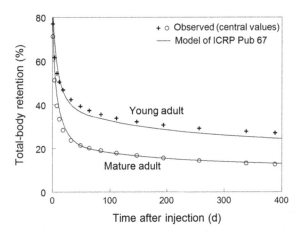

FIGURE 6.26   Comparison of predictions (curves) of the systemic model for strontium recommended in ICRP Publication 67 (ICRP 1993) with total-body retention data (+, o) reported for adult
human subjects administered radio-strontium. Mature adult defined as age ≥25 y at injection.
Young adult defined as age 21 y at injection.

TABLE 6.10   Typical Dietary Iodine in the United States (μg d^{-1})

Group	Age (y)			
	6–11	12–19	20–39	40–59
Males	180	280	280	280
Females	160	210	160	130
Total population	170	250	210	200

The following typical values for dietary intake of iodine by adults are based on worldwide survey data for iodine intake:

130 μg d^{-1} for women,

190 μg d^{-1} for men, and

160 μg d^{-1} as a gender-averaged value.

Table 6.10 lists estimated typical daily intakes of iodine as a function of age and gender in the U.S. population. The estimates are based on 50 percentile values for urinary iodine (μg/liter) in each group as determined in the National Health and Nutrition Survey, 2001–2002, together with reference values for the daily volume of urine in each group (ICRP 2002).

The following overview of the systemic biokinetics of iodine in adult humans is excerpted from a review by Leggett (2010), which provides an extensive bibliography.

6.4.3.2.1.2 Absorption and Distribution of Inorganic Iodide

Iodine occurs in foods mainly as inorganic iodide. Other forms of iodine in foods are reduced to iodide in the alimentary tract before absorption. Absorption is primarily from the small intestine but may occur to some extent from the stomach and other sites along the alimentary tract. Absorption normally is rapid and nearly complete.

Absorbed iodide is distributed quickly throughout the extracellular fluids (ECF). Most of the iodide that leaves blood is recycled to blood within 1–2 h and much of it is recycled within a few minutes.

The iodide ion is largely excluded from most cells but rapidly traverses the red blood cell (RBC) membrane. Equilibration between plasma iodide and RBC iodide occurs in minutes. The equilibrium concentration of iodide in RBC is about the same as in plasma.

A substantial portion of iodide entering blood is concentrated in the salivary glands and stomach wall by active transport. It is subsequently secreted into the alimentary tract contents in saliva and gastric juice and nearly completely reabsorbed to blood. As a central estimate, the rate of clearance of plasma iodide in saliva plus gastric secretions is 43 mL/min. The concentration of iodine in these secretions is about 30 times its concentration in plasma. There is a delay of about 20 min between uptake of iodine by the salivary glands and stomach wall and appearance in the stomach contents, and a delay of about 30 min between the peak concentration in plasma and the peak concentration in secretions into the alimentary tract.

The thyroid and kidneys are in competition for blood iodide and hence for the body's supply of iodide due to the rapid recycling of total-body iodide through blood. Normally more than 90% of the loss of iodine from the body is due to renal clearance of iodide. Little inorganic iodide is lost in feces. Sweat does not appear to be an important mode of loss of iodide, except perhaps in hot climates or during intense exercise.

Iodide in blood plasma is filtered by the kidneys at the glomerular filtration rate. About 70% of the filtered iodide is reabsorbed to blood, and the rest enters the urinary bladder contents and is excreted in urine. Renal clearance expressed as the volume of plasma iodide or blood iodide cleared per unit time is nearly constant over a wide range of plasma concentrations for a given age and gender. As a central estimate, renal clearance is about 37 mL plasma/min for euthyroid adult males. Renal clearance of iodide expressed as plasma volumes per unit time appears to be about 25%–30% lower on average in women than in men, but fractional loss of total-body iodide in urine per unit time is similar for men and women.

The concentration of radioiodide in the kidneys may exceed that in most extrathyroidal tissues for a brief period after acute input into blood. The liver typically accumulates a few percent of radioiodide soon after ingestion or intravenous administration but much less per gram of tissue than the kidneys.

6.4.3.2.1.3 *Behavior of Iodide and Organic Iodine in the Thyroid*
The basic unit of cellular organization within the thyroid is the follicle (Figure 6.27), a spherical structure typically a few hundredths of a millimeter in diameter. Each follicle is composed of a single layer of epithelial cells enclosing a lumen filled with a viscous material called colloid. The colloid consists mainly of thyroglobulin, a protein synthesized by follicular cells and secreted into the lumen. Thyroglobulin serves as a matrix for production and storage of the thyroid hormones T_4 and T_3.

Iodide is actively transported from blood plasma into thyroid follicular cells at the plasma membrane. A normal thyroid can concentrate the iodide ion to 20–40 times its concentration in blood plasma. Some of the trapped iodide leaks back into blood, but most of it diffuses across the follicular cell and enters the follicular lumen, where it is converted to organic iodine.

The kinetics of trapping and binding of intravenously injected ^{131}I by the thyroid has been studied in hyperthyroid and euthyroid (normal thyroid) subjects, first with no inhibition of binding, and later with administration of a drug that inhibited binding. The results indicate that the rate of binding of trapped iodide is much greater than the rate of return of trapped iodide to blood. When iodide binding was blocked before administration of ^{131}I, activity in the thyroid reached a peak at times varying from several minutes to an hour or more after injection. Typically, the rate of loss of trapped ^{131}I from the blocked thyroid was 2%–3% per minute. The rate of binding of trapped ^{131}I by the thyroid was nearly three times its rate of loss back to blood.

Iodide is transported across the luminal membrane of the follicular cell into the lumen and oxidized at the cell-colloid interface. The neutral iodine atoms formed by oxidation of iodide are bound (organified) within the lumen to specific residues of the amino acid

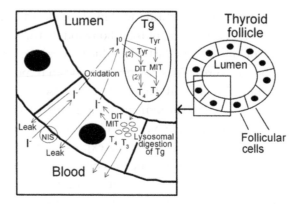

FIGURE 6.27 Diagram of a thyroid follicle, showing main steps in synthesis and secretion of T_4 and T_3 (steps described in main text). I^- = iodide, I^0 = neutral iodine, NIS = sodium-iodide symporter, Tg = thyroglobulin, Tyr = tyrosine, MIT = monoiodotyrosine, DIT = diiodotyrosine.

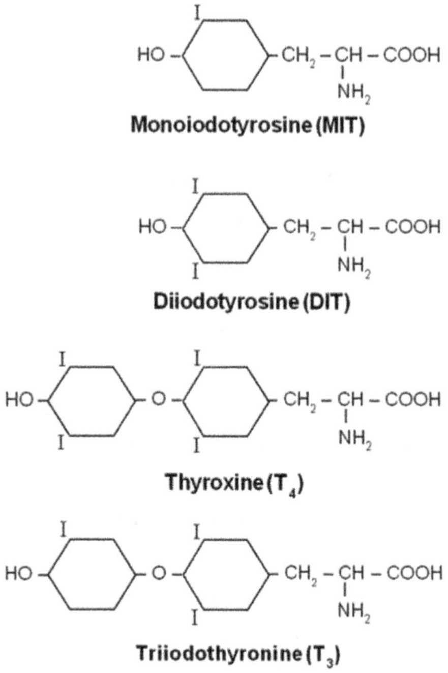

Monoiodotyrosine (MIT)

Diiodotyrosine (DIT)

Thyroxine (T_4)

Triiodothyronine (T_3)

FIGURE 6.28 Structure of the thyroid hormones and their precursors.

tyrosine. Some tyrosine residues gain one iodine atom, forming monoiodotyrosine (MIT), and others gain two iodine atoms, forming diiodotyrosine (DIT) (Figures 6.27 and 6.28). T_4 is formed within the lumen by the coupling of two DIT molecules and hence has four iodine atoms, and T_3 is formed within the lumen by coupling of one MIT molecule to one DIT molecule and hence has three iodine atoms. The lumen typically contains 10–15 times more T_4 than T_3.

The thyroid adapts to prolonged reductions or increases in iodine intake by adjusting its rate of uptake of iodide from blood. Adaptation of thyroidal clearance of iodide to dietary intake results in an inverse relation between net 24-h thyroidal uptake of ingested ^{131}I and average 24-h urinary excretion of stable iodine. Results of a number of studies of dietary iodine and thyroidal uptake of ^{131}I in the same populations indicate that thyroid uptake of ^{131}I is about 14%–15% for stable iodine intake approaching or moderately exceeding 400 µg d^{-1}, 16%–27% for intake of 250–330 µg d^{-1}, 40%–45% for intake of 80–85 µg d^{-1}, 54%–59% for intake of 40–55 µg d^{-1}, and about 90% for intake of 5–10 µg d^{-1}.

In adults with iodine sufficient diet, the thyroid typically stores 5–15 mg of hormonal iodine. Estimates of the rate S of secretion of hormonal iodine by the thyroid (µg I d^{-1}) in individual normal adult subjects range from less than 30 µg d^{-1} to more than 150 µg d^{-1}. Typical values for adults given in reviews and textbooks are generally in the range 55–85 µg d^{-1}. There is a decline in thyroid hormone secretion with increasing adult age after the fifth or sixth decade (Figure 6.29). The secretion rate appears to be about one-third lower on average in women than in men, although there is some overlap is measurements for the two sexes. The following reference values of S for adults are based on collected data on thyroidal secretion of iodine as T_4 for ages 18–65 y, and the assumption that T_4 represents 90% of total secretion of hormonal iodine:

52 µg d^{-1} for females,

76 µg d^{-1} for males, and

64 µg d^{-1} as a gender-averaged value.

Fractional transfer of iodine from thyroid stores to blood per unit time depends on the size of current thyroid stores, the rate of secretion of thyroid hormones, and the extent of

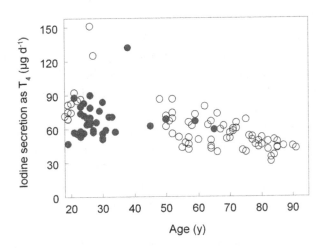

FIGURE 6.29 Rate of secretion of hormonal iodine as T_4 from the thyroid in healthy adult male subjects. Open circles represent data from a single study involving a large number of subjects. Shaded circles represent data collected from several smaller studies.

leakage of iodide to blood from MIT and DIT deiodinated in follicular cells (which may be large in persons with iodine-rich diets). For example, assuming first-order kinetics and negligible leakage of iodide from MIT and DIT to blood, thyroidal stores of 5 mg and a secretion rate of hormonal iodine of 64 μg d^{-1} correspond to a half-time of about 54 d; stores of 10 mg and secretion of 76 mg correspond to a half-time of about 91 d; and stores of 15 mg and a secretion rate of 80 μg d^{-1} correspond to a half-time of about 130 d. A biological half-time of 90 d appears to be a reasonable central estimate for normal adults based on observed values, that is, half-times estimated from external measurement of radioiodine retention in the thyroid.

6.4.3.2.1.4 Behavior of Extrathyroidal T_4 and T_3

Upon secretion by the thyroid into blood, T_4 and T_3 are rapidly and almost completely bound to plasma proteins. Little, if any, enters the RBC. As a result of protein binding, clearance of organic iodine from the circulation is slower than removal of the iodide ion from the circulation. Reported concentrations of protein-bound iodine in blood plasma of euthyroid subjects generally are in the range 3–8 μg/100 mL and cluster about 5–6 μg/100 mL.

A number of investigators have studied the kinetics of radiolabeled T_4 after its intravenous administration to human subjects. The removal half-time from blood plasma typically increases from about 1 h at 20–60 min after injection to about 1 wk at equilibrium. Early disappearance from plasma may represent mainly distribution throughout the extracellular fluids plus uptake by hepatocytes. The slower decline at later times may represent uptake by cells and binding to intracellular proteins throughout the body, reduction to inorganic iodide due to use of the hormones by cells, and biliary secretion, followed by fecal excretion of part of the organic iodine entering the liver. External measurements together with liver biopsy data indicate that the liver accumulates roughly 35% (22%–52%) of injected T_4 during the first 3–4 hours after administration and contains roughly 25% (14%–40%) of extrathyroidal T_4 at equilibrium.

The kinetics of labeled T_3 has been difficult to determine with much precision, in large part due to interference of iodoproteins generated by metabolism of the injected trace material. Human studies indicate high initial uptake of labeled T_3 by the liver, but a shorter retention time than T_4 in the liver. The liver content at equilibrium has been estimated as 5%–21% of the total extrathyroidal T_3 pool.

A portion of T_4 or T_3 entering the liver is secreted into the small intestine in bile. The secreted form is poorly absorbed to blood and is largely excreted in feces. This accounts for about one-fifth of the loss of organic iodine from extrathyroidal tissues, and reduction to iodide and return to the blood iodide pool accounts for the rest. Endogenous fecal excretion of organic iodine can become a major source of loss of iodine during periods of low intake of iodine.

Most estimates of the mass of extrathyroidal organic iodine at equilibrium are in the range 500–1000 μg. Most estimates of the biological half-life of T_4 in normal subjects are in the range 5–9 d. The half-life of T_3 is about 1 d, and that of an inactive variant of T_3 called reverse T_3 (rT_3, an inactive variant of T_3) is a few hours. Extrathyroidal conversion of T_4 to

T_3 or rT_3 results, in effect, in an extension of the half-life of T_4. Measurements on euthyroid adult males of ages 18–91 y indicate that the rate of T_4 production as well as its turnover rate, representing the combined rate of deiodination and fecal excretion, decrease with age starting sometime before age 50 y. Measured rates of deiodination of T_4 are similar in male and female subjects in the same age groups.

6.4.3.2.2 Biokinetic Models for Systemic Iodine
6.4.3.2.2.1 ICRP Publication 2
The systemic model for iodine applied in ICRP Publication 2 (ICRP 1960) is unusually detailed for that document. The organs of reference for iodine are total body, thyroid, kidneys, liver, spleen, testes, and bone. Deposition fractions for these organs are 1.0, 0.3, 0.04, 0.12, 0.005, 0.005, and 0.07, respectively. Biological half-times are 138 d for total body and thyroid; 7 d for kidneys, liver, spleen, and testes; and 14 d for bone.

6.4.3.2.2.2 The Riggs Model and Its Variations Used in Radiation Protection
A number of physiological systems models have been developed from results of radioiodine studies on human subjects to describe quantitative aspects of the metabolism of iodine as an essential element in humans. A relatively simple three-compartment biokinetic model of iodine developed by Riggs (1952) for applications in physiological and clinical studies has been used, sometimes with modified parameter values, by the ICRP for many years as the basis of its biokinetic models for occupational or environmental intake of radioiodine. The Riggs model was adopted for application to workers in ICRP Publication 30 (ICRP 1979) and carried over to ICRP Publication 68 (ICRP 1994b). The Riggs model with age-specific parameter values was used in the ICRP Publication 72 series on age-specific doses to members of the public from intake of environmental radionuclides (ICRP 1990, 1993, 1995a,b,c).

The Riggs model with parameter values applied to workers in relatively recent ICRP reports (ICRP 1994b, 1997) is shown in Figure 6.30. The compartments and paths of transfer represent absorption of dietary iodine to blood as inorganic iodide; competition between thyroidal and renal clearance for circulating inorganic iodide; production, storage, and secretion of hormonal iodine by the thyroid; deiodination of most of the secreted hormonal iodine and recycling of inorganic iodide; and loss of the remainder of secreted hormonal iodine in feces.

Variations of the Riggs model and some more detailed iodine models have been developed for specific applications in radiation protection including: age-specific dosimetry of internally deposited radioiodine for application to environmental exposures (Stather, Greenhalgh, and Adams 1983; Johnson 1987; ICRP 1990); estimation of doses to patients from medical applications of radioiodine (Committee on Medical Internal Radiation Dose 1975; Robertson and Gorman 1976; McGuire and Hays 1981; Johansson et al. 2003); dose to the embryo/fetus or nursing infant from intake of radioiodine by the mother (Berkovski 1999b, Berkovski 2002, ICRP 2002); and reduction of radioiodine dose by administration of potassium iodide (Adams and Bonnell 1962, Ramsden et al. 1967, Zanzonico and Becker 2000). The model of Berkovski (Berkovski 1999a,b, 2002) for the pregnant or nursing

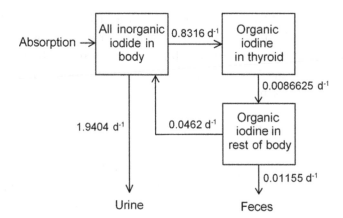

FIGURE 6.30   Biokinetic model for iodine introduced by Riggs (1952) and widely used in radiation protection. The transfer coefficients are those used in relatively recent ICRP reports (ICRP 1994b, 1997). In those reports the compartments labeled "All inorganic iodide in body," "Organic iodine in thyroid," and "Organic iodine in rest of body" are called "Blood," "Thyroid," and "Rest of body," respectively.

mother, and the model of Johannsson et al. (Johansson et al. 2003) designed for applications in nuclear medicine, provide relatively detailed descriptions of the early biokinetics of inorganic iodide to allow improved dosimetry of short-lived radioiodine.

6.4.3.2.2.3 An Updated Systemic Model for Radioiodine Intake in the Workplace
Recently the ICRP adopted an updated model for iodine (Leggett 2010) for use in upcoming revisions of ICRP documents on occupational intake of radioiodine. The updated model is intended to describe the iodine cycle in the human body in sufficient detail to provide improved dose estimates for short-lived isotopes of iodine that may be encountered in and around nuclear facilities or used in nuclear medicine.

The structure of the model, including connections with the ICRP's Human Alimentary Tract Model (ICRP 2006), is shown in Figure 6.31. Baseline parameter values for workers are listed in Table 6.11. Each of the values is in the form of a transfer coefficient, defined as fractional transfer of the contents of the donor compartment per unit time.

The updated ICRP model is a consolidation of three sub-models describing three physiological systems that determine the iodine cycle in the human body:

(1) A sub-model describing the behavior of extrathyroidal inorganic iodide. This sub-model is an extension of a model developed by Hays and Wegner (1965) from bioassay and external measurements of intravenously administered ^{131}I in young adult males during the early hours after administration. The model of Hays and Wegner was modified mainly by the addition of compartments and transfer coefficients representing inorganic iodide kinetics in the kidneys and liver, and by adjustment of flow rates to other compartments to account for this change in model structure. The following compartments shown in Figure 6.31 are used to describe the

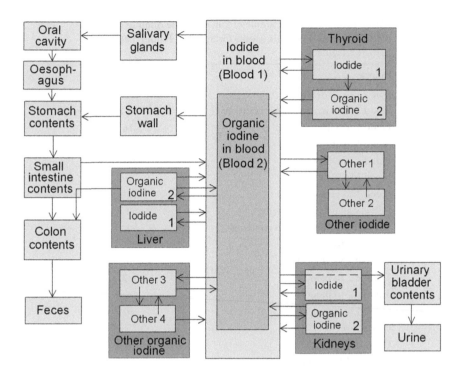

FIGURE 6.31 Structure of the updated ICRP biokinetic model for systemic iodine (Leggett 2010; ICRP 2018).

behavior of extrathyroidal inorganic iodide: a compartment representing iodide in blood (Blood 1); Salivary glands; Stomach wall; Liver 1, representing iodide in liver; Kidneys 1, representing iodide in kidneys; Other 1, representing rapidly exchangeable iodide in extracellular fluids of extrathyroidal tissues other than kidneys and liver; Other 2, representing slowly exchangeable iodide in extrathyroidal tissues other than kidneys and liver; and a series of compartments representing different segments of the alimentary tract as represented in the ICRP's Human Alimentary Tract Model.

(2) A sub-model describing the behavior of iodine in the thyroid. The structure and parameter values of this sub-model were developed independently of existing models. The thyroid is divided into compartments representing inorganic iodide (Thyroid 1) and organic iodine (Thyroid 2). Thyroid 1 receives iodide from Blood 1, feeds iodide to Thyroid 2, and leaks some iodide back to Blood 1. Thyroid 2 converts iodide to organic iodine and transfers organic iodine into the blood organic iodine pool (Blood 2). An arrow representing leakage of activity from Thyroid 2 into Blood 1 is included for application to subjects with unusually high dietary iodine, but the baseline transfer coefficient from Thyroid 2 to Blood 1 is set to zero.

(3) A sub-model describing the behavior of extrathyroidal organic iodine. This sub-model is an extension of a model of extrathyroidal T_4 kinetics developed by Nicoloff and Dowling (1968) from measurements of ^{131}I-labeled T_4 in healthy human subjects.

TABLE 6.11   Baseline Parameter Values for the Adult in the ICRP's Updated Biokinetic Model for Systemic Iodine (Leggett 2010, ICRP 2018)

Pathway	Transfer Coefficient (d^{-1})
Blood 1 to Thyroid 1	7.26[a]
Blood 1 to Urinary bladder contents	11.84
Blood 1 to Salivary gland	5.16
Blood 1 to Stomach wall	8.60
Blood 1 to Other 1	600
Blood 1 to Kidneys 1	25
Blood 1 to Liver 1	15
Salivary gland to Oral cavity	50
Stomach wall to Stomach contents	50
Thyroid 1 to Thyroid 2	95
Thyroid 1 to Blood 1	36
Thyroid 2 to Blood 2[b]	0.0077
Thyroid 2 to Blood 1	0[c]
Other 1 to Blood 1	330
Other 1 to Other 2	35
Other 2 to Other 1	56
Kidneys 1 to Blood 1	100
Liver 1 to Blood 1	100
Blood 2 to Other 3	15
Other 3 to Blood 2	21
Other 3 to Other 4	1.2
Other 4 to Other 3	0.62
Other 4 to Blood 1	0.14
Blood 2 to Kidneys 2	3.6
Kidneys 2 to Blood 2	21
Kidneys 2 to Blood 1	0.14
Blood 2 to Liver 2	21
Liver 2 to Blood 2	21
Liver 2 to Blood 1	0.14
Liver 2 to Right colon contents	0.08

[a] Depends on the ratio Y/S, where Y (μg d^{-1}) is dietary intake of stable iodine and S (μg d^{-1}) is the rate of secretion of hormonal stable iodine by the thyroid.

[b] For high intake of stable iodine the outflow from Thyroid 2 is split between Blood 2 and Blood 1 as described by Leggett (2010).

[c] Non-zero only for high intake of stable iodine (Leggett 2010).

The model of Nicoloff and Dowling was modified mainly by the addition of a compartment representing organic iodine in the kidneys that is assumed to have the same rate of exchange with blood plasma per gram of tissue as does the liver. The following compartments shown in Figure 6.31 are used to describe the behavior of extrathyroidal organic iodine: Blood 2, representing thyroid hormones bound to plasma proteins; Liver 2, representing organic iodine in liver; Kidneys 2, representing organic iodine in kidneys; Other 3, representing rapidly exchangeable organic iodine in extracellular fluids of extrathyroidal tissues other than kidneys and liver; and Other 4, representing slowly exchangeable organic iodine in extrathyroidal tissues other than kidneys and liver.

In the full model defined in Figure 6.31 and Table 6.11, iodine is assumed to be removed from the body only through urinary and fecal excretion. Iodide moves to Urine after transfer from Blood 1 into Urinary bladder contents. This represents the net result of glomerular filtration of iodide, reabsorption of much of the filtered iodide to blood, and transfer of the remainder to the urinary bladder contents followed by excretion in urine. Organic iodine is excreted in feces after transfer from Liver 2 to Right colon, representing the net result of secretion into the small intestine and the transfer of unabsorbed organic iodine to the right colon followed by excretion in feces.

Assuming that stable iodine intake and excretion are in balance, the transfer coefficient λ from Blood iodide to Thyroid iodide can be estimated in terms of the dietary stable iodine Y (μg d^{-1}) and the rate S of secretion of stable iodine by the thyroid (μg d^{-1}) using reference (typical) values for Y and S:

$$\lambda = \frac{16.34}{0.98\left(\dfrac{Y}{S}\right) - 0.2} (d^{-1}) \tag{6.1}$$

Thus, λ depends on the ratio Y:S. For example, the ratio Y:S based on the reference values Y = 190 μg d^{-1} and S = 76 μg d^{-1} for a male worker is 190:76 = 2.5. The same ratio is derived from reference values for a female worker: Y:S = 130 μg d^{-1}:52 μg d^{-1} = 2.5. The resulting sex-independent transfer coefficient based on Equation (6.1) is 7.26 d^{-1}.

Equation (6.1) is applicable to any combination of Y and S that gives a transfer coefficient of at least 2.5 d^{-1}. For lower derived values, the transfer coefficient is set at 2.5 d^{-1} based on indications that, with iodine-rich diets, Equation (6.1) does not apply, and thyroid uptake becomes increasingly difficult to model as dietary iodine increases. The transfer coefficient 2.5 d^{-1}, together with baseline values for other coefficients in the model, gives a 24-h thyroid content of about 12% of the ingested amount. This appears to be a reasonable average value for dietary iodine between 400 and 2000 μg d^{-1}, although considerable variability in thyroid uptake is seen between individual subjects at these levels.

Predictions of the systemic iodine in the adult human defined by Figure 6.31 and Table 6.11 are compared with observations in Figures 6.32 through 6.34 and Table 6.12.

Figures 6.32 and 6.33 show observations (symbols) and model predictions (curves) of the distribution of radioiodine in the first few hours after intravenous injection into adult humans. The open circles in these figures represent means for healthy young adult males. The close agreement in Figure 6.32 between predictions and the open circles is to be expected because the parameter values dominating model predictions were based in part on these data. The triangles in Figure 6.32 represent median values for individual euthyroid patients. The values represented by plus signs in Figures 6.32 and 6.33 represent mean

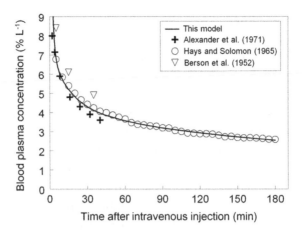

FIGURE 6.32 Model predictions of clearance of intravenously injected radioiodine from plasma compared with central values determined in three studies.

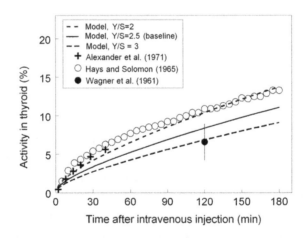

FIGURE 6.33 Model predictions of thyroidal uptake of intravenously injected ^{131}I compared with mean values of external measurements for three study groups.

values for several euthyroid subjects. The model predictions shown in Figure 6.32 are for total blood iodide.

In Figure 6.33, the observations are compared with model-generated curves based on three different values of the transfer coefficient from blood to thyroid. This transfer coefficient is derived from Equation (6.1) and depends on the ratio Y/S, where Y is dietary stable iodine ($\mu g\ d^{-1}$) and S is daily secretion of hormonal iodine by the thyroid ($\mu g\ d^{-1}$). Estimates of Y and S were not reported for the three study groups addressed in the figure. The group represented by plus signs was from a region with relatively low dietary iodine, suggesting a ratio Y/S less than the baseline value 2.5. The transfer coefficient based on the ratio Y/S = 2 yields reasonable agreement with thyroidal uptake data for that group, as well as data for a group of healthy young adult male subjects represented by open circles. Short-term urinary data for the third group, represented by the single closed circle, indicate mean

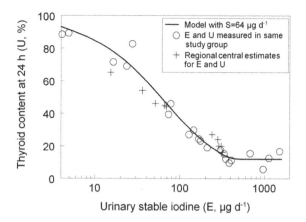

FIGURE 6.34   Model predictions and observations of 24-h uptake of radioiodine by thyroid (U) as a function of daily urinary excretion of stable iodine (E).

TABLE 6.12   Model Predictions of Mass or Concentration of Iodine in Tissues and Fluids at Equilibrium

Quantity	Dietary Iodine (μg d^{-1})/ Thyroidal Secretion of Organic Iodine (μg d^{-1})			
	130/52[a]	160/64[a]	190/76[a]	300/100[b]
Iodine in thyroid (μg)	6750	8310	9870	13,000
Iodine in blood plasma (μg dl^{-1})	0.22	0.27	0.32	0.51
Total extrathyroidal inorganic iodide (μg)	58	71	84	135
Organic iodine in blood plasma (μg dl^{-1})	4.3	5.2	6.2	8.2
Total extrathyroidal organic iodine (μg)	520	640	760	1000

[a] Baseline transfer coefficient describing thyroidal uptake (7.26 d^{-1}) is applied because the ratio of daily intake of iodine Y to daily thyroidal secretion S is 2.5.

[b] Transfer coefficient from blood iodide to thyroid iodide is 5.96 d^{-1} based on Equation (6.1).

iodine intake on the order of 200 μg d^{-1}, suggesting a ratio Y/S greater than the baseline value 2.5. The transfer coefficient based on the ratio Y/S = 3 is consistent with mean 2-hour thyroidal uptake for that group.

Model predictions of the percentage U of ingested radioiodine in the thyroid at 24 h after intake, assuming no radioactive decay, are compared in Figure 6.34 with observed values for subjects with different levels E of stable iodine in urine. Model predictions are based on the transfer coefficients in Table 6.11, except that the transfer coefficient from Blood 1 to Thyroid 1 was varied with E, as described by Equation (6.1), down to a minimum value of 2.5 d^{-1}. For this comparison, the value S was set (kept fixed) at a gender-averaged reference value of 64 μg d^{-1}.

The model with baseline parameter values (Table 6.11) predicts that the thyroid contains about 29% of ingested or intravenously injected iodine at 24 h after intake, assuming no radioactive decay. The content of the thyroid is predicted to peak at about 30% of the ingested or injected amount during the period 24–48 h after intake.

Model predictions of the equilibrium content of iodine in the thyroid, concentration of inorganic iodide and organic iodine in blood, and total extrathoracic contents of inorganic iodide and organic iodide are listed in Table 6.12 for different combinations of dietary iodine Y and thyroidal secretion rate S. The predicted values for each of these quantities based on reference values for dietary stable iodine Y and secretion rate of hormonal iodine S for women, total adult population, and men (see the first three columns of model predictions) are within the ranges of reported values for euthyroid subjects. For example, predictions of the mass of iodide in the thyroid at equilibrium are 6.75–9.87 g, compared with typical values of 5–15 mg. Predictions of the concentration of organic iodine in blood plasma are 4.3–6.2 μg/dl, compared with commonly reported values of 3–8 μg/dl.

6.4.3.2.2.4 Extension of Transfer Coefficients of the Updated Iodine Model to Pre-Adult Ages
The iodine model for adults described in the previous section (Leggett 2010) was extended to pre-adult ages (Leggett 2017). Age-specific parameter values are listed in Table 6.13. These values were developed for use by the ICRP in upcoming documents on: (a) doses to members of the public from environmental radionuclides, and (b) doses to patients from administered radiopharmaceuticals.

A transfer coefficient developed for adults was applied to children, unless there was clear evidence in the literature of age dependence. The only transfer coefficients for which variation with age is clearly indicated by reported data are the value describing transfer of organic iodine from Thyroid 2 to Blood Organic Iodine and values describing movement of extrathyroidal organic iodine. The transfer coefficients for these paths are calculated as follows: (1) the transfer coefficient from Thyroid 2 to Blood Organic Iodine at a given age is $\ln(2)/T_{1/2}$, where $\ln(2)$ = natural logarithm of 2 = 0.69315 and $T_{1/2}$ refers to the biological half-times of iodine in the thyroid (e.g., 10 d in infants and 50 d at age 10 y); (2) all parameter values describing the movement of extrathyroidal organic iodine at a given age (the last 12 values in each column in Table I-4) are (7/T) times the corresponding value for adults, where T is the age-specific turnover time of extrathyroidal organic iodine discussed earlier, (e.g., 4 d in infants and 5.5 d at age 10 y).

The following paragraphs summarize the results of the review of the age-specific behavior of iodine in the human body.

Thyroid uptake of iodine: Thyroidal uptake of iodine typically is much higher during the first week or two of life than at later times. For example, uptake of ^{131}I during the first 2–3 d of life averages about 70% of the administered amount. The thyroidal hyperactivity observed soon after birth is not considered in the development of the age-specific transfer coefficients listed in Table 6.13, which are for intake ages of 100 d or higher.

Regional studies of radioiodine uptake by the thyroid in different age groups suggest that there is little, if any, age dependence in uptake beyond early infancy, except perhaps for a modest decline after the fifth or sixth decade. Age-specific uptake values determined in one relatively large set of euthyroid subjects (60 subjects ages 2.5 mo to 18 y and 64 adults) are shown in Figure 6.35.

TABLE 6.13 Baseline Transfer Coefficients for the Updated Iodine Model (Figure 6.31) (Leggett 2017)

Pathway[a]	100 d	1 y	5 y	10 y	15 y Male	15 y Female	Adult Male	Adult Female
Blood Iodide to Thyroid 1	7.26E+00	7.26E+00	7.26E+00	7.26E+00	7.26E+00	7.26E+00	7.26E+00	7.26E+00
Blood Iodide to UB Cont	1.184E+01	1.184E+01	1.184E+01	1.184E+01	1.184E+01	1.184E+01	1.184E+01	1.184E+01
Blood Iodide to Salivary	5.16E+00	5.16E+00	5.16E+00	5.16E+00	5.16E+00	5.16E+00	5.16E+00	5.16E+00
Blood Iodide to St Wall	8.60E+00	8.60E+00	8.60E+00	8.60E+00	8.60E+00	8.60E+00	8.60E+00	8.60E+00
Blood Iodide to Other 1	6.00E+02	6.00E+02	6.00E+02	6.00E+02	6.00E+02	6.00E+02	6.00E+02	6.00E+02
Blood Iodide to Kidneys 1	2.50E+01	2.50E+01	2.50E+01	2.50E+01	2.50E+01	2.50E+01	2.50E+01	2.50E+01
Blood Iodide to Liver 1	1.50E+01	1.50E+01	1.50E+01	1.50E+01	1.50E+01	1.50E+01	1.50E+01	1.50E+01
Salivary to St Cont	5.00E+01	5.00E+01	5.00E+01	5.00E+01	5.00E+01	5.00E+01	5.00E+01	5.00E+01
St Wall to St Cont	5.00E+01	5.00E+01	5.00E+01	5.00E+01	5.00E+01	5.00E+01	5.00E+01	5.00E+01
Thyroid 1 to Thyroid 2	9.50E+01	9.50E+01	9.50E+01	9.50E+01	9.50E+01	9.50E+01	9.50E+01	9.50E+01
Thyroid 1 to Blood Iodide	3.60E+01	3.60E+01	3.60E+01	3.60E+01	3.60E+01	3.60E+01	3.60E+01	3.60E+01
Thyroid 2 to Blood Organic	6.93E-02	4.62E-02	2.31E-02	1.39E-02	1.07E-02	1.07E-02	7.70E-03	7.70E-03
Thyroid 2 to Blood Iodide	0.00E+00	0.00E+00	0.00E+00	0.00E+00	0.00E+00	0.00E+00	0.00E+00	0.00E+00
Other 1 to Blood Iodide	3.30E+02	3.30E+02	3.30E+02	3.30E+02	3.30E+02	3.30E+02	3.30E+02	3.30E+02
Other 1 to Other 2	3.50E+01	3.50E+01	3.50E+01	3.50E+01	3.50E+01	3.50E+01	3.50E+01	3.50E+01
Other 2 to Other 1	5.60E+01	5.60E+01	5.60E+01	5.60E+01	5.60E+01	5.60E+01	5.60E+01	5.60E+01
Kidneys 1 to Blood Iodide	1.00E+02	1.00E+02	1.00E+02	1.00E+02	1.00E+02	1.00E+02	1.00E+02	1.00E+02
Liver 1 to Blood Iodide	1.00E+02	1.00E+02	1.00E+02	1.00E+02	1.00E+02	1.00E+02	1.00E+02	1.00E+02
Blood Organic to Other 3	2.63E+01	2.33E+01	2.10E+01	1.91E+01	1.75E+01	1.75E+01	1.50E+01	1.50E+01
Other 3 to Blood Organic	3.68E+01	3.27E+01	2.94E+01	2.67E+01	2.45E+01	2.45E+01	2.10E+01	2.10E+01
Other 3 to Other 4	2.10E+00	1.87E+00	1.68E+00	1.53E+00	1.40E+00	1.40E+00	1.20E+00	1.20E+00
Other 4 to Other 3	1.09E+00	9.64E-01	8.68E-01	7.89E-01	7.23E-01	7.23E-01	6.20E-01	6.20E-01
Other 4 to Blood Iodide	2.45E-01	2.18E-01	1.96E-01	1.78E-01	1.63E-01	1.63E-01	1.40E-01	1.40E-01
Blood Organic to Kidneys 2	6.30E+00	5.60E+00	5.04E+00	4.58E+00	4.20E+00	4.20E+00	3.60E+00	3.60E+00
Kidneys 2 to Blood Organic	3.68E+01	3.27E+01	2.94E+01	2.67E+01	2.45E+01	2.45E+01	2.10E+01	2.10E+01
Kidneys 2 to Blood Iodide	2.45E-01	2.18E-01	1.96E-01	1.78E-01	1.63E-01	1.63E-01	1.40E-01	1.40E-01
Blood Organic to Liver 2	3.68E+01	3.27E+01	2.94E+01	2.67E+01	2.45E+01	2.45E+01	2.10E+01	2.10E+01
Liver 2 to Blood Organic	3.68E+01	3.27E+01	2.94E+01	2.67E+01	2.45E+01	2.45E+01	2.10E+01	2.10E+01
Liver 2 to Blood Iodide	2.45E-01	2.18E-01	1.96E-01	1.78E-01	1.63E-01	1.63E-01	1.40E-01	1.40E-01
Liver 2 to Right Colon	1.40E-01	1.24E-01	1.12E-01	1.02E-01	9.33E-02	9.33E-02	8.00E-02	8.00E-02

[a] Other 1 and Other 2 are the inorganic iodide pools in "Other tissue." Other 3 and Other 4 are the organic iodine pools in "Other tissue."

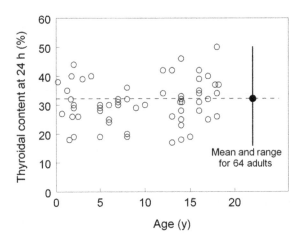

FIGURE 6.35   Comparison of thyroidal uptake of ^{131}I at 24 h in euthyroid children and adults from the same region (Oliner et al. 1957).

Biological half-time in the thyroid: The biological half-time of radioiodine in the thyroid as a function of age at intake has been measured in several studies. Results for any one age group are highly variable. The collective data suggest a sizable increase in the half-time between birth and about age 5–6 y, and then at most a modest increase to early adulthood. There appears to be little, if any, change in the half-time from early adulthood until at least the fifth or sixth decade, after which there may be a moderate decline. Based on a review and analysis of values reported in the literature, Dunning Jr and Schwarz (1981) estimated mean half-times of 16 d (range, 6–23 d) in infants; 13 d (4–39 d) for ages 0.5–2 y; 50 d (19–118 d) for ages 6–16 y; and 85 d (21–372 d) for ages >18 y. Results of four experimental studies involving pre-adult subjects are summarized in Figure 6.36. Selected baseline biological half-times for use in the updated iodine model (Figure 6.31) are 10 d in infants (age 100 d), 15 d at age 1 y, 30 d at age 5 y, 50 d at age 10 y, 65 d at age 15 y, and 90 d in young or middle-aged adults.

Rate of thyroidal secretion of T_4 and T_3: Results of clinical and experimental studies indicate that the mass of organic iodine secreted daily by the thyroid increases with age from infancy to early adulthood, then remains steady through the fifth or sixth decade of life, and declines thereafter. Representative values are shown in Figure 6.37.

Rate of degradation of extrathyroidal organic iodine: The biological half-time of extrathyroidal T_4 increases with age throughout life. Central half-times estimated from collected data are 4 d in infants, 5 d in children, 6 d in adolescents, 7 d in young adults, 8 d in middle-aged adults, and 9 d in elderly adults. The half-time of extrathyroidal T_4 essentially determines the half-time of extrathyroidal hormonal iodine. The following reference values are used to develop baseline transfer coefficients describing the behavior of extrathyroidal organic iodine at different ages: 4 d in infants, 4.5 d at age 1 y, 5 d at age 5 y, 5.5 d at age 10 y, 6 d at age 15 y, and 7 d in adults.

Adjustment of thyroid uptake to account for atypical dietary iodine levels: As is the case for adults, balance considerations imply that the transfer coefficient (d^{-1}) from the blood

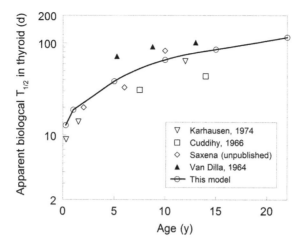

FIGURE 6.36   Measured and modeled biological half-time of iodine in the thyroid.

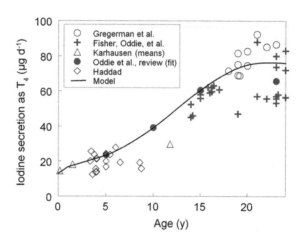

FIGURE 6.37   Measured and modeled rate of secretion of T_4 by the thyroid in males from birth to early adulthood. In the model, secretion of iodine as T_4 is assumed to be 93% of total secretion of hormonal iodine by the thyroid.

iodide pool to the thyroid can be estimated from Equation (6.1), where Y (µg) is daily dietary intake of stable iodine, and S (µg) is daily secretion of hormonal iodine by the thyroid. Changes with age in the hormonal iodine secretion rate S are reasonably well established (Figure 6.37). Thus, the transfer coefficient from blood to thyroid can be adjusted for children when average stable iodine intake Y is known.

6.4.3.3 Cesium

6.4.3.3.1 Summary of the Database

Cesium is chemically similar to the essential element potassium and is a qualitative physiological analog of K, but important quantitative differences in the biokinetics of K and Cs arise from different rates of transport of these elements across biological membranes. The

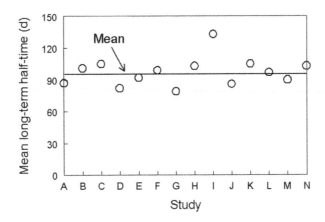

FIGURE 6.38 Mean long-term half-times (d) for total-body retention of Cs in 14 studies involving healthy adult male subjects. The number of subjects in individual studies ranged from 2 to >100.

most important factor affecting the residence time of ^{137}Cs in the body appears to be the mass of K in the body, which is related to the muscle mass and hence the total-body weight.

Whole-body retention of Cs has been studied in many human subjects, some exposed to elevated concentrations of radiocesium in the environment and others administered Cs isotopes under controlled conditions. Retention over the first year or two after intake usually can be closely approximated by a sum of two exponential terms, with the long-term component representing most of the systemic deposit. The half-time of the long-term component has been found to vary with age, gender, diet, muscle mass, pregnancy, and elevation above sea level. Autopsy studies on environmentally exposed humans, as well as experimental studies on laboratory animals, indicate that Cs is fairly uniformly distributed in the body.

Results of 14 studies of the long-term half-time of cesium in healthy adult males from different regions of the world are summarized in Figure 6.38. These include both controlled and environmental studies. The number of subjects per study varied from 2 to more than 100. Differences in average retention times in different study groups may result in part from differences in measurement techniques but are attributable in part to variation of Cs biokinetics from one population to another. It has been found, for example, that the retention half-time in Japanese adult males is shorter on average than that in adult males from North America or Europe. Mean half-times in the 14 studies ranged from 79 to 133 d and averaged 97 ± 13 d. Inter-subject variability within a given study generally was small, with a typical coefficient of variation of about 20%, and a typical geometric standard deviation of 1.2.

In eight of the studies indicated in Figure 6.38, retention half-times were determined in adult females as well as adult males. Although there was some overlap in individual half-times for males and females, the mean half-time for females was 15%–34% lower than that for males in each of these eight studies (Figure 6.39). Other studies have shown that the long-term half-time of Cs generally is reduced during pregnancy to about two-thirds of the half-time in the same woman before or after pregnancy.

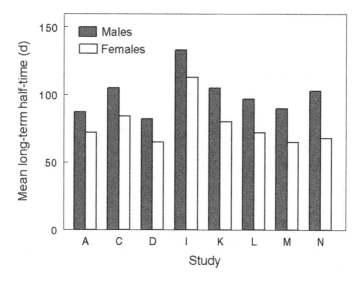

FIGURE 6.39   Comparative total-body half-times of Cs in adult males and adult females in eight studies.

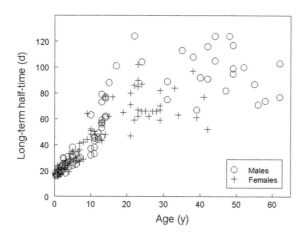

FIGURE 6.40   Measured Cs retention half-times at different ages.

Findings from a recent review of differences with age in the long-term biological half-time of Cs are summarized in Figure 6.40. The data in this figure are from controlled studies of Cs retention, follow-up studies of persons exposed due to the Chernobyl accident, and subjects exposed during an incident in Brazil in which a ^{137}Cs source was found and opened. According to the collective data, the retention half-time for the first year of life is generally in the range 15–20 d. Between age 1 y and adulthood, the average half-time increases about five-fold in the male and about four-fold in the female. Differences with gender are not apparent until age 13–15 y, after which higher average values are determined for males than for females. These data, combined with a number of other studies of Cs retention in adults, indicate a gradual decline in the retention time of Cs in the body after about age 50 y.

6.4.3.3.2 Systemic Biokinetic Models
6.4.3.3.2.1 ICRP Publication 2
The systemic biokinetic model for cesium recommended in ICRP Publication 2 (ICRP 1960) appears to have been based largely on whole-body retention of cesium determined in a controlled study involving a few healthy human subjects, together with results of biokinetic studies of cesium in rats. The organs of reference assigned in the model are total body, muscle, lungs, kidneys, spleen, liver, and bone. Deposition fractions for these organs are 1.0, 0.4, 0.003, 0.01, 0.005, 0.07, and 0.04, respectively. Biological half-times are 70 d, 140 d, 140 d, 42 d, 98 d, 90 d, and 140 d, respectively.

6.4.3.3.2.2 ICRP Publication 30
An updated cesium model based on a considerably expanded biokinetic database was introduced in ICRP Publication 30 (ICRP 1979). In that model, cesium is assumed to be uniformly distributed in the body at all times after uptake to blood. Whole-body retention at time t (days) is represented as a sum of two exponential terms:

$$R(t) = a \cdot e^{-0.693t / T_1} + (1-a) \cdot e^{-0.693t / T_2},$$

where T_1 and T_2 are biological half-times for short-term and long-term components of retention, respectively. Parameter values a = 0.1, T_1 = 2 d, and T_2 = 110 d were applied to the worker. The underlying data were primarily whole-body retention data from controlled studies on human subjects administered ^{137}Cs by ingestion or intravenous injection, and information on the gross distribution of cesium in the body derived from animal studies and autopsy data for human subjects.

6.4.3.3.2.3 ICRP Publication 68 and the Publication 72 Series
In the ICRP's series of reports on environmental intake of radionuclides (ICRP 1990, 1993, 1995a,b,c) (called the Publication 72 series) the parameter values in the cesium model of Publication 30 (a, T_1, and T_2) were extended to pre-adult ages based on age-specific retention data for ^{137}Cs in human subjects. For those reports completed after 1991, the model structure used in Publication 30 was modified to depict explicit excretion pathways in order to improve dose estimates to the urinary bladder and colon, which are given tissue weighting factors in the effective dose quantity defined in ICRP Publication 60 (ICRP 1991a). For ages 100 d and 1 y, the cesium model as applied in the Publication 72 series has only one term, that is, it is assumed that T_1 = T_2 at these early ages. The assigned half-time is 16 d for infants and 13 d for age 1 y. For ages 5, 10, and 15 y, the coefficient a is 0.45, 0.30, and 0.13, respectively; the short-term half-time T_1 is 9.1, 5.8, and 2.2 d, respectively; and the long-term half-time T_2 is 30, 50, and 93 d, respectively. Activity lost from tissues is excreted in urine after residence in the urinary bladder contents and in feces after residence in the upper and lower large intestines. A urinary to fecal excretion ratio of 4:1 is assigned. The cesium model for adults used in the Publication 72 series is applied to workers in ICRP Publication 68 (ICRP 1994b).

6.4.3.3.2.4 An Updated Model for Use in Future ICRP Reports

A more detailed systemic biokinetic model for cesium based on updated information was developed by Leggett et al. (2003). That model is constructed around a blood flow model that describes the fraction of cardiac output received by different compartments (Leggett and Williams 1995; ICRP 2002). The model structure is shown in Figure 6.41. The distribution of cesium is assumed to be determined by the distribution of cardiac output, modified by tissue-specific "extraction fractions," that is, fractions of cesium extracted from blood by different tissues as the blood moves through the tissue. Experiments with laboratory animals and in vitro material have established that the extraction fractions for cesium and other alkali metals (e.g., potassium and rubidium) vary considerably from one tissue to another.

The cesium model shown in Figure 6.41 was modified for use by the ICRP (Leggett 2013). The original model structure (Leggett et al. 2003) was revised for three reasons: (1) for greater consistency with the ICRP's typical biokinetic modeling scheme; (2) to provide a more detailed skeletal model for dosimetric purposes; and (3) to describe exchanges between the systemic model and alimentary tract in terms of compartments of the ICRP's updated Human Alimentary Tract Model (the HATM). The method of development of

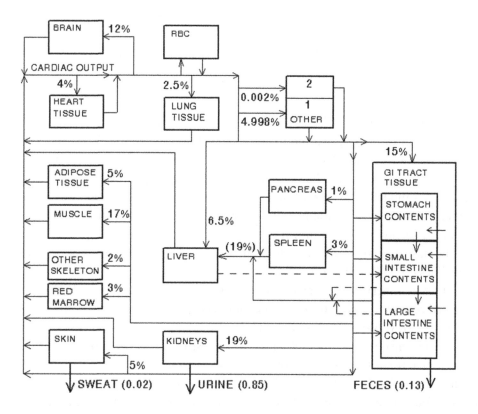

FIGURE 6.41 Structure of a physiologically based biokinetic model for cesium in the human body (Leggett et al. 2003). Solid arrows represent plasma flow and broken arrows represent flow not involving plasma. Percentages indicate a reference distribution of cardiac output in the resting adult male. Numbers beside SWEAT, URINE, and FECES are fractions of cumulative excretion.

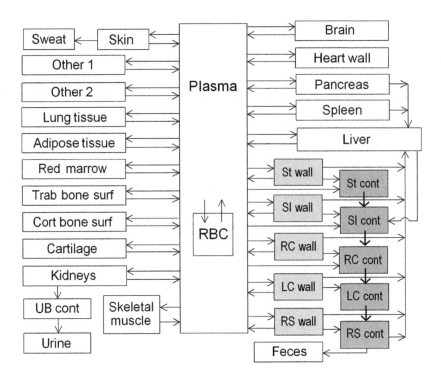

FIGURE 6.42 Structure of the ICRP OIR model for cesium, modified from the model shown in Figure 6.41.

parameter values of the ICRP OIR model is the same as in the original model; in fact, parameter values of the original model were carried over to the ICRP OIR model insofar as allowed by the two model structures. Also, predictions of the time-dependent cesium content in organs based on the ICRP OIR model were required to be consistent with those of the original model. Thus, for all practical purposes, the modification of the model of Leggett et al. (2003) was essentially only a change to the model structure. In the following, the modified version of the cesium model (Figure 6.42) is referred to as the "ICRP OIR model," because its first use by the ICRP is in a series of reports on occupational intake of radionuclides (OIR) (ICRP 2018).

Except where otherwise indicated, the following description applies to both versions of the updated cesium model. Movement of cesium is depicted as a system of first-order processes in the OIR version. The transfer rate (fractional transfer per unit time, also called a transfer coefficient) from plasma into a tissue T is estimated as the product of the plasma flow rate to that tissue (1766 plasma volumes per day as a reference value for the adult male, given in ICRP Publication 89 [ICRP 2002]) and a tissue-specific extraction fraction, E_T. The extraction fraction for a tissue is defined as the fraction of cesium atoms extracted by that tissue during passage of cesium from arterial to venous plasma.

Data on tissue-specific extraction fractions for cesium and its physiological analogues potassium and rubidium include comparative studies of these three elements. For example, extraction by the myocardium in dogs was estimated as 0.71 (range, 0.64–0.80) for potassium, 0.65 (0.58–0.76) for rubidium, and 0.22 (0.09–0.30) for cesium. More information on extraction fractions was found for potassium and rubidium than for cesium. Data for potassium and rubidium were extrapolated to cesium by applying modifying factors as indicated by comparative data on selectivity of these elements. Initial selections of extraction fractions were modified in some cases after testing the model against reported cesium distributions in the early minutes or hours after administration to laboratory animals or human subjects. For example, an initially selected extraction fraction of 0.003 for brain had to be reduced to 0.002 to reproduce the extremely slow buildup of cesium observed for the brain. The final selections of extraction fractions for cesium are as follows: 0.2 for kidneys, walls of the gastrointestinal tract, and heart muscle; 0.05 for liver and skin; 0.002 for brain; and 0.1 for all other tissues.

The transfer rate from tissue T to plasma is estimated from the relative contents of cesium in plasma and T at equilibrium. If T exchanges cesium only with plasma, if A and P are the fractions of total-body cesium in the tissue and plasma at equilibrium, and if R_1 is the transfer rate from plasma to T, then the transfer rate R_2 from T to plasma is determined as $R_2 = R_1 \times P/A$.

The equilibrium distribution of cesium in the adult male (Table 6.14) is based on reference masses of tissues, together with relative concentrations of stable cesium or radiocesium in tissues determined mainly from autopsy data for chronically exposed persons.

The use of extraction fractions and the equilibrium distribution for cesium to derive transfer rates between plasma and tissues is illustrated for skeletal muscle. The transfer rate from plasma to skeletal muscle is estimated as $0.1 \times 0.17 \times 1766 \text{ d}^{-1} = 30.022 \text{ d}^{-1}$, where 0.1 is the estimated extraction fraction for skeletal muscle, 0.17 is the fraction of cardiac output going to skeletal muscle, and 1766 d^{-1} is cardiac output in plasma volumes per day. The transfer rate from skeletal muscle back to plasma is $0.002 \times 30.022 \text{ d}^{-1}/0.8 = 0.0751 \text{ d}^{-1}$, where 0.002 and 0.8 are, respectively, fractions of total-body cesium in plasma and skeletal muscle at equilibrium.

The concept of an extraction fraction does not apply to red blood cells (RBC). The transfer rates between plasma and RBC are derived from estimates for potassium and comparative data on potassium and cesium. The transfer rate for potassium from plasma to RBC is estimated from data from several experimental studies as 6 d^{-1}, and the rate from RBC to plasma is estimated as 0.38 d^{-1}. The rate of transfer of cesium into RBC is roughly 0.3 times that of potassium in humans, rabbits, and rats and therefore is approximately $0.3 \times 6 \text{ d}^{-1} = 1.8 \text{ d}^{-1}$. As is the case for other tissues that exchange cesium with plasma, the transfer rate from RBC to plasma can be determined from the cesium inflow rate (1.8 d^{-1}) and the equilibrium fractions of cesium in plasma and RBC, respectively. Based on the equilibrium fractions given in Table 6.14, the transfer rate of cesium from RBC to plasma is estimated as $1.8 \text{ d}^{-1} \times 0.002/0.014 = 0.257 \text{ d}^{-1}$. This is consistent with the finding of Forth et al. (1963) that the outflow rate of cesium from rabbit RBC is about two-thirds that of potassium ($2/3 \times 0.38 \text{ d}^{-1} = 0.253 \text{ d}^{-1}$).

TABLE 6.14   Tissue Masses, Cesium Equilibrium Distribution, and Tissue Blood Flow for a Reference Adult Male[a], Used to Derive Transfer Coefficients of the ICRP OIR Model

Compartment	Mass[b] (g)	Equilibrium Cesium Content[b] (Fraction of Total-Body Cesium)	Blood Flow (Fraction of Cardiac Output)
Adipose tissue[c]	12,000	0.01	0.05
Brain	1450	0.01	0.12
GI contents	900	0.004	–
GI tract tissue	1170	0.012	0.15
Heart	330	0.0035	0.04
Kidneys	310	0.004	`0.19
Liver	1800	0.02	0.065 (arterial) (0.19) (portal)
Lungs	500	0.006	0.025
Skeletal muscle	29,000	0.8	0.17
Plasma	3100	0.002	–
Red blood cells (RBC)	2500	0.014	–
Skeleton	10,500	0.07	0.05
Red marrow		(0.015)	(0.03)
Bone and other tissue		(0.055)	(0.02)
Skin	3300	0.01	0.05
Spleen	150	0.002	0.03
Pancreas	140	0.002	0.01
Other	5850	0.0305	0.05
Totals	73,000	1.00	1.00

[a] Tissue masses from ICRP 2002; cesium equilibrium distribution based on autopsy studies of Yamagata (1962) and other investigators (see review by Williams and Leggett 1987). Tissue blood flow from a model of Leggett and Williams (1995); also see ICRP 2002.
[b] Without blood.
[c] Separable adipose tissue excluding yellow bone marrow.

For a compartment T that receives cesium from plasma, but loses cesium to multiple compartments, one can still estimate the total outflow rate from the inflow rate and equilibrium contents of T and plasma, but additional information is required to divide the total outflow rate from T into separate transfer rates representing different paths of movement. For example, the derived rate of loss R from skin was divided into transfer rates R_1 and R_2 representing the rate of loss from skin to plasma and the rate of loss from skin to sweat, respectively. The value for R_2 was set for consistency with data of Yamagata et al. (1966) on appearance of ^{132}Cs in sweat after ingestion of this radionuclide by a human subject, and R_1 was determined as $R—R_2$.

Several of the transfers of cesium depicted in the model do not involve exchange between plasma and tissues and therefore had to be derived by methods other than those described above. In general, these other transfers were derived from physiological considerations combined with empirical data. For example, the rate of transfer of cesium into the gastrointestinal tract in liver bile could be estimated from data on the rate of bile flow in man and observed concentration ratios for cesium in liver and bile. Similar considerations were applied to transfer of cesium in other secretions into the contents of the gastrointestinal tract.

Urinary excretion of cesium is depicted in the model as transfer from plasma to a well-mixed kidney compartment and division of outflow from that compartment to plasma and the contents of the urinary bladder. Transfer from plasma to kidneys is represented as an effective extraction fraction times the blood flow rate to kidneys, where the effective extraction fraction includes atoms temporarily retained in the tubules after filtration at the glomerulus, as well as atoms entering kidney tissue directly from blood plasma. The division of kidney outflow between plasma and urinary bladder contents is set for consistency with short-term urinary excretion data for healthy adult males. It is assumed that the renal deposit represents the only source of urinary cesium. That is, it is assumed that none of the urinary cesium arises from filtered or secreted atoms that pass immediately to the urinary bladder without being retained in the kidney tissues.

Endogenous fecal excretion is accounted for by transfer of cesium into the contents of the alimentary tract in saliva, gastric juices, pancreatic secretions, liver bile, and other secretions. It is assumed that 99% of the secreted activity that reaches the small intestine is reabsorbed to blood and that absorption occurs only in the small intestine.

The model depicts a small component of very long-term retention observed in human subjects involved in an incident in Goiânia, Brazil involving cesium intake by members of the public, and in experimental studies on the behavior of ^{137}Cs in rats. In adult human subjects, this small component of retention had an estimated half-time on the order of 500 d and represented an estimated 0.01%–0.25% of uptake to blood, with estimates falling between 0.04% and 0.07% for five of the eight subjects. In rats, this component represented less than 0.01% of injected ^{137}Cs and had a half-time of 150–200 d. Because the physiological basis for this long-term retention component is not known, it is represented in the model as uptake and retention in a compartment called "Other 2," rather than in an explicitly identified tissue. This long-term retention component does not represent an important contribution to dose per unit intake of radiocesium but can be important with regard to interpreting bioassay data collected long after exposure.

Parameter values for the adult male are given in Table 6.15 for the ICRP OIR model. As indicated earlier, parameter values were carried over from the original model (Leggett et al. 2003) insofar as allowed by the two model structures. The remaining parameter values of the ICRP OIR model were set to reproduce as closely as practical the retention, distribution, and excretion of cesium predicted by the original model. Figures 6.43 through 6.45 compare model predictions with observations for human subjects. In these cases, model predictions (curves) are virtually the same for original and OIR versions of the model.

The ICRP OIR model, as well as the original version of that model (Leggett et al. 2003), can be used to simulate the effect of binding of cesium to Prussian Blue (PB) or other unabsorbed material in the gut. The simulation is carried out by changing the relative fractions of cesium assumed to move from the small intestine contents to blood and to the contents of the large intestine. If it is assumed that all cesium entering the small intestine is taken to the large intestine contents and subsequently to feces, the long-term retention half-time for the adult male decreases by about 60%. Studies on adult male subjects exposed to high

TABLE 6.15   Transfer Coefficients for the Adult Male in the ICRP OIR Model for Cesium (ICRP 2018)

From	To	Transfer Coefficient (d^{-1})
Plasma	Red blood cells	1.8
Plasma	Skeletal muscle	30.0
Plasma	Liver	19.5
Plasma	Kidneys	67.1
Plasma	Spleen	5.30
Plasma	Pancreas	1.77
Plasma	Skin	4.42
Plasma	Adipose tissue	8.83
Plasma	Brain	0.424
Plasma	Heart wall	14.1
Plasma	Lung tissue	4.42
Plasma	Red marrow	5.3
Plasma	Cartilage	3.0
Plasma	Trabecular bone surface	1.59
Plasma	Cortical bone surface	1.06
Plasma	Stomach wall	3.53
Plasma	Stomach content	4.52
Plasma	Small intestine wall	35.3
Plasma	Small intestine content	1.05
Plasma	Right colon wall	5.65
Plasma	Right colon content	0.02
Plasma	Left colon wall	5.65
Plasma	Rectosigmoid colon wall	2.83
Plasma	Other 1	9.71
Plasma	Other 2	0.00353
Red blood cells	Plasma	0.257
Muscle	Plasma	0.0751
Liver	Plasma	2.14
Liver	Small intestine content	0.113
Kidneys	Urinary bladder content	1.68
Kidneys	Plasma	31.9
Spleen	Plasma	5.03
Spleen	Liver	0.265
Pancreas	Plasma	1.68
Pancreas	Liver	0.0883
Skin	Plasma	0.867
Skin	Excreta	0.0159
Adipose tissue	Plasma	1.77
Brain	Plasma	0.0848
Heart wall	Plasma	8.07
Lung tissue	Plasma	1.47
Red marrow	Plasma	0.706
Cartilage	Plasma	0.2

(Continued)

TABLE 6.15 (CONTINUED)   Transfer Coefficients for the Adult Male in the ICRP OIR Model for Cesium (ICRP 2018)

From	To	Transfer Coefficient (d⁻¹)
Bone surface compartments	Plasma	0.212
Stomach wall	Plasma	4.16
Stomach wall	Liver	0.219
GI wall compartments[a]	GI content compartments[a]	0.21
Small intestine wall	Plasma	9.87
Small intestine wall	Liver	0.519
Colon wall compartments[b]	Plasma	6.86
Colon wall compartments[b]	Liver	0.361
Other 1	Plasma	0.762
Other 2	Plasma	0.00141

[a] Stomach, SI, right colon, left colon, or rectosigmoid colon wall to corresponding contents.

[b] Right colon, left colon, or rectosigmoid colon wall.

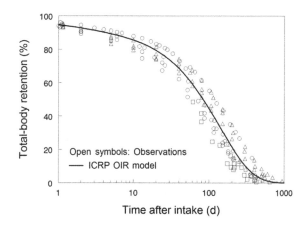

FIGURE 6.43   Comparison of model predictions of total-body retention of cesium in adult males based on the ICRP OIR model with observed values.

levels of ^{137}Cs indicate that PB typically reduced the long-term retention half-time by about 35%–85%, depending on the time between the exposure and the start of PB treatment, and presumably also on inter-subject variability in the biokinetics of cesium.

Age- and sex-specific parameter values for the ICRP OIR model are derived by adjusting the organ sizes and blood flow rates to organs. Shorter total-body residence times of cesium are predicted by this method for women and children than for adult males, mainly due to the smaller muscle masses in women and children. For example, parameter values for a reference adult female are based on the assumption of lower transfer of cesium from plasma to skeletal muscle in women than in men, balanced by uniformly higher transfer from plasma to the remaining outlets from plasma. The age- and gender-specific total-body retention of cesium predicted by this method is consistent with central values derived from reported observations.

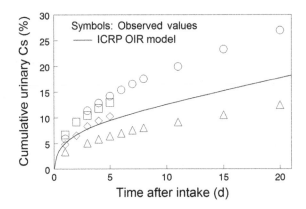

FIGURE 6.44 Comparison of model predictions of cumulative urinary excretion of cesium in adult males with predictions of the original model (Leggett 2003) and observed values (data sources given in Leggett 2003).

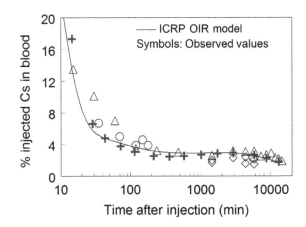

FIGURE 6.45 Comparison of model predictions of disappearance of cesium from blood in adult males following intravenous injection with predictions of the original model (Leggett 2003) and with observed values (data sources given in Leggett 2003).

Dose coefficients for intravenous injection of selected cesium isotopes into the reference adult male based on ICRP OIR model are compared in Table 6.16 with values based on the ICRP's current systemic model for cesium (the two-exponential model used in ICRP Publication 68 (ICRP 1994b), described above). The comparisons are intended to show dosimetric implications of the ICRP OIR model, which is best achieved by restricting attention to the case of direct input of the parent radionuclides into blood. The dosimetric methods used for both sets of calculations are those of ICRP Publication 68. The differences between dose coefficients based on the Publication 68 model and the ICRP OIR model result largely from the different time-dependent distributions of cesium depicted in the models. The ICRP OIR model predicts an initially highly heterogeneous distribution of cesium that changes to a moderately heterogeneous distribution over the first day after intake. This results in relatively wide distributions of tissue dose coefficients for short-lived

TABLE 6.16 Comparison of Dose Coefficients for Intravenous Injection of Cesium Isotopes Based on the ICRP OIR Model with Values Based on the Cesium Model of ICRP Publication 68[a]

Tissue	Ratio of Coefficient Based on Proposed Models to Coefficient Based on Pub. 68 Models							
	Cs-130 (29.2 min)	Cs-134m (2.90 h)	Cs-129 (32.1 h)	Cs-131 (9.69 d)	Cs-136 (13.2 d)	Cs-134 (2.06 y)	Cs-137 (30.2 y)	Cs-135 (2.3 × 10⁶ y)
Bone surface	1.0	2.3	1.2	1.3	1.3	0.8	1.9	3.7
Colon wall	7.4	5.0	1.5	1.0	1.0	0.8	1.4	0.6
Kidneys	25	9.5	2.1	1.4	1.2	0.9	0.9	0.8
Liver	3.0	2.9	2.0	1.3	1.1	0.8	0.9	0.7
Muscle	0.6	0.8	1.0	1.3	1.1	1.1	1.3	1.8
Pancreas	4.3	4.2	2.2	1.7	1.3	1.0	1.1	1.3
Red marrow	1.3	1.5	1.2	1.2	1.1	0.9	1.1	0.9
Spleen	6.4	4.8	1.9	1.3	1.1	0.9	0.9	0.7
SI[b] wall	9.4	5.7	1.4	1.0	1.0	0.8	0.9	0.7
Stomach wall	6.5	4.8	1.8	1.3	1.1	0.9	0.9	0.7
Testes	0.5	0.6	0.8	0.7	0.8	0.7	0.5	0.2
Thyroid	0.5	0.6	0.8	0.8	0.8	0.7	0.5	0.2
Effective dose	3.2	2.3	1.3	0.97	0.96	0.80	0.87	0.55

[a] The dosimetric models of ICRP Publication 68 (ICRP 1994b) and decay data of ICRP Publication 107 (ICRP 2008) were used in both sets of calculations. Differences in the two sets of dose coefficients arise only from differences in the systemic models applied.

[b] Small intestine.

cesium isotopes and more narrow distributions for cesium isotopes with intermediate to long half-lives. The model of ICRP Publication 68 depicts a uniform distribution of cesium in the body and hence yields more nearly uniform tissue dose coefficients than those based on the proposed models, regardless of the half-life of the cesium isotope.

6.4.3.4 Plutonium

Many different models of the distribution, retention, and excretion of systemic plutonium have been developed since the early 1940s to assess doses to workers or members of the public from intake of plutonium. The following paragraphs summarize the evolution of biokinetic data and systemic models for plutonium used over the years, with emphasis on models recommended in reports of the ICRP. Except where otherwise referenced, the material in this section is taken from the following reports and papers: ICRP (ICRP 1993), Leggett (2003), and Leggett et al. (2005).

Plutonium production in the United States began in the early 1940s. Initially, biokinetic data from plutonium tracer studies on rats were used to relate urinary excretion of plutonium to its body burden in workers and to examine its distribution and retention in the body. In studies conducted in 1945 and 1946 by Wright Langham and colleagues at the Los Alamos Scientific Laboratory, 18 seriously ill persons were injected with tracer amounts of [239]Pu in order to gain more direct information on the fate of internally deposited plutonium in man. For the next 50 years, the results of the Langham studies represented the primary data source for modeling the biokinetics of plutonium in workers (Leggett 2003).

Extensive urinary and fecal excretion measurements were made on the Langham subjects during the first few months after injection, and limited measurements were made at later times. At least three of the subjects lived for many years after the study, and plutonium excretion measurements were made on these subjects in the mid-1970s. Others died from their diagnosed illnesses during the first few months after injection, and tissue samples collected at autopsy were analyzed for plutonium. The investigators found no major differences from the distribution of plutonium in rats, with the exception that the liver appeared to accumulate more plutonium in man than in rats. Excretion data for the human subjects over the first 138 d after exposure, supplemented with longer-term measurements on occupationally exposed persons, were used to derive power-function predictors of the urinary and fecal excretion rates of plutonium as a function of time:

$$\text{Urine}: Y_u(t) = 0.20t^{-0.74} \tag{6.2}$$

$$\text{Feces}: Y_f(t) = 0.63t^{-1.09} \tag{6.3}$$

where Y_u and Y_f are percentages of injected plutonium in urine and feces, respectively, during day t (>1) after administration. On the basis of these excretion functions, the time to eliminate half of the administered plutonium was estimated as 84–200 years.

The conclusions of Langham and coworkers formed the basis for the biokinetic model for plutonium used in ICRP Publication 2 (ICRP 1960) (Figure 6.46). The organs of reference (see the section on evolution of ICRP systemic biokinetic models) for absorbed plutonium are total body, bone, liver, and kidneys, and total body, which are assumed to receive 100%, 80%, 15%, and 2%, respectively, of the absorbed amount. The assumed biological removal half-times from the organs of reference are 178 y, 200 y, 82 y, and 88 y, respectively.

In 1965, the ICRP formed a task group to review information on the biological behavior of plutonium and related elements and update biokinetic models for these elements. By this time plutonium biokinetics had been studied in several animal species, and considerably more information on occupational intakes had accumulated. Interspecies extrapolation of biokinetic data was used heavily in the construction of the model for plutonium recommended in the task group report (ICRP 1972) and applied a few years later in ICRP Publication 30 (Part 1) (ICRP 1979). In that model, it is assumed that plutonium leaves the transfer compartment (blood) with a half-time of 0.25 d, with 45% depositing on bone surfaces, 45% in the liver, 0.001% g^{-1} in gonads, and the remaining ~10% is removed in excreta (Figure 6.47). Plutonium is assumed to move from bone surfaces and liver to excretion with half-times of 100 y and 40 y, respectively, and to be permanently retained in gonads. Parameter values were based in part on data for man, but the estimates relied heavily on information for different animal species. The biological half-time in liver was based mainly on an apparent relationship between body weight and hepatic retention derived from animal data. The biological half-time in the skeleton was based mainly on a relationship between lifespan and skeletal retention derived from animal data. The task group found relatively little information on related elements and concluded: "Until contrary information is available, it is reasonable to assume that there is general similarity

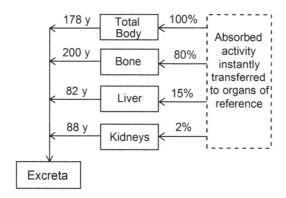

FIGURE 6.46 Systemic model for plutonium used in ICRP Publication 2 (ICRP 1960).

in the physiological behavior of plutonium and americium and of other actinides and the lanthanides."

In 1981, another ICRP task group was formed to address updated information on the biokinetics of plutonium and other actinide elements. This group relied heavily on autopsy data on the distribution of plutonium in occupationally and environmentally exposed persons. They found that the partition of plutonium between liver and skeleton varies widely in individual cases but concluded that the most likely average deposition is 50% in the skeleton and 30% in the liver. In view of the high inter-subject variability and the fact that available data represented mainly long-term retention, the task group recommended that the assumption of ICRP Publication 30 of equal distribution between skeleton and liver (45% and 45%) continue to be used for radiation protection. They recommended, however, that the previously assumed half-times of 40 y for the liver and 100 y for the skeleton be replaced with values of 20 y and 50 y, respectively. The recommended model, that is, the model of Part 1 of ICRP Publication 30 with the removal half-times for liver and skeleton reduced to 20 y and 50 y, respectively, was published in ICRP Publication 48, and also used in Part 4 of Publication 30 to replace the plutonium model described in Part 1 of the Publication 30 series.

Durbin (1972) reanalyzed the human injection data in view of the specific illnesses of the subjects, the samples of bone collected, similarities between plutonium and iron, material balance, occupational data, and animal data. She used information for subjects judged to have normal excretory function, supplemented with animal data, to derive the following representations of the urinary (U) and fecal (F) excretion rates (% d^{-1}) as a function of time t (d):

$$U(t) = 0.41e^{-0.58t} + 0.12e^{-0.13t} + 0.013e^{-0.017t} + 0.003e^{-0.0023t} + 0.0012e^{-0.00017t} \quad (6.4)$$

$$F(t) = 0.60e^{-0.35t} + 0.16e^{-0.11t} + 0.012e^{-0.012t} + 0.002e^{-0.0018t} + 0.0012e^{-0.00017t} \quad (6.5)$$

The long-term removal rate of 0.00017 d^{-1}, corresponding to a half-time of about 4000 d, was based on data for dogs. These functions were adopted as plutonium bioassay models in ICRP Publication 54 (ICRP 1989).

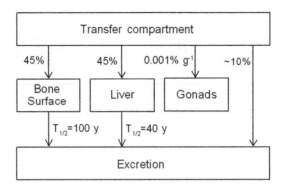

FIGURE 6.47 Systemic biokinetic model for plutonium recommended in ICRP Publication 19 (ICRP 1972), as applied in ICRP Publication 30 (Part 1) (ICRP 1979). In Part 4 of Publication 30 (ICRP 1988), the removal half-times from liver and bone were revised to 20 y and 50 y.

By the 1970s, it had been clearly established from animal and human studies that plutonium tends to accumulate on bone surfaces, but gradually becomes buried in bone volume due to bone restructuring processes that normally occur throughout life. Some investigators had begun to model the burial of plutonium in bone volume, called volumization, in an effort to sharpen dose estimates to radiosensitive cells of bone surface and red marrow. In the mid-1980s, a bioassay and dosimetry model for plutonium was developed at ORNL in an effort to extend these bone volumization models and connect skeletal kinetics to a more comprehensive and physiologically realistic description of the systemic behavior of plutonium. The model structure (Figure 6.48) depicted recycling of activity from tissues to blood, bone restructuring as a controlling factor in the behavior of skeletal activity, and three physically identifiable compartments within the liver with different rates and directions of transfer of activity. The kidneys were included in the urinary tract tissue shown in Figure 6.48. Parameter values were based on estimated rates of physiological processes and plutonium-specific data for human subjects or laboratory animals, particularly dogs. Data on dogs were used, for example, to estimate a removal half-time from bone marrow of 0.25 y, which was considerably lower than previous estimates. Parameter values were based on physiological processes where feasible (e.g., bone remodeling rates) and otherwise on fits to plutonium-specific data. Some parameter values changed with age during adulthood, including those describing the initial division between the skeleton and liver and translocation of skeletal plutonium. At all ages, the skeleton was assumed to have higher initial uptake than the liver as indicated at that time by essentially all animal data, as well as the results of Langham's human injection studies in the 1940s. Studies of plutonium workers through the mid-1980s had revealed the extent to which Langham's equation for urinary excretion (Equation 6.2) underestimates the urinary excretion rate at times remote from intake, so that the long-term urinary excretion rate could be estimated reasonably well. Also, information had been published on long-term excretion rates (at ~10,000 d) in two subjects injected with plutonium in the 1940s. It was recognized that the one-compartment blood model underestimated blood plutonium at intermediate times and that the

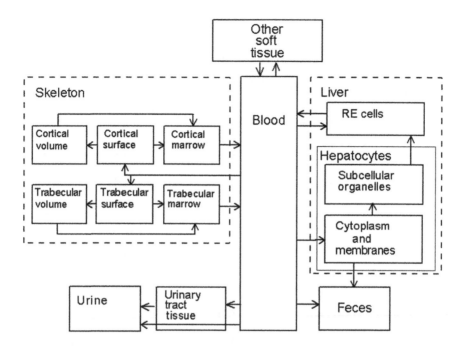

FIGURE 6.48   Precursor to the current ICRP model for plutonium (Leggett 1985).

method of treating the time dependence in urinary clearance of plutonium resulted in an overestimate of accumulation of plutonium in the urinary tract tissues, but it was not evident how these problems could be resolved without complicating the model considerably, particularly the representation of plutonium in blood.

A simplified version of the ORNL model (Figure 6.49) was adopted for use in ICRP Publication 56 (ICRP 1990), the first part of a series of ICRP documents on doses to members of the public from environmental exposure to radionuclides. The ICRP had not yet moved toward physiological realism in its models but attempted to provide best available dose estimates using minimal, easily solved mathematical representations of the distribution of systemic activity.

The plutonium model of ICRP Publication 56 was intended for dosimetric purposes only and did not depict biologically realistic excretion pathways. The bone model was carried over from the ORNL model, but the representations of liver and soft tissues were changed for the purposes of reducing the number of compartments and pathways, eliminating the apparent overestimate of plutonium accumulation in urinary tract tissues, and addressing gonads separately from other soft tissues. The accumulation of plutonium in urinary tract tissues assumed in the ORNL model was shifted to "Other soft tissues."

The model of ICRP Publication 56 was revised in ICRP Publication 67 (Figure 6.50), the second part of the series on doses to members of the public. The revisions were made mainly to reflect updated information on the long-term distribution of plutonium in workers and address recent modifications of the ICRP's effective dose concept (ICRP 1991b). Also, preliminary, unpublished data on the early excretion of intravenously injected plutonium by

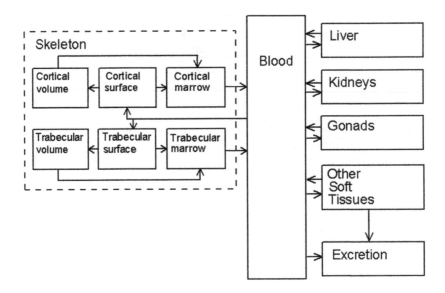

FIGURE 6.49   Structure of the systemic model for plutonium used in ICRP Publication 56 (1989).

two healthy human subjects were used to modify initial urinary and fecal excretion rates slightly. The revised model has been applied to workers as well as members of the public. For example, it was used in ICRP Publication 68 (ICRP 1994b) to derive dose coefficients for intake of plutonium isotopes by workers, and in ICRP Publication 78 (ICRP 1997) as the basis for interpretation of bioassay data.

The model adopted in ICRP Publication 67 (Figure 6.50) moved back in the direction of physiological realism and restored specific excretion pathways, which were needed to address the new recommendations of ICRP Publication 60 regarding tissues at risk. However, it retained one of the biologically unrealistic shortcuts used in ICRP Publication 56, that is, a transfer from a soft tissue compartment directly to excretion pathways (specifically, to the urinary bladder in the modified model). This transfer was used to account for an apparent difference in urinary clearance from blood of initially absorbed plutonium and recycled plutonium. No changes were made to correct the recognized underestimate of blood plutonium at intermediate times. Nearly all animal and human data still indicated that the skeleton had greater uptake than the liver at all ages.

Information developed since the completion of ICRP Publication 67 (ICRP 1993) includes the results of two studies involving intravenous injection of plutonium isotopes into healthy volunteers. One of the studies, initiated at the Harwell Laboratory in Great Britain, involved six adult males and six adult females. The other, conducted at the National Radiological Protection Board (NRPB) in Great Britain, involved five adult males. These studies provide data on blood retention, total-body retention, and urinary and fecal excretions rates, and (for the Harwell study) externally measured plutonium in the liver over several years following administration.

Much additional excretion and autopsy data for plutonium workers in the United States, Great Britain, and Russia has been published since the appearance of ICRP Publication 67.

The updated occupational data provide information on the time-dependent distribution of plutonium in systemic tissues, daily urinary excretion of plutonium as a percentage of the systemic burden, and dependence of plutonium biokinetics on the health of the workers.

Comparisons of predictions of the model of ICRP Publication 67 with updated information show consistency with regard to total-body retention, daily urinary excretion, and the long-term distribution of plutonium. Inconsistencies are seen regarding some aspects of the early behavior of plutonium, most notably the initial division between the liver and skeleton. In the Harwell subjects, peak estimates of the liver content based on external counts averaged more than 70% of the administered activity, compared with a model prediction of about 30%. The external measurements involve uncertainties that may lead to modest overestimates of the liver content but provide credible evidence of substantially greater uptake of plutonium by the liver than depicted in previous systemic models for plutonium.

Autopsy data for plutonium workers show that there is considerable variability in the division of activity between the liver and skeleton at all times, with the skeleton containing more plutonium than the liver in some cases and less in others. Statistical analysis of the data indicates, however, that the liver typically is the more important repository soon after exposure, and that there is a gradual shift of activity to the skeleton (Figure 6.51).

Two other discrepancies between observations and predictions of the plutonium model of ICRP Publication 67 are evident from later data for early or intermediate times after injection.

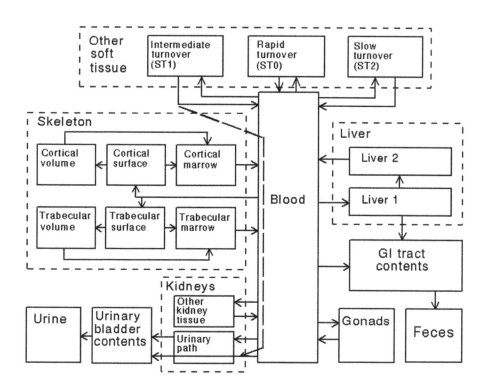

FIGURE 6.50 Structure of the systemic biokinetic model for plutonium adopted in ICRP Publication 67 (ICRP 1993).

First, the predicted level of plutonium in blood falls below observed values by about 2 weeks after injection and remains too low for an extended period. Second, the fecal excretion rate falls below the central tendency of observed values at about 2 weeks after injection and remains too low for a few weeks. The underestimate in the blood content may stem from an underestimate of the feed from liver back to blood. The underestimate in fecal excretion may stem from inaccuracies associated with both of the assumed sources of fecal excretion, namely, a feed from liver to the intestinal contents representing biliary secretion, and a feed from blood to intestinal contents representing all other secretions into the intestines.

A proposed update of the model for systemic plutonium applied in ICRP Publication 67 was developed to resolve the discrepancies between that model and biokinetic data for plutonium developed since 1993. The structure of the proposed model is shown in Figure 6.52. Parameter values are listed in Table 6.17.

The proposed model structure is an extension of the generic model structure for plutonium and other actinides used in ICRP Publication 67 and subsequent ICRP documents. The following primary changes to the model of Publication 67 were made:

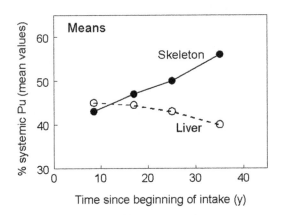

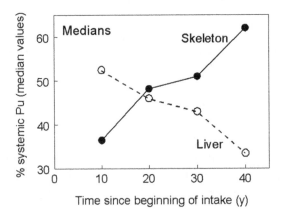

FIGURE 6.51 Shift with time in the mean or median systemic distribution of plutonium, based on data for plutonium workers at the Mayak facility in Russia (Suslova et al. 2002).

TABLE 6.17   Baseline Parameter Values for the Adult, in the Updated Systemic Model for Plutonium: The Initial Input to Blood via Absorption or Injection Is Assumed to Distribute Rapidly between Blood 1 (70%) and ST0 (30%)

Source	Destination	Transfer Coefficient (d^{-1})
Blood 1	Liver 0	4.6200×10^{-1}
Blood 1	Cortical surface	8.7780×10^{-2}
Blood 1	Cortical volume	4.6200×10^{-3}
Blood 1	Trabecular surface	1.2474×10^{-1}
Blood 1	Trabecular volume	1.3860×10^{-2}
Blood 1	Urinary bladder contents	1.5400×10^{-2}
Blood 1	Renal tubules	7.7000×10^{-3}
Blood 1	Other kidney	3.8500×10^{-4}
Blood 1	Upper large intestine contents	1.1550×10^{-2}
Blood 1	Testes	2.6950×10^{-4}
Blood 1	Ovaries	0.8470×10^{-4}
Blood 1	ST1	1.8511×10^{-2}
Blood 1	ST2	2.3100×10^{-2}
ST0	Blood 1	9.9000×10^{-2}
Blood 2	Urinary bladder contents	3.5000×10^{0}
Blood 2	Blood 1	6.7550×10^{1}
Blood 2	ST0	2.8950×10^{1}
Renal tubules	Urinary bladder contents	1.7329×10^{-2}
Other kidney	Blood 2	1.2660×10^{-4}
ST1	Blood 2	1.3860×10^{-3}
ST2	Blood 2	1.2660×10^{-4}
Liver 0	Small intestine contents	9.2420×10^{-4}
Liver 0	Liver 1	4.5286×10^{-2}
Liver 1	Blood 2	1.5200×10^{-3}
Liver 1	Liver 2	3.8000×10^{-4}
Liver 2	Blood 2	1.2660×10^{-4}
Testes	Blood 2	3.8000×10^{-4}
Ovaries	Blood 2	3.8000×10^{-4}
Cortical surface	Cortical marrow	8.2100×10^{-5}
Cortical surface	Cortical volume	2.0500×10^{-5}
Cortical volume	Cortical marrow	8.2100×10^{-5}
Trabecular surface	Trabecular marrow	4.9300×10^{-4}
Trabecular surface	Trabecular volume	1.2300×10^{-4}
Trabecular volume	Trabecular marrow	4.9300×10^{-4}
Cortical marrow	Blood 2	7.6000×10^{-3}
Trabecular marrow	Blood 2	7.6000×10^{-3}

- A second blood compartment is added to provide a physiologically meaningful approach to modeling the observed phenomenon that urinary clearance of plutonium from blood is higher for recycled plutonium atoms than for plutonium atoms that initially enter blood. This is accomplished by assigning initially entering atoms to one of these blood compartments (Blood 1) and recycled atoms (i.e., atoms returning from tissues to blood) to the other (Blood 2), and then assigning Blood 1 and Blood 2 different fractional losses to Urinary bladder content. The rationale is that the form of plutonium initially entering blood may be different from plutonium returning from tissues to blood. For lack of better information, the portion of activity leaving Blood 2 that does not go directly to the Urinary bladder content is assumed to distribute proportionally to the original input to blood. Higher deposition in liver (60%, compared with 30% in Publication 67) and lower deposition in bone (30%, compared with 50% in Publication 67) are assumed.

- Another compartment is added to the liver and assigned a moderately high rate of loss (half-time of 1 y, with 80% going to Blood 2 and 20% to the long-term liver compartment Liver 2) to reproduce an observed gradual shift in the systemic burden from liver to skeleton (Figure 6.51), despite the higher uptake by liver assumed in the updated model.

The bone model was also changed slightly to depict different sites of deposition in bone, that is, to change the assumption that bone surface is the only initial bone repository to the assumption that bone surface is the primary repository, but there is also some direct entry into bone volume. This change has little effect on dose estimates for plutonium isotopes.

Predictions of the model of ICRP Publication 67 and the proposed update are compared in Figures 6.53 through 6.57 with data from recent plutonium injection studies on healthy

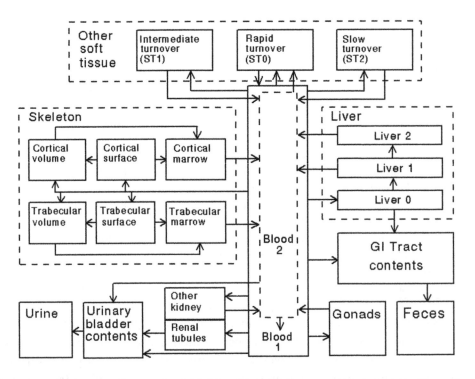

FIGURE 6.52 Structure of a proposed biokinetic model for systemic plutonium (Leggett et al. 2005) intended to resolve discrepancies between the model of ICRP Publication 67 and later experimental and occupational data.

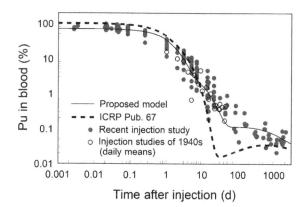

FIGURE 6.53   Time-dependent blood content of plutonium as predicted by the proposed model (see Figure 6.52 and Table 6.17) and the model of ICRP Publication 67, and as measured in human injection studies (Langham et al. 1950, Newton et al. 1998).

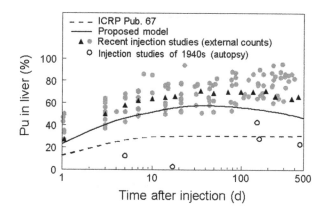

FIGURE 6.54   Time-dependent content of plutonium in the liver as predicted by the proposed model (see Figure 6.52 and Table 6.17) and the model of ICRP Publication 67, and as measured in human injection studies (Langham et al. 1950, Durbin 1972, Newton et al. 1998). The diamonds show estimates for a selected subject of the most recent study (Newton et al. 1998).

human subjects and from the plutonium injection studies of the 1940s involving seriously ill subjects. Data from the studies of the 1940s are reasonably consistent with recent observations with the main exception of the initial content of the liver. The early human studies have the advantage that the liver content was based on tissue analysis, but the disadvantage that liver uptake and retention may have been affected by the poor health of the subjects. The more recent studies have the advantage that the subjects were healthy, but the disadvantage that the liver content was based on external measurements, which could involve non-trivial errors associated with calibration factors and assumptions concerning the distribution of non-liver plutonium. Nevertheless, potential errors in more recent estimates seem too small to explain the large discrepancies between results of the two studies. In the

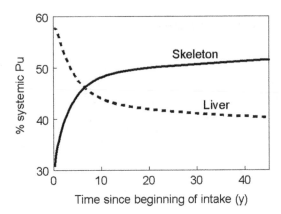

FIGURE 6.55 Contents of liver and skeleton (% systemic burden) as a function of time after the start of chronic uptake to blood at a constant rate, as predicted by the proposed update of the plutonium model of ICRP Publication 67 (see Figure 6.52 and Table 6.17).

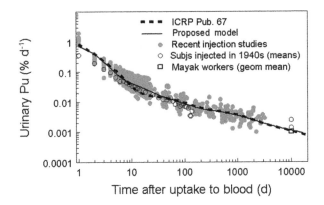

FIGURE 6.56 Time-dependent urinary excretion of plutonium as predicted by the proposed model (see Figure 6.52 and Table 6.17) and the model of ICRP Publication 67, and measured in human injection studies and in Mayak workers (Langham et al. 1950, Durbin 1972, Rundo et al. 1975, Talbot, Newton, and Warner 1993, Talbot, Newton, and Dmitriev 1997, Popplewell et al. 1994, Warner, Talbot, and Newton 1994, Khokhryakov et al. 1994, Khokhryakov et al. 2000, Newton et al. 1998, Ham and Harrison 2000).

selection of parameter values for the liver, data from recent human injection studies were given greater weight than data from the human studies of the 1940s. The time-dependent systemic distribution of activity in plutonium workers was also a major consideration.

The proposed model eliminates apparent underestimates of the blood content (Figure 6.53), liver content (Figure 6.54), and fecal excretion rate (Figure 6.57) at early or intermediate times after exposure. Predicted relative contents of liver and skeleton as a function of time after start of chronic intake (Figure 6.55) depict a gradual shift of plutonium from liver to skeletal, as indicated by data for Mayak workers (Figure 6.51).

Predictions of the urinary excretion rate produced by the present model are nearly identical to those produced by the model of ICRP Publication 67, except that the present model

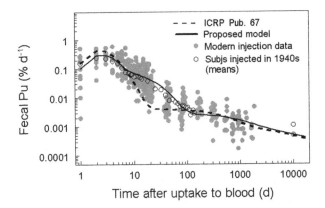

FIGURE 6.57 Time-dependent fecal excretion of plutonium as predicted by the proposed model (see Figure 6.52 and Table 6.17) and the model of ICRP Publication 67, and measured in human injection studies (Langham et al. 1950, Durbin 1972, Rundo et al. 1975, Talbot, Newton, and Warner 1993, Talbot, Newton, and Dmitriev 1997, Newton et al. 1991, Newton et al. 1998).

yields slightly higher estimates at 10–30 d after injection (Figure 6.56). Predictions of both models are reasonably consistent with experimental and occupational data on retention in kidneys, testes, and other soft tissues. For example, the proposed model predicts that the content of other soft tissues is about 8% of the systemic burden at times remote from exposure, compared with an average of about 7% of the systemic burden in 120 Mayak workers, and about 9.5% in nine USTUR subjects.

6.4.4 Biokinetic Models for Radionuclides Produced in Vivo by Decay of Parent Nuclides

6.4.4.1 General Considerations

Radioactive decay results in emission of radiation and transformation of an atom, called the parent, into a different type of atom, called the progeny or daughter radionuclide. The progeny radionuclide may also be radioactive, in which case it will also eventually decay and emit radiation. This leads to a sequence of different radionuclides and decay events, eventually producing a stable nuclide. The parent radionuclide together with the sequential set of radionuclides produced by this process is called a decay chain. The members of a decay chain excluding the original parent radionuclide are referred to collectively as the radioactive progeny of that parent radionuclide.

To estimate tissue doses following intake of a parent radionuclide that heads a decay chain involving one or more radioactive progeny, it is necessary to provide a biokinetic model, not only for the parent radionuclide, but also for each of its radioactive progeny of potential dosimetric significance. A radioactive progeny should be regarded as a potentially significant contributor to dose, unless it can be demonstrated that the radiations potentially emitted in the body by the progeny represent negligible additions to the dose delivered by the parent. This is often demonstrable, for example, in cases in which the radiological half-life of the progeny is sufficiently long that there would be relatively few decays of the progeny in the body during a person's lifetime, even if there were no biological removal of the progeny.

One of two general assumptions concerning the behavior of radionuclide progeny has usually been applied by the ICRP in the assignment of biokinetic models to radioactive progeny produced in the body, for the purpose of calculating dose coefficients for the parent:

Assumption of shared kinetics: Chain members produced in the body following intake of a parent radionuclide adopt the biokinetic model of the parent radionuclide.

Assumption of independent kinetics: Chain members produced in the body following intake of the parent follow their own characteristic behavior, i.e., the same biokinetics as if entering the body as a parent radionuclide.

The assumption of shared kinetics of radioactive progeny typically is much easier to implement than the assumption of independent kinetics, essentially because a single model applies to the entire chain. As described below, a number of technical difficulties often must be overcome in order to implement the assumption of independent kinetics of chain members. Because it is the more convenient assumption, shared kinetics traditionally has been applied by the ICRP for purposes of deriving dose coefficients for radionuclides with dosimetrically significant progeny.

Shared kinetics of chain members produced in the respiratory tract following inhalation of a radionuclide has some logical support, in that the kinetics of a deposited radionuclide and its chain members usually is presumed to be determined largely by properties of the carrier (e.g., particulate material) rather than properties of the radionuclides. Shared kinetics of chain members in the contents of the alimentary tract (excluding absorption to blood) also appears to be a reasonable assumption, in that the transfer of the parent and progeny through the tract presumably is controlled mainly by mass movement of the contents.

On the other hand, data collected from experimental studies on laboratory animals and follow-up studies of accidentally exposed human subjects indicate that radioactive progeny produced in the body following injection or absorption of a parent radionuclide to blood usually tend to migrate from the parent and follow their characteristic biological behavior (Leggett, Dunning, and Eckerman 1984). This excludes radionuclides produced in bone volume, which tend to remain with the parent radionuclide in bone volume, presumably because retention in bone volume is determined by the rate of bone remodeling rather than chemical properties of the chain members. The noble gases are exceptions, in that experimental studies indicate that they migrate from bone volume to blood with a half-time of a few days, compared with a typical bone turnover time of several years.

Some movement toward the assumption of independent kinetics of chain members produced in systemic pools is seen in the ICRP's series of reports on environmental intake of radionuclides by members of the public, called the Publication 72 series after the final, summary report (ICRP 1990, 1993, 1995a,b,c), and in ICRP Publication 68 (ICRP 1994b) on occupational intake of radionuclides. In those documents, the assumption of independent kinetics is applied to radioactive progeny produced in systemic compartments excluding bone volume compartments following intake of radioisotopes of selected elements (e.g., lead, radium, thorium, or uranium). In updated ICRP documents on occupational or environmental intake of radionuclides (ICRP 2015, 2016, 2018), the assumption of independent kinetics (excluding production in bone volume) is the standard assumption rather than the exception.

In practice, implementation of the assumption of independent kinetics of chain members often is not straightforward, due to structural differences in the systemic models for many parent and progeny combinations. The resolution of this problem requires modifications of the biokinetic models for some chain members. Generally, the model for the parent is not changed, but the models for the subsequent chain members may require modification in order to solve the system of equations represented by the collective biokinetic models. For example, a radionuclide may be born in an explicitly designated tissue T in the parent's model that is not an explicitly designated tissue in the progeny radionuclide's characteristic model. When this happens, the rate of removal of the progeny radionuclide from T and the destination of the removed activity must be defined before the models can be solved.

Even if the progeny radionuclide is produced in a tissue that is an explicitly designated source organ in the progeny radionuclide's characteristic model, a meaningful method of implementation of the assumption of independent kinetics may not be evident if the progeny radionuclide's model divides the tissue into compartments that are not identifiable with compartments in the parent's model. For example, this may occur if the division of the tissue into compartments is based on physiological or anatomical considerations for the parent and on a kinetic basis for the progeny, or vice versa.

The selection of a specific method of implementation of independent kinetics of chain members (i.e., a specific approach to modifying systemic models for chain members to produce a solvable set of models) for a given parent element may depend on: the availability of specific information (e.g., experimental data) on the behavior of the given chain members following production in the body; the sensitivity of dose estimates to uncertainties in the behavior of chain members; the lengths of radionuclide chains for that element; and the complexity and consistency of the characteristic systemic models for chain members. In some cases, primarily involving progeny radionuclides with short radiological or biological half-times, the characteristic systemic model for the progeny might be replaced by a simpler model judged as adequate for practical purposes in view of the uncertainties in its short-term behavior following in vivo production. For example, short-lived progeny radionuclides might be assumed to decay at their site of production, in lieu of specific information. As a second example, it seems reasonable to apply a simplistic model involving direct and relatively fast removal of noble gas progeny from the body, because it would rarely be worthwhile to attempt to provide a detailed description of the behavior of noble gas progeny over the generally short retention time following their production in the body.

In the development of models for progeny radionuclides, it is important that the systemic model applied to an element X as a progeny of a parent element Y be the same for all chains headed by Y as the parent, to help keep the modeling effort to a manageable level. For example, the systemic model applied to ^{224}Ra produced in a systemic pool following intake of ^{228}Th should also be applied to ^{223}Ra produced in a systemic pool following intake of ^{227}Th.

6.4.4.2 Case Studies of Treatment of Radioactive Progeny Produced in Vivo

The following examples illustrate approaches to implementing the assumption of independent kinetics. The reader is referred to Leggett (2013) for more detailed discussions.

These case studies all involve cesium as a parent radionuclide. Four different parent radionuclides are considered: ^{125}Cs, ^{127}Cs, ^{134m}Cs, and ^{137}Cs. In all cases the model applied to the cesium parent is the model defined in Figure 6.42 and Table 6.15 in the earlier section on the systemic behavior of cesium.

Case 1: ^{125}Cs as a Parent (the Chain ^{125}Cs/^{125}Xe/^{125}I)

Cesium-125 ($T_{1/2}$ = 45 m) decays to ^{125}Xe (16.9 h), which decays to ^{125}I (59.4 d). Studies of the fate of inhaled xenon in human subjects indicate multiple components of retention of xenon following absorption to blood and uptake by tissues (Susskind et al. 1977). The half-times of these retention components vary from less than 1 minute to several hours. The longest half-time presumably is associated with retention in fatty tissues with relatively low blood perfusion rates. Based on the sizes and half-times of the various retention components, the mean residence time in the adult human is on the order of 30 min. In the present model for xenon as a progeny radionuclide, xenon produced in soft tissue compartments transfers to plasma at the rate 50 d^{-1}, corresponding to a half-time of about 20 min and a mean residence time in the body of about 30 min. Xenon produced on bone surface transfers to plasma at the rate 100 d^{-1}, a value estimated for radon as a progeny of radium and used in ICRP Publication 67 (ICRP 1993). Xenon entering plasma is exhaled at the rate 1000 d^{-1} (corresponding to a half-time of 1 min), a default value used in recent ICRP documents to represent extremely fast removal from a compartment. Iodine produced by decay of xenon is assumed to follow the model for iodine as a parent radionuclide defined by Figure 6.31 and Table 6.11. Iodine produced in a compartment of the cesium model that is not identifiable with a compartment in the iodine model is assumed to transfer to the blood iodide of the iodine model (Figure 6.31) at the rate 330 d^{-1}, the highest transfer rate to the blood iodide compartment in the iodine model. Iodine produced in red blood cells is assumed to transfer rapidly (at 1000 d^{-1}) to the blood iodide pool.

Case 2: ^{127}Cs as a Parent (the Chain ^{127}Cs/^{127}Xe)

Cesium-127 (6.25 h) decays to ^{127}Xe (36.4 d). The model for xenon as a progeny of cesium is described above (for ^{125}Cs as a parent radionuclide). As indicated earlier, the systemic model applied to an element X as a progeny of a parent element Y is the same for all chains headed by Y as the parent.

Case 3: ^{134m}Cs as a Parent (the Chain ^{134m}Cs/^{134}Cs)

Cesium-134m (2.9 h) decays to ^{134}Cs (2.06 y). The model for cesium as a parent radionuclide is applied to ^{134}Cs produced in systemic compartments. In this case, there is no distinction between the assumptions of shared and independent kinetics of chain members.

Case 4: ^{137}Cs as a Parent (the Chain ^{137}Cs/^{137m}Ba)

Cesium-137 (30.2 y) decays to ^{137m}Ba (2.55 min, 94.4% yield) and ^{137}Ba (stable, 5.6% yield), and ^{137m}Ba decays to ^{137}Ba. The model for ^{137m}Ba produced in systemic compartments is adapted from the barium model for adults applied in ICRP Publication 67 (ICRP 1993) to adult members of the public, and ICRP Publication 68 (ICRP 1994b) to workers. The structure of that model is the same as the structure of the systemic model for strontium described in an earlier section. The parameter values of that model are listed below in Table 6.18.

The barium model of Publications 67/68 was not designed for application to very short-lived isotopes, such as ^{137m}Ba. The model is consistent with results of human studies indicating that bone and colon are the primary early repositories following entry of barium into blood (Korsunskii, Tarasov, and Naumenko 1981), but it understates the rate at which barium initially leaves plasma and deposits at these sites (Leggett 1992). The model is modified for application to ^{137m}Ba as ^{137}Cs progeny by depicting the early, rapid phase of plasma clearance, and eliminating features not relevant to the fate of the short-lived isotope ^{137m}Ba. As indicated in the earlier discussion of strontium biokinetics, kinetic studies with radioisotopes of the alkaline earth elements indicate that upon entry into blood,

TABLE 6.18   Transfer Coefficients in the Biokinetic Model for Systemic Barium (ICRP 1993, 1994b) Used as the Starting Point for the Model for Barium as a Progeny of Systemic Cesium[a]

Pathway	Transfer Rate (d^{-1})
Plasma to urinary bladder contents	2.24
Plasma to right colon	20.16
Plasma to trabecular bone surface	9.72
Plasma to cortical bone surface	7.78
Plasma to ST0	23.0
Plasma to ST1	7.0
Plasma to ST2	0.14
Trabecular bone surface to plasma	0.578
Trabecular bone surface to exchangable volume	0.116
Cortical bone surface to plasma	0.578
Cortical bone surface to exchangeable volume	0.116
ST0 to Plasma	7.67
ST1 to Plasma	0.693
ST2 to Plasma	0.00038
Exchangeable trabecular bone volume to surface	0.0097
Exchangeable to non-exchangeable trabecular bone volume	0.0042
Exchangeable cortical bone volume to surface	0.0097
Exchangeable to non-exchangeable cortical bone volume	0.0042
Non-exchangeable cortical bone volume to plasma	0.0000821
Non-exchangeable trabecular bone volume to plasma	0.000493

[a] The model structure and compartment names are the same as in the systemic model for strontium of ICRP Publication 67 (ICRP 1993) and 68 (ICRP 1994b), described in an earlier section.

these elements equilibrate rapidly with an extravascular pool about three times the size of the plasma pool. The rapid equilibration of plasma with a threefold larger extravascular pool, as well as plasma clearance data for barium in healthy human subjects (Newton et al. 1991), are reproduced by: increasing the transfer coefficients of the Publication 67/68 model for barium in adults by a factor of two for transfers from plasma to bone surface and excretion pathways; increasing the transfer coefficient λ from plasma to a rapid-turnover soft tissue compartment (called ST0) by a factor of 8; and setting the transfer coefficient from ST0 to plasma at $\lambda/3$.

Rates of removal of ^{137m}Ba from its sites of production to plasma are set for reasonable agreement with experimental data of Wasserman, Twardock, and Comar (1959), who demonstrated considerable dissociation of ^{137m}Ba from ^{137}Cs in rats at 4–7 d after intraperitoneal administration of ^{137}Cs/^{137m}Ba. Those investigators found that ^{137m}Ba exceeded equilibrium proportions in bone, whole blood, and plasma by factors of 3.3, 3.9, and 14, respectively. Some soft tissues were moderately deficient in ^{137m}Ba, while others showed little or no deviation from equilibrium. Wasserman and coworkers concluded nevertheless that soft tissues likely were the main source of the excess ^{137m}Ba in plasma and that red blood cells probably also contributed to the excess. Skeletal muscle, which was not sampled, seems likely to have been the main contributor to the excess ^{137m}Ba in plasma and bone, as muscle has been found to contain the preponderance of systemic cesium by a few days after its acute uptake to blood. Based on the findings of Wasserman and coworkers, the rates of transfer of ^{137m}Ba from red blood cells (of the cesium model) to plasma and from muscle to plasma are set at 1000 d^{-1} (half-time of 1 min), an ICRP default value representing extremely fast removal from these pools. The rate of transfer of ^{137m}Ba from all other soft tissue compartments of the cesium model (including cartilage) to blood is set at 200 d^{-1} (half-time of 5 min), chosen to yield at most a moderate deficiency of ^{137m}Ba in these tissues compared with equilibrium values. Barium-137m produced in bone in the cesium model is assumed to decay at the site of production. Barium-137m produced in or entering the urinary bladder contents is removed to urine at the rate 12 d^{-1}, which is the ICRP's reference value for the rate of urinary bladder emptying in workers (ICRP 1994b). The movement of ^{137m}Ba produced in or entering the gastrointestinal contents is determined by the generic transit rates in the ICRP's Human Alimentary Tract Model (HATM).

The resulting model for ^{137m}Ba produced by decay of systemic ^{137}Cs is shown in Figure 6.58. The numbers next to the arrows are transfer coefficients (d^{-1}) for ^{137m}Ba. The box in Figure 6.58 labeled "All other soft tissue [compartments] of Cs model" represents ^{137m}Ba produced in the cesium model compartments representing liver, kidneys, gastrointestinal tract walls, spleen, pancreas, heart, brain, skin, lung tissue, other 1, and other 2, plus cartilage. Plasma, RBC, trabecular bone surface, cortical bone surface, urinary bladder contents, and the five gastrointestinal content compartments are common to the cesium and barium models. ST0 represents ^{137m}Ba that enters extravascular spaces (excluding bone) from plasma; it is assumed that no decays of ^{137}Cs occur in ST0. Barium-137m produced in a bone surface compartment or entering that compartment from blood is assumed to decay there.

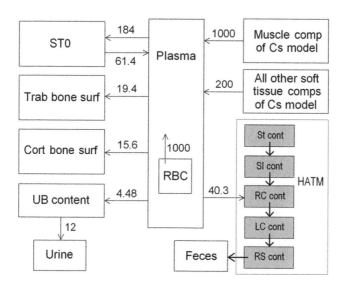

FIGURE 6.58 Model for ^{137m}Ba produced by decay of ^{137}Cs following absorption or injection of ^{137}Cs into blood. The values next to the arrows are transfer coefficients (d^{-1}) for ^{137m}Ba. ST0 is a soft tissue compartment with rapid turnover, Trab = trabecular, Cort = cortical, surf = surface, UB = urinary bladder, RBC = red blood cells, Comp(s) = compartment(s), cont = content, St = stomach, SI = small intestine, RC = right colon, LC = left colon, RS = rectosigmoid colon, HATM = Human Alimentary Tract Model.

By design, the model for ^{137m}Ba shown in Figure 6.58 is consistent with plasma clearance data (Newton et al. 1991) for 6 healthy adult men following intravenous administration of ^{133}Ba (Figure 6.59). The model for ^{137m}Ba as a progeny of ^{137}Cs is also designed for reasonable consistency with findings of Wasserman et al. (1959) regarding the dissociation of ^{137m}Ba in plasma following administration of ^{137}Cs in rats. The proposed models for cesium and ^{137m}Ba as a progeny of ^{137}Cs predict that the plasma content of ^{137m}Ba at 4–7 d after injection of ^{137}Cs to blood is 13–16 times the equilibrium value, compared with an average ratio of 14 determined by Wasserman and coworkers for that period. The bone content of ^{137m}Ba at 4–7 d is predicted to be about 2 times the equilibrium value compared with the ratio 3.3 determined by Wasserman and coworkers. The high rate of migration of ^{137m}Ba from its sites of production to bone indicated by the findings for rats could not be reproduced closely while remaining consistent with reported biokinetic data for barium (as a parent) in human subjects.

Dose calculations based on the models for cesium and progeny described above indicate that the contribution of progeny to dose coefficients is substantial for the ^{137}Cs and ^{134m}Cs chains, of marginal importance for the ^{125}Cs chain, due mainly to a roughly twofold increase in the dose coefficient for the thyroid due to growth of ^{125}I in the body, and negligible for ^{127}Cs. The contributions to dose estimates from progeny of ^{134m}Cs and ^{137}Cs based on the models described above as well as the models of ICRP Publication 68 (ICRP 1994b) are shown in Table 6.19. The contribution of ingrowing ^{134}Cs to tissue dose coefficients for ^{134m}Cs based on the proposed model for cesium varies considerably (4%–53%) from one

TABLE 6.19   Contribution of Progeny to Injection Dose Coefficients for ^{134m}Cs and ^{137}Cs Based on Proposed Models and Models of ICRP Publication 68 (ICRP 1994b)

| | Contribution of Progeny to Injection Dose Coefficient (%) | | | |
| | Cs-134m | | Cs-137 | |
Tissue	Proposed Models	Pub. 68	Proposed Models	Pub. 68
Bone surface	13	38	32	61
Colon wall	9	42	79	58
Kidneys	4	41	65	60
Liver	12	42	68	61
Muscle	53	39	42	58
Pancreas	11	44	59	63
Red marrow	24	41	64	60
Small intestine wall	6	43	73	62
Spleen	7	41	67	60
Stomach wall	8	40	67	59
Testes	49	39	84	58
Thyroid	51	40	85	59
Effective dose	15	41	74	59

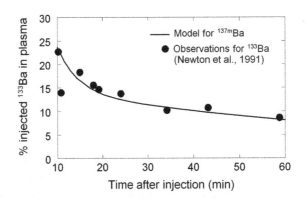

FIGURE 6.59   Predictions of plasma clearance of intravenously injected barium based on the model for barium as cesium progeny, compared with observations for healthy adult human subjects.

tissue to another. Based on the model of ICRP Publication 68, decays of ingrowing ^{134}Cs represent roughly 40% of the dose coefficient for each tissue because both the parent and progeny are assumed to be uniformly distributed in tissues at all times. For the ^{137}Cs/^{137m}Ba chain, the contribution of ingrowing ^{137m}Ba to the dose coefficient is predicted to vary from 32 to 85% based on the proposed models and to be roughly 60% for all tissues based on the model of ICRP Publication 68. The large contribution of ^{137m}Ba to the dose coefficient for colon based on the proposed models (79%) results in a significantly different dose coefficient for colon compared with the coefficient generated from the systemic model for cesium used in ICRP Publication 68.

REFERENCES

Adams, C. A., and T. A. Bonnell. 1962. "Administration of Stable Iodide as a Means of Reducing Thyroid Irradiation Resulting from Inhalation of Radioactive Iodine". *Health Physics* 7 (3–4):127–149.

Bates, D. V., B. R. Fish, T. F. Hatch, T. T. Mercer, and P. E. Morrow. 1966. "Deposition and Retention Models for Internal Dosimetry of the Human Respiratory Tract. Task Group on Lung Dynamics". *Health Physics* 12 (2):173.

Berkovski, V. 1999a. "Radioiodine biokinetics in the mother and fetus. Part 1. Pregnant woman". In G. Thomas, A. Karaoglou, and E. D. Williams (eds), *Radiation and Thyroid Cancer*. Singapore: World Scientific Publishing Co., 319–325.

Berkovski, V. 1999b. "Radioiodine biokinetics in the mother and fetus. Part 2. Fetus". In G. Thomas, A. Karaoglou, and E. D. Williams (eds), *Radiation and Thyroid Cancer*. Singapore: World Scientific Publishing Co., 327–332.

Berkovski, V. 2002. "New Iodine Models Family for Simulation of Short-Term Biokinetics Processes, Pregnancy and Lactation". *Food and Nutrition Bulletin* 23 (3 Suppl):87–94.

Committee on Medical Internal Radiation Dose. 1975. "MIRD Dose Estimate Report No 5. Summary of Current Radiation Dose Estimates to Humans from 123i, 124i, 125i, 126i, 130i, 131i, and 123i as Sodium Iodide". *The Journal of Nuclear Medicine* 16:857–860.

Dunning Jr, D. E., and G. Schwarz. 1981. "Variability of Human Thyroid Characteristics and Estimates of Dose from Ingested 131i". *Health Physics* 40 (5):661–675.

Durbin, Patricia W. 1972. *Plutonium in Man: A New Look at the Old Data*. Berkeley, CA: Univ. of California.

Forth, W., Oberhausen, E., Pfleger, K., and Weske, G. 1963. Beitrag zur Klarung der Ursachen der Anriecherung von Cesium-137 im Organismus. *Experienta* 19:25–26.

Eve, I. S. 1966. "A Review of the Physiology of the Gastrointestinal Tract in Relation to Radiation Doses from Radioactive Materials". *Health Physics* 12 (2):131–161.

Ham, G. J., and J. D. Harrison. 2000. "The Gastrointestinal Absorption and Urinary Excretion of Plutonium in Male Volunteers". *Radiation Protection Dosimetry* 87 (4):267–272.

Hays, Marguerite T., and Louis H. Wegner. 1965. "A Mathematical and Physiological Model for Early Distribution of Radioiodide in Man". *Journal of Applied Physiology* 20 (6):1319–1328.

International Commission on Radiological Protection. 1960. *ICRP Publication 2: Recommendations of the ICRP, Report of Committee II on Permissible Dose for International Radiation*. Oxford, U.K.: Pergamon Press.

International Commission on Radiological Protection. 1968. *ICRP Publication 10: Evaluation of Radiation Doses to Body Tissues from Internal Contamination Due to Occupational Exposure*. Oxford, U.K.: Pergamon Press.

International Commission on Radiological Protection. 1972. *ICRP Publication 19: The Metabolism of Compounds of Plutonium and Other Actinides*. Oxford: Pergamon Press.

International Commission on Radiological Protection. 1973. *ICRP Publication 20: Alkaline Earth Metabolism in Adult Man*. Oxford: Pergamon Press.

International Commission on Radiological Protection. 1979. "ICRP Publication 30 (Part 1): Limits for Intakes of Radionuclides by Workers". *Ann ICRP* 2 (3–4).

International Commission on Radiological Protection. 1980. "ICRP Publication 30 (Part 2): Limits for Intakes of Radionuclides by Workers". *Ann ICRP* 4 (3–4).

International Commission on Radiological Protection. 1981. "ICRP Publication 30 (Part 3): Limits for Intakes of Radionuclides by Workers". *Ann ICRP* 6 (2–3).

International Commission on Radiological Protection. 1988. "ICRP Publication 30 (Part 4): Limits for Intakes of Radionuclides by Workers". *Ann ICRP* 19 (4).

International Commission on Radiological Protection. 1989. "ICRP Publication 54: Individual Monitoring for Intakes of Radionuclides by Workers". *Ann ICRP* 19 (1–3):1–137.

International Commission on Radiological Protection. 1990. "ICRP Publication 56: Age-Dependent Doses to Members of the Public from Intake of Radionuclides - Part 1". *Ann ICRP* 20 (2):1–22.

International Commission on Radiological Protection. 1991a. "ICRP Publication 60: 1990 Recommendations of the International Commission on Radiological Protection". *Ann ICRP* 21 (1–3).

International Commission on Radiological Protection. 1991b. "ICRP Publication 60: 1990 Recommendations of the International Commission on Radiological Protection". *Ann ICRP* 21 (1–3).

International Commission on Radiological Protection. 1993. "ICRP Publication 67: Age-Dependent Doses to Members of the Public from Intake of Radionuclides: Part 2 - Ingestion Dose Coefficients". *Ann ICRP* 23 (3–4):1–167.

International Commission on Radiological Protection. 1994a. "ICRP Publication 66: Human Respiratory Tract Model for Radiological Protection". *Ann ICRP* 24 (1–3):1–482.

International Commission on Radiological Protection. 1994b. "ICRP Publication 68: Dose Coefficients for Intakes of Radionuclides by Workers". *Ann ICRP* 24 (4).

International Commission on Radiological Protection. 1995a. "ICRP Publication 69: Age-Dependent Doses to Members of the Public from Intake of Radionuclides - Part 3 Ingestion Dose Coefficients". *Ann ICRP* 25 (1):1–167.

International Commission on Radiological Protection. 1995b. "ICRP Publication 71: Age-Dependent Doses to Members of the Public from Intake of Radionuclides - Part 4 Inhalation Dose Coefficients". *Ann ICRP* 25 (3-4):1–405.

International Commission on Radiological Protection. 1995c. "ICRP Publication 72: Age-Dependent Doses to Members of the Public from Intake of Radionuclides - Part 5 Compilation of Ingestion and Inhalation Coefficients". *Ann ICRP* 26 (1).

International Commission on Radiological Protection. 1997. "ICRP Publication 78: Individual Monitoring for Internal Exposure of Workers". *Ann ICRP* 27 (3–4).

International Commission on Radiological Protection. 2002. "ICRP Publication 89: Basic Anatomical and Physiological Data for Use in Radiological Protection - Reference Values". *Ann ICRP* 32 (3–4):1–277.

International Commission on Radiological Protection. 2006. "ICRP Publication 100: Human Alimentary Tract Model for Radiological Protection". *Ann ICRP* 36 (1–2):1–366.

International Commission on Radiological Protection. 2007. "ICRP Publication 103: The 2007 Recommendations of the International Commission on Radiological Protection". *Annals of the ICRP* 37.

International Commission on Radiological Protection. 2008. "ICRP Publication 107: Nuclear Decay Data for Dosimetric Calculations". *Ann ICRP* 38 (3):1–26.

International Commission on Radiological Protection. 2015. "ICRP Publication 130: Occupational Intakes of Radionuclides: Part 1". *Ann ICRP* 44 (2):1–188.

International Commission on Radiological Protection. 2016. "ICRP Publication 134: Occupational Intakes of Radionuclides: Part 2". *Ann ICRP* 45(3–4):1–352.

International Commission on Radiological Protection. 2018. "ICRP Publication 137: Occupational Intakes of Radionuclides: Part 3". *Ann ICRP* 46 (3–4):1–486.

Johansson, L., S. Leide-Svegborn, S. Mattsson, and B. Nosslin. 2003. "Biokinetics of Iodide in Man: Refinement of Current ICRP Dosimetry Models". *Cancer Biotherapy and Radiopharmaceuticals* 18 (3):445–450.

Johnson, J. R. 1987. "A Review of Age Dependent Radioiodine Dosimetry". In G. B. Gerber (ed), *Age-Related Factors in Radionuclide Metabolism and Dosimetry*. Dordrecht, Netherlands: Martinus Nijhoff.

Khokhryakov, V. F., Z. S. Menshikh, K. G. Suslova, T. I. Kudryavtseva, Z. B. Tokarskaya, and S. A. Romanov. 1994. "Plutonium Excretion Model for the Healthy Man". *Radiation Protection Dosimetry* 53 (1–4):235–239.

Khokhryakov, V. F., K. G. Suslova, R. E. Filipy, J. R. Alldredge, E. E. Aladova, S. E. Glover, and V. V. Vostrotin. 2000. "Metabolism and Dosimetry of Actinide Elements in Occupationally-Exposed Personnel of Russia and the United States: A Summary Progress Report". *Health Physics* 79 (1):63–71.

Korsunskii, V. N., N. F. Tarasov, and A. Z. Naumenko. 1981. "Clinical Evaluation of Ba-133m as an Osteotropic Agent". *Meditsinskaya Radiologiya* 10:45–48.

Langham, Wright H., Samuel H. Bassett, Payne S. Harris, and Robert E. Carter. 1950. "Distribution and Excretion of Plutonium Administered Intravenously to Man". *Health Physics* 38 (6):1031–1060.

Leggett, R. W. 1985. "A Model of the Retention, Translocation and Excretion of Systemic Pu". *Health Physics* 49 (6):1115–1138.

Leggett, R. W. 1992. "A Generic Age-Specific Biokinetic Model for Calcium-Like Elements". *Radiation Protection Dosimetry* 41 (2–4):183–198.

Leggett, R. W. 2003. "Reliability of the ICRP's Dose Coefficients for Members of the Public: III. Plutonium as a Case Study of Uncertainties in the Systemic Biokinetics of Radionuclides". *Radiation Protection Dosimetry* 106 (2):103–120.

Leggett, R. W. 2010. "A Physiological Systems Model for Iodine for Use in Radiation Protection". *Radiation Research* 174 (4):496–516.

Leggett, R. W. 2013. "Biokinetic Models for Radiocaesium and Its Progeny". *Journal of Radiological Protection*. 33:123–140.

Leggett, R. W. 2017. "An Age-Specific Biokinetic Model for Iodine". *Journal of Radiological Protection* 37:864–882.

Leggett, R. W., D. E. Dunning, and K. F. Eckerman. 1984. "Modelling the Behaviour of Chains of Radionuclides inside the Body". *Radiation Protection Dosimetry* 9 (2):77–91.

Leggett, R. W., K. F. Eckerman, and L. R. Williams. 1982. "Strontium-90 in Bone: A Case Study in Age-Dependent Dosimetric Modeling". *Health Physics* 43 (3):307–322.

Leggett, R. W., K. F. Eckerman, V. F. Khokhryakov, K. G. Suslova, M. P. Krahenbuhl, and S. C. Miller. 2005. "Mayak Worker Study: An Improved Biokinetic Model for Reconstructing Doses from Internally Deposited Plutonium". *Radiation Research* 164 (2):111–122.

Leggett, R. W., and L. R. Williams. 1995. "A Proposed Blood Circulation Model for Reference Man". *Health Physics* 69 (2):187–201.

Leggett, R. W., L. R. Williams, D. R. Melo, and J. L. Lipsztein. 2003. "A Physiologically Based Biokinetic Model for Cesium in the Human Body". *Science of the Total Environment* 317 (1):235–255.

McGuire, R. A., and M. T. Hays. 1981. "A Kinetic Model of Human Thyroid Hormones and Their Conversion Products". *The Journal of Clinical Endocrinology & Metabolism* 53 (4):852–862.

National Council on Radiation Protection and Measurements. 2007. *NCRP Report 156: Development of a Biokinetic Model for Radionuclide-Contaminated Wounds and Procedures for Their Assessment, Dosimetry and Treatment*. Bethesda, MD: National Council on Radiation Protection and Measurements.

Newton, D., G. E. Harnson, C. Kang, and A. J. Warner. 1991. "Metabolism of Injected Barium in Six Healthy Men". *Health Physics* 61 (2):191–201.

Newton, D., R. J. Talbot, C. Kang, and A. J. Warner. 1998. "Uptake of Plutonium by the Human Liver". *Radiation Protection Dosimetry* 80 (4):385–395.

Nicoloff, J. T., and J. T. Dowling. 1968. "Estimation of Thyroxine Distribution in Man". *Journal of Clinical Investigation* 47 (1):26.

Oliner, Leo, Robert M. Kohlenbrener, Theodore Fields, Ralph Kunstadter, H., and Dolores R. Goldstein. 1957. "Thyroid Function Studies in Children: Normal Values for Thyroidal I131 Uptake and Pbi131 Levels up to the Age of 18". *The Journal of Clinical Endocrinology & Metabolism* 17 (1):61–75.

Popplewell, D. S., G. J. Ham, W. McCarthy, and C. Lands. 1994. "Transfer of Plutonium across the Human Gut and Its Urinary Excretion". *Radiation Protection Dosimetry* 53 (1–4):241–244.

Ramsden, D., F. H. Passant, C. O. Peabody, and R. G. Speight. 1967. "Radioiodine Uptakes in the Thyroid Studies of the Blocking and Subsequent Recovery of the Gland Following the Administration of Stable Iodine". *Health Physics* 13 (6):633–646.

Riggs, D. S. 1952. "Quantitative Aspects of Iodine Metabolism in Man." *Pharmacological Reviews* 4 (3):284–370.

Robertson, James S., and Colum A. Gorman. 1976. "Gonadal Radiation Dose and Its Genetic Significance in Radioiodine Therapy of Hyperthyroidism". *Journal of Nuclear Medicine: Official Publication, Society of Nuclear Medicine* 17 (9):826–835.

Rundo, J., P. M. Starzyk, J. Sedlet, R. P. Larsen, R. D. Oldham, and J. J. Robinson. 1975. Excretion Rate and Retention of Plutonium 10,000 Days after Acquisition. Argonne National Lab., IL, USA.

Shagina, N. B., E. I. Tolstykh, V. I. Zalyapin, M. O. Degteva, V. P. Kozheurov, E. E. Tokareva, L. R. Anspaugh, and B. A. Napier. 2003. "Evaluation of Age and Gender Dependences of the Rate of Strontium Elimination 25–45 Years after Intake: Analysis of Data from Residents Living Along the Techa River". *Radiation Research* 159 (2):239–246.

Stather, J. W., J. R. Greenhalgh, and N. Adams. 1983. "The Metabolism of Iodine in Children and Adults". *Radiological Protection Bulletin* 54:9–15.

Suslova, K. G., V. F. Khokhryakov, Z. B. Tokarskaya, A. P. Nifatov, M. P. Krahenbuhl, and S. C. Miller. 2002. "Extrapulmonary Organ Distribution of Plutonium in Healthy Workers Exposed by Chronic Inhalation at the Mayak Production Association". *Health Physics* 82 (4):432–444.

Susskind, Herbert, Harold L. Atkins, Stanton H. Cohn, Kenneth J. Ellis, and Powell Richards. 1977. "Whole-Body Retention of Radioxenon". *Journal of Nuclear Medicine: Official Publication, Society of Nuclear Medicine* 18 (5):462–471.

Talbot, R. J., D. Newton, and S. N. Dmitriev. 1997. "Sex-Related Differences in the Human Metabolism of Plutonium". *Radiation Protection Dosimetry* 71 (2):107–121.

Talbot, R. J., D. Newton, and A. J. Warner. 1993. "Metabolism of Injected Plutonium in Two Healthy Men". *Health Physics* 65 (1):41–46.

Warner, A. J., R. J. Talbot, and D. Newton. 1994. "Deposition of Plutonium in Human Testes". *Radiation Protection Dosimetry* 55 (1):61–63.

Wasserman, R. H., A. R. Twardock, and C. L. Comar. 1959. "Metabolic Dissociation of Short-Lived Barium-137m from Its Cesium-137 Parent". *Science* 129 (3348):568–569.

Williams, L. R., and R. W. Leggett. 1987. "The Distribution of Intracellular Alkali Metals in Reference Man". *Physics in Medicine & Biology* 32 (2):173.

World Health Organization. 2001. *Assessment of Iodine Deficiency Disorders and Monitoring Their Elimination*. 2nd edn. Edited by United National Children's Fund Joint Publication of International Council for Control of Iodine Deficiency Disorders, and World Health Organization. Geneva: World Health Organization.

Yamagata, Noboru. 1962. "The Concentration of Common Cesium and Rubidium in Human Body". *Journal of Radiation Research* 3 (1):9–30.

Yamagata, Noboru, Kiyoshi Iwashima, Teruo Nagai, Kazuo Watari, and Takeshi A. Iinuma. 1966. "In Vivo Experiment on the Metabolism of Cesium in Human Blood with Reference to Rubidium and Potassium". *Journal of Radiation Research* 7 (1):29–46.

Zanzonico, Pat B., and David V. Becker. 2000. "Effects of Time of Administration and Dietary Iodine Levels on Potassium Iodide (Ki) Blockade of Thyroid Irradiation by 131i from Radioactive Fallout". *Health Physics* 78 (6):660–667.

CHAPTER 7

Dosimetric Models

John R. Ford, Jr. and John W. Poston, Sr.

CONTENTS

7.1	Introduction and General Overview	307
7.2	Source and Target Organs	313
7.3	Nuclear Decay Data	317
	7.3.1 Sources of Nuclear Decay Data	318
	7.3.2 Applications and Uncertainties of Nuclear Data	321
	7.3.3 Most Recent Updates to Nuclear Data	321
7.4	Specific Effective Energy	322
7.5	Specific Absorbed Fractions for Photons	327
7.6	Specific Absorbed Fractions for Neutrons	329
7.7	Specific Effective Energy and Absorbed Fractions for Charged Particles	330
References		332

7.1 INTRODUCTION AND GENERAL OVERVIEW

In the context of this chapter, the term "dosimetric models" encompasses a number of approaches to determining the absorbed dose to humans exposed to ionizing radiation from both internal and external sources. These determinations for internal sources have been almost exclusively computational. The approaches began very simply but over time, as the computational technology evolved, so did the capabilities to produce more realistic models of the human body. Early in their development, the models were called "anthropomorphic models" (having human characteristics) but soon the term "phantom" was applied to these models. The term "phantom" will be used almost exclusively in this chapter.

The purpose of a dosimetric model (either simple or complex) is to provide a framework in which an estimate of the absorbed dose* to an organ or the entire body can be obtained for a particular exposure situation. Models are used when both external and internal radiation sources are being considered. The hope is that, the more realistic the model, the more accurate the dose estimate. This may or may not be true, because there are many parameters that must be considered in the dose calculation and the model (i.e., geometry) is just

---

* As reinforced in Chapter 2, the absorbed dose is a fundamental dosimetric quantity describing the energy deposited per unit mass of a volume or region of interest. The units are J/kg in the SI system and the special unit is the gray (Gy).

one, but not the only one, of the many major considerations. A more realistic phantom does not always assure a more reliable dose estimate.

Advances in computational techniques over more than half a century have both simplified and complicated dose assessment techniques. Initially, calculations were based simply on the physics of the situation. For example, Berger published the results of his calculations for point sources of electrons and beta particles in water and other media (Berger 1971). He considered monoenergetic, point sources of electrons with nine energies between 25 keV and 4 MeV. In addition, he calculated the distribution of absorbed dose for 75 beta-emitting radionuclides.

Early calculations focused on simple models and the well-understood transport of radiation through matter; generally, in tissue, and later in bone and lung tissue. Early models were simple shapes (spheres, ellipsoids, disks, and cylinders). For example, Ellett and Humes reported Monte Carlo calculations of the absorbed fraction of energy in unit density spheres and ellipsoids surrounded by a scattering medium (Ellett and Humes 1971). These calculations were an improvement on those published earlier by Brownell, Ellett, and Reddy (Brownell, Ellett, and Reddy 1968), in that the volume containing the radioactivity was embedded in a large scattering medium of the same composition. In 1959, the International Commission on Radiological Protection (ICRP) in Publication 2 modeled the total body of an adult male as a sphere with an effective radius of 30 cm (ICRP 1960a). Inside this sphere were specifications for the muscle, the respiratory system, and the gastrointestinal tract. No justification for the assumption of a spherical geometry was given in their publication. Some dosimetrists often used a very simple representation of Standard Man in their lectures on internal dose calculations. One approach was to represent the body as a sphere with a mouth, a single lung, and a gastrointestinal tract. Even though this model was quite rudimentary, it represented the level of knowledge and the simplifying assumptions embodied in the calculations.

Later, the ICRP produced a more complex, but still incomplete model, outlining the principal metabolic pathways of radionuclides taken into the body (ICRP 1968). This model is shown in Figure 7.1. The concept of a "standard man" for use in internal dose calculations originated more than 60 years ago. When early dosimetrists compared their dose estimates for inhaled or ingested radionuclides (or their estimates of permissible levels in air and water), they found that agreement was not always good. This lack of agreement was due primarily to the use of widely varying biological data in the dose calculations. For this reason, a "Standard Man" was proposed to be used in all dose calculations. This standard man was assumed to be an adult male with a number of parameters, such as organ masses, total-body mass, respiration rate, ingestion rate, excretion rates, and other parameters specified for use in dose calculations. The first agreements on a standard man were formulated by the National Council on Radiation Protection and Measurements (NCRP) at a conference held at Chalk River, Canada in 1949 (Taylor 1984). It should be clear that these data were never intended to represent an adult human in all aspects. The main purpose was to specify only those characteristics needed for purposes of dosimetry. As stated above, in these early calculations, that is, for maximum concentrations in air and water, it was assumed that the

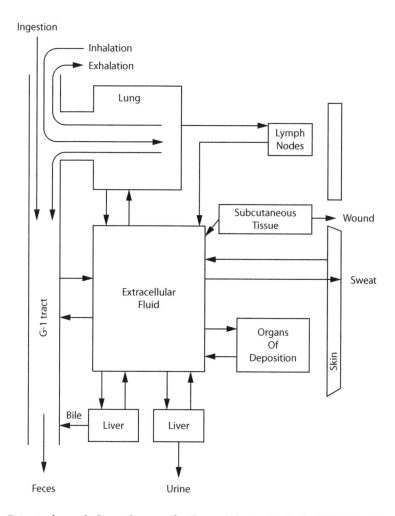

FIGURE 7.1 Principal metabolic pathways of radionuclides in the body (ICRP 1968).

"whole body" of an adult male human could be represented as a 30-cm diameter sphere. In addition, the respiratory and gastrointestinal models were significantly simplified.

The ICRP Report of Committee II on Permissible Dose for Internal Radiation (ICRP 1960a) provided data for use with "Standard Man." These data included 46 major elements in the total body as well as the elemental distributions in 36 organs and structures associated with the human body. The mass of the total body was assumed to be 70 kg and the masses and assumed dimensions of the organs and systems in the body were specified. Intake and excretion of Standard Man are shown in Tables 7.1 and 7.2, in terms of water and air balance. Assumptions associated with the respiratory tract and the gastrointestinal tract are given in Tables 7.3 and 7.4, respectively. These data give the impression that the early calculations to control the intake of radioactive materials were much more sophisticated than was actually the case. In reality, calculations of maximum permissible concentrations (MPC) in air and water and internal dose assessments were accomplished using mechanical machines, or very early computers that only were capable of relatively simple

TABLE 7.1   Intake and Excretion of the
Standard Man—Water Balance (ICRP 1960a)

Intake (cm³/day)		Excretion (cm³/day)	
Food	1000	Urine	1400
Fluids	1200	Sweat	600
Oxidation	300	From lungs	300
		Feces	200
Total	**2500**	**Total**	**2500**

TABLE 7.2   Intake and Excretion of the Standard Man—Air Balance (ICRP 1960a)

	O₂ (vol. %)	CO₂ (vol. %)	N₂ + others (vol. %)
Inspired air	20.94	0.03	79.03
Expired air	16	4.0	80
Alveolar air (inspired)	15	5.6	–
Alveolar air (expired)	14	6.0	–
Vital capacity of lungs:		3–4 liters (men)	
			2–3 liters (women)
Air inhaled during 8-hour work day			10^7 cm³/day
Air inhaled during 16-hour not at work			10^7 cm³/day
		Total	2×10^7 cm³/day
Interchange area of lungs			50 m²
Area of upper respiratory tract, trachea, bronchi			20 m²
Total surface area of respiratory tract			70 m²

TABLE 7.3   Particulates in the Respiratory Tract of Standard Man (ICRP 1960a)

Distribution	Readily Soluble Compounds (%)	Other Compounds (%)
Exhaled	25	25
Deposited in the upper respiratory passages and subsequently swallowed	50	50
Deposited in the lungs (lower respiratory passages)	25 (This is taken up into the body)	25[a]

[a] Of this, half is eliminated from the lungs and swallowed in the first 24 hrs, making a total of 62.5% swallowed. The remaining 12.5% is retained in the lungs with a half-life of 120 days, it being assumed that this portion is taken into body fluids.

mathematical manipulations. Remember the organs of the body, as well as the total body, were assumed to be spherical.

As computer techniques improved in the early 1960s, more realistic representations of an adult human male were soon being described by Fisher and Snyder (1966). The first phantom consisted of three distinct regions: the tissue, the lungs, and the skeleton. The exterior surfaces of the phantom were represented by three simple regions based on simple shapes. The trunk of the body was represented by an elliptical cylinder representing

TABLE 7.4 The Gastrointestinal Tract of the Standard Man (ICRP 1960a)

Portion of GI Tract That Is Critical Tissue	Mass of Contents (g)	Time Food Remains, τ (day)	Fraction from Lung to GI Tract, f_a	
			(sol.)	(insol.)
Stomach (S)	250	1/24	0.50	0.625
Small intestine (SI)	1100	4/24	0.50	0.625
Upper large intestine (ULI)	35	8/24	0.50	0.625
Lower large intestine (LLI)	150	18/24	0.50	0.625

the torso and enclosing the arms and the hips. The head and neck was a separate elliptical cylinder, and the legs of the phantom were enclosed in a truncated ellipsoidal cone (see Figure 7.2). The tissues in the phantom were composed primarily of hydrogen, carbon, nitrogen, and oxygen. A limited number of trace elements were also included to more realistically model the tissue. For example, in the skeleton, about 18% of additional elements were included. These were primarily calcium and phosphorous. The composition of the lungs was slightly different from that of other soft tissues. The lungs were assumed to contain little fat and a much larger fraction of blood than other organs. The densities of the skeletal region (bone plus marrow), lungs, and the remainder (soft tissue) of the phantom were 1.4862, 0.2958, and 0.9869 g/cm³, respectively.

The next significant step in standardizing the parameters to be used in dose calculations was the publication of ICRP Publication 23 in 1975 (ICRP 1975). This publication was an encyclopedia of information on the adult human. In general, the data contained in this volume were focused on the adult male, but the publication also included information on the adult female and, in many cases, children of several ages. In addition, there were data on the chemical (elemental) composition of the individual organs in the human body. Later, the ICRP suggested a name change from "Standard Man" to "Reference Man." The intent was to extend and revise the concept to provide a more adequate basis for the assessment of exposure of all groups of the population. But, Snyder cautioned the dosimetry community of the weaknesses in the concept:

> The real need of the health physicist is not merely one Standard Man; for, however carefully he may be defined; he will still be representative of only a small fraction of the population the health physicist must consider. Thus, the concept of Standard Man should not merely define an individual but should include ranges of variations about this norm and provide procedures for taking these individual differences into account when they are significantly altering the dose estimate.
>
> (Snyder 1966)

It seems that Snyder's admonition has been ignored by many current-day dosimetrists as they rush to formulate new phantoms that are based, not on the large database used in

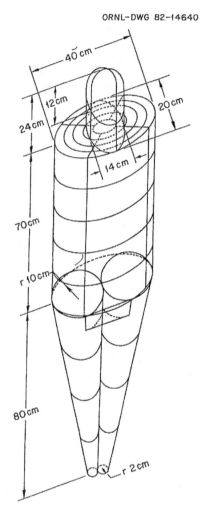

ORNL–DWG 82–14640

FIGURE 7.2   The Snyder–Fisher adult human phantom (Poston 1983).

Reference Man, but on data obtained from one male and one female, and/or a very limited data set.

Later, organs with realistic shapes and locations were added. A total of 20 source organs and 19 target organs were ultimately described in a more realistic phantom. A source organ was an organ in the body that contains a radioactive material uniformly distributed in the organ. A target organ was the organ in the body for which a "dose" calculation is required. Usually the "dose" was called an "absorbed fraction of energy" (usually simply the absorbed fraction or AF). The AF was the ratio of the energy deposited (or absorbed) in the target organ to the total energy emitted by the source organ. It must be realized that a source organ was also a target organ, because radiation energy may be absorbed in the organ without escaping the organ. Thus, the source organ was a target organ but a target organ may not necessarily be a source organ.

This representation was incorporated into a Monte Carlo computer code originally developed at the Oak Ridge National Laboratory. Snyder, Ford, Warner, and Watson

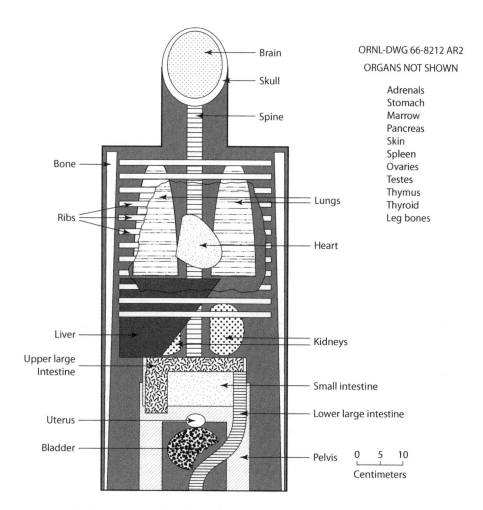

FIGURE 7.3   Anterior view of principal organs in the head and trunk of the phantom (Poston 1983).

provided estimates of the absorbed fractions for monoenergetic photon sources uniformly distributed in various organs of a heterogeneous phantom (Snyder et al. 1974). The principal organs of the head and trunk are shown in Figure 7.3, and the idealized model of the skeleton is shown in Figure 7.4. Later, a "family" of phantoms was designed which included both an adult male and female, and children of ages newborn, 1 year, 5 years, 10 years, and 15 years old. Current computer codes are much more sophisticated and the computer models (i.e., the phantoms) were based on actual measurements from a single male and a single female. More detail on the development of anthropomorphic phantoms can be found in Sections 7.2 and 7.3.

7.2 SOURCE AND TARGET ORGANS

Prior to World War II, ^{226}Ra was essentially the only radionuclide used in medicine. After peace was restored in 1945, man-made radionuclides were developed in the late 1940s for use in medical applications. At the same time, methods of calculating the absorbed dose to the patients were being developed. The first formalizations of these procedures were

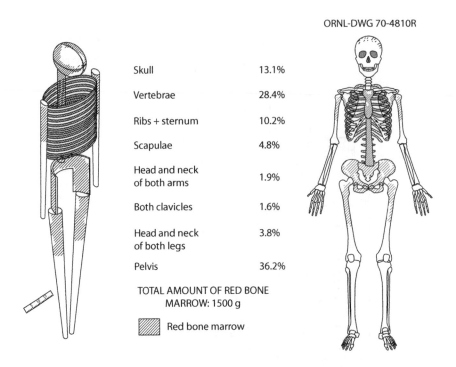

FIGURE 7.4   Idealized model of the skeleton with percentages of red bone marrow (Poston 1983).

published in 1948. L.D. Marinelli and his colleagues published two articles that reviewed the status of the field and provided a general approach to radionuclide dosimetry (Marinelli 1942; Marinelli, Quimby, and Hine 1948). This general dosimetric approach was accepted quickly and became the standard approach. Generally, these papers were considered to represent the beginning of modern radiation dosimetry in nuclear medicine.

A major contribution was made to Marinelli's approach radiation dosimetry by Ellett and his colleagues (Ellett and Humes 1971). These authors defined a term called the absorbed fraction, that is, the fraction of energy emitted by the radiation source (originally called s) absorbed in a specific volume of tissue (originally called v). These concepts were incorporated into an early Monte Carlo transport code and calculations were performed for photon sources of a variety of energies and for tissue volumes of many sizes and shapes. This research led Loevinger and Berman to conclude that equations for use in internal dosimetry could be derived in simple terms, which were independent on the characteristics of the radiation (Loevinger and Berman 1968). This approach was called the "MIRD Schema."

The general equation used to calculate the average dose, D, delivered to a target mass, m, is:

$$D = \frac{A}{m} \sum_i \varphi_i \Delta_i,$$ (7.1)

where A is the cumulated activity, φ_i is the absorbed fraction of energy for a given energy E_i target configuration, and source geometry; and Δ_i is the equilibrium dose constant.

Dillman published decay schemes and equilibrium dose constants for use in these calculations (Dillman 1969). The above equation was ultimately modified, simply by redefining the parameters, for use in calculations published by the International Commission on Radiological Protection in ICRP Publication 23 and ICRP Publication 30 (ICRP 1979).

The data comprising what was called Reference Man were collected from Western Europe and North America data sources. In addition, a large appendix to this publication contained from Monte Carlo calculations of the "specific absorbed fractions of energy" for monoenergetic photon sources over a range of energies (from 0.01 to 4 MeV) and a number of organs defined in Reference Man. This publication was perhaps one of the first to introduce the terms "source organ" and "target organ." The source organ sizes and masses were based on the Reference Man data. The source organ contained the photon emitter and the target organ was the region of interest in which the photon energy is deposited. There is a difference between the "absorbed fraction" and the "specific absorbed fraction." The specific absorbed fraction is the absorbed fraction of energy divided by the organ mass (in grams).

As stated above, the initial phantom, used especially in internal dose calculations, was assumed to be an adult male, 20 to 30 years of age, residing in North America or Western Europe. Later, there was demand for phantoms of other sizes to represent the adult female and children of several ages. Snyder and his colleagues answered this need by introducing a "family" of phantoms (Fischer and Snyder 1966). Initially, the phantoms of other ages were based on the design of the adult male. These were called "similitude phantoms." That is, the design of the adult was arranged in a Cartesian coordinate system and the phantoms representing other ages were obtained simply by "shrinking" the dimensions along the x-, y-, and z-axes. This approach produced an adult female, a newborn, a 1-year-old, a five-year-old, a 10-year-old, and a 15-year-old. The adult female model also served for the 15-year-old phantom. The similitude approach was the first attempt to provide phantoms other than the adult male. This approach, while extending the dosimetric capabilities, was not completely acceptable because organs internal to the body were "shrunk" by the same factors as the external dimensions. The result was that specific organs in the younger phantoms were not always of the correct size, mass, and location. For example, the organ in the child may be larger and shrink as the child approaches adulthood, for example, the thymus gland. It was very clear that a child was not simply a "shrunken adult." Another limitation of the similitude phantoms was that, even though the organ masses had to be known to calculate the absorbed dose, the organ masses were never published in the open literature.

These weaknesses led to an effort to design additional phantoms independently of the adult Reference Man. There were several design efforts conducted at ORNL during this period. For example, the ORNL research group reported on the design of two phantoms representing a one-year-old and a five-year-old child (Hwang et al. 1976b). Later, she and her colleagues undertook the design of a newborn child (Hwang et al. 1976a). A separate effort to design a 15-year-old was reported by Jones and his colleagues (Jones et al. 1976). These phantoms were similar to the "similitude phantoms" in many ways, but the dimensions of the phantom and the size and location of the internal organs were established using data from the literature. Table 7.1 presents a summary of the organ masses for four of the individualized phantoms. Later, Deus and Poston reported on the redesign of the 10-year-old (Deus and Poston 1976). This latter design was intended to "point the way" to the next generation of phantoms. However, the design was complicated and there were other approaches that proved to be superior. Basically, this phantom was too advanced for the time, and it was never used in calculations for the ICRP or for the MIRD Committee.

In 1978, Snyder, Ford, and Warner published results using a revised adult male phantom which became known as a "heterogeneous phantom" (Snyder et al. 1978). As in the past, these calculations of the specific absorbed fractions assumed monoenergetic photons uniformly distributed in the source organs of the phantom. Twelve photon energies were considered, ranging from 0.01 to 4.0 MeV. In addition, major modifications to the phantom included:

1. The head section was modified by topping the right elliptical cylinder with half an ellipsoid.

2. The leg region was separated into two legs by assuming this region could be represented by frustums of two circular cones.

3. A region representing the male genitalia was established outside the trunk, on the front of the revised model of the legs.

4. Modifications of the skeleton were included by providing a detailed description of the scapulae and the clavicles.

5. The gastrointestinal tract was modified significantly. The stomach, small intestine, and the upper and lower large intestine were designed to be organs with walls and contents.

6. The skin was added to the phantom as both a source region and target region.

The modifications described above brought the source/target organs (or regions) to a total of twenty each. In order to provide increased reliability, the number of photon histories for each calculation was increased to 60,000. However, for the low-energy photons, the results remained unreliable for some source/target combinations. Other methods were instituted to attempt to remedy the situation. For example, the absorbed fractions of energy for some organs were determined by plotting the absorbed fractions for a specific organ as a function of photon energy and extrapolating to lower energies assuming a smooth curve with shapes similar to those of other, larger organs.

As computer techniques improved, other models were designed and incorporated into the Monte Carlo codes. For example, more complex structures were added to specific organs of interest. A model of the kidneys was developed which consisted of three regions rather than the original "solid organ" (Patel 1988). Another example of specific organ or region modeling is the revised model of the adult head and brain published by a MIRD Committee Task Group (Bouchet et al. 1999). This model allowed the calculation of absorbed fractions and S-values for both monoenergetic photons and electrons over twelve energies ranging from 0.01 to 4.0 MeV. Target regions included the total brain, the caudate nuclei, the cerebellum, the cerebral cortex, the cranial CSF, the cranium, the lateral ventricles, the lentiform nuclei, the spinal CSF, the spinal skeleton, the thalami, the third ventricle, the white matter, and the thyroid gland. More than ten years after Patel, Bouchet and his colleagues introduced age-dependent, multiregion models of the kidneys (Bouchet et al. 2002).

Another MIRD Committee Task Group reported on the development of a dynamic urinary bladder for radiation dose calculations (Thomas et al. 1992). The bladder was modeled as an expanding sphere with a volume range (i.e., bladder contents) of 10 to 770 mL. The mass of the wall was 45 g and the volume of the wall was 45 cm³. It was assumed that there was uniform thinning of the bladder wall as the bladder filled, while the mass of the wall remained constant. Rather than perform calculations for monoenergetic radiations, the calculations focused on only a few radionuclides tagged to specific compounds such as ^{18}F, ^{99m}Tc, ^{123}I, ^{124}I, ^{131}I, ^{111}In, and ^{89}Sr.

Finally, as mentioned earlier, adding a region to the "phantom" to represent the skin of the adult human was another advancement. This region allowed more realistic results to be acquired for exposure situations such as submersion in a cloud of noble gas, and the skin also serving as another source organ.

In the 1980s, Cristy and Eckerman, at the Oak Ridge National Laboratory, designed a new series of the stylized phantoms with a range of ages. This series included an adult and phantoms representing a newborn and children of 1, 5, 10, and 15 years of age. The designs were based on data from ICRP Publication 23 (ICRP 1975). However, in some cases, the similitude approach used by Snyder and Fisher was employed to determine organ shapes and locations. Nevertheless, these "new" phantoms were a significant step forward. Later, the 15-year-old was modified to serve a dual purpose. This phantom represented a 15-year-old male as well as the adult female. The breasts, ovaries, and the uterus of the 15-year-old were redesigned to represent those of an adult female. Other contributions to this group of phantoms included a more stylized breast region, a new model of heart developed by Coffey and colleagues (Coffey, Cristy, and Warner 1981), and an improved model of the thyroid gland. Other minor modifications in organ size and location were made to avoid "over-lapping" organs.

Table 7.5 lists important dimensions of the "family" of phantoms. A more detailed discussion of these phantoms can be found in Eckerman et al. (1999) (Tables 7.6 through 7.10).

7.3 NUCLEAR DECAY DATA

In order to calculate the absorbed dose to any tissue or organ, one must first determine the energies of the incident radiation(s) and the amount of energy absorbed by the tissue or organ of interest. Then, to calculate a dose equivalent (early ICRP publications) or equivalent dose, defined in ICRP 60 and used in subsequent publications (ICRP 1991), the types of radiation must be known and the absorbed dose modified accordingly. That is, modified originally by the Relative Biological Effectiveness (RBE) in the original ICRP recommendations, later by Q from about Publication 6 (ICRP 1960b), and finally until the adoption of the radiation weighting factor and the associated equivalent dose in the 1990s).

TABLE 7.5   Gastrointestinal Tract (GI) Model for Reference Man (ICRP 1975)

Section of Gastrointestinal Tract	Mass of Walls (g)	Mass of Contents (g)	Mean Residence Time (day)	λ (day⁻¹)
Stomach (ST)	150	250	1/24	24
Small intestine (SI)	640	400	4/24	6
Upper large intestine (ULI)	210	220	13/24	1.8
Lower large intestine (LLI)	160	135	24/24	1

TABLE 7.6   Source and Target Regions and Their Masses (Watson and Ford 1980)

Source Organ	Mass (g)	Target Organ	Mass (g)
Adrenals	14	Adrenals	14
Bladder contents	200	Bladder wall	45
Stomach contents	250	Stomach wall	150
Small intestine contents	400	Small intestine wall	640
Upper large intestine contents	220	Upper large intestine wall	210
Lower large intestine contents	135	Lower large intestine wall	160
Kidneys	310	Kidneys	310
Liver	1800	Liver	1800
Lungs	1000	Lungs	1000
Muscle	28000	Muscle	28000
Ovaries	11	Ovaries	11
Pancreas	100	Pancreas	100
Cortical bone	4000	Thymus	20
Trabecular bone	1000	Bone surfaces	120
Red bone marrow	1500	Red marrow	1500
Skin	2600	Skin	2600
Spleen	180	Spleen	180
Testes	35	Testes	35
Thyroid	20	Thyroid	20
Total body	70000	Uterus	80

To accomplish these calculations, the ICRP and other radiation protection agencies have relied on nuclear decay data from a variety of sources.

7.3.1 Sources of Nuclear Decay Data

In the first major set of recommendations for internal dose calculation, the ICRP used the 1953 Table of Isotopes by Hollander, Perlman, and Seaborg, along with the 1957 version of Sullivan's Trilinear Chart of the Nuclides supplemented with compilations of data from the Nuclear Data Group of the NRC combed from National Bureau of Standards circulars and supplements (1950–1952) and annual compilations of Nuclear Science Abstracts from 1952–1957. This effort provided the information needed to formulate effective energies for a large number of radionuclides and chains of parent radionuclides and their daughters (Cook 1960).

TABLE 7.7   Elemental Composition (% by Weight) of Tissues of the Phantom (ICRP 1975)

Element	Skeletal Tissue[a]	Lung Tissue[b]	Total Body Minus Skeleton and Lungs[c]
H	7.04	10.21	10.47
C	22.79	10.01	23.02
N	3.87	2.8	2.34
O	48.56	75.96	63.21
Na	0.32	0.19	0.13
Mg	0.11	7.4×10^{-3}	0.015
P	6.94	0.082	0.24
S	0.17	0.23	0.22
Cl	0.14	0.27	0.14
K	0.15	0.2	0.21
Ca	9.91	7.0×10^{-3}	0
Fe	8.0×10^{-3}	0.037	6.3×10^{-3}
Zn	4.8×10^{-3}	1.2×10^{-3}	3.2×10^{-3}
Rb	0	3.7×10^{-4}	5.7×10^{-4}
Sr	3.2×10^{-3}	5.9×10^{-4}	3.4×10^{-5}
Zr	0	0	8.0×10^{-4}
Pb	1.1×10^{-3}	4.1×10^{-5}	1.6×10^{-5}

[a] Density 1.4862 g/cm^3.
[b] Density 0.2958 g/cm^3.
[c] Density 0.9869 g/cm^3.

TABLE 7.8   Distribution of Red Bone Marrow in the Idealized Skeleton (Poston 1983)

Skeletal Region	Percent of Total Mass
Skull	13.10%
Vertebrae	28.40%
Ribs and sternum	10.20%
Scapulae	4.80%
Head and neck of both arms	1.90%
Both clavicles	1.60%
Head and neck of both legs	3.80%
Pelvis	36.20%
Total red bone marrow	**1500 g**

These sources provided the principal types of radiations emitted, the physical half-lives, the branching ratios for different decay pathways, and the energies of the emitted radiations. In addition, the compilations considered contributions from X-rays, beta particles, conversion electrons, annihilation photons, gamma rays, alpha particles, and recoil nuclei from alpha emission. For the RBEs, a value of 1 was used for photons and most beta-particles and electrons, a value of 1.7 was used if the beta particles or electrons had a maximum energy of less than 30 keV, a value of 10 for alpha particles, and finally, a value of 20 for the recoil nuclei (ICRP 1959).

TABLE 7.9   Weight and Vertical Dimensions of the Age-Specific Phantoms (Poston 1983)

Age (years)	Weight (kg)	Trunk (cm)	Head (cm)	Legs (cm)
0	3.15	23	13	16
1	9.11	33	16	28.8
5	18.12	45	20	46
10	30.57	54	22	64
15	53.95	65	23	78
Adult	69.88	70	24	80

TABLE 7.10   Masses of Red and Yellow Marrow and Bone in the Phantom (Poston 1983)

Skeletal Region	Red Marrow (g)	Bone (g)	Yellow Marrow (g)
ARMS			
Upper	28.5	474	9.5
Lower	0	520	389
CLAVICLES	24	49	8
LEGS			
Upper	57	2036	19
Lower	0	1588	461
PELVIS	543	177	181
RIBS	153	677	201
SCAPULAE	72	206	24
SKULL			
Cranium	178.5	557	59.5
Mandible	18	439	6
SPINE			
Upper	51	130	17
Middle	211.5	533	70.5
Lower	163.5	87	54.5
TOTAL	**1500**	**7473**	**1500**

A few years later, the ICRP defined the dose equivalent as the absorbed dose times a quality factor (QF now Q) based on the unrestricted linear energy transfer (LET) of the incident radiation. The values of Q ranged from 1 to 20, and, in general, the ICRP recommended values that were equal to the original RBE values for different types of radiation. Neutrons were considered for the first time, and if neutron energy and LET data were not available, the ICRP suggested the use of a value of 10 for Q (ICRP 1960b).

When the next major update to the ICRP recommendations was in preparation, the ICRP turned to the Nuclear Data Project, led by Snyder at Oak Ridge National Laboratory and the work of Dillman. Dillman developed a computer program that could be used to calculate decay schemes and energy yields of different types of radiation from Evaluated Nuclear Structure Data Files (ENSDF) (Dillman 1980).

In ICRP 30 Parts I–III and the supplements, only minimal decay schemes were published and bremsstrahlung calculations were performed only for the noble gas radionuclides (ICRP 1979, 1980, 1981). The full data and decay schemes were published in ICRP Publication 38, which provided a brief description of the methods, and tables, and included graphic representations of the decay schemes, tables of data listing yields (branching ratios), and energies of radiations (ICRP 1983).

A select set of decays schemes (some updated from the earlier publication) for medically important radionuclides were also published in a MIRD publication (Weber et al. 1989).

7.3.2 Applications and Uncertainties of Nuclear Data

The nuclear data in the original publications of the ICRP and other advisory bodies were used to calculate effective energies for each of the critical organs identified in ICRP Publication 2. They used very simple assumptions, for example, that all of the energy for most radiations was absorbed in the critical organ. As the system of radiation protection developed, the nuclear data was used for the determination of effective half-lives, chains of daughter radionuclides to be included with the dose resulting from the intake of a parent radionuclide, and the determination of the absorbed fractions and specific effective energy for dose calculations in later publications.

An analysis of the uncertainties associated with dose estimates from ICRP methods was performed at Oak Ridge National Laboratory and it was found that the nuclear data used for most of the ICRP calculations currently in print was generally sound. For very long-lived radionuclides, some significant differences were found between the half-lives reported in Publication 38 and the most current NUBASE values (ICRP 1983). For the most commonly encountered radionuclides of interest in radiation protection or medical applications, the difference between published values and more recent NIST measurements were less than 1%. There were some notable differences in the decay schemes, particularly for ^{80}Sr and for some positron emitting radionuclides where an uncertainty in the yield could lead to quite a difference in the total energy released (Leggett, Eckerman, and Meck 2008).

7.3.3 Most Recent Updates to Nuclear Data

In 2008, data from ICRP Publication 38 were superseded by the new information in ICRP Publication 107. It takes the form of a compact disk that contains information including beta spectra and neutron spectra of over twelve hundred radionuclides from 97 different elements with atomic numbers less than 101. For practitioners of nuclear medicine, 333 of the radionuclides that are included in this report have been published separately in a monograph by the Society of Nuclear Medicine. The database was compiled with an updated version of the EDISTR code (EDISTR04) reading ENSDF files for input. It is currently the most exhaustive database available for nuclear data to be used for radiation protection purposes. More short-lived radionuclides are included, and where space limits restricted some of the radiations that were considered in earlier versions, the full decay schemes are now available. The most recent recommendations incorporate this new source of nuclear data into their dosimetry calculations (ICRP 1991, 1994, 2006, 2007, 2008, 2009).

7.4 SPECIFIC EFFECTIVE ENERGY

In ICRP Publication 30, an entirely new system of internal dose assessment was established. For the most part, this system has been adopted into law and has been in use in the United States since 1990. The committed dose equivalent was defined as the total dose equivalent to an organ or tissue over the 50 years after intake of a radioactive material. The dose equivalent (or the committed dose equivalent) is proportional to the product of the total number of nuclear transformations occurring in the source tissue over the time period of interest and the energy absorbed per gram of target tissue per nuclear transformation of the radionuclide, modified by the appropriate quality factor. In the symbolism used by the ICRP, this becomes:

$$H_{T,50} = kU_S SEE(T \leftarrow S)$$ (7.2)

where, U_S is the total number of spontaneous nuclear transformations of a radionuclide in the source organ (S) over a period of 50 years after the intake and SEE is the Specific Effective Energy imparted per gram of target tissue, from a transformation occurring in a source tissue.

The total number of transformations in a source organ is obtained by integrating (or summing) over time an equation that describes the way material is retained in the organ. This retention equation includes losses due to radioactive decay as well as losses due to the biokinetics of the radionuclide and resulting biological elimination. In the most current ICRP formulations, U_S has units of transformations per Becquerel. The use of unit intakes allows for the calculation of dose conversion coefficients, which can then be multiplied by a known intake to arrive at a total committed dose for a tissue or organ.

The specific effective energy is obtained from a consideration of the radiological characteristics of the radionuclide deposited in the organ. All of these parameters, except for one, may be obtained from a review of the decay scheme of the particular radionuclide. The equation for SEE is:

$$SEE(T \leftarrow S) = \sum_i \frac{Y_i E_i AF(T \leftarrow S)_i Q_i}{M_T}$$ (7.3)

where:

Y_i	= the yield of the radiations of type i per transformation of the radionuclide j;
E_i	= the average or unique energy of radiation i in units of MeV;
$AF(T \leftarrow S)_i$	= the fraction of energy absorbed in target organ T per emission of radiation i in source organ S;
Q_i	= the appropriate quality factor for radiation of type i; and
M_T	= the mass of the target organ in units of grams.

The factor $AF(T \leftarrow S)$ is called the absorbed fraction of energy and is the ratio of the energy absorbed in a target organ (T) to the total energy emitted by the radionuclide in the source organ (S). For alpha and beta radiation, all energy is assumed to be absorbed in the organ containing the radionuclide. In this case, the absorbed fraction in the source organ is equal to 1.0, that is, when the source "irradiates" itself $(S = T)$. The absorbed fraction in all other target organs is assumed to be zero (i.e., $S \neq T$). There are two exceptions to this general rule for alpha and beta radiation. These are special situations in which the source is the skeleton or the contents of the GI tract, and the targets are the cells on bone surfaces and the red marrow or those cells in the walls of the GI tract. These situations will be discussed in more detail when the specific absorbed fractions for different types of radiation are discussed.

For penetrating radiations, generally X-rays and gamma rays, but in later recommendations, energetic beta particles may deposit energy in organs some distance away from the source organ. The absorbed fractions can be calculated by using radiation transport codes to determine the fraction of emitted energy deposited in other target organs. In the original publication of data for penetrating photons (energies greater than 10 keV), the ICRP did not publish data which gives the photon absorbed fractions of energy; instead, they provided tables of specific absorbed fractions (SAF). The SAF is defined as the absorbed fraction divided by the mass of the target organ; in other words

$$\text{SAF} = \frac{AF(T \leftarrow S)}{M_T} \tag{7.4}$$

The committed dose equivalent has units of sieverts per unit intake of activity. Therefore, the quantities on the right-hand side of the equation must be multiplied by a constant to bring both sides into agreement. Therefore, to bring both sides into agreement, it is only necessary to multiply by 1000 g/kg and 1.6 × 10–13 J/MeV. Equation (7.2) becomes:

$$H_{T,50}(T \leftarrow S) = 1.6 \times 10^{-10} U_S \text{SEE}(T \leftarrow S) \tag{7.5}$$

The subscript "$T,50$" on the committed dose equivalent is intended as a reminder that the calculation of the committed dose equivalent is for a particular target organ and the time period of concern is 50 years.

Strangely enough, at about the same time the United States was adopting the recommendations of the ICRP, the organization was promulgating a new set of recommendations. Of the other dozen or so dosimetric quantities mentioned in ICRP Publication 60, two are of particular concern for internal dose assessment purposes. The "committed equivalent dose" is the time-integrated equivalent dose rate in a tissue from time t_0, the age at the time of intake, to age t, and is calculated by the following expression:

$$H_T(t - t_0) = \int_{t_0}^{t} \dot{H}_T(t, t_0) dt \tag{7.6}$$

where

$$\dot{H}_T(t, t_0) = c \sum_{S} \sum_{j} q_{S,j}(t, t_0) \text{SEE}(T \leftarrow S; t)_j \tag{7.7}$$

with c (sometimes C) being a conversion factor to obtain the appropriate units, $q_{S,j}(t,t_0)$ represents the activity of a given radionuclide j present at age t after an intake at age t_0 in organ or tissue S. The term $\text{SEE}(T \leftarrow S;t)_j$ is the specific effective energy deposited in the target organ T (ICRP 1991). For occupational exposures, the integration time is fifty years, and for the dose coefficients provided by the ICRP, the intake is assumed to have occurred at age twenty. This assumption also holds true for environmental exposures of adults. For children, the age at the time of intake is considered and the integration is carried out to age seventy. Since the data tables only consider a single nuclide at a time, the equation is often simplified to the following:

$$H_T(\tau) = \sum_S U_S(\tau)\text{SEE}(T \leftarrow S;t) \tag{7.8}$$

where τ is the integration time after an intake in years. If this is not specified, it should be taken to be 50 years for an adult, or to the age of 70 years for a minor. For occupational exposures, we obtain the expression:

$$H_T(50) = \sum_S U_S(50)\text{SEE}(T \leftarrow S;t) \tag{7.9}$$

On the right side of the equation, $U_S(50)$ represents the number of nuclear transformations in 50 years in a source region, S. This is equivalent to the earlier term, U_S, of ICRP Publication 30. Of course, this term would be calculated using more recent biokinetic models. Similarly, the "specific effective energy," $\text{SEE}(T \leftarrow S)$, is the equivalent dose in the target per transformation in the source region, and is now expressed as:

$$\text{SEE}(T \leftarrow S) = \sum_R \frac{Y_R E_R w_R AF(T \leftarrow S;t)_R}{m_T(t)} \tag{7.10}$$

where:

Y_R	= the yield of the radiations of type R per transformation of the radionuclide;
E_R	= the average or unique energy of radiation R in units of MeV;
$AF(T \leftarrow S;t)_R$	= the fraction of energy absorbed in target organ T per emission of radiation R in source organ S;
w_R	= the appropriate radiation weighting factor for radiation of type R; and
m_T	= the mass of the target organ in units of grams.

From the above it can be seen that primarily the radiation-weighting factor has been substituted for Q in the earlier form of the expression and age dependence is now included (compare with Equation (7.3) above). In many ICRP publications, dose conversion coefficients are provided for the committed equivalent doses in the principal tissues for a unit intake of radionuclide. The term $h_T(\tau)$ is used to denote a dose conversion coefficient.

The counterpart to the effective dose for internal exposures is the committed effective dose, which is defined as follows (ICRP 1991):

$$E(t-t_0) = \sum_T w_T H_T(t-t_0) + w_{\text{Remainder}} H_{\text{Remainder}}(t-t_0) \tag{7.11}$$

Again, in the case of the tabulated values, this can be simplified to:

$$E(\tau) = \sum_T w_T H_T(\tau) = \sum_T w_T \sum_S U_S(\tau) \text{SEE}(T \leftarrow S;t) \tag{7.12}$$

And the expression for the 50-year committed effective dose for workers is given as:

$$E(50) = \sum_T w_T H_T(50) = \sum_T w_T \sum_S U_S(50) \text{SEE}(T \leftarrow S;t) \tag{7.13}$$

In a number of the latest ICRP publications, $e(50)$ is the dose coefficient that represents the occupational committed effective dose over 50 years due to 1 Bq of intake of a particular radionuclide (ICRP 1995). For environmental exposures, the committed effective dose is computed to age 70, which is designated in the publications as $e(70)$. With these new definitions, the Commission has recommended that in cases in which previous dosimetric quantities need to be combined with the present quantities, no effort should be made to correct old internal dose values. The committed effective dose equivalent and the committed effective dose should be summed directly. The same applies for all the other new quantities and their ICRP Publication 30 counterparts (ICRP 1991).

Although there were changes in the radiation and tissue weighting factors with the publication of ICRP Publication 103, there were no further changes in terminology or notation. Care must be taken to know which particular combination of phantoms, biokinetic models, nuclear data, and weighting factors are used to produce a particular set of recommendations. Some of the confusion may be alleviated in the near future. The ICRP has just started to release its latest round of recommended dose conversion coefficients with specific absorbed fractions calculated using the newest voxelized phantoms and nuclear data (ICRP 2016).

In Publication 60, and most recently in Publication 103, the ICRP has identified additional tissues and organs at risk and have further defined which tissues should be included in the Remainder for effective dose calculations. The most current recommended tissue weighting factors (w_T) are given in Table 7.11 and explained in detail in Section 2.7.3 of Chapter 2.

The addition of new biokinetic models for the respiratory tract and the alimentary tract (see Chapter 7, this volume) necessitated changes in the way the colon and lung effective doses are calculated. For the colon, this is relatively simple using the older gastrointestinal tract model: the tissue-weighting factor should be applied to the mass average of the equivalent dose in the wall of the upper (ULI) and lower large intestine (LLI).

TABLE 7.11   Tissue Weighting Factors[a]

Organ/Tissue	w_T
Bone surface	0.01
Bladder	0.04
Brain	0.01
Breast	0.12
Colon	0.12
Gonads	0.08
Liver	0.04
Lungs	0.12
Esophagus	0.04
Red bone marrow	0.12
Salivary glands	0.01
Skin	0.01
Stomach	0.12
Thyroid	0.04
Remainder[b]	0.12

[a] Adapted from Table B.2 (ICRP 2007).
[b] Adrenals, extrathoracic tissue, gallbladder, heart, kidneys, lymphatic nodes, muscle, oral mucosa, pancreas, prostate (male), small intestine, spleen, thymus, uterus/cervix (female).

As it turns out the relative masses are largely age independent, so the equivalent dose in the colon is simply,

$$H_{\text{Colon}} = 0.57 H_{\text{ULI}} + 0.43 H_{\text{LLI}} \tag{7.14}$$

For the new model, introduced in Publication 100 (ICRP 2006), the ICRP suggests that the equivalent dose be averaged according to the following formula:

$$H_{\text{Colon}} = \frac{\left(m_{rc} h_{rc} + m_{lc} h_{lc} + m_{rs} h_{rs} \right)}{\left(m_{rc} + m_{lc} + m_{rs} \right)} \tag{7.15}$$

where
m designates the mass and
h designates the equivalent dose of a particular compartment and
       the subscripts rc, lc, and rs designate the right, left, and rectosigmoid sections of the colon.

For the new respiratory model, a more complicated scheme is used. The equivalent dose to the extrathoracic region is given by the expression:

$$H_{ET} = H_{ET_1} A_{ET_1} + H_{ET_2} A_{ET_2} + H_{LN_{ET}} \tag{7.16}$$

TABLE 7.12   Weighting Factors for the Partition of Detriment
among Respiratory Tissues[a]

Tissue	A_i
Extrathoracic region (a remainder tissue):	
ET_1	0.001
ET_2	1
LN_{ET}	0.001
Thoracic region (lung):	
BB	0.333
Bb	0.333
AI	0.333
LN_{TH}	0.001

[a] Adapted from Table 31 of ICRP Publication 66 (ICRP 1994).

And for the thoracic region (lungs in all ICRP tables)

$$H_{TH} = H_{BB}A_{BB} + H_{bb}A_{bb} + H_{LN_{TH}}A_{LN_{TH}} \tag{7.17}$$

where H_i is the equivalent dose to a particular region, A_i is the weighting factor for the radiosensitivity of that region, and the subscripts designate the compartments of the model (Table 7.12).

7.5 SPECIFIC ABSORBED FRACTIONS FOR PHOTONS

Treatment of internal photon sources has evolved with the ability of computers to carry out more complex radiation transport calculations and with the availability of increasingly complex anatomical phantoms. In the earliest days of radiation protection, empirical formulas were used to estimate the fraction of energy deposited in a critical organ. Of course, this was an unsatisfactory approach, as it was clearly important to consider the contribution to the dose of surrounding tissues and organs, not just the organ where the photon source resided. The first real attempt to address the "crossfire" problem of internal penetrating radiation emitters was provided in an Appendix to ICRP Publication 23 on Reference Man (ICRP 1975). Radiations from individual radionuclides were not tabulated, but the data for twelve monoenergetic photon sources with energies in the range 0.01 to 4.0 MeV were provided with Specific Absorbed Fractions for a combination of source and target organs and tissues. The values were obtained by Monte Carlo methods with at least 60,000 source photons in the original Reference Man phantom. One major assumption made was that the sources were uniformly distributed in the source organs. For some organs and tissues, the statistics of the estimates exceeded 50%, and the estimate was modified with the use of a buildup factor formula. For the gastrointestinal tract compartments, the absorbed fractions were determined in the walls of the organs and the bladder was treated as solid, that is, no fluid volume. For the red bone marrow, regions of the skeleton were used as targets but not as a source.

With the development of age-dependent phantoms and more complex respiratory tract and alimentary tract models, the specific absorbed fractions were determined in particular locations within some of the compartments of the biokinetic models. In particular, the ICRP recommended specific layers within the epithelial tissues where adult stem cells are known to reside. Since these are the putative origin cells of radiation-induced cancers, determining the doses for these cells should improve the accuracy of the risk estimates. For example, in some respiratory compartments, the energy deposition was determined in a tissue layer that corresponded to the depth in the epithelium for secretory and basal cell nuclei. Similar calculations were done at depths appropriate for the basal cells (esophagus) or crypt cells (small intestine and colon) in various compartments of the alimentary tract. This approach resulted in smaller specific absorbed fractions than those obtained with earlier, cruder methods. Unfortunately, there may still be some overestimate of the risk associated with radiation exposure in these tissues, as it is clear that not all of the cells in the selected target layers actually have the potential to give rise to a neoplasm. There are distinct subpopulations in these tissues that have varying degrees of sensitivity to carcinogens and varying proliferative potentials.

For external exposures, a similar evolution has occurred throughout the development of ICRP recommendations. In the earliest recommendations, the emphasis was on the use of sufficient shielding to reduce the exposure or dose in the air to acceptable levels. In Publication 21, the ICRP first introduced general absorbed fractions for photons in tissue-like phantoms. The phantoms were crude and cuboidal or cylindrical of roughly adult torso dimensions (ICRP 1971). In the late 1980s and early 1990s, more realistic Monte Carlo calculations were being performed and these formed the basis of regulations promulgated by the EPA. Eckerman and others used a hermaphroditic phantom in the calculation of specific absorbed fractions for a range of photon energies for infinite planar sources in soil, semi-infinite clouds in the air, and immersion in water (Eckerman and Ryman 1993). For planar sources in soil, calculations were performed at depths from 0 to 4 mean free paths in soil for photons of select energies ranging between 0.1 and 5.0 MeV. The code ALGAMP was used for these calculations, and organ doses were determined for 24 organs and tissues in the phantom. Similar calculations were performed for air and water submersion and dose equivalent coefficients for the target tissues from ICRP Publication 30 (ICRP 1979). This approach was in keeping with the current regulatory requirements of the United States. These calculations were subsequently used to develop the cancer risk coefficients published in Federal Guidance Report No. 13 (Eckerman et al. 1999), but some of the biokinetic models used were from more contemporaneous ICRP recommendations.

In ICRP Publication 74 (ICRP 1996), more complex phantoms in a variety of geometries were used and the absorbed doses per air kerma, in all the target organs described in ICRP Publication 60 (ICRP 1991), were calculated using a suite of Monte Carlo radiation transport codes. For the most part, the MIRD adult male and female phantoms were used to obtain equivalent dose conversion factors for incident, monoenergetic photons for select energies between 0.01 and 10 MeV.

The ICRP, in its latest set of recommendations that are in preparation or in press, make use of the standardized voxelized Reference Male and Reference Female (ICRP 2009). The Task Group has opted for some loss of resolution due to the size of the voxels in favor of more realistic placement of the organs and tissues in the phantoms. In the latest version, Version 4-2-3-0 (Rogers et al. 2003) is used for the calculation of photon and electron specific absorbed fractions for each of the organs and tissues of interest. The calculations examined energies from 10 keV to 10 MeV and were computed separately for each phantom. The dose conversion coefficients will result from an average of the values obtained in the male and female phantom. Presumably, these voxelized phantoms will also be employed to update the equivalent dose conversion factors for external exposures in the near future.

7.6 SPECIFIC ABSORBED FRACTIONS FOR NEUTRONS

Neutron emitters were generally ignored for internal dose assessment purposes in the early recommendations of the ICRP and other regulatory bodies. However, starting in ICRP Publication 38 (ICRP 1983), the decay schemes used a method to approximate the average energy and yields of radiations accompanying each spontaneous fission for a particular radionuclide. This information was updated in ICRP Publication 107 in which the neutron spectra were expressed by a Watt spectrum (ICRP 2008). The average energies of the neutrons, fission fragments, prompt and delayed gamma rays, and delayed beta particles and their yields could be expressed by relatively simple functions that depended on the average number of neutrons emitted, the branching fraction for spontaneous fission, the A and Z of the parent nuclide, and the parameters of the Watt spectrum.

Fission fragments and most of the beta particles are commonly treated as non-penetrating radiations that can only deposit energy in the source organ or tissue. The largest fraction of the energy from the fission is deposited locally. It is expected that in the new voxelized phantoms, some of the heterogeneity of dose deposition from spontaneous fission sources could be observed in the calculations and the contribution from the penetrating radiations could be more fully accounted for as well.

For exposures from external sources of neutrons, guidance and calculations followed the course that was seen with photons. Real absorbed dose/kerma fractions became available for monoenergetic neutrons in ICRP Publication 51(ICRP 1987). These equivalent dose conversion coefficients were calculated using radiation weighting factors for the incident neutron energies. These coefficients could conceivably be used to calculate the total dose due to a known spectrum of neutrons. The tables provide equivalent dose conversion coefficients for monoenergetic neutron beams incident on an anthropomorphic phantom from a variety of directions. For these calculations, the neutrons had a selection of energies that ranged from 1 eV to 13.5 MeV.

In Publication 74, which superseded Publication 51, a similar set of calculations for photons, neutrons, and electrons over a broader range of energies were performed (ICRP 1996). The neutron energies ranged from 0.001 eV to 180 MeV. Monoenergetic, parallel beams of neutrons were incident on Adult MIRD phantoms from a variety of directions. These calculations included an isotropic case and a rotational case. Although the calculations used idealized, broad parallel beams for most of the calculations, these results still represent some of the most useful data for determining external neutron doses.

7.7 SPECIFIC EFFECTIVE ENERGY AND ABSORBED FRACTIONS FOR CHARGED PARTICLES

From the inception of radiological protection, alpha particles and low-energy beta particles have been considered only an internal threat. In the first set of recommendations, all alpha-emitting radionuclides were considered to deposit all of the energy of the transformation in the critical organ. This deposited energy included the recoil energy of the nucleus after alpha emission. For beta-emitting radionuclides, an empirical formula, based on the maximum beta energy, was used to determine the total amount of energy deposited in the critical organ, and contributions to the dose in nearby organs or tissues were ignored.

In ICRP Publication 30, the same general approach applied (ICRP 1979, 1980, 1981). For alpha and beta radiation, all energy was assumed to be absorbed in the organ containing the radionuclide. In this case, the absorbed fraction in the source organ is equal to 1.0, that is, when the source "irradiates" itself ($S = T$). The absorbed fraction in all other target organs is assumed to be 0 (i.e., $S \neq T$). There are two exceptions to this general rule for alpha and beta radiation. These are special situations in which the source is the bone or the contents of the GI tract, and the targets are cells on bone surfaces and the red marrow, or those cells in the walls of the GI tract.

For bone, a table of average absorbed fractions for non-penetrating radiations was published that divided the types of radiation and their source locations and assigned absorbed fractions based on the target tissues (see Table 7.13). These absorbed fractions are used in the standard Specific Effective Energy equation (discussed previously) to calculate the dose equivalent. For the gastrointestinal tract, all the sources are considered to be in the contents of the gut, a not entirely correct simplifying assumption. The targets of the radiation damage are the stem cells lining the mucosal layer of the inner surface of the different gastrointestinal tract compartments. In order to determine the dose equivalent, an empirical relation for the specific absorbed fraction was developed for non-penetrating radiations.

TABLE 7.13 Recommended Absorbed Fractions for Dosimetry of Radionuclides in Bone

Source Organ	Target Organ	Class of Radionuclide				
		Alpha Emitter Uniform in Volume	Alpha Emitter on Bone Surfaces	Beta Emitter Uniform in Volume	Beta Emitter on Bone Surfaces $\bar{E}_\beta \geq 0.2 Me$	Beta Emitter on Bone Surfaces $\bar{E}_\beta < 0.2 Me$
Trabecular bone	Bone surfaces	0.025	0.25	0.025	0.025	0.25
Cortical bone	Bone surfaces	0.01	0.25	0.015	0.015	0.25
Trabecular bone	Red bone marrow	0.05	0.5	0.35	0.5	0.5
Cortical bone	Red bone marrow	0	0	0	0	0

This equation is:

$$SAF = 0.5 \frac{1}{M_c} v \qquad (7.18)$$

where:

M_c is the mass of the contents of the gastrointestinal tract compartment in question and

v is the penetration factor.

The penetration factor was defined as one for betas, zero for recoil nuclei, and 0.01 for alpha particles and fission fragments.

In the additional Parts and Supplements for ICRP Publication 30, calculations began to take into account some of the doses to nearby organs and tissues due to bremsstrahlung from energetic beta particles and for energetic beta particles that could reach adjacent organs (ICRP 1979, 1980, 1981). When the new respiratory tract and alimentary tract models were introduced, the use of target cell layers in the lung and digestive tract models at particular depths led to most alpha particles and some low-energy beta particles failing to reach those target depths. Therefore, for the stomach and some other compartments, the absorbed fraction for alpha particles is again zero. This result is less of a problem in the lung, where most of the target cell layers are within 30 micrometers of the surface.

Variations in age for those cases resulted in calculated specific absorbed fractions or absorbed fractions, which were generally higher, as the target layers were closer to the surface for younger models. With the new voxelized models, the problem of resolution leads to uncertainty in the calculations. Most voxel sizes are larger than the dimensions of the target layers, and so precise energy calculations will be unlikely. This is particularly true for skeletal and lung compartments where the resolution is not sufficient to depict the different areas of sources and targets.

For external exposures, the general practice has been to treat all beta particles and energetic electrons as equal and the other types of charged particle are not considered a general protection concern, except for the case of astronauts. The first extensive calculation for external energetic electrons and protons were published in ICRP Publication 21 (ICRP 1971). In Publication 51, there were data for pions and quality factors with respect to LET for a variety of ions. The first really useful dose conversion coefficients for monoenergetic electrons with selected energies between 70 keV and 10 MeV are found in Publication 74. The dose conversion coefficients shown in the tables were calculated for three different skin thicknesses from 0.07 to 10 mm (ICRP 1996).

In the most recent update to all the dose conversion factors for external beams, the widest set of energies and particle types were examined. The authors used an ensemble of radiation transport codes to calculate fluence to dose conversion coefficients for both effective dose and organ absorbed doses. These coefficients were calculated using the Reference Adult Male and Reference Adult Female voxelized phantoms. The incident radiations and the energy ranges considered were external beams of monoenergetic photons, electrons, positrons, neutrons, muons, protons all up to energies of 10 GeV, pions (negative/positive) of 1 MeV–200 GeV and helium ions of 1 MeV/u–100 GeV/u. Once again, the phantoms were exposed to broad parallel beams of incident radiation from multiple directions. Fluence to effective dose conversion coefficients was derived from the organ dose conversion coefficients, the radiation weighting factor w_r, and the tissue-weighting factor w_t, according to the recommendations of ICRP Publication 103 (ICRP 2007).

Over the years, a number of investigators have used MRI or CT images captured from volunteers to produce "voxelized" data sets for geometry inputs into radiation transport codes. These all had limitations, mainly due to the fact that the individuals imaged did not correspond well to the Reference Persons, which were the bases of dose calculations. In an attempt to rectify this limitation, two sets of data (originally named "Golem" and "Laura") were used as templates to produce mathematical phantoms for a Reference Male and a Reference Female. Scaling was employed to obtain the desired dimensions, and for most tissues and organs a more realistic spatial arrangement of the organs was obtained. However, the voxel size limited the resolution of the phantom in some key areas.

The location of source and target tissues in the skeleton could not be fully rendered. Similarly, the smaller airways could not be segmented and the skin epithelium was slightly more massive in the phantoms. The gallbladder was larger in both the male and female than the given reference values and, where there were excess voxels, these were treated as adipose tissue. These phantoms are now in use by the ICRP in calculations of absorbed fractions and dose coefficients for more recent publications (ICRP 2009).

Data files for the two phantoms can be obtained online as a supplement to Publication 110 at http://www.icrp.org/publication.asp?id=ICRP%20Publication%20110. More details can be found in Chapter 5.

REFERENCES

Berger, Martin J. 1968. "MIRD Pamphlet No. 2: Energy Deposition in Water by Photons from Point Isotopic Sources." *Journal of Nuclear Medicine, Supplement* No.1:15–25.

Bouchet, L. G., W. E. Bolch, H. P. Blanco, B. W. Wessels, J. A. Siegel, D. A. Rajon, I. Clairand, and G. Sgouros. 2002. "MIRD Pamphlet No. 19: Absorbed Fractions and Radionuclide S Values for Six Age-Dependent Multiregion Models of the Kidney." *Journal of Nuclear Medicine* 44 (7):1113–1147.

Bouchet, L. G., W. E. Bolch, D. A. Weber, H. L. Atkins, and J. W. Poston Sr. 1999. "MIRD Pamphlet No. 15: Radionuclide S Values in a Revised Dosimetric Model of the Adult Head and Brain." *Journal of Nuclear Medicine* 40 (3):62S–101S.

Brownell, G. L., W. H. Ellett, and A. R. Reddy. 1968. "Absorbed Fractions for Photon Dosimetry." *Journal of Nuclear Medicine* 9 (Suppl 1):29–39.

Coffey, J. L., M. Cristy, and G. G. Warner. 1981. "MIRD Pamphlet No. 13: Specific Absorbed Fractions for Photon Sources Uniformly Distributed in the Heart Chambers and Heart Wall of a Heterogeneous Phantom." *Journal of Nuclear Medicine* 22 (1):65–71.

Cook, M. J. 1960. " Bibliography for Biological, Mathematical and Physical Data." *Health Physics* 3 (1): 235–378.

Dillman, L. T. 1969. "MIRD Pamphlet No. 4: Radionuclide Decay Schemes and Nuclear Parameters for Use in Radiation-Dose Evaluation." *Journal of Nuclear Medicine* 10 (Suppl. 2):1–32.

Dillman, L. T. 1980. *EDISTR-a Computer Program to Obtain a Nuclear Decay Database for Radiation Dosimetry* (ORNL/TM-6689). Oak Ridge, TN: Oak Ridge National Laboratory.

Eckerman, K. F., R. W. Leggett, C. B. Nelson, J. S. Pushkin, and A. C. B. Richardson. 1999. *Federal Guidance Report No. 13: Cancer Risk Coefficients for Environmental Exposure to Radionuclides.* Washington, DC: U.S. Environmental Protection Agency.

Eckerman, K. F., and J. C. Ryman. 1993. *Federal Guidance Report No. 12: External Exposure to Radionuclides in Air, Water, and Soil.* Washington, DC: U.S. Environmental Protection Agency.

Ellett, W. H., and R. M. Humes. 1971. "MIRD Pamphlet No. 8: Absorbed Fractions for Small Volumes Containing Photon-Emitting Radioactivity." *Journal of Nuclear Medicine* 12 (Suppl. 5): 35–32.

Hwang, J. M. L., R. L. Shoup, G. G. Warner, and J. W. Poston. 1976a. *Mathematical Description of a Newborn Human for Use in Dosimetry Calculations* (ORNL/TM-5453). Oak Ridge, TN: Oak Ridge National Laboratory.

Hwang, J. M. L., R. L. Shoup, G. G. Warner, and J. W. Poston. 1976b. *Mathematical Descriptions of a One- and Five-Year Old Child for Use in Dosimetry Calculations* (ORNL/TM-5293). Oak Ridge, TN: Oak Ridge National Laboratory.

International Commission on Radiological Protection. 1960a. *ICRP Publication 2: Permissible Dose for Internal Radiation.* New York, NY: Pergamon Press.

International Commission on Radiological Protection. 1960b. *ICRP Publication 6: Recommendations of the ICRP.* Oxford, U.K.: Pergamon Press.

International Commission on Radiological Protection. 1968. *ICRP Publication 10: Evaluation of Radiation Doses to Body Tissues from Internal Contamination Due to Occupational Exposure.* Oxford, U.K.: Pergamon Press.

International Commission on Radiological Protection. 1975. ICRP Publication 23: Report on the Task Group on Reference Man. *Ann ICRP.* 23 (1).

International Commission on Radiological Protection. 1971. *ICRP Publication 21: Data for Protection against Ionizing from External Sources: Supplement to ICRP Publication 15.* Oxford, U.K.: Pergamon Press.

International Commission on Radiological Protection. 1979. "ICRP Publication 30 (Part 1): Limits for Intakes of Radionuclides by Workers." *Ann ICRP* 2 (3–4).

International Commission on Radiological Protection. 1980. "ICRP Publication 30 (Part 2): Limits for Intakes of Radionuclides by Workers." *Ann ICRP* 4 (3–4).

International Commission on Radiological Protection. 1981. "ICRP Publication 30 (Part 3): Limits for Intakes of Radionuclides by Workers." *Ann ICRP* 6 (2–3).

International Commission on Radiological Protection. 1983. "ICRP Publication 38: Radionuclide Transformations - Energy and Intensity of Emissions." *Ann ICRP* (11–13):1–1250.

International Commission on Radiological Protection. 1987. "ICRP Publication 51: Data for Use in Protection from External Radiation." *Ann ICRP* 17 (2–3).

International Commission on Radiological Protection. 1991. "ICRP Publication 60: 1990 Recommendations of the International Commission on Radiological Protection." *Ann ICRP* 21 (1–3).

International Commission on Radiological Protection. 1994. "ICRP Publication 66: Human Respiratory Tract Model for Radiological Protection." *Ann ICRP* 24 (1–3):1–482.

International Commission on Radiological Protection. 1995. "ICRP Publication 72: Age-Dependent Doses to Members of the Public from Intake of Radionuclides - Part 5 Compilation of Ingestion and Inhalation Coefficients." *Ann ICRP* 26 (1).

International Commission on Radiological Protection. 1996. "ICRP Publication 74: Conversion Coefficients for Use in Radiological Protection against External Radiation." *Ann ICRP* 26 (3–4).

International Commission on Radiological Protection. 2006. "ICRP Publication 100: Human Alimentary Tract Model for Radiological Protection." *Ann ICRP* 36 (1–2):1–366.

International Commission on Radiological Protection. 2007. "ICRP Publication 103: The 2007 Recommendations of the International Commission on Radiological Protection." *Ann ICRP* 37 (2–4).

International Commission on Radiological Protection. 2008. "ICRP Publication 107: Nuclear Decay Data for Dosimetric Calculations." *Ann ICRP* 38 (3):1–26.

International Commission on Radiological Protection. 2009. "ICRP Publication 110: Adult Reference Computational Phantoms." *Ann ICRP* 39 (2):1–165.

International Commission on Radiological Protection. 2016. "ICRP Publication 133: The ICRP Computational Framework for Internal Dose Assessment for Reference Adults: Specific Absorbed Fractions." *Ann ICRP* 45 (2):1–74.

Jones, R. M., J. W. Poston, J. L. Hwang, T. D. Jones, and G. G. Warner. 1976. *The Development and Use of a Fifteen Year Old Equivalent Mathematical Phantom for Internal Dose Calculations* (ORNL/TM-5278). Oak Ridge, TN: Oak Ridge National Laboratory.

Leggett, R. W., K. F. Eckerman, and R. A. Meck. 2008. *Reliability of Current Biokinetic and Dosimetric Models for Radionuclides: A Pilot Study* (ORNL/TM-2008/131). Oak Ridge, TN: Oak Ridge National Laboratory.

Loevinger, R., and M. Berman. 1968. "MIRD Pamphlet No. 1: A Schema for Absorbed-Dose Calculations for Biologically-Distributed Radionuclides." *Journal of Nuclear Medicine* 9 (Suppl. 1):7–14.

Marinelli, L. D. 1942. "Dosage Determination with Radioactive Isotopes I." *American Journal of Roentgenology and Radium Therapy* 47 (210).

Marinelli, L. D., E. H. Quimby, and G. J. Hine. 1948. "Dosage Determination with Radioactive Isotopes II: Practical Considerations in Therapy and Protection." *American Journal of Roentgenology and Radium Therapy.* 59 (260).

Patel, J. S. 1988. "A Revised Model of Kidney for Medical Internal Radiation Dose." M.S., Department of Nuclear Engineering, Texas A&M University.

Poston, J. W. 1983. "Reference Man: A System for Internal Dose Calculation." In J. E. Till and H. R. Meyer (eds) *Radiological Assessment: A Textbook on Environmental Dose Analysis.* Washington, DC: U.S. Nuclear Regulatory Commission.

Rogers, D. W. O., I. Kawrakow, J. P. Seuntjens, B. R. B. Walters, and E. Mainegra-Hing. 2003. NRC User Codes for Egsnrc. In *NRCC Report PIRS-702 (Rev. B)*.

Snyder, W. S. 1966. "The Standard Man in Relation to Internal Radiation Dose Concepts." *American Industrial Hygiene Association Journal* 27 (6):539–545.

Snyder, W. S., M. R. Ford, G. G. Warner, and H. L. Fisher Jr. 1978. "MIRD Pamphlet No. 5: Revised. Estimates of Absorbed Fractions for Monoenergetic Photon Sources Uniformly Distributed in Various Organs of a Heterogeneous Phantom." *Journal of Nuclear Medicine* 19.

Snyder, W. S., M. R. Ford, G. G. Warner, and S. B. Watson. 1974. *A Tabulation of Dose Equivalent Per Microcurie-Day for Source and Target Organs of an Adult for Various Radionuclides.* Oak Ridge, TN: Oak Ridge National Laboratory.

Taylor, Lauriston S. 1984. *Tripartite Conferences on Radiation Protection: Canada, United Kingdom, United States (1949–1953).* Washington, DC: National Bureau of Standards.

Thomas, S. R., M. G. Stabin, C. T. Chen, and R. C. Samaratunga. 1992. "MIRD Pamphlet No. 14: A Dynamic Urinary Bladder Model for Radiation Dose Calculations." *Journal of Nuclear Medicine* 33 (5):783–802.

Watson, S. B., and M. R. Ford. 1980. *User's Manual to the ICRP Code: A Series of Computer Programs to Perform Dosimetric Calculations for the ICRP Committee 2 Report* (ORNL/TM-6980). Oak Ridge, TN: Oak Ridge National Laboratory.

Weber, D. A., K. E. Eckerman, L. T. Dillman, and Ryman. J. C. 1989. *MIRD: Radionuclide Data and Decay Schemes.* New York, NY: Society of Nuclear Medicine.

CHAPTER 8

Dose Coefficients

Nolan E. Hertel and Derek Jokisch

CONTENTS

8.1 Computation of Dose Coefficients External Irradiation 337
    8.1.1 Transport Methods 339
        8.1.1.1 Transport Equation 339
    8.1.2 Discrete Ordinates Method 342
    8.1.3 Adjoint Transport Equation 344
    8.1.4 Monte Carlo Methods 347
        8.1.4.1 The Kerma Approximation versus Tracking of Charged Particles 348
        8.1.4.2 Variance Reduction Techniques 350
        8.1.4.3 Reciprocity Method 351
    8.1.5 External Irradiation Dose Coefficients 352
        8.1.5.1 Photons 352
        8.1.5.2 Neutrons 354
8.2 Computation of Dose Coefficients for Internal Emitters 358
    8.2.1 Description of Quantities Used in Internal Dose Calculations 358
        8.2.1.1 Effective Dose 358
        8.2.1.2 Equivalent Dose 359
        8.2.1.3 S-Coefficient 360
        8.2.1.4 Number of Nuclear Transformations 361
        8.2.1.5 Committed Equivalent Dose Coefficient 362
        8.2.1.6 Committed Effective Dose Coefficient 363
        8.2.1.7 Specific Absorbed Fraction 363
    8.2.2 Calculation Details 364
        8.2.2.1 Accounting for Activity in Systemic "Other" 364
        8.2.2.2 Interpolation Techniques 365
        8.2.2.3 Integration Techniques 366
        8.2.2.4 Decay Chain Considerations 366
        8.2.2.5 Target Tissues with Constituents 366
    8.2.3 Derived Quantities 367
        8.2.3.1 Annual Limit on Intake 367
        8.2.3.2 Derived Air Concentration 368

8.3   Computation of Dose Coefficients for External Environmental Radiation Fields   368
       8.3.1   Dose Rate Coefficients for Contaminated Soil   369
               8.3.1.1   Geometry and Compositions   371
               8.3.1.2   Variance Reduction and Other Speedups   373
               8.3.1.3   Dose Rate Coefficients for Electrons in Soil   374
       8.3.2   Submersion in Contaminated Air   374
               8.3.2.1   Submersion Dose Due to Photons   374
               8.3.2.2   Organ Dose from Electrons in Air   375
       8.3.3   Immersion in Contaminated Water   376
               8.3.3.1   Photon Sources in Water   376
               8.3.3.2   Electron Sources in Water   377
       8.3.4   Example of Environmental Dose Rate Coefficients   377
               8.3.4.1   Computational Phantoms   378
               8.3.4.2   Calculating Absorbed Dose for Contaminated Soil   378
               8.3.4.3   Calculation Absorbed Dose for Submersion in Contaminated Air   383
               8.3.4.4   Calculation Absorbed Dose for Submersion in Contaminated
                         Water   384
               8.3.4.5   Dose Rate Coefficients for Radionuclides   384
8.4   Sample Calculations Based on Dose Coefficients   384
       8.4.1   Ingestion of ^{90}Sr   384
       8.4.2   Ingestion of ^{131}I   385
       8.4.3   Ingestion of ^{137}Cs   386
       8.4.4   Inhalation of ^{210}Po   387
       8.4.5   Inhalation of ^{239}Pu   388
References   390

SINCE STATE-OF-THE-ART DOSE COEFFICIENTS are discussed in this chapter, some introductory and historical information will be beneficial to the reader. The generation of reference dose coefficients, or, in the case of dose reconstruction, the generation of individual-specific dose coefficients are addressed by computational dosimetry. Siebert and Thomas (1997) define computation dosimetry as the sub-discipline of computation physics that is devoted to radiation metrology. They relate it to the process of connecting and ordering known data, theories, and models to create new data. The "new" data, in this case, dose coefficients, can be used to generate reference values for regulations to limit dose and provide insights into the dose occurring due to irradiation of a target system, in this work, the human body or an adequate representation of it. A large set of data is required, including radionuclide decay data and emissions, as well as neutron, gamma ray, and other particle interaction data. A complete theory of radiation dosimetry would predict the reaction of organs and cells in a living target when irradiated by ionizing radiation and not just the physical deposition of dose. Such a theory does not fully exist. However, computational dosimetry is based on sound principles.

In the present, the Monte Carlo method is the principal workhorse for performing the radiation transport and energy deposition required in computational dosimetry and very sophisticated models of the human body are available and becoming more detailed every day. However, the use of both computational and anthropomorphic models always contains at least some approximations to reality, particularly if the dose to a specific individual is desired. The Monte Carlo method finds widespread use, as it is the state-of-the-art approach to track the secondary charged particles from neutral particle interactions and primary charged particles as they deposit energy in the media of interest through Coulomb interactions.

This chapter will be divided into four major sections. The first section will address the computation of dose coefficients for the external irradiation of the human body. These coefficients are used, among other things, in operational dose determinations and shielding analyses. The second section will address the computation of dose coefficients for internal emitters. The third section will present dose coefficient methods and results for external irradiation by environmental sources of radiation, namely by submersion in contaminated air, immersion in contaminated water, and irradiation by contaminated soil. These coefficients find use in the analyses of releases of radionuclides into the environment and are useful in accident analyses, emergency response, and radiological assessments. The fourth section demonstrates the use of the internal emitter dose coefficients and their application in computing dose.

8.1 COMPUTATION OF DOSE COEFFICIENTS EXTERNAL IRRADIATION

Following the approach of Shultis and Faw (2000), to address the use of external dose coefficients (traditionally referred to fluence-to-dose conversion coefficients), in practice, the most complete computation of a dosimetric quantity (absorbed dose, effective dose, etc.) represented by R at some location in a nuclear facility or in a contaminated environment is given by the following integral over the target volume V by

$$R = \int_0^\infty dE \int_{4\pi} d\Omega \int_V dV \, \mathcal{R}(\mathbf{r}, E, \hat{\Omega}) \Phi(\mathbf{r}, E, \hat{\Omega}) \tag{8.1}$$

The function $\mathcal{R}(\mathbf{r}, E, \hat{\Omega})$ is the contribution to the quantity being computed in the target (human body, dose-measuring instrument, etc.) to a particle of energy E which is traveling in the direction $\hat{\Omega}$ at the location defined by the vector $\mathbf{r}$ per differential length of travel. The quantity $\Phi(\mathbf{r}, E, \hat{\Omega})$ is commonly referred to as the angular- and energy-dependent fluence, and it would be known throughout the target volume to evaluate the integral. In the International Commission on Radiation Units and Measurements (ICRU) recommended terminology, it is called the particle radiance and would be represented by $\Phi_{\Omega, E}$ and defined as $\Phi_{\Omega, E} = \dfrac{d\Phi}{dE d\Omega}$. In this discussion, the more common terminology and notation will be used.

For most applications, a phantom is too complicated to be placed directly into models of nuclear systems, and so on. One step toward simplification of the required computation would be performing the appropriate surface integral using a surface response function $\mathcal{R}_s(\mathbf{r}_s, E, \hat{\Omega}_s)$. In the section on the adjoint method (Section 8.1.3) and in Section 8.3, variations of this approach are referenced, and their use discussed. This approach would be the equivalent to solving the following integral to obtain the dosimetry quantity value.

$$R = \int_0^\infty dE_s \int_{4\pi} d\Omega_s \int_s dV \mathcal{R}_s(\mathbf{r}_s, E_s, \hat{\Omega}_s) \Phi_s(\mathbf{r}_s, E_s, \hat{\Omega}_s) \tag{8.2}$$

In many cases where a dose quantity is to be computed, particularly when the dose outside a shield is desired, an idealized dose coefficient is used. The computation is reduced to the use of the energy-dependent fluence at a point of interest, and an energy-dependent conversion coefficient is applied which was computed independently for idealized radiation field geometries, namely,

$$R = \int_0^\infty dE \mathcal{R}(E)\Phi(\mathbf{r}_0, E) \tag{8.3}$$

where $\Phi(\mathbf{r}_0, E)$ is the energy-dependent fluence at the location of interest defined by the vector $\mathbf{r}_0$. This approach relies on the use of dose coefficients $\mathcal{R}(E)$ which were computed by groups such as the International Commission on Radiological Protection (ICRP) for what are referred to as reference geometries. In general, the use of such dose coefficients can be judiciously used to compute doses in a variety of applications in practice so that at worst a conservative overestimate of the actual dose at a location is obtained. Dose coefficients were recommended in ICRP Publication 116 (ICRP 2010) for the six reference fields shown in Figure 8.1. The ICRP recommended reference dose coefficients are listed in from left to right: antero-posterior, postero-anterior, left lateral, right lateral, rotational, and isotropic. The rightmost set of idealized geometries are the caudal and cranial irradiation geometries of Veinot, Eckerman, and Hertel (2015) and Veinot et al. (2016). The commonly used abbreviations for the idealized radiation field geometries are given as part of the figure. Note that all the coefficients are based on parallel beam irradiation of the phantoms, except for the isotropic radiation field which is based on the irradiation of the individual from all directions. The rotational geometry (ROT) assumes that the phantom is rotated in

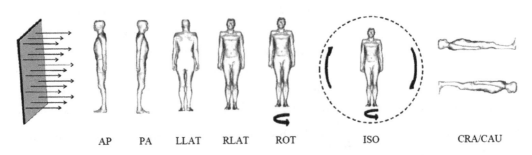

| AP | PA | LLAT | RLAT | ROT | ISO | CRA/CAU |

FIGURE 8.1 Idealized irradiation geometries. The first six irradiation fields are those used in ICRP Publication 116 and the caudal and cranial irradiations are shown at the far right (Veinot et al. 2016).

a parallel beam. The caudal and cranial geometries address the need for dose coefficients when the radiation source is overhead or below the feet of the phantom.

The discussion in this section will cover some basic concepts in the computation of external dose coefficients but will by no means be an exhaustive treatment. Here, the techniques are more important than presenting the values since the values will likely change over time. In recent years, the computational sophistication and increasing detail/complexity in the anthropomorphic phantoms used in the calculation of dose coefficient tracks closely with the development of high-speed, large memory computers and, more recently, with the development of parallelized high-performance computing clusters. Approximations used in dose coefficients will be briefly enumerated. The Monte Carlo method will be qualitatively reviewed and a few samples of state-of-the art external dose coefficients will be presented. The current anthropomorphic phantoms were previously discussed in other chapters of this book, so they are only discussed as is necessary to assist in providing insight into the origin of reported dose coefficients.

8.1.1 Transport Methods

The starting point of computational dosimetry consists of emitting the particles from sources. This is true for both internal and external dosimetry modeling. In the former case, the sources are radionuclides deposited in various organs and tissues of the body and, in the latter case, the sources can vary from radiation fields generated at nuclear and radiation-generating facilities to fields due to environmental contamination. To state the obvious, external dosimetry refers to the radiation emanating from sources outside the person. For external computational dosimetry, as previously mentioned, standard idealized irradiation geometries are used to generate reference dose coefficients. When doses need to be calculated in different fields, the practice is to compute the fluence at a location and fold it with a reference dose coefficient. In some cases, the actual field may not have the angular dependence of the field used to generate the dose coefficients. However, judicious selection of the coefficients used can provide acceptable dose values. So as a first stage in the application of dose coefficients, the emitted particles must be transported to, or possibly measured at, a location of interest from the source.

8.1.1.1 Transport Equation

The following discussion is rather brief but is included for completeness. There are a large number of books and papers that address the derivation of the transport equation and the approximations used to solve them. The interested reader is referred to discussions in Shultis and Faw (2000), as well as Schaeffer (1973), for the discussion of the transport equation in the solution of fixed source and shielding problems. These are likely the most relevant to the present discussions, although there is a plethora of textbooks and references books that address the derivation and solution of the transport equation through analytical and numerical techniques.

This section follows a pedestrian approach compared to the more detailed approach used by Schaeffer (1973). The reader interested in a more detailed discussion, including the accompanying derivations, is directed to that reference. The Boltzmann transport equation was formulated to calculate the coefficient of self-diffusion for a gas in which molecules are assumed to scatter as elastic spheres. This is equivalent to the transport of radiation, particularly neutral particle transport, with the exception that, for transport in a medium, particle–particle collisions are ignored due to their rarity compared to particle interactions with atoms in the medium through which they are transported.

The equation is largely an accounting statement for the particle balance in a given increment of phase space. In its full-blown description, the angular- and energy-dependent fluence rate is a function of seven independent variables (three that define the location in space, two that define the direction of the particle motion, one that corresponds to the energy of the particle, and one that describes time dependence). In the present discussion, the time dependence for external and internal irradiation of the body need not be described to obtain dose coefficients, as it can be brought in to the dose calculation subsequently. The differential fluence obtained from the solution of the transport equation completely describes the radiation field throughout a system subject to the boundary conditions used in the solution process.

The quantity $\Phi(\bar{r}, E, \hat{\Omega})d\Omega dE$, which has units of particles per unit area, is the fluence of particles at the location defined by vector $\bar{r}$ which travels in the solid angle $d\Omega$ about direction $\hat{\Omega}$ with energies between E and $E + dE$. Performing a balance on a differential volume in six-dimensional phase space $dVd\Omega dE$, illustrated in Figure 8.2, leads to the linear Boltzmann transport equation. It is linear since it incorporates no term for particle–particle interactions but accounts only for particle interactions in the medium in which they are traveling. Particles can be added and subtracted to the phase space volume $dVd\Omega dE$ through the following processes:

- Additions

  o Source particles can be born in dV into the appropriate energy and direction intervals.

  o Particles can flow into dV with the appropriate energy and directions from adjacent spatial regions.

  o Particles with other directions and/or energies can undergo interaction in dV and be transferred into the appropriate energy and direction.

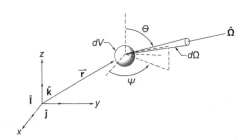

FIGURE 8.2   Volume element in phase space after Schaeffer 1973.

- Losses

  o Particles can be absorbed in dV or undergo interactions that change their energy and/or direction.

  o Particles can flow out of dV into adjacent spatial regions.

The flow in and out of the volume dV is combined into one term called the net leakage and the particle balance for the volume in phase space can be stated in words as:

$$\text{Net Leakage} + \text{Interactions} = \text{Inscattering} + \text{Source}$$

Here, "interactions" means that the particles in $d\Omega dE$ react by multiple processes and are removed from $d\Omega dE$. The three-dimensional form of the linearized Boltzmann transport equation is shown below. Again "linearized" refers to the fact that no accounting is taken of the low probability of particle–particle interactions compared to particle–atom interactions since the densities of the media in which the particles are being transported are much higher than the densities of the transported particles.

$$\hat{\Omega} \cdot \vec{\nabla}\Phi(\vec{r}, E, \hat{\Omega}) + \mu(\vec{r}, E)\Phi(\vec{r}, E, \hat{\Omega}) = S(\vec{r}, E, \hat{\Omega})$$

$$+ \int_0^\infty dE' \int_{4\pi} d\Omega' \mu_s(\vec{r}, E' \to E, \hat{\Omega}' \to \hat{\Omega})\Phi(\vec{r}, E', \hat{\Omega}') \tag{8.4}$$

where

$\mu(\vec{r}, E)$	= interaction coefficient at position $\vec{r}$ at energy E;
$S(\vec{r}, E, \hat{\Omega})$	= emission of particles of energy E in direction $\hat{\Omega}$ from a source at position $\vec{r}$;
$\mu_s(\vec{r}, E' \to E, \hat{\Omega}' \to \hat{\Omega})$	= interaction coefficient at position $\vec{r}$ for the transfer of a particle with energy E' and direction $\hat{\Omega}'$ to energy E and direction $\hat{\Omega}$.

In its simplest form, it is the scattering interaction transfer coefficient, but it can take on more complicated forms, for example, to include two neutrons being produced in an (n,2n) reaction.

When solved, the transport equation provides a complete mathematical description of neutral particle radiation fields, that is, photons and neutrons, throughout a system. Its use in the transport of charged particles is complicated by their continuous slowing down (energy loss) by Coulomb reactions with the medium in which they are being transported.

The analytical solution of the equation can only be done for very simple and largely idealized cases. Often these problems are used to benchmark the accuracy of codes which employ numerical solution. Approximate solutions can be obtained either by making simplifying assumptions which results in an equation that can be solved analytically, for example, diffusion theory (although even diffusion theory equations cannot always be solved analytically), or by using numerical methods to solve the equation or a simplified form of it. Given today's high-speed computers with large memory capabilities, rather large systems can be modeled and solved, for example, an entire nuclear power plant.

The numerical solutions of the transport equation include the following steps in one form or another:

- The equation is normally not solved in a continuous energy manner and is simplified to be solved as a coupled set of equations over energy ranges. The first step is generating an energy multigroup formulation of the equation and the accompanying cross sections (interaction coefficients).

- Selecting a method to represent the differential scattering cross sections and the angular dependence of the fluence.

- Approximating the equations by relating the spatial derivatives of the angular and energy fluence as functions of $\bar{r}$ at points on a spatial mesh.

- Solving the resulting system of algebraic equations, usually by an iterative method until the multigroup fluences converge to values which solve the set of equations.

Here again, the interested reader can find the description of numerical methods, such as the straight-ahead method, the method of moments, the integral transport method, and the diffusion approximation in a number of textbooks and reference books, for example, Lewis and Miller 1984.

8.1.2 Discrete Ordinates Method

For many years, the available photon dose conversion coefficients found in the American National Standard ANSI/ANS 6.1.1-1977 entitled *Neutron and gamma-ray flux-to-dose-rate factors* (ANSI 1977) were used extensively in shielding and design applications. The majority of the values in that standard were computed using the discrete ordinates method to solve the transport equation in a slab phantom. A brief review of that method and the results is presented by Claiborne and Trubey (1970). First, the equation is reduced to its one-dimensional version and integrated over each of the G energy group ranges to form G equations that are coupled. Since, in one dimension, the angular fluence is azimuthally symmetric and, therefore, the equation is averaged over all azimuthal angles to obtain the following equation

$$\omega \frac{\partial \Phi_g(z,\omega)}{\partial z} + \mu_g \Phi_g(z,\omega)$$

$$= \sum_{g'=1}^{G} \int_{-1}^{1} d\omega' \mu_{g' \to g}(z, \omega' \to \omega) \Phi_{g'}(z, \omega') + S_g(z, \omega); \ g = 1, 2, \ldots, G \tag{8.5}$$

where

$\omega = \cos\theta$ and θ is the polar angle with respect to the positive z-axis.

The equation yields the solution for the multigroup angular fluence as a function of distance into the slab. The angular fluence in each energy group is defined by $\Phi_g = \int_{E_{g+1}}^{E_g} dE\Phi(z,E,\omega)$. By convention, the highest energy group is $g = 1$ and the lowest energy group is $g = G$. As in the three-dimensional and the continuous energy version of the transport equation, the interaction coefficient $\mu_{g'\to g}(z,\omega'\to\omega)$ represents the probability of in-scatter to group g and direction ω from another energy group g' and particle direction ω'. This transfer term is represented using a truncated Legendre polynomial expansion.

In this method, the angular variable ω is represented by a finite number of discrete directions along which the particles can stream. The spatial variables are then discretized to form intervals or meshes, finite differencing of the derivatives is performed, the integral is evaluated using a quadrature, and the resulting set of equations are algebraic equations which can be solved. The reader is directed to Shultis and Faw (2000) for the derivation of the discrete ordinates equations in one dimension.

In the work of Claiborne and Trubey (1970), a one-dimensional, tissue-equivalent slab, which was 30 cm thick and consisted of the 11-element composition of standard man at unit density, was employed for computation of photon dose coefficients using the method of discrete ordinates. In this case, the geometry is that shown in Figure 8.3 with the slab being divided into 36 increments: 8 increments of 0.25 cm at the source side of the slab followed by 28 1 cm increments. Sixteen discrete directions were employed, referred as an S_{16} quadrature. The cross-section expansion was included up through the 5th order term, referred to as a P_5 expansion. Although not specifically stated in the article, the tissue kerma was folded by the multigroup fluences computed in each interval to obtain the average dose coefficient in each interval.

Two sets of 23 energy groups were used: one to cover the energy range from 16 MeV down to 0.01, and the other to cover the lower energy range from 2.72 MeV down to 0.01 MeV allowing for a more detailed circulation at lower energies. Computations were done for both an isotropic fluence on the surface and a parallel beam hitting the surface. The dose coefficients for a given energy group were then obtained by taking the value in the spatial interval having the highest absorbed dose and normalizing it to a unit fluence incident on the surface of the slab phantom. This was common practice for obtaining dose

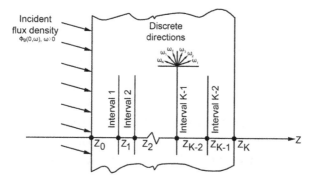

FIGURE 8.3   Discrete ordinates method in a slab.

coefficients for simple geometric phantoms (cylinder, slabs, spheres) before the advent of the ICRU operational quantities and anthropomorphic phantoms. It is frequently referred to as the Maximum Absorbed Dose Equivalent (MADE). These values were used in ANSI/ANS6.1.1-1977 (ANSI 1977) and were tabulated for interpolation using the energy of the midpoint of each of the groups.

8.1.3 Adjoint Transport Equation

In the calculation of responses involving the angular- and energy-dependent fluence or the total energy-dependent fluence, such as a detector response or, in a similar vein, the dose in a region or organ of a phantom, some portions of phase space are more "important" than others in contributing to the desired response (e.g., interactions and dose). For example, in the computation of photon absorbed dose in a region of a phantom, the secondary electron trajectories which intersect the region are obviously more important than those that do not. As a consequence, some photon paths are more important to the absorbed dose than others which are far from the region of interest.

As another example of importance, in high-energy hadrons-nucleus collisions, the more energetic secondary particles emitted at high energy in the forward direction are more important than the lower energy secondary particles emitted at large angles with respect to the original hadron direction. The concept of importance is used in the Monte Carlo method of radiation transport to increase the number of particles heading toward the volume in which dose is to be calculated. Using such importances to perform variance reduction leads to a greater number of particle histories contributing to the absorbed dose tally. Modern Monte Carlo codes allow the assignment of importance parameters to regions of the problem to generate more particles in regions that contribute to the dose tally. For computations in a simple phantom geometry, physical insight into the problem may be straightforward enough so that the code user can assign appropriate importance values to special locations to accomplish variance reduction in the quantity being tallied. In more complicated geometries, the adjoint transport equation can be used to determine the importance of particle trajectories and input into the Monte Carlo code (O'Brien 1980).

The adjoint transport equation corresponding to the forward transport equation given in Equation (8.4) is the following.

$$-\hat{\Omega}\cdot\bar{\nabla}\Phi^{+}(\bar{r},E,\hat{\Omega})+\mu(\bar{r},E)\Phi^{+}(\bar{r},E,\hat{\Omega})=X(\bar{r},E,\hat{\Omega})$$

$$+\int_{0}^{\infty}dE'\int_{4\pi}d\Omega'\mu_{s}(\bar{r},E\to E',\hat{\Omega}\to\hat{\Omega}')\Phi^{+}(\bar{r},E',\hat{\Omega}') \tag{8.6}$$

Here, $\Phi^{+}(\bar{r},E,\hat{\Omega})$ is the adjoint angular- and energy-dependent fluence and $X(\bar{r},E,\hat{\Omega})$ is an adjoint source which can be arbitrarily selected for a given problem.

Many textbooks derive this equation, although most do it for multiplying systems rather than for fixed source problems.[*] If the adjoint source is selected to be the cross section in

---

[*] See Bell and Glasstone (1970) and Lewis and Miller (1984).

the volume of a detector, then Φ^+ have units of counts-cm² and if the kerma factor in a given volume, it would have units of Gy-cm². O'Brien states that the adjoint Boltzmann equation describes particles starting at low energy from a target region which gain energy with each collision and then emerge at the actual source. In that regard, the term $\mu_s(\bar{\mathbf{r}}, E \rightarrow E', \hat{\Omega} \rightarrow \hat{\Omega}')$ is the transpose of the forward transport term $\mu_s(\bar{\mathbf{r}}, E' \rightarrow E, \hat{\Omega}' \rightarrow \hat{\Omega})$, at least in matrix form. The paper by Hansen and Sandmeier (1965) is instructive in the use of adjoint computations to determine the response of a detector and that application can be similarly applicable to a dose computation in a volume of a system.

O'Brien's discussion on the use of the adjoint in a dosimetry calculation is now presented. Hansen and Sandmeier showed that when the adjoint source is taken to be a collision cross section for some effect in a volume, the adjoint fluence on the outer surface of a finite medium in a vacuum is the energy response for the system of interest. For the incident fluence on that surface S, namely $\Phi(\bar{\mathbf{r}}_S, E, \hat{\Omega})$, where $\bar{\mathbf{r}}_S$ defines the surface in space, the reaction rate in the detector is then

$$R = \int_S dS \int_0^\infty dE \int_{4\pi} d\Omega \left[\Phi(\bar{\mathbf{r}}_S, \hat{\Omega}, E)\Phi^+(\bar{\mathbf{r}}_S, \hat{\Omega}, E)\hat{\Omega} \cdot \hat{\mathbf{n}} \right] \tag{8.7}$$

where $\hat{\mathbf{n}}$ is the inwardly directed surface normal on surface S. The power in this approach when performed in a Monte Carlo simulation is that the adjoint particles can be started at the desired point or in a desired volume (organ) and followed to the source, with all particles making a contribution. This contrasts with the forward Monte Carlo computation where the particles emitted from a source largely would not make it to the target region and contribute to the desired score. This approach has the power of starting with the appropriate energy-dependent kerma in an organ in the anthropomorphic phantom and computing the dose response on a coupling surface surrounding the phantom for all energies and angles in one calculation. If individual organ doses are required, then the computation must be repeated for each organ. However, the resulting dose response function on the coupling surface can be used in multiple irradiation geometries by computing the angular- and energy-dependent fluence on the surface and folding it with the response to obtain an organ dose. The drawback here is that secondary charged particles cannot be used as the adjoint organ source term; in application, the resulting dose assumes that the kerma approximation is valid.

A slight modification of this approach was used in the calculation of organ dosimetry in DS86 Dosimetry System 1986: U.S.–Japan Joint Reassessment of Atomic Bomb Radiation Dosimetry in Hiroshima and Nagasaki (Roesch 1987). The adjoint solutions were carried forward to DS02: Dosimetry System 2002 (Young and Kerr 2002). The approach was to compute these adjoint solutions on a surface, surrounding phantoms in standing, kneeling, and lying down positions. Then the neutron and gamma ray forward angular- and energy-dependent fluences (discrete ordinates solution) at various locations could be folded with the response functions over the coupling surfaces. For the DS86 response computations, a Monte Carlo code was used to compute the adjoint fluences (referred to in those reports as the adjoint leakage) on the desired coupling surface. Since the researchers wanted to allow for updates to nuclear data that might occur and lead to modifications in the kerma coefficients, they used the energy-dependent fluence as their adjoint source. The computations were performed using multigroup energy representations with a value of unity entered for each energy group fluence in the organ to start the computation, that is, as the adjoint source. This resulted in conversion coefficients, the adjoint fluences, which were folded with the forward computed energy- and angular-dependent fluences from the source, in this case, the bomb, to obtain the multigroup fluences in each of the organs. This fluence was then multiplied by the kerma coefficient for the organ to obtain the organ doses. In equation form, this is written as

$$\Phi_{\text{organ}}(E') = \int_S dS \int_0^\infty dE \int_{4\pi} d\Omega [\Phi(\bar{\mathbf{r}}_S, \hat{\mathbf{\Omega}}, E) \Phi^+(\bar{\mathbf{r}}_S, \hat{\mathbf{\Omega}}, E) \hat{\mathbf{\Omega}} \cdot \hat{\mathbf{n}}] \tag{8.8}$$

and is shown schematically in Figure 8.4. The forward tracks serve to form $\Phi(\bar{\mathbf{r}}_S, \hat{\mathbf{\Omega}}, E)$ on the surface and the adjoint tracks (illustration of lung as adjoint source) serve to form $\Phi^+(\bar{\mathbf{r}}_S, \hat{\mathbf{\Omega}}, E)$. This was a rather elegant approach for solving the problem, given the computational resources available in the 1980s, and, in fact, it was a stretch in that time period. If the energy-dependent kerma coefficient had been used to perform the adjoint calculation, the value of $\Phi^+(\bar{\mathbf{r}}_S, \hat{\mathbf{\Omega}}, E)$ would be a dose coefficient that computed absorbed doses in the organs of the phantom after folding with the angular- and energy-dependent fluence on the coupling surface, that is,

$$D_{\text{organ}}(E') = \int_S dS \int_0^\infty dE \int_{4\pi} d\Omega [\Phi(\bar{\mathbf{r}}_S, \hat{\mathbf{\Omega}}, E) \Phi^+(\bar{\mathbf{r}}_S, \hat{\mathbf{\Omega}}, E) \hat{\mathbf{\Omega}} \cdot \hat{\mathbf{n}}] \tag{8.9}$$

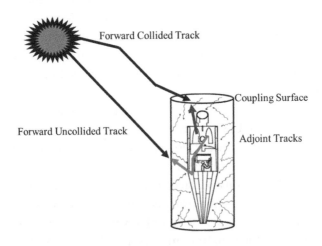

FIGURE 8.4 Forward-adjoint coupling similar to that in DS86 (1987).

8.1.4 Monte Carlo Methods

The Monte Carlo method is now the principal tool used in computational dosimetry because of the ability to represent complex three-dimensional geometries and track secondary charged particles. Lux and Koblinger (1991) define the Monte Carlo method as follows:

> A stochastic model is constructed in which the expected value of a certain random variable is equivalent to the physical quantity to be determined. The expected value is then estimated by the average of several independent samples representing the random variable introduced above. For the construction of the series of independent samples, random numbers follow the distributions of the variable to be estimated are used.

In this discussion, the physical quantity desired is absorbed dose.

Monte Carlo simulation has become a widely used technique in the solution of radiation transport problems as computer speed and power have evolved to their present state. Although some would say that the Monte Carlo method is using a stochastic simulation to generate solutions to the Boltzmann equation, the present authors liken it more to doing a particle-by-particle experiment and tallying the score in regions of interest. The method requires probability density functions to be sampled randomly to provide detailed particle transport and interaction histories. The greatest strength of the method is that it allows the user to perform simulations in very geometrically complicated systems. However, since the approach is stochastic, the challenge is to generate statistically significant results in what may be a small volume within a large computational model. To do so may take excessive computational time, even with today's capabilities, to obtain statistically significant results. There is a rather extensive set of texts, book chapters, and articles on the Monte Carlo method with emphasis on radiation transport; the ones most used by the authors, in no specific order, are chapters in Lewis and Miller (1984), Shultis and Dunn (2012), Haghighat (2016), Schaeffer (1973), Turner, Wright, and Hamm (1985), Profio (1979), Lux and Koblinger (1991), Shultis and Faw (2000), and Carter and Cashwell (1975).

A very abbreviated flowchart illustrating the basic flow of the Monte Carlo process is shown in Figure 8.5.

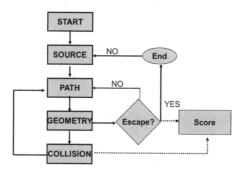

FIGURE 8.5   Simplified flowchart for a Monte Carlo simulation.

The process in what is known as classical analog Monte Carlo includes these steps:

- Selection of a particle from the source to begin a particle history:

    o Unless the particle is emitted from a fixed-point source, a particle emission location must be selected randomly over an area or in a volume from probability distributions.

    o The direction of the emitted particle must be selected randomly from probability distributions. Sometimes the direction of the emitted particle is fixed, e.g., in the case of a parallel beam for the reference radiation geometries previously discussed.

    o If the source is not monoenergetic, the energy of the particle must be chosen randomly from a probability distribution.

    o There are instances where the particle emission in time must be sampled as well. This is generally not the case in dose coefficient computations.

- The pathlength to the first interaction is randomly sampled from a probability distribution using the cross section or interaction coefficient.

- The location of this interaction in the geometry (and consequently material composition*) must be determined. The location of the interaction requires a fair amount of effort to locate. If the particle has exited the system, its history is stopped. In some cases, the code user wants to score the number of particles exiting the system, e.g., escaping a shield in a shielding problem.

- The collision type is determined and the particle, if not absorbed, continues to be tracked going through the process again. Secondary particles may be generated by the collision and must be tracked so that their entire histories are determined.

- Scoring (tallying) depends on the type of particle tally used and the quantity desired in the computation. If one wants the fluence to be determined in a volume, the track length travelled by particles in that volume must be summed and scored. In the case of dose computations, one may either use the fluence multiplied by a kerma coefficient in the kerma approximation or compute the energy deposited by secondary charged particle interactions in the volume.†

8.1.4.1 The Kerma Approximation versus Tracking of Charged Particles

For most of the twentieth century, the kerma approximation was employed in computations of absorbed dose in phantoms (the reader is referred to Chapter 2 for the definition of kerma). The kerma per unit fluence is termed the kerma coefficient and is

---

* For photons, this requires determining with what atomic element the photon interacted; for neutrons, this requires determining with what nuclide the neutron interacted.

† There are multiple ways to tally fluence and other quantities. The scope of the discussion in this chapter does not allow an in-depth treatment covering the details of the method.

represented K/Φ. Although kerma is defined in terms of the initial transfer of energy to matter, it has been used as an approximation to absorbed dose. This approximation approaches the true value of absorbed dose to the extent that charged particle equilibrium exists. ICRU Report 85a states that "charged-particle equilibrium exists at a point if the distribution of the charged-particle radiance with respect to energy is constant within distances equal to the maximum charged-particle range" (ICRU 2011). Collisional kerma is used in the kerma approximation to absorbed dose when radiative losses by the secondary charged particles are not negligible. The reader is directed to ICRU Report 85a for additional discussion.

Over the last two decades, the capabilities for tracking secondary charged particles in Monte Carlo codes have greatly improved and become readily available. In today's highly sophisticated Monte Carlo code environment, one can track the secondary charged particles and deposit their energy in volumes of interest. In organs or volumes of the phantoms sufficiently far from the surface of the phantom (more than the range of the most energetic secondary charged particle), the values obtained by doing secondary charged particle transport are not significantly different from using the kerma approximation, at least at energies where kerma is defined. As the energy of photons increases, production of bremsstrahlung by the secondary electrons causes the kerma approximation to fail. In addition, as the energy and consequently the range of the secondary charged particles increases, obtaining charged particle equilibrium at the surface of the body, at the interfaces of organs, and in small organs is more difficult to obtain.

In Figure 8.6, ambient dose equivalent at a depth of 10 mm computed using the kerma approximation is plotted, as are the values computed for the absorbed dose at several depths using secondary electron tracking. The kerma approximation differs from the calculated absorbed dose with secondary particle tracking occurring at low energies, as few

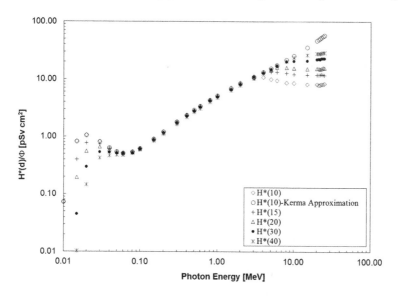

FIGURE 8.6   Ambient dose equivalent at various depths in the ICRU sphere as a function of energy calculated by the kerma approximation and by secondary particle tracking (Shannon 2009).

electrons reach the depths of the tally volumes. At higher energies, the departure shows the inadequacy of kerma approximation as radiative energy transfer from the secondary electrons becomes more common and the range of the electrons increases. The same would be true for neutrons where secondary protons and alpha particles would be depositing the energy in the sphere, but the kerma approximation for neutrons would hold at shallower depths and higher energies as the ranges of the protons would be shorter at a given energy than for an electron.

8.1.4.2 Variance Reduction Techniques

The primary challenge in the use of the analog Monte Carlo method in obtaining statistically significant scores (tallies) in the region of interest is the execution time of the code in complex geometries. Variance reduction techniques are employed to increase the number of particles that contribute to the tally. There are a number of techniques that are used. In such techniques, biased distributions are often applied and in order to continue playing a fair game, the weight of the particle must be reduced to compensate. Nonanalog procedures can be applied at each particle collision, boundary crossing, or other event during a particle history to reduce the variance in the answer. The goal is to increase the fraction of particles that contribute to the tally and to compensate for the modification in sampling, the weight of the particle is adjusted.

Analog Monte Carlo calculations consist of generating particle histories for N independent particle histories. The result of this calculation is a sample mean, where x is some desired property of the histories, for example, energy deposited in a volume for an absorbed dose computation,

$$\hat{x} = \frac{1}{N} \sum_{n=1}^{N} x_n \tag{8.10}$$

The value for x_n can be zero for a large portion of the particle histories. Using variance reduction, the number of particles interacting in the volume can be increased, but the weight of the particles must be adjusted to maintain the value of the sample mean, which is the estimate of the true mean in the problem computation, namely,

$$\hat{x} = \frac{1}{N} \sum_{n=1}^{N} w_n \tilde{x}_n \tag{8.11}$$

where w_n is the weight of the particle making the contribution $\tilde{x}_n$ after the application of the variance reduction technique.

The most common technique applied in Monte Carlo codes to increase the transport of particles to the volumes of interest is absorption suppression. In analog Monte Carlo simulation, the particle history would end when it was absorbed. In absorption suppression, a fraction of the particle is absorbed at each collision site, namely $\left(\mu_a \big/ \mu_t \right)$ multiplied by its current weight, and a particle continues on with its weight reduced by a factor of

$\left(1 - \mu_a \middle/ \mu_t\right)$. This serves the purpose of more particles penetrating deeper into a system, but at a reduced weight.

Some other variance reduction techniques worth mentioning in the computation of absorbed dose are enumerated below:

- *Importance sampling* is when the sampling procedure uses modified distributions to ensure that the most important elements of phase space are sampled. In addition to required particle weight adjustments, the history of a particle may be terminated below a certain weight value. This often involves the a process referred to as "Russian roulette." When the particle weight goes below a certain threshold value, a random number is uniformly generated to terminate the history *(k-1)* out of *k* times. The particle continues to be tracked *1/k* of the time with its weight increased by a factor of *k*. Included in this approach are geometric splitting and the use of alternative sampling distributions, e.g., sampling the more important energies of the source spectrum. One can also use the adjoint fluence of the problem to determine regions of importance (Hiller et al. 2016).

- The *exponential transform* is an approach which stretches the distances between collisions in the forward direction. This is essentially performed by reducing the interaction coefficient of the particle and adjusting the weight accordingly. It can be applied in preferred directions or throughout the system. As an example, it would serve to drive low-energy photons deeper into a phantom before interacting.

- Another approach is to *force collisions*. This serves to increase the number of collisions in a volume; this can be particularly useful if the volume has small dimensions and low density, both of which lead to low interaction rates. In this approach, the particle is split into two smaller weight particles; the first passes through without interaction and the second is forced to collide. The weights are adjusted to reflect the collision probability. This can be applied in the volume where the dose is desired, as well as in adjacent volumes in the problem in which interactions would be important contributors to the dose.

8.1.4.3 Reciprocity Method

In some instances, the source region may be quite large, and the target organ may be small in comparison, so that the tally of absorbed dose in the target organ may suffer from large statistical uncertainties. In such cases, the source and target region may be reversed in the calculation to obtain a better estimate of the dose in the target region, that is, the particles may be emitted from the target region and the dose tallied is the source region. This is still a forward transport computation, but the radiation emissions now emanate from what was the target volume.

Shultis and Faw (2000) states that if the source strength per unit mass is kept constant when the source and target regions are reversed, the total energy absorption in the source region from particles being emitted from the target region is equal to the total energy absorption in the target region, if the particles had been emitted from the source region. However, unless the materials involved have similar interaction coefficients, this approach may not hold for scattered radiations. It is strictly satisfied in an infinite homogeneous medium, or for the uncollided radiation in a heterogeneous medium in which the interaction coefficients are proportional to the local density. Attix (2008) has a discussion of its use in dose computations. Its use in external irradiation dose coefficient calculations is discussed by Hiller et al. (2016), and its use in internal radiation dosimetry is discussed in a recent article by Wayson et al. (2012), although the authors in the latter article (mistakenly) equate their use of the reciprocity theory as being an adjoint computation.

8.1.5 External Irradiation Dose Coefficients

This section only deals with neutrons and photon dose coefficients and, at that, is only a sampling of what could be reported and discussed concerning such coefficients. It is an attempt to give the flavor of computational results for those coefficients over the years. In addition, energies of interest at high-energy facilities are not covered in this section and dose coefficients for charged particles also are not presented.

8.1.5.1 Photons

In Figure 8.7, several sets of dose coefficients (fluence to dose) for photons are displayed. For the Claiborne and Trubey multigroup, discrete ordinates computations in a 30-cm thick slab using the kerma approximation are plotted at the midpoint of each energy group (Claiborne and Trubey 1970). These values are maximum absorbed dose coefficients and were the recommended set in ANSI/ANS6.1.1-1977, which was used for many years to convert photon fluence to dose for shielding and a variety of other applications (ANSI 1977).

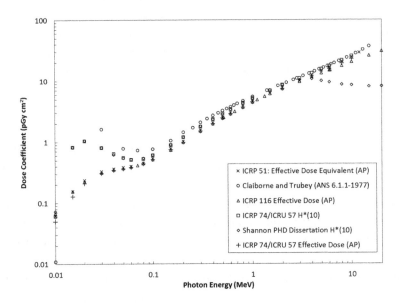

FIGURE 8.7 Photon dose coefficients to convert from fluence to absorbed dose.

The effective dose equivalent coefficients based on ICRP Publication 26 (ICRP 1977) recommendations were taken from the values reported in ICRP Publication 51 (ICRP 1988) and plotted in the figure along with values of its replacement quantity effective dose. The values of effective dose based on ICRP Publication 60 (ICRP 1991) recommendations are taken from ICRP Publication 74 (ICRP 1996)/ICRU Report 57 (ICRU 1998), and the ones based on ICRP Publication 103 (ICRP 2007) recommendations were taken from ICRP Publication 116 (ICRP 2010). For these three sets, the AP irradiation geometry is presented; the reader is directed to those documents for other irradiation geometries which were omitted here to keep the graphs less congested. The values for effective dose equivalent from ICRP Publication 51 (ICRP 1988) and effective dose from ICRP Publication 74 (ICRP 1996) were generated using the kerma approximation, while the effective dose values from ICRP Publication 116 (ICRP 2010) were computed using secondary electron transport. The differences between the effective doses and the effective dose equivalent follow from different tissue weighting factors, possibly improved photon interaction data, phantom differences, and the use of secondary particles tracking rather than the kerma approximation for those from ICRP Publication 116 (ICRP 2010).

Two sets of the ambient dose equivalent, $H^*(10)$, values are shown in the figure; recall that ambient dose equivalent is computed at a depth of 10 mm in the ICRU sphere by aligning and expanding the energy-dependent fluence at a point, that is, making a parallel beam to irradiate one side of the ICRU sphere. Those from ICRP Publication 74 (ICRP 1996)/ICRU Report 57 (ICRU 1998) are the latest recommended values at a 1 cm (10 mm) depth in the ICRU sphere on the principal axis and were computed using the kerma approximation. Ambient dose equivalent is intended for use in area monitoring and, as with all operational quantities, was intended to be a conservative estimate of the limiting quantities. Worthy of mention is the strong overestimate of the effective dose and effective dose equivalent at low energies by the ambient dose equivalent. However, it does overestimate effective dose over the entire energy range plotted. At the present, the ICRP Publication 116 (ICRP 2010) coefficients are the recommended values for effective dose and the ICRP Publication 74 (ICRP 1996)/ICRU Report 57 (ICRU 1998) are the recommended values for ambient dose equivalent.

The $H^*(10)$ values from Shannon (2009) were computed using secondary electron tracking rather than the kerma approximation. At approximately 3 MeV, the ICRP Publication 74 (ICRP 1996)/ICRU Report 57 (ICRU 1998) values of $H^*(10)$ depart from those of Shannon due to the use of the kerma approximation rather than performing energy deposition by secondary electron tracking. The $H^*(10)$ based on secondary particle tracking is not a conservative estimate of the effective dose above that energy and the difference worsens as the energy increases.* Using a deeper depth in the ICRU sphere to compute a value of ambient dose equivalent would delay the onset of this departure (Figure 8.7) from the effective dose; it also would also decrease the over-response of ambient dose equivalent at low energies. Data for higher energies are available in the literature.

---

* See ICRP Publication 116 (2010) for details at higher energies.

TABLE 8.1   Percent Differences between ICRP Publication 116 (ICRP 2010), Effective Dose the Effective Dose Equivalent from ICRP Publication 51 (ICRP 1988), and the Effective Dose from ICRP Publication 74 (ICRP 1996)

Energy (MeV)	H_E (ICRP 51)/E (ICRP 116)	E (ICRP 74)/E (ICRP 116)
0.01	1.00	0.72
0.1	1.03	1.00
0.5	1.03	1.00
1	1.02	1.00
4	1.07	1.02
6	1.11	1.05
8	1.17	1.08
10	1.20	1.12

For the range of interest between 0.1 and 10 MeV, the differences between the effective dose (regardless of the recommendations) and effective dose equivalent are shown in Table 8.1 for a sampling of energies over the range plotted in Figure 8.7. There are differences in tissue weighting factors between the values as well as the organs and the protocol for organ selection for use in the weighted sum to compute effective dose. However, at 4 MeV, the impact of tracking secondary electrons rather than using the kerma approximation in the computation of effective dose is evident, although it is not nearly as great as the magnitude of the difference with the ambient dose equivalent. This is due to the averaging nature of the effective dose formalism. Using the kerma approximation, as opposed to tracking secondary electrons, leads to an absorbed dose overestimate that can be significant for organs near the surface of the phantom, while for larger organs and for organs deeper in the body, the impact is almost negligible. As an example of that, in Figure 8.8, the organ absorbed doses to the female breast is shown from ICRP Publication 74 (ICRP 1996)/ICRU Report 57 (ICRU 1998) (kerma approximation) and ICRP Publication 116 (ICRP 2010) (secondary electron tracking). Above approximately 3 MeV, the kerma approximation starts to overestimate the dose, and the difference increases significantly with increasing energy. The differences at low energies between the two sets of data are likely due to the differences in phantoms. Other organs near the surface of the illuminated side of the phantom would follow similar trends, although the departure of the kerma approximation from the absorbed dose may occur at different energies, that is, for the skin it occurs below 1 MeV given in ICRP Publication 116. In Figure 8.9, the values from those two publications are plotted for the liver, a much larger organ which is further from the surface of the body. In that case, the kerma approximation is an excellent estimate of the absorbed dose in the liver as charged particle equilibrium is established.

8.1.5.2 Neutrons

Seven sets of neutron dose coefficients for external irradiation are shown in Figure 8.10. The oldest ones track back to the early 1970s. All of them are for parallel beams incident on the phantom. Those taken from ICRP Publication 21 (ICRP 1971) were based on

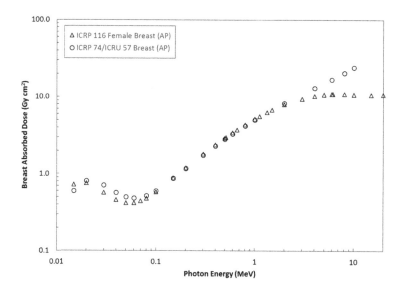

FIGURE 8.8 The breast absorbed dose for AP irradiation geometry for ICRP Publication 74 (ICRP 1996)/ICRU Report 57 (ICRU 1998) (kerma approximation) compared to ICRP Publication 116 (ICRP 2010) (secondary electron tracking).

the maximum absorbed dose equivalent in a semi-infinite slab phantom 30 cm thick and those from National Council on Radiation Protection (NCRP) Report 38 (NCRP 1971) were based the maximum absorbed dose equivalent in a cylinder that was 30 cmØ × 60 cm. Both these two sets were based on the kerma approximation.

Values for the effective dose equivalent were taken from ICRP Publication 51 (ICRP 1988) which used the ICRP Publication 26 recommendations (ICRP 1977). To compute effective dose equivalent, absorbed doses were computed as a function of L and folded with the Q-L relationship from ICRP Publication 26 to arrive at organ dose equivalent; these doses were then summed using the organ dose formalism of ICRP Publication 26 (ICRP 1977). Also plotted are these same calculated values multiplied by a factor of 2, as per the statement for the Paris meeting in 1985 on the quality factor (ICRP 1985). The computations were performed with the kerma approximation.

The ambient dose equivalent coefficients at a depth in the ICRU sphere of 10 mm, $H^*(10)$, are also plotted as are the recommended values taken from ICRP Publication 74 (ICRP 1996)/ICRU Report 57 (ICRU 1998). The neutron dose coefficients from this publication used the kerma approximation below 20 MeV and tracking charged particles was performed for neutron interactions occurring above 20 MeV. Since this set is based on several sets of calculations, phantoms, approaches, and codes, the reader should go to the publication to gain a full understanding of the differences in techniques and phantoms. These computations used the Q-L relationship recommended in ICRP Publication 60 (ICRP 1991) to obtain dose equivalent.

The effective doses from ICRP Publication 116 (ICRP 2010) and ICRP Publication 74 (ICRP 1996)/ICRU Report 57 (ICRU 1998) are also plotted. Most of the differences below 1 MeV between these two sets are due to the reduction of the radiation weighting factor by

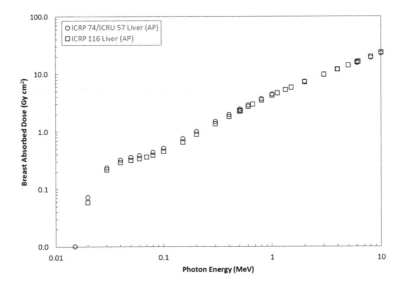

FIGURE 8.9   The liver absorbed dose for AP irradiation geometry for ICRP Publication 74 (ICRP 1996)/ICRU Report 57 (ICRU 1998) (kerma approximation) compared to ICRP Publication 116 (ICRP 2010) (secondary electron tracking; female phantom).

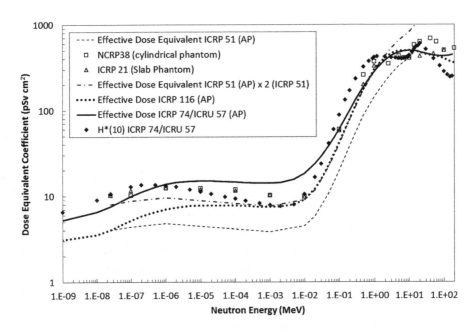

FIGURE 8.10   Neutron dose coefficients to convert from fluence-to-dose equivalent quantities.

close to a factor of 2 in ICRP Publication 103 (ICRP 2007), as compared to ICRP Publication 60 (ICRP 1991). Above 20 MeV, the differences are attributable in large part to the differences in the nuclear data employed as the ICRP Publication 74/ICRU Report 57 results, which employed codes that used models above approximately 20 MeV and the ICRP 116 results used evaluated cross-section data to higher energies and then models. Both the

ICRP Publication 74/ICRU Report 57 and the ICRP Publication 116 recommended values used the kerma approximation below 20 MeV. The kerma approximation is applicable in this energy range, with the exception that secondary particles should be transported to get a better value of the absorbed dose in the skin (Chen and Chilton 1979). In the range from 1 MeV to 20 MeV, the two sets agree well. There are data in ICRP Publication 116 to much higher energies, but that discussion would require a more comprehensive discussion than fits the scope of this chapter.

The impact of resonances in the neutron cross section for tissue and organs is generally not discussed in the publications containing recommended fluence-to-dose coefficients. For example, the dose coefficients for neutrons in ICRP Publication 116 for the energy range 0.1 to 5 MeV were computed for the following neutron energies: 0.1, 0.15, 0.2, 0.3, 0.5, 0.7, 0.9, 1.0, 1.2, 1.5, 2.0, 3, 4 and 5 MeV. To demonstrate the effect of resonances on the structure of the dose coefficients that can be missed by selecting an energy grid at random, computations were performed in 0.1 MeV energy increments for 0 to 5 MeV in the ICRU sphere for aligned and expanded fields of monoenergetic neutrons. The ICRU sphere was chosen since it provides fewer geometric complications and serves to demonstrate the effect in a straightforward fashion. Secondary particle tracking was used, and the absorbed dose was computed at 0.07, 10, and 105 mm depths into the sphere along the principal axis, that is, ambient absorbed dose $D^*(d)$, values at those depths. In the vicinity of resonances, the energy grid was further refined to better define the structure in the absorbed dose. Some of the very narrow resonances were not specifically studied, as they would have required an even finer local energy grid, and the point of this discussion is to demonstrate the effect of resonances. The resulting values are plotted in Figure 8.11. At the 0.07 mm and

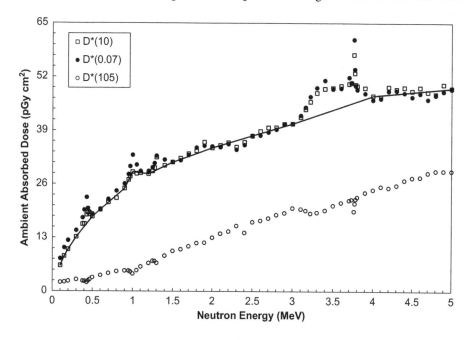

FIGURE 8.11 Neutron absorbed dose at 0.07, 10, and 105 mm depths in the ICRU spherical phantom. The solid line connects the values at the energies used in ICRP Publication 116 energy grid.

10 mm depths, the impact of the cross-section resonance structure serves to increase the reaction rate at, or in the vicinity of, the resonances, leading to associated increases in the absorbed dose. At the 105 mm depth, the resonances result in depressions, albeit small at this depth, in the absorbed dose, due to the preferential removal of neutrons with energies in or near the resonance energies as they penetrate the sphere. The latter effect would be more pronounced at deeper depths into a phantom.

The solid line shown connects the values that would have been available if the ICRP Publication 116 energy grid were the only energies at which the absorbed doses were computed. It serves to highlight structure that would be omitted by not considering the cross-section variation with neutron energy before choosing the energy grid for computation. The question that arises is if any of this missed structure in the absorbed dose coefficients are of consequence when used in practice, especially since monoenergetic neutron fields are rare. It is likely that the polyenergetic spectra encountered in practice might only be impacted by the missed region of higher dose magnitude between 3 and 4 MeV. The past use of multigroup cross sections, if properly weighted in their preparation, would give estimates that would more likely be average over energy ranges rather than point estimates.

8.2 COMPUTATION OF DOSE COEFFICIENTS FOR INTERNAL EMITTERS

When radionuclides enter the body prior to undergoing radioactive decay, the resulting dose is typically computed based on an estimate of the activity which was taken into the body. Chapter 10 details the methods used to estimate such intakes. Once this activity is determined, internal dose coefficients are applied to convert the intake to a committed effective dose. This section describes the methods and details used in the computation of the internal dose coefficients.

8.2.1 Description of Quantities Used in Internal Dose Calculations

This section describes and defines the quantities used in computing internal dose coefficients. The names and definitions of the quantities presented here represent the current terminology used by the ICRP (ICRP 1979, 1991, 2007). Most of these quantities can be traced to either equivalent or analogous quantities used historically in nuclear medicine (Loevinger, Budinger, and Watson 1991; ICRP 2015a) or radiation protection (ICRP 1979, 1991). Bolch et al. (2009) contains a detailed description of these historical quantities and an approach for unifying the quantity nomenclature.

8.2.1.1 Effective Dose

ICRP Publication 103 (ICRP 2007) defines an effective dose as the tissue and gender-weighted average of the equivalent doses to a set of radiosensitive target tissues. The concept of effective dose is discussed in detail in Chapter 2 but is highlighted in the following discussion.

Effective dose is represented mathematically in Equation (8.12).

$$E = \sum_T w_T \frac{H_T^{\text{female}} + H_T^{\text{male}}}{2} \tag{8.12}$$

TABLE 8.2   ICRP Publication 103 Tissue Weighting Factors (ICRP 2007)

Tissue	w_T	$\sum w_T$
Bone marrow (red), colon, lung, stomach, breast, remainder tissues[a]	0.12	0.72
Gonads	0.08	0.08
Bladder, esophagus, liver, thyroid	0.04	0.16
Bone surface, brain, salivary glands, skin	0.01	0.04
	Total	1.00

[a] Remainder tissues: Adrenals, extrathoracic (ET) region, gallbladder, heart, kidneys, lymphatic nodes, muscle, oral mucosa, pancreas, prostate, small intestine, spleen, thymus, uterus/cervix.

where

w_T is the tissue weighting factor for target tissue T and

H_T is the equivalent dose to the same target in the female and male.

The tissue weighting factors are also defined in ICRP Publication 103 (ICRP 2007) and are presented here in Table 8.2.

Note that there are 13 tissues comprising the Remainder tissue. Each of the thirteen is weighted equally in the calculation. In other words, the equivalent dose to the Remainder tissue is simply the arithmetic mean of the 13 tissues, as given in Equation (8.13):

$$H_{\text{Remainder}} = \frac{1}{13} \sum_T H_T \tag{8.13}$$

8.2.1.2 Equivalent Dose

Equivalent dose is defined in ICRP Publication 103 (ICRP 2007) as the radiation weighted dose to a target tissue. The concept of equivalent dose is discussed in detail in Chapter 2, but highlighted in the following discussion. Equivalent dose is computed as the product of the radiation weighting factor and the mean absorbed dose as shown in Equation (8.14):

$$H_T = \sum_R w_R D_{R,T} \tag{8.14}$$

where w_R is the radiation weighting factor and $D_{R,T}$ is the mean absorbed dose to target tissue T for radiation type R. The radiation weighting factors from ICRP Publication 103 are presented in Table 8.3.

To compute the equivalent dose due to a source inside the body, it is helpful to separate the calculation into a source term and an energy deposition term. These two quantities are multiplied to give the equivalent dose rate as shown in Equation (8.15):

$$\dot{H}_T = \sum_{r_S} A_S S_w \left(r_T \leftarrow r_S \right) \tag{8.15}$$

TABLE 8.3   ICRP Publication 103 (ICRP 2007) Radiation Weighting
Factors

Radiation Type	w_R
Photons	1
Electrons and muons	1
Protons and charged pions	2
Alpha particles, fission fragments, heavy ions	20
Neutrons	A continuous function of neutron energy (see ICRP Publication 103)

where A_S, the source term, is the activity in source region r_S, and $S_w(r_T \leftarrow r_S)$, the energy deposition term, is the S-coefficient for radiations emitted from source region r_S depositing energy in target tissue r_T. These two terms are described in more detail in the next two sections.

8.2.1.3 S-Coefficient

The S-coefficient is defined as the radiation weighted dose to a target tissue per nuclear transformation occurring in a source region (ICRP 2015b). It is computed as shown in Equation (8.16):

$$S_w(r_T \leftarrow r_S) = \sum_R w_R \sum_i E_{R,i} Y_{R,i} \Phi(r_T \leftarrow r_S, E_{R,i}) \tag{8.16}$$

where
  $E_{R,i}$ is the energy of the ith emission of type R;
  $Y_{R,i}$ is its corresponding yield per nuclear transformation; and
  $\Phi(r_T \leftarrow r_S, E_{R,i})$ is the specific absorbed fraction corresponding to this radiation type and energy for emissions in source region r_S depositing energy in target tissue r_T.

The specific absorbed fraction is the fraction of source energy emitted which is absorbed in the target per mass of the target. The specific absorbed fraction is discussed further in Section 8.2.1.7.

The energies, yields, and radiation types associated with a particular radionuclide decay are tabulated in ICRP Publication 107 (ICRP 2008). This publication includes mathematical descriptions of the beta energy spectrum. For a beta-emitting radionuclide, the summation over i in Equation (8.16) turns into an integral, as shown in Equation (8.17):

$$S_{w-\text{beta}}(r_T \leftarrow r_S) = \int_{i=0}^{i\text{max}} E_{R,i} Y_{R,i} \Phi(r_T \leftarrow r_S, E_{R,i}) \tag{8.17}$$

As a practical matter, when performing the calculation, it is helpful to separately calculate the beta contribution (given by Equation 8.17) to the S-coefficient from the other radiation types contribution (given by Equation 8.16). They can be simply summed later to give the S-coefficient for all radiation types.

8.2.1.4 Number of Nuclear Transformations

The intake of radionuclides results in an exposure to the individual which occurs over a period of time after the time of intake. The time of this exposure can be short or long, depending on the physical half-life and the biological removal rate of the radioactive specie. It is desirable, then, to compute the integrated dose contribution from an intake over a time period.

If the S-coefficient is constant over the time of integration, it is efficient to integrate the activity over this time and therefore compute the total number of nuclear transformations occurring in each source region over this time (commitment period), as shown in Equation (8.18).

$$\tilde{A}(r_S,\tau) = \int_{t_0}^{t_0+\tau} A_S(t)\,dt \tag{8.18}$$

where

t_0 is the time of intake,

τ is the commitment period,

$A_S(t)$ is the activity as a function of time in the source region, and

$\tilde{A}(r_S,\tau)$ is the total number of transformations occurring in this source region over the commitment period.

This quantity has also been referred to as the integrated activity or the cumulated activity (Loevinger, Budinger, and Watson 1991; Bolch et al. 2009).

An internal dose coefficient is the dose per unit activity intake. To compute a dose coefficient, the activities in each source region are divided by the intake activity as shown in Equation (8.19).

$$\tilde{a}(r_S,\tau) = \frac{\tilde{A}(r_S,\tau)}{A_{\text{intake}}} = \int_{t_0}^{t_0+\tau} a_S(t)\,dt$$

$$a_S(t) = \frac{A_S(t)\,dt}{A_{\text{intake}}} \tag{8.19}$$

where

A_{intake} is the total intake activity at the time of intake (t_0),

$a_S(t)$ is the activity in a source region per unit activity intake, and

$\tilde{a}(r_S,\tau)$ is the number of nuclear transformations in a source region per unit activity intake.

When performing the calculation, it becomes practically useful to consider an intake of 1 Bq, therefore making the division by the intake activity trivial.

8.2.1.5 Committed Equivalent Dose Coefficient

For workers and adult members of the public, the S-coefficient remains constant over the commitment period, making use of the integrated activity in Equation (8.18) helpful. When considering intakes in children, however, the S-coefficient varies as a function of time since the specific absorbed fraction time varies with age. In this case, the integration must be performed over the product of the activity as a function of time and the S-coefficient.

Equation (8.15) can now be updated using the activity terms described in Equation (8.19). Equation (8.20) gives the equivalent dose rate coefficient:

$$\dot{h}_T(t) = \sum_{rs} a_S(t) S_w(r_T \leftarrow r_S, t) \tag{8.20}$$

where $\dot{h}_T(t)$ is the equivalent dose rate coefficient to a target tissue as a function of time. Integration of this quantity over the commitment period gives the committed equivalent dose coefficient as shown in Equation (8.21):

$$h_T = \sum_{rs} \int_{t_0}^{t_0+\tau} a_S(t) S_w(r_T \leftarrow r_S, t) dt \tag{8.21}$$

where h_T is the committed equivalent dose coefficient to a target tissue.

If computing the coefficient for intake occurring as an adult, then the S-coefficient is constant over time and can be removed from the integration. Since the activity is the only term remaining inside the integral, the number of nuclear transformations (integrated activity) can be used instead, giving Equation (8.22):

$$h_T = \sum_{rs} S_w(r_T \leftarrow r_S) \int_{t_0}^{t_0+\tau} a_S(t) dt$$

$$h_T = \sum_{rs} \tilde{a}(r_S, \tau) S_w(r_T \leftarrow r_S) \tag{8.22}$$

The commitment period depends on the age of the reference individual at intake. If calculating the dose coefficient for intake as an adult, the commitment period is 50 years. If the intake takes place as a child, the commitment period extends from the time of intake to an age of 70 years, as given in ICRP Publication 103 (ICRP 2007).

8.2.1.6 Committed Effective Dose Coefficient

Once the committed equivalent dose coefficients have been computed for each target tissue defined in Table 8.1 and for each gender, the committed effective dose coefficient can be computed using the relationship in Equation (8.12). Equation (8.23), then, gives the committed effective dose coefficient:

$$e = \sum_T w_T \frac{h_T^{\text{female}} + h_T^{\text{male}}}{2}$$

$$e = \frac{1}{2} \left(\sum_T w_T h_T^{\text{female}} + \sum_T w_T h_T^{\text{male}} \right) \tag{8.23}$$

where e is the committed effective dose coefficient. Note that while the second form of Equation (8.23) is mathematically equivalent, care should be taken not to report a gender-specific effective dose coefficient. By ICRP Publication 103 definition, effective dose is a gender-averaged quantity (ICRP 2007).

8.2.1.7 Specific Absorbed Fraction

Returning to the S-coefficient and Equation (8.16), the specific absorbed fraction quantity warrants further explanation. Equation (8.24) gives the definition of the specific absorbed fraction.

$$\Phi\left(r_T \leftarrow r_S, E_{R,i}\right) = \frac{\phi\left(r_T \leftarrow r_S, E_{R,i}\right)}{m_T} \tag{8.24}$$

where $\phi\left(r_T \leftarrow r_S, E_{R,i}\right)$ is the absorbed fraction, the fraction of energy emitted from a source region which is absorbed in a target tissue of mass, m_T.

Specific absorbed fractions for the reference adult are tabulated in ICRP Publication 133 (ICRP 2016a). The tabulations are specific to each type of radiation (electrons, alpha particles, photons, neutrons) and each gender. For each gender and type of radiation, except neutrons, the table contains specific absorbed fractions for each source and target combination at many (greater than 20) discrete energies. Neutron specific absorbed fractions are listed by radionuclide rather than energy and have already been fission-spectrum weighted.

The specific absorbed fractions in ICRP Publication 133 (ICRP 2016a) are computed for various geometries using a variety of computational phantoms all designed to represent reference adults. The reference adult is defined in ICRP Publication 89 (ICRP 2002). The majority of the specific absorbed fractions were derived from Monte Carlo radiation transport using the computational phantoms described in ICRP Publication 110 (ICRP 2009) (Zankl and Wittmann 2001; Zankl et al. 2005; Zankl et al. 2012). Additional dosimetry models are taken from ICRP Publication 66 (ICRP 1994) and ICRP Publication 130 (ICRP 2015b) for the respiratory tract while new values appear in ICRP Publication 133 (ICRP 2016a) for the alimentary tract and skeleton (Hough et al. 2011; Jokisch et al. 2011; Bahadori et al. 2011; Johnson et al. 2011; O'Reilly et al. 2016). The list of reference source regions and target tissues with their masses are found in tables in Annex A of Publication 133 (ICRP 2016a).

To compute dose coefficients resulting from intakes in children, it is necessary to have specific absorbed fractions derived from computational phantoms representing a variety of ages. Such data is currently in preparation for publication by the ICRP.

8.2.2 Calculation Details

A series of steps is followed, using the equations in Section 8.2.1 to compute the committed effective dose coefficient. These steps can be summarized below.

1. Calculation of the activity in each source region per unit intake as a function of time (see Chapter 6 discussing biokinetic models).

2. Determination of the gender-specific absorbed dose per unit intake to each target tissue.

3. Application of radiation weighting to determine the committed equivalent dose coefficient for each target tissue.

4. Application of gender-averaging and tissue-weighting to obtain committed effective dose coefficient.

There are several details involved in these steps which warrant further description. The subsequent sections provide some of the non-trivial details associated with these calculations.

8.2.2.1 Accounting for Activity in Systemic "Other"

As described in Chapter 6, the biokinetic models give a series of differential equations which, when solved, provide the activity in each source region as a function of time. The systemic biokinetic models make use of one or more "Other" compartments. These compartments represent all systemic tissues not otherwise specified within the systemic model. The constituent source regions comprising the "Other" compartment, therefore, vary by element. When performing the dosimetry calculation, the contribution from activity in "Other" must be appropriately handled. There are two mathematically equivalent approaches for handling the activity in "Other."

First, as described in ICRP Publication 133 (ICRP 2016a), specific absorbed fractions can be computed in each case for a source region of "Other" based on a source-mass-weighted average of the "Other" constituents specific absorbed fractions. Equation (8.25) is used to derive a specific absorbed fraction for the "Other" source region irradiating each possible target.

$$\Phi\left(r_T \leftarrow \text{Other}, E_{R,i}\right) = \frac{1}{M_{\text{Other}}} \sum_{r_S} M_{r_S} \Phi\left(r_T \leftarrow r_S, E_{R,i}\right) \qquad (8.25)$$

where

M_{r_S} is the mass of each source region constituting "Other,"

M_{Other} is the total mass of "Other" (computed on a case-by-case basis), and

$\Phi\left(r_T \leftarrow \text{Other}, E_{R,i}\right)$ is the specific absorbed fraction for source material in "Other" irradiating a particular target tissue.

A second approach is to take the activity in "Other" and distribute it to its constituents by source-mass fraction, as shown in Equation (8.26):

$$\tilde{a}\left(r_S, \tau\right) = \frac{M_{r_S}}{M_{\text{Other}}} \tilde{a}\left(\text{Other}, \tau\right) \tag{8.26}$$

where

$\tilde{a}\left(\text{Other}, \tau\right)$ is the number of transformations occurring in "Other," and

$\tilde{a}\left(r_S, \tau\right)$ is the number of transformations occurring in a constituent of "Other."

While mathematically equivalent to the first, this approach allows use of the specific absorbed fractions as tabulated. Finally, while Equation (8.26) is given in terms of the number of nuclear transformations, the same approach can be applied to the activity in "Other" as a function of time.

Note that the definition of "Other" tissues is based on those source regions explicit in the systemic biokinetic model and a list of eligible soft tissues. It is not impacted by source regions invoked as a result of the alimentary tract or respiratory tract model. Confusion can result, for example, for an inhalation case. The biokinetic model associated with the respiratory tract will result in source activity in the extrathoracic and thoracic lymph nodes (LN-ET and LN-Th). If, however, these lymphatic source regions do not appear in the systemic biokinetic model, then some of the systemic "Other" activity must be assigned to these source regions. In such a case, the total activity in the LN-ET region would be the sum of the activity assigned to it from the respiratory tract model and the activity it receives as a result of being some fraction of the systemic "Other."

8.2.2.2 Interpolation Techniques

As mentioned earlier, the specific absorbed fractions provided in ICRP Publication 133 (ICRP 2016a) are tabulated at discrete energies. It is necessary to interpolate between these energies when seeking the specific absorbed fraction at a specific energy of a nuclear emission. The specific absorbed fraction tables contain values at more than 20 energies. As a result, linear interpolation techniques can provide reasonable estimates of the effective dose coefficient. If more precision is desired, other interpolation algorithms can be implemented.

One such algorithm achieves a monotone interpolation using a piecewise cubic Hermite spline. This method was developed by Fritsch and Carlson (1980). This algorithm uses all known data points and will return multiple interpolants in single call.

When computing dose coefficients for intakes in children, the calculation can be slow given the large number of interpolations required over multiple time steps and radiation energies. The efficiency of the calculation will be improved if the S-coefficients are computed at each of the phantom ages (newborn, 1-year-old, 5-year-old, 10-year-old, 15-year-old, adult), and then interpolating to obtain S-coefficients at each desired time step. While mathematically equivalent, interpolating the specific absorbed fractions with respect to time prior to the S-coefficient calculation will prove much slower, given the large number of specific absorbed fractions (one for each radiation emission) that go into a single S-coefficient.

8.2.2.3 Integration Techniques

Integration is necessary over the beta energy spectrum as described in Equation (8.17). If computing a dose coefficient for intake as a child, integration will be necessary in Equation (8.21). The crudest approximation to the integral would be to apply a trapezoidal summation of the discrete points in the data set. To improve the quality of the integration, however, the Fritsch and Carlson interpolation algorithm can be invoked. Rather than integrating over n data points, the interpolation algorithm can create many interpolants (100, for example) between each pair of data points before the trapezoidal summation is applied.

8.2.2.4 Decay Chain Considerations

Intake of radionuclides which decay into radioactive progeny need to be treated appropriately when computing dose coefficients. ICRP Publication 130 (ICRP 2015b) describes the manner in which radioactive progeny produced post intake are modeled. It is important for the user of internal dose coefficients to understand that the value represents the dose resulting from intake of a pure radioactive parent, but includes dose contribution from the parent's progeny produced by radioactive decay while inside the body. Dose contribution from any radioactive progeny generated outside the body and then ingested or inhaled must be treated as a separate intake. For example, if considering the ingestion of a mixed ^{90}Sr/^{90}Y source material, the ^{90}Sr dose coefficients could be used to estimate dose to the reference individual resulting from the amount of ^{90}Sr originally ingested, and would include dose contribution from ^{90}Y produced while the ^{90}Sr was inside the body. Dose coefficients for ^{90}Y would be needed to consider the dose contribution from ^{90}Y which existed at the time of ingestion.

8.2.2.5 Target Tissues with Constituents

ICRP Publication 133 (ICRP 2016a) presents four target tissues whose doses are computed as weighted sums of multiple constituent regions. For example, to compute dose to the colon, the doses to the right colon, left colon, and rectosigmoid colon are summed using the weighting factors shown in Table 8.3. Equation (8.27) is used to compute the committed equivalent dose coefficients to these target tissues (Table 8.4).

$$h_T = \sum_{r_T} f(r_T, T) h(r_T) \qquad (8.27)$$

TABLE 8.4   Target Region Fractional Weights Given in ICRP Publication 133 (ICRP 2016a)

Target Tissue, T	Constituent Tissue, r_T	Abbreviation	$f(r_T, T)$
Extrathoracic region	ET_1 basal cells	ET1-bas	0.001
	ET_2 basal cells	ET2-bas	0.999
Lung	Bronchi basal cells	Bronch-bas	1/6
	Bronchi secretory cells	Bronch-sec	1/6
	Bronchiolar secretory cells	Bchiol-sec	1/3
	Alveolar-interstitial	AI	1/3
Colon	Right colon	RC-stem	0.4
	Left colon	LC-stem	0.4
	Rectosigmoid colon	RS-stem	0.2
Lymphatic nodes	Extrathoracic lymph nodes	LN-ET	0.08
	Thoracic lymph nodes	LN-Th	0.08
	Systemic lymph nodes	LN-Sys	0.84

8.2.3 Derived Quantities

Part 20 of Title 10 of the Code of Federal Regulations (U.S. Nuclear Regulatory Commission 2014) defines a couple of operational quantities used in radiation protection from intake of radionuclides. The Annual Limit on Intake (ALI) and derived air concentration (DAC) are specific to each radionuclide and mode of intake. Both quantities are defined specific to the internal doses associated with an adult reference worker.

8.2.3.1 Annual Limit on Intake

The ALI is defined as the amount of radioactive material which, if taken into the body will result in the reference worker receiving either the effective dose limit (0.05 Sv) or the equivalent dose limit to an organ or tissue (0.5 Sv). In either case, the committed dose over 50 years post intake is compared to the annual dose limit. Calculation of the ALI can be described with Equation (8.28), as follows:

$$ALI = MIN\left(\frac{0.05}{e}, \frac{0.5}{h_{max}}\right) \tag{8.28}$$

where the committed effective dose coefficient, e, and the maximum equivalent dose coefficient, h_{max}, are in Sv/Bq, yielding an ALI in Bq. Note that an ALI can be separately calculated for each type of intake and for each chemical form of a radionuclide. It is worth mentioning that the latest approach described in this chapter, and presented in the most recent ICRP publications, utilizes gender-specific phantoms rather than anthropomorphic phantoms. Therefore, the calculation gives unique gender-specific committed equivalent dose coefficients to each target. A decision will need to be made by regulators if the ALI is continued to be used in the same manner. Should h_{max} be defined as the largest dose to any target tissue in any gender, or should it be defined as the maximum gender-averaged quantity to any target? Note that the latter quantity is not defined by the ICRP for any meaningful use. For the examples given later in this chapter, the authors have elected to use the largest dose to a target tissue in any gender.

8.2.3.2 Derived Air Concentration

The DAC is defined for inhalation cases only and is the air concentration in a room that would result in the intake of an ALI to a reference worker breathing 1.1 m^3/h of air for a period of 2000 hours. The DAC in Bq per m^3 is then mathematically calculated as:

$$DAC = \frac{ALI}{2200} \qquad (8.29)$$

where the ALI is given in Bq.

8.3 COMPUTATION OF DOSE COEFFICIENTS FOR EXTERNAL ENVIRONMENTAL RADIATION FIELDS

In this section, dose rate coefficients for external exposure that relate the dose rate to organs and tissues of the body to the concentrations of radionuclides in an environmental media are presented. The reader should keep in mind that "environmental" radionuclides in the usage in this chapter *should not be misconstrued* to be naturally occurring radionuclides; rather, it refers to radionuclides released to the environment and consequently being suspended in air and water, as well as deposited and infiltrating the soil. The radiation dose depends strongly on the temporal and spatial distribution of the radionuclide and the duration of the exposure. Since this is a difficult problem, it is common to use idealized source-to-receptor geometries in which the radionuclide is considered uniformly distributed and effectively infinite or semi-infinite in extent. These idealizations provide dose coefficients that are adequate for setting radiation protection standards in terms of addressing the dose due to releases of radioactivity into the air and deposition on the ground and water. These infinite or semi-infinite source representations can result in conservative overestimates of the actual doses when the environmental contamination is finite is extent.

For a uniform concentration $C(t)$ of a radionuclide, at time t, on a surface (units of Bq/m^2) or in a volume (Bq/m^3) which is infinite or semi-infinite infinite in extent, the equivalent dose $H_T(T_A, T_E)$ in tissue T of an individual of age T_A for an exposure time T_E is

$$H_T(T_A, T_E) = \int_0^{T_E} C(\tau)\dot{h}_T(T_A + \tau)d\tau \qquad (8.30)$$

where $\dot{h}_T(t)$ is the time-dependent dose rate coefficient for external exposure. The coefficient $\dot{h}_T$ represents the dose rate to tissue T of the body of an individual of initial age T_A per unit time-integrated exposure expressed in terms of the time-integrated concentration of the radionuclide. The dose rate coefficient depends on the age of the individual since reference pre-adult individuals have less stature and body mass than reference adults. As a consequence, the attenuation properties between the surface of the body and the organs of the body are different for different statured individuals, but for most regulatory and accident response purposes reference phantoms (see discussion in Chapter 5) are used in the computations of $\dot{h}_T(t)$. It might be argued that even during the adult life, the attenuation properties of the body changes, but this is normally not accounted for. In the case of the adult years, the dose rate coefficient is assumed to be independent of time.

Environmental dose coefficients are most frequently constructed by interpolating in energy between the dose rate coefficients computed for monoenergetic particle energies and folding the interpolated particle energies with the emission probabilities from the radionuclide decay scheme and summing to form the radionuclide-specific dose rate coefficients. The radiations of primary concern in computing environmental dose coefficients for radionuclides are photons, electrons, and positrons. Beta and positron dose rate coefficients can be generated in an analogous manner from monoenergetic electron and positron dose rate coefficients by folding them with the energy distributions of the emitted electrons or positrons.

This discussion will be limited to the computation of monoenergetic photon and electron/positron dose rate coefficients which can be used to generate radionuclide-specific dose rate coefficients from the monoenergetic data. The calculation of organ doses from irradiation of the human body by photon emitters distributed in the environment requires the solution of a complex radiation transport problem which can involve the transport of particles through large distances in the air portion and through many mean free paths in soil for infinite or semi-infinite sources. In theory, the dose rate coefficients for environmental sources should be computed for infinite soil, water, and air media. However, in practice, the computations need only be concerned with dimensions that are several mean free paths in extent. However, the transport of particles through several mean free paths of material with Monte Carlo methods is challenging. As a result, Monte Carlo methods require the use of variance reduction techniques to ensure acceptably small uncertainties in the computed doses. Such methods will be briefly discussed, but only in a general sense. The reader is directed to the literature for greater detail on the application of Monte Carlo sampling and techniques in radiation transport problems, as well as for a deeper understanding of variance reduction methods. Not all variance reduction techniques may be appropriate for dose computations, but the ones that can be used serve to greatly reduce computation times.

8.3.1 Dose Rate Coefficients for Contaminated Soil

The two most frequently computed environmental dose rate coefficients used for soil are an infinite planar source on the ground surface and a volumetric source to some depth in the soil. In these cases, the computed dose rate coefficients are almost entirely dependent on photon emission from the radionuclide. To compute a volume source, the approach has been to perform dose rate computations for uniformly contaminated planes at different depths in the soil. This approach is readily extended to generate dose rate constants for volumetric sources for various subsurface distributions of radionuclides by performing a weighted numerical integration of the doses due to a series of plane sources at different depths. The number of planar sources and their spacing may limit the accuracy of describing the variation with depth if it must be folded with complicated depth distribution functions. However, the usual case would be to perform such an integration for soil uniformly contaminated to a desired depth. An example of a distribution that might be used for such an integration would be a Gaussian distribution, which would approximate the depth distribution for an instantaneous or short-term deposition on the surface, followed by infiltration of the contamination into the ground. The use of the planar source on the ground surface is an obvious approximation to fallout or deposition from a plume, while the use of a planar source at some depth below the ground surface could approximate the covering of a layer of contamination by a layer of soil to prevent the spread of surface contamination or provide shielding. Although the capability to perform computations for other phantom positions is possible, reference dose rate coefficients for exposure to contaminated ground source planes at various depths have to date been performed with the phantom, regardless of age, standing upright on the ground surrounded by air.

To be used in the integration process described above to obtain dose rate constants for volumetric sources in the soil, planar source depths are chosen to facilitate an accurate integration to determine continuous source-depth profiles. For a soil contamination of infinite depth, one can use a maximum depth of four times the mean free path of the photon. At this depth, less than 2% of the photons initially emitted directly upward (parallel to the positively directed surface normal) escape from this depth without interaction. Since the photon emission from this plane is isotropic, the fraction of photons from a plane at this depth is far less than 2%. Therefore, using four mean free paths for the depth of a planar source at infinite depth is acceptable.

The air–ground interface (0 mm depth) source is an idealized flat plane without anything other than air above the soil surface. In truth, there are a variety of conditions of sources, including the presence of vegetation, surface irregularities, and migration of the deposited activity downward that provide some shielding of the surface contamination. Rather than a planar source on the surface of the soil, a planar source at a depth of 3 mm has been used to account for this effect, usually referred to as the ground roughness effect (e.g., see Bellamy et al. 2018). The effect of ground roughness by placing the planar source at 3 mm is applicable in the first months after a wet deposition (Jacob et al. 1986; ICRU 1994; Jacob et al. 1994). The reader is referred to the previous work related to surface roughness by Jacob et al. (1986), Saito et al. (1990), Petoussi et al. (1991), and Petoussi-Henss et al. (2012). Other approaches to account for the effect of surface roughness include multiplying the dose rate constant computed on the air–ground interface by a factor ranging from ½ to 2/3 (Spencer, Chilton, and Eisenhauer 1980).

A direct Monte Carlo simulation involves the combination of a deep penetration (i.e., transport through many mean free paths of air and/or soil) and a complex geometry (the human phantom). Since the dose rate coefficients may be desired for different aged, and therefore sized, phantoms, the computations are generally performed in two steps rather than repeatedly running the environmental transport portion of the simulation.

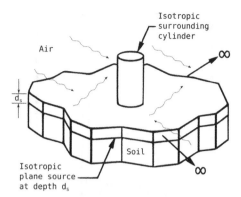

FIGURE 8.12 Radiation field due to a contaminated ground plane, on a cylinder surrounding the phantom (Bellamy et al. 2018).

(1) Calculation of the incident radiation on the phantom (angular and energy fluence) recorded on a closed surface surrounding the phantom (Figure 8.12) is first performed. This surface serves as a coupling surface for the second step in the simulations. The energy and angular data of the photons hitting the coupling surface are stored as a function of position on the coupling surface for step (2).

(2) The photons from the data stored in step (1) are then transported from the coupling (here a cylinder) surface and the doses are tallied in the organs of the phantom (Figure 8.12). This approach permits different sized phantoms to be placed in the cylinder without redoing the entire computation that was performed in step (1). The effective dose rates for the reference persons are computed from the tissue dose rates using the tissue weighting factors; see Table 8.2 for the latest recommended factors from ICRP Publication 103 (ICRP 2007) (Figure 8.13).

This two-step computational approach reduces the complexity of the human phantom from the calculation of the incident radiation field for the infinite or semi-infinite sources. This approach may be employed if the presence of the phantom does not significantly perturb the incoming angular flow rate across the coupling cylinder. The phantom can affect the incoming directions of the radiation field only for those photons which, having interacted in the phantom, pass out of the surrounding surface, scatter in the surrounding media, and return across the closed surface. This is at most a second-order effect, as demonstrated by Saito et al. (1990), and that component of the radiation can be picked up by keeping the air and ground medium in place during the second step of the computation, allowing scatter by the phantom into the environment.

8.3.1.1 Geometry and Compositions

To allow the broadest implementation, the dose rate coefficients should be computed per unit source strength (1 Bq m^{-2}) for each photon energy in a set of monoenergetic photon sources. The number of energies is a compromise between computational time and having

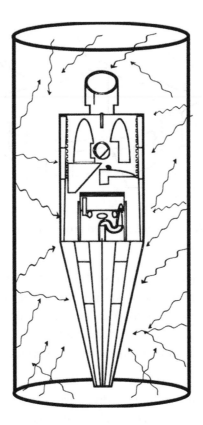

FIGURE 8.13   Angular current source on the coupling cylinder surrounding the phantom (Bellamy et al. 2018).

sufficient detail in energy to allow adequate interpolation. The dose rate values at these photon energies can be interpolated to obtain the dose rate constants for the energies corresponding to the emissions at other photon energies. For reference dose rate constants, the photons are emitted isotropically and uniformly across the plane source. The results of these plane source calculations are used to determine the tissue dose coefficients for a reference individual standing at the air–ground interface using the two-step process outlined above.

Although one could use any of a number of soil compositions, the use of reference soil compositions has largely been the approach undertaken. As an example, one frequently used soil composition, given in Table 8.5, is typical of silty soil (ICRU 1994) containing 30% water and 20% air by volume. Generally, a generic soil density of 1.6×10^3 kg m^{-3} is employed. For most radionuclides, the radiation field above the air–ground interface can be scaled to account for differences in soil density (Beck and De Planque 1968; Chen 1991). Below 100 keV, where the photoelectric effect is more likely to be the dominate interaction channel, photons are readily absorbed in the ground and density scaling is not appropriate (ICRU 1994). Detailed information on the chemical composition of various soils can be found in Helmke (Helmke and Sumner 2000). Changes in soil density have a much larger effect on the dose rate coefficient than changes to the elemental composition.

TABLE 8.5 Soil Composition (ICRU 1994)

Element	Mass Fraction
H	0.021
C	0.016
O	0.577
Al	0.050
Si	0.271
K	0.013
Ca	0.041
Fe	0.011
Total	1.000

The air medium above the ground must be accounted for in the dose rate coefficients to obtain the photon data on the coupling cylinder (step 1). It is sufficient to use a three mean free path thickness of air above the ground surface for the photon source energy. Photons scattered in the atmosphere at a height greater than three mean free paths will have travelled a total distance of more than six mean free paths from the soil surface to reach the coupling surface, and therefore will not make a discernable contribution to the organ doses in the phantom.

8.3.1.2 Variance Reduction and Other Speedups

To reinforce concepts of variance reduction presented in Section 8.1, variance reduction methods are employed to speed up the computations for large problems. in the computation of dose coefficients. In a variance reduction method, a biased probability distribution is sampled in the place of the true probability distribution. The requirement to play a fair game is met by adjusting the weight of the particle to conserve the normalization required not to obtain a biased result. Most Monte Carlo transport codes perform absorption suppression by absorbing a fraction of the particle at each collision and allowing the remainder of the particle to undergo other interacts and continue to be transported. In this case, the fraction of the particle that is absorbed is the ratio of the absorption coefficient to the total attenuation coefficient $\left(\mu_a/\mu_t\right)$ while $\left(1-\mu_a/\mu_t\right)$ of the particle continues to be transported.

As reinforced in Section 8.1, exponential transform or pathlength stretching are other variance reduction methods that artificially reduce the attenuation coefficient in materials and adjust the particle weight to continue playing a fair game; particles can be pushed deeper into a medium without biasing the final result. In the case of soil in the computation of dose coefficients, this can serve to push more photons through the air-soil interface, albeit at reduced weight. It may be useful to employ a forced collision technique as well. In this variance reduction technique, the particles are forced to interact throughout a volume or volumes of materials in the problem, and their weights are adjusted by the probability that they would have interacted at that position. The reader is referred to more detailed literature on variance reduction techniques in Section 8.1.

8.3.1.3 Dose Rate Coefficients for Electrons in Soil

Due to their short range, the dose to tissues other than skin by directly electrons and positrons does not need to be considered. The bremsstrahlung produced by electrons and positrons slowing down in environmental media can deliver dose to other tissues and must be considered. When electrons slow down, a relatively small fraction of their initial kinetic energy (only about 0.2% for a 1.0 MeV electron in air) is converted to bremsstrahlung. Bremsstrahlung energy is distributed from zero up to the initial electron energy. For pure beta emitters, it is the only source of radiations sufficiently penetrating to cause dose to tissues below the skin.

Additionally, the dose due to annihilation photons from positron annihilation must be considered. In the case of a uniformly contaminated depth of soil with a positron emitter, it can be assumed that the 0.511 MeV photon source term is twice the positron emission rate per decay and the data previously computed for planar sources of photons in the soil can be used to compute this component of the dose rate constant.

8.3.2 Submersion in Contaminated Air

Air submersion dose rate coefficients are dose coefficients to an individual exposed to uniformly contaminated air. Here the approach would be to calculate the submersion dose rate coefficients for a set of monoenergetic photon or electron emitters which are uniformly distributed in the air. If one desires to compute the dose rate using multiple phantoms, a two-step procedure analogous to that used in soil contamination computations is applicable:

(1) The energy-angle information as a function of position on the surface of a coupling cylinder is tallied from photons emitted in air without the phantom present.

(2) The tissue-equivalent dose rate coefficients are then computed by transporting the photons from the coupling surface using the data acquired in step (1) after placing a phantom inside the cylinder. This step can be repeated for various sized and age phantoms using the data from step (1), thereby saving redoing the air transport portion of the problem over and over for different sized phantoms.

8.3.2.1 Submersion Dose Due to Photons

The air submersion exposure geometry involves an individual standing on an uncontaminated flat surface of infinite area in uniformly contaminated air. The source is then a semi-infinite cloud containing a uniformly distributed monoenergetic photon emitter of strength (1 Bq m^{-3}) surrounding a phantom standing on the air–ground interface.

The air composition given in Table 8.6 for conditions of 40% relative humidity, a pressure of 760 mm Hg, a temperature of 20°C, and a density of 1.2 kg m^{-3}, has generally been used for reference dose rate coefficient computations (e.g., see Bellamy et al. 2018). The dose rate coefficients for air submersion can be scaled to account for different air densities for application.

TABLE 8.6   Air Composition

Element	Mass Fraction
H	0.00064
C	0.00014
N	0.75086
O	0.23555
Ar	0.01281
Total	1.00000

Most frequently the dose rate coefficients for submersion in a semi-infinite volume of contaminated air have been arrived at by taking one-half the dose rate coefficient due to immersion in an infinite cloud source, see Dillman (1974), Poston and Snyder (1974), and Kocher (1980, 1981). Ryman et al. (1981) has shown this to be a good approximation for air dose (within 20%) at energies of 20 keV or greater. At lower energies, there would be an increase in dose with increasing height along the phantom if the phantom were truly modeled as standing on uncontaminated ground (Bellamy et al. 2018).

One way to speed up the submersion computation in a Monte Carlo code is to employ a reflective boundary condition. In this approach, as performed in the U.S. Environmental Protection Agency's Federal Guidance Report 15 (FGR 15) (Bellamy et al. 2018), one can consider a cube with a side length of 10 m containing an emitter of strength 1 Bq m^{-3}. Note that the dimensions are not unique, but the cube must be larger than the coupling cylinder. By setting reflective boundary conditions (Chilton, Shultis, and Faw 1984) on the six walls of the cube, an infinite volume is represented. Photons contacting the wall are reflected with the angle of incidence equal to the angle of reflection, thus providing the same results as a cylinder placed in an infinite air medium. The cylindrical coupling surface is placed in the middle of the cube and its location is illustrated in Figure 8.14.

8.3.2.2 Organ Dose from Electrons in Air

For an exposed individual standing at the boundary of a semi-infinite, uniformly contaminated atmospheric cloud, the electron or positron direct dose again would only affect the skin dose of the phantoms. The air volume only out to the range of the electron/positron energy from the surface of the phantom is all that has to be considered to represent the effects of an infinite source medium in terms of the direct deposition of energy in the phantom by electrons and positrons. This skin dose can be computed by modeling the effect of that volume outside the phantom in a Monte Carlo code. However, it also lends itself to the use of stopping powers or range-energy relations. This can be performed analytically with the help of existing codes (Shultis and Faw 2000; Bellamy et al. 2018).

The bremsstrahlung production in air must be included in the computations and the production of annihilation photons from positron annihilation must be modeled. In the case of positron emitters, a uniform source of 0.511 MeV photons can be emitted throughout the air medium normalized to twice the positron emission intensity per decay of the

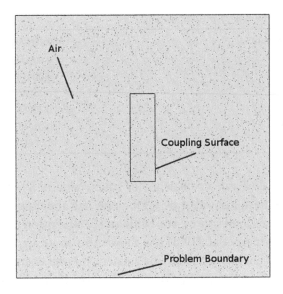

FIGURE 8.14   Geometry used for generating the air submersion coupling surface. The dots denote source particle generated randomly within the cube (Bellamy et al. 2018).

radionuclide. In the case of bremsstrahlung, a multistep process is more productive than modeling a semi-infinite airspace contaminated with a beta emitter and tracking each individual electron to produce the bremsstrahlung. A Monte Carlo computation of the bremsstrahlung production in a box of air sufficient in dimension to fully stop the electrons and positrons can be performed for a set of monoenergetic electrons that encompass the range of energies from radionuclides. This bremsstrahlung production calculation can be used to tally the bremsstrahlung energy resulting from the slowing down of that energy electron. This leads to a set of bremsstrahlung spectra for each electron energy in the monoenergetic set. The bremsstrahlung spectrum due to a particular beta emitter in air can be generated by performing a properly weighted integral of these spectra. Photons can then be emitted uniformly in the air medium using this generated spectrum to compute the bremsstrahlung contribution to the dose rate coefficient. This is beneficial, as the tracking of electrons in Monte Carlo codes is quite time consuming. One may also use the literature to find such sets of bremsstrahlung spectra due to monoenergetic electrons.

8.3.3 Immersion in Contaminated Water

8.3.3.1 Photon Sources in Water

Dose rate coefficients for water immersion are calculated under the assumption that an individual is completely immersed in an infinite volume of uniformly contaminated water. It is possible to simulate an effectively infinite pool using relatively small dimensions (Figure 8.15) in comparison to the contaminated air and soil computations; this is possible since the linear attenuation coefficient of water is much greater than that of air. The phantom can be placed in a relatively small container of water and the Monte Carlo simulation can be performed in a single run for each phantom by generating source photons in the water. This greatly reduces computational time, since the extent of the source is

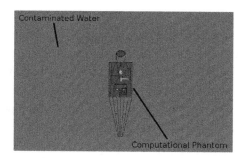

FIGURE 8.15   Computational geometry used for water immersion for the estimating adult organ dose rate coefficients (Bellamy et al. 2018).

no longer infinite. As an example, in FGR 15 the computational phantoms were placed in a cube with 400 cm long sides, corresponding to 11 mean free paths for a 5 MeV photon. Monoenergetic photons can be generated uniformly in the contaminated water. To form radionuclide-specific dose rate coefficients, the monoenergetic photon data can be interpolated, weighted by the emission probabilities, and summed to obtain the radionuclide-specific photon dose rate coefficient.

For lower photon energies (below approximately 30 keV), it is difficult to get reasonable statistical uncertainties for the doses in smaller organs at depths in the body. In such cases, the theory of reciprocity can be applied to improve the statistical uncertainties in the results for a given runtime (King 1912, Mayneord 1940, Bell and Glasstone 1970, Shultis and Faw 2000, and Attix 2008). In this approach, the source photons are generated in the organs and the dose to the water is now tallied. This approach works because the water and tissue compositions have very similar atomic numbers, and at these low energies, virtually all the photons only undergo absorption reactions.

8.3.3.2 Electron Sources in Water

In this case, the amount of water surrounding the phantom only needs to extend to a distance equal to the electron (or positron) range to determine the direct energy deposition. In this small volume of water, a set of monoenergetic electron (or positron) dose rate coefficients can be obtained and used as previously mentioned for the air and soil dose coefficients to form radionuclide specific values. As discussed in the corresponding section on electron dose coefficients in air, these computations can also be performed in an analytic approach using point kernel methods. The bremsstrahlung and annihilation photon dose contributions can be computed in the same manner as for the air contamination methods.

8.3.4 Example of Environmental Dose Rate Coefficients

Although there are multiple sources of environmental dose rate coefficients, a sampling of FGR 15 (Bellamy et al. 2018) is presented herein. The ICRP is also producing a new set of such coefficients, but they were not released before this book was written. The coefficients presented here will be limited to monoenergetic photon values presented in that report. The report contains dose coefficients for over 1200 radionuclides.

8.3.4.1 Computational Phantoms

This work employed a series of stylized phantoms originally developed at ORNL in the early 1980s (Cristy and Eckerman 1987) and modified by Han, Bolch, and Eckerman (2006). The reader is referred to Appendix B of that report, as well as Chapter 5 of the present text, for more information on the history and development of the phantoms. It was noted by the authors of FGR 15 that several organ volumes in the updated phantoms were found to be imprecisely declared when comparing the volume obtained by the ray tracing capability within the Monte Carlo n-Particle radiation transport code, MCNP6 (Pelowitz 2013). In that work, corrections were made whenever the disparity in volume was greater than 3%. The family of phantoms consisted of newborn, 1-year-old, 5-year-old, 10-year-old, 15-year-old, and adult phantoms (Figure 8.16).

8.3.4.2 Calculating Absorbed Dose for Contaminated Soil

8.3.4.2.1 Photons

In FGR 15 the radiation field due to isotropic infinite plane sources was computed for 13 photon energies and for planar sources at six depths in the soil. The monoenergetic photon energies ranged from 0.01 to 5.0 MeV. Those include the ground surface plane, the surface roughness approximation plane at 3 mm, and planar sources at depths corresponding to 0.2, 1.0, 2.5, and 4.0 mean free paths. The authors of that report used a total soil thickness of 3 mean free paths for planar sources up to 1 mean free path deep while for the source planes at depths of 2.5 and 4 mean free paths, the total soil thicknesses are 3.5 and 5 mean free paths, respectively.

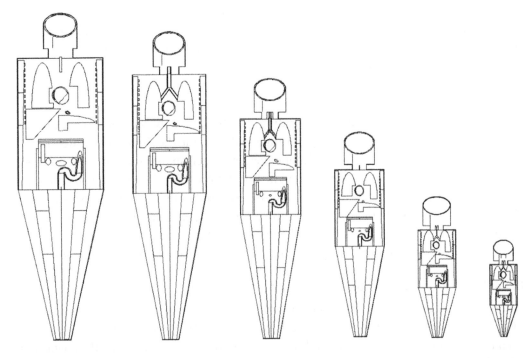

FIGURE 8.16 "Family" of phantoms including newborn through 15 years, plus adults used in the derivation of external dose rate coefficient.

The first step of computing the required data on the cylindrical coupling surface was performed using the "surface source write" feature of the MCNP6 code (Pelowitz 2013). This code option records the position, angle, and energy of photons incident on the coupling surface cylinder due to emission of monoenergetic photons from the ground plane source. In that study, electrons liberated by photon interactions were not transported, but bremsstrahlung photons were generated and transported using the thick target assumption model available in MCNP6.

For all photon energies and source depths, coupling cylinders 200 cm high with 30 cm radii were used. These coupling cylinders were chosen to be as small as possible while still completely enclosing the largest phantom. The distance between the base of the coupling cylinder and the air-soil interface was 0.01 cm. The coupling cylinder served as a passive detector/recorder of incident photons, thus the presence of the coupling surface did not affect the photon transport. Absorbed doses to tissues were computed using the kerma approximation using track length estimators. Tissue doses were calculated as the product of photon fluence and tissue-specific kerma coefficients. Absorbed dose rate coefficients for active marrow and bone surface were based on a track-length estimate of skeletal fluence combined with ICRP skeletal fluence-to-dose response functions (Cristy and Eckerman 1987; ICRP 2010).

After the coupling surface computations were performed, the phantoms were placed individually inside the coupling cylinder with the volume between the cylinder and phantom filled with air. The distance between the cylinder base and the phantom was less than 0.1 cm. The resulting dose rate coefficients for the monoenergetic planar source on the surface of the ground are tabulated in Table 8.7 for the six phantoms, as is the air kerma rate at 1 meter above the ground. By and large, the dose rate coefficients in that table decrease with the phantom age for a given energy. This is qualitatively explained by the fact that that the dose rate at a point above the surface decreases with distance above the surface, and

TABLE 8.7 Effective Dose Rate Coefficient: Monoenergetic Ground Plane Source

Photon Energy (MeV)	Air Kerma (Gy m²/Bq s)	Effective Dose Rate Coefficient (Sv m²/Bq s)					
		Newborn	1 y	5 y	10 y	15 y	Adult
0.01	2.94E–16	4.70E–18	2.99E–18	1.85E–18	1.77E–18	6.91E–19	6.40E–19
0.015	2.75E–16	1.46E–17	9.95E–18	8.09E–18	7.76E–18	3.10E–18	2.83E–18
0.02	2.19E–16	2.75E–17	1.68E–17	1.42E–17	1.30E–18	7.30E–18	6.80E–18
0.03	1.39E–16	5.43E–17	3.70E–17	2.94E–17	2.60E–17	1.93E–17	1.84E–17
0.05	8.90E–17	7.56E–17	6.33E–17	5.55E–17	5.10E–17	4.26E–17	4.13E–17
0.07	9.02E–17	8.93E–17	8.13E–17	7.42E–17	7.00E–17	6.14E–17	5.98E–17
0.1	1.18E–16	1.18E–16	1.10E–16	1.03E–16	9.85E–17	8.83E–17	8.65E–17
0.15	1.82E–16	1.78E–16	1.65E–16	1.54E–16	1.49E–16	1.35E–16	1.32E–16
0.2	2.52E–16	2.42E–16	2.25E–16	2.09E–16	2.02E–16	1.83E–16	1.80E–16
0.5	6.67E–16	6.19E–16	5.76E–16	5.38E–16	5.19E–16	4.74E–16	4.65E–16
1	1.25E–15	1.18E–15	1.10E–15	1.04E–15	1.00E–15	9.28E–16	9.12E–16
2	2.17E–15	2.10E–15	1.98E–15	1.87E–15	1.83E–15	1.71E–15	1.68E–15
5	4.25E–15	4.17E–15	3.98E–15	3.82E–15	3.74E–15	3.56E–15	3.51E–15

the organs for the older phantoms are higher above the ground used in the computation of effective dose and are higher above the surface, as well as shielded by more intervening body mass.

In FGR 15, dose rate coefficients for volumetric sources were obtained by first interpolating the dose rate coefficients for the six planes over soil depth and then integrating over the source volume. In this manner, the coefficients were obtained for uniformly contaminated soil to thicknesses of 1, 5, and 15 cm, and "infinite" in extent (4 mean free paths).

If is the dose rate coefficient (Sv m² Bq⁻¹ s⁻¹) for tissue T for a plane isotropic source P at energy E and depth τ (mean free paths), then the dose rate coefficient for a volumetric source extending from the air–ground interface to depth L (cm) is

$$\hat{h}_{T,L}(E) = \frac{1}{\mu}\int_0^{\mu L}\hat{h}_{T,P}(E,\tau)\,d\tau \qquad (8.31)$$

where μ is the linear attenuation coefficient (including coherent scattering) for soil at energy E (Berger et al. 1998). In FGR 15, the dose rate coefficients for each organ at the six source depths were interpolated on a fine grid using a log-linear Hermite cubic spline (Fritsch and Carlson 1980) and then integrated.

The monoenergetic effective dose rate coefficients as computed and reported in FGR 15 are plotted in Figure 8.17 for the six reference age groups considered (Figures 8.18 through 8.22).

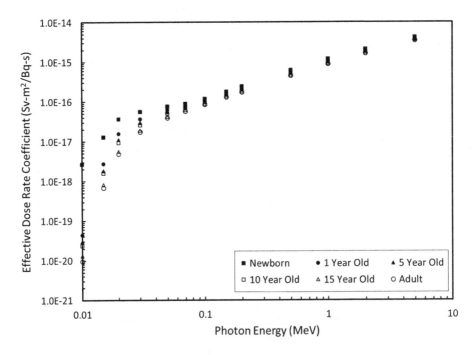

FIGURE 8.17 Effective dose rate coefficients (Sv m² Bq⁻¹ s⁻¹) for an infinite planar source at the ground surface.

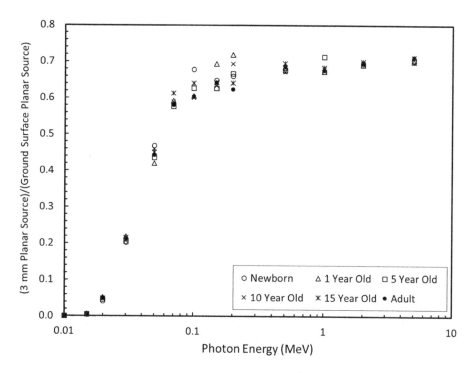

FIGURE 8.18 Ratio of the effective dose rate due to a planar source at 3 mm to account for ground roughness to the ground surface planar source of Figure 8.17.

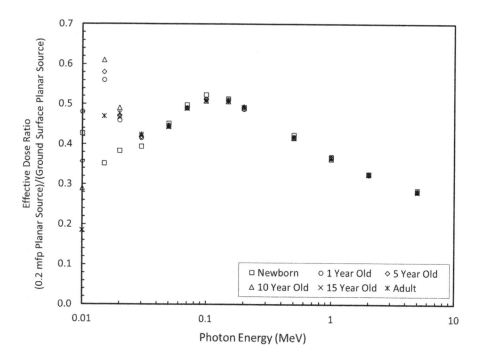

FIGURE 8.19 Ratio of the effective dose rate due to a planar source at 0.2 mean free path to the ground surface planar source of Figure 8.17.

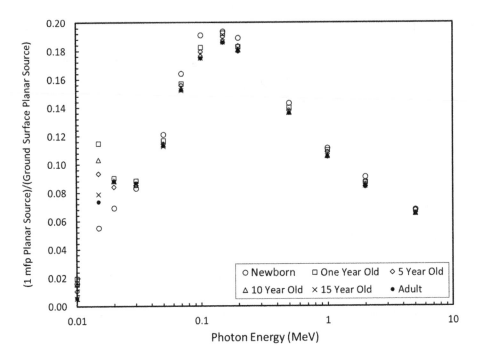

FIGURE 8.20   Ratio of the effective dose rate due to a planar source at 1 mean free path to the ground surface planar source of Figure 8.17.

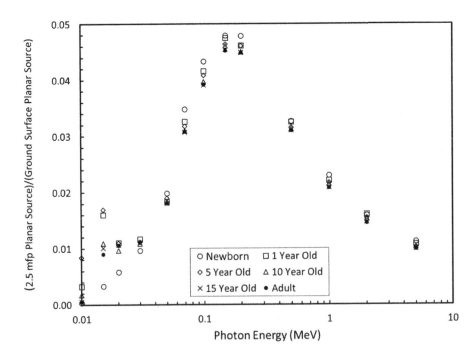

FIGURE 8.21   Ratio of the effective dose rate due to a planar source at 2.5 mean free paths to the ground surface planar source of Figure 8.17.

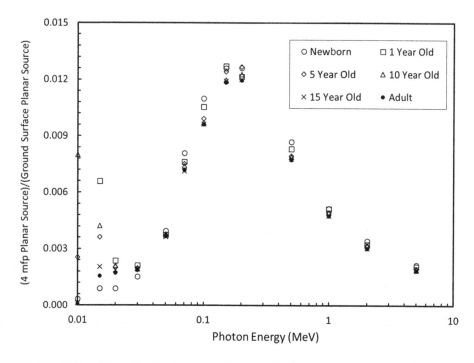

FIGURE 8.22 Ratio of the effective dose rate due to a planar source at 4.0 mean free paths to the ground surface planar source of Figure 8.17.

8.3.4.2.2 Electrons

The DOSFACTER code of Kocher (1981) was used to calculate skin dose rate coefficients for a set of monoenergetic electron emissions. The resulting coefficients were convoluted for radionuclides using the energy and intensity of beta and electron emissions from ICRP Publication 107 (ICRP 2008). The reader is directed to Kocher (1981) for the details of the code, and to FGR 15 for a more in-depth discussion of the generation of the dose coefficients. The radionuclide beta spectra from the ICRP Publication 107 (ICRP 2008) dosimetric data file were used to evaluate the contribution of the beta particles to the skin dose and also in the determination of the bremsstrahlung yield.

8.3.4.3 Calculation Absorbed Dose for Submersion in Contaminated Air

In FGR 15, the tissue dose rate coefficients in each phantom are computed in a similar approach to the surface soil source calculations using MCNP6. Again, the tissue doses were calculated as the product of a track-length fluence estimator and tissue-specific kerma coefficients. The resulting dose coefficient is divided by 2 to account for the semi-infinite geometry.

In FGR 15, the maximum electron continuous slowing down range in air is 1 m. Therefore, the air source region is effectively infinite in extent and the DOSFACTER code was again used to compute the skin dose due to electrons. The reader again is referred to FGR 15 for a more detailed discussion.

8.3.4.4 Calculation Absorbed Dose for Submersion in Contaminated Water

The radionuclide-specific contribution of electrons to skin dose is derived using a point kernel method as in the case of air submersion. The photon dose rate coefficients were determined according to the manner previously described, as was the bremsstrahlung.

8.3.4.5 Dose Rate Coefficients for Radionuclides

In FGR 15, the first step in the development of external dose rate coefficients for radionuclides was to derive dose rates to tissues for monoenergetic photon sources at 13 energies ranging from 0.01 to 5.0 MeV. Radionuclide-specific dose rate coefficients were constructed from the monoenergetic coefficients using the energy spectrum of the desired radionuclide. For nearly all radionuclides addressed in FGR 15, external dose is due entirely to gamma and electron emissions. Prompt and delayed emissions for photons and β^+ and β^- particles following spontaneous fission are included in the decay data tabulations of ICRP Publication 107 (ICRP 2008) and thus were included in the calculations of dose rate coefficients in FGR 15. The contribution from neutrons accompanying spontaneous fission was not included in the computations.

The energies and intensities of the radiations emitted by the radionuclide decays were from ICRP Publication 107 (ICRP 2008). The equivalent dose rate coefficient h_T^S for tissue T and exposure mode S was expressed as

$$h_T^S = \sum_{j=e,\gamma}\left[\sum_i y_j(E_i)\,\hat{h}_{T,j}^S(E_i) + \int_0^\infty y_j(E)\,\hat{h}_{T,j}^S(E_i)dE\right] \tag{8.32}$$

where $\hat{h}_{T,j}^S(E_i)$ is the equivalent dose rate coefficient for tissue T irradiated in exposure mode S by monoenergetic radiation of type j, and energy E_i, $y_j(E_i)$ is the yield of discrete radiations of type j, and energy E_i, and $y_j(E)$ denotes the yield of continuous radiations per nuclear transformation with energy between E and $E + dE$.

8.4 SAMPLE CALCULATIONS BASED ON DOSE COEFFICIENTS

The methods described in Section 8.2 have been used to compute dose coefficients which appear in the ICRP Occupational Intake of Radionuclides Series of publications (2015b, 2016b, 2017). In the following sections, example dose coefficients are presented for five different examples of radionuclide intake. In each case, the most important source regions are presented along with the largest committed equivalent dose coefficients. Table 8.8 summarizes the committed dose coefficients for each example and Table 8.9 gives the ALI and, for inhalations, the DAC. In both inhalation cases, the ICRP default particle size for workers of 5 μm AMAD has been used.

8.4.1 Ingestion of ^{90}Sr

Due to its similar valence electron structure to calcium and 29-year half-life, strontium-90 will be taken up in the mineral portion of the skeleton. Figure 8.23 shows that the dominant majority of nuclear transformations over the commitment period occur in the bone

TABLE 8.8   Example Dose Coefficients

Radionuclide and Intake Mode	Committed Effective Dose Coefficient (Sv/Bq)	Target Tissue Receiving Largest Equivalent Dose	Committed Equivalent Dose Coefficient (Sv/Bq)
^{137}Cs ingestion $f_A = 1.0$	1.4×10^{-8}	Female kidneys	3.1×10^{-8}
^{90}Sr ingestion $f_A = 0.25$	2.4×10^{-8}	Female bone surface	2.7×10^{-7}
^{131}I ingestion	1.6×10^{-8}	Female thyroid	4.4×10^{-7}
^{210}Po inhalation type M	1.1×10^{-6}	Female extrathoracic	2.0×10^{-5}
^{239}Pu inhalation type S	1.8×10^{-5}	Female lymph nodes	2.6×10^{-4}

TABLE 8.9   Example ALIs and DACs

Radionuclide and Intake Mode	Annual Limit on Intake (Bq)	ALI Based on Effective or Equivalent Dose	DAC (Bq/m³)
^{137}Cs ingestion $f_A = 1.0$	3.7×10^6	Effective	NA
^{90}Sr ingestion $f_A = 0.25$	1.8×10^6	Equivalent	NA
^{131}I ingestion	1.1×10^6	Equivalent	NA
^{210}Po inhalation type M	2.6×10^4	Equivalent	12
^{239}Pu inhalation type S	2.0×10^3	Equivalent	0.89

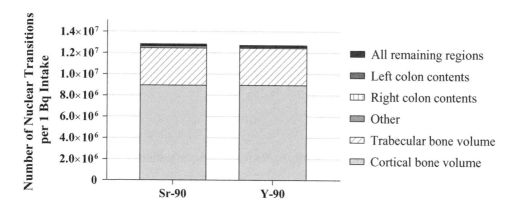

FIGURE 8.23   Number of nuclear transformations per 1 Bq ingestion of ^{90}Sr stacked by source region.

volume. The progeny of strontium-90, yttrium-90, is much shorter lived and therefore does not move appreciably far away in the body from the source region it is born in before decaying. Therefore, the number of transformations for parent and progeny are roughly the same in all source regions.

Since the beta particles accompanying the decay of ^{90}Sr and ^{90}Y primarily originate from the bone volumes, the sensitive tissues receiving the largest equivalent doses are the bone surface (endosteum) and the red, or active marrow. This is apparent when viewing Figure 8.24.

8.4.2 Ingestion of ^{131}I

The thyroid uses iodine to produce hormones required for the regulation of metabolic processes, therefore ingested iodine will quickly find itself in the thyroid (ICRP 2017). Figure 8.25

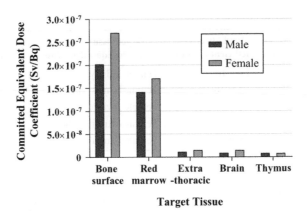

FIGURE 8.24   The largest committed equivalent dose coefficients due to ingestion of ^{90}Sr.

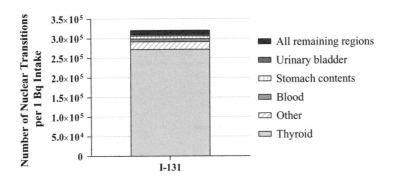

FIGURE 8.25   Number of nuclear transformations per 1 Bq ingestion of ^{131}I stacked by source region.

shows that the thyroid is the dominant source region for ingestion of iodine-131. Correspondingly, the emitted radiations (primarily beta) deposit the majority of their energy in the thyroid as seen in Figure 8.26.

8.4.3 Ingestion of ^{137}Cs

Due to its valence electron structure, cesium exhibits biochemistry similar to potassium. Due to the body's widespread use of potassium, cesium tends to accumulate in a variety of tissues but will favor skeletal muscle, as shown in Figure 8.27 (ICRP 2017). Cesium-137's progeny is barium-137m whose half-life (2.25 min) is short, but long enough to allow for migration out of the tissue it is created in and back to the bloodstream. Given the distribution of muscle throughout the body, and the long-range nature of ^{137m}Ba's 661.6 keV gamma ray, the dose to tissues ends up more uniformly distributed than the other examples in this chapter, as seen in Figure 8.28.

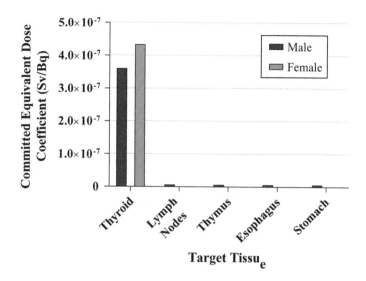

FIGURE 8.26　The largest committed equivalent dose coefficients due to ingestion of ^{131}I.

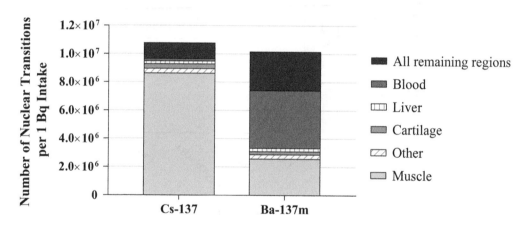

FIGURE 8.27　Number of nuclear transformations per 1 Bq ingestion of ^{137}Cs stacked by source region.

8.4.4 Inhalation of ^{210}Po

The inhalation of polonium-210, type M predictably results in a large number of nuclear decays taking place in the lung, primarily in the alveolar interstitial tissue. Since ^{210}Po is an alpha emitter, the most important target tissues include the lungs. Figure 8.29 also shows a dominating dose to the sensitive tissues in the extrathoracic region (nose). While the number of transitions occurring in the nose are not high enough to show up separately in Figure 8.30, the physical proximity between the extrathoracic source regions and the small nasal target tissues result in a large specific absorbed fraction for alpha particles, and therefore a dominating committed equivalent dose coefficient for the extrathoracic region.

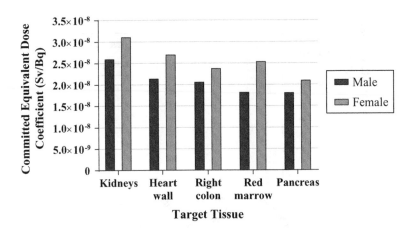

FIGURE 8.28 The largest committed equivalent dose coefficients due to ingestion of [137]Cs.

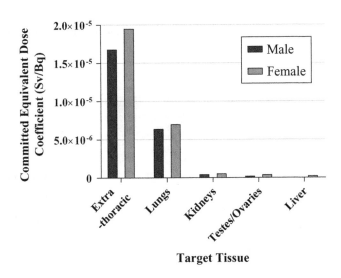

FIGURE 8.29 The largest committed equivalent dose coefficients due to inhalation of [210]Po.

8.4.5 Inhalation of [239]Pu

Similar to the prior example, inhalation of Type S plutonium-239 results in a significant number of decays taking place in the lung (alveolar interstitium). However, due to the long half-life of [239]Pu (24,110 years), it has time to move out of the lung into other tissues. Figure 8.31 shows that significant activity will be present in the liver, the bone surfaces, and the thoracic lymph nodes, which are involved in clearing material deposited in lung tissue. Since [239]Pu is an alpha emitter, significant doses are seen in the corresponding target tissues to the source regions identified in Figure 8.31. Figure 8.32 indicates significant committed equivalent dose coefficients in the extrathoracic tissue, lymphatic nodes, lungs, liver, and bone surfaces. Note that the dose to the lymphatic nodes is driven by dose to those nodes in the thoracic region which are invoked in transportation of plutonium out of the lung tissue.

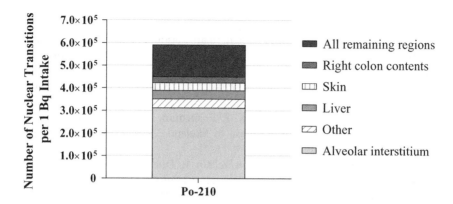

FIGURE 8.30 Number of nuclear transformations per 1 Bq inhalation of ^{210}Po stacked by source region.

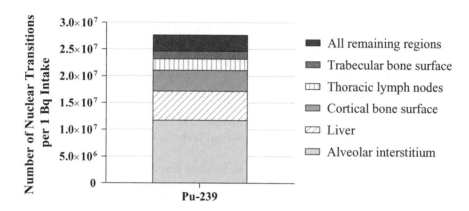

FIGURE 8.31 Number of nuclear transformations per 1 Bq inhalation of ^{239}Pu stacked by source region.

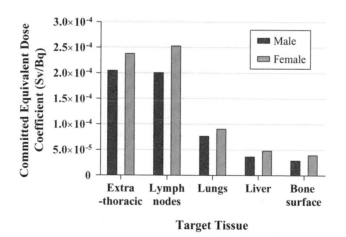

FIGURE 8.32 The largest committed equivalent dose coefficients due to inhalation of ^{239}Pu.

REFERENCES

American National Standards Institute. 1977. American National Standard: *Neutron and Gamma-Ray Flux-to-Dose-Rate Factors (ANSI/ANS 6.1.1-1977)*. La Grange Park, IL: American Nuclear Society.

Attix, F. H. 2008. *Introduction to Radiological Physics and Radiation Dosimetry*. New York, NY: John Wiley & Sons.

Bahadori, A. A., P. Johnson, D. W. Jokisch, K. F. Eckerman, and W. E. Bolch. 2011. "Response Functions for Computing Absorbed Dose to Skeletal Tissues from Neutron Irradiation". *Physics in Medicine & Biology* 56 (21):6873.

Beck, H., and G. De Planque. 1968. *The Radiation Field in Air Due to Distributed Gamma-Ray Sources in the Ground*. New York, NY: New York Operations Office (Atomic Energy Commission).

Bell, G. I., and S. Glasstone. 1970. *Nuclear Reactor Theory*. Washington, DC: US Atomic Energy Commission.

Bellamy, M. B., S. A. Dewji, R. W. Leggett, M. M. Hiller, K. G. Veinot, R. P. Manger, K. F. Eckerman et al. 2018. *Federal Guidance Report No. 15: External Exposure to Radionuclides in Air, Water and Soil*. Edited by U.S. Environmental Protection Agency. Washington, DC: U.S. Environmental Protection Agency.

Berger, M. J., Hubbell, J. H. Seltzer, S. M., Chang, J., Coursey, J. S., Sukumar, R., Zucker D. S., and Olson, K. 1998. *XCOM: Photon Cross Sections Database, NBSIR 87-3597*. Washington, DC: National Institute of Standards and Technology.

Bolch, W. E., K. F. Eckerman, G. Sgouros, and S. R. Thomas. 2009. "MIRD Pamphlet No. 21: A Generalized Schema for Radiopharmaceutical Dosimetry-Standardization of Nomenclature". *Journal of Nuclear Medicine* 50 (3):477.

Carter, L. L., and E. D. Cashwell. 1975. *Particle-Transport Simulation with the Monte Carlo Method*. Edited by U.S. Department of Commerce National Technical Information Service. Springfield, VA: Los Alamos Scientific Laboratory.

Chen, S. Y. 1991. "Calculation of Effective Dose-Equivalent Responses for External Exposure from Residual Photon Emitters in Soil". *Health Physics* 60 (3):411–426.

Chen, S. Y., and A. B. Chilton. 1979. "Depth-Dose Relationships near the Skin Resulting from Parallel Beams of Fast Neutrons". *Radiation Research* 77 (1):21–33.

Chilton, A. B., J. K. Shultis, and R. E. Faw. 1984. *Principles of Radiation Shielding*. Old Tappan, NJ: Prentice Hall Inc.

Claiborne, H. C., and D. K. Trubey. 1970. "Dose Rates in a Slab Phantom from Monoenergetic Gamma Rays". *Nuclear Applications and Technology* 8 (5):450–455.

Cristy, M., and K. F. Eckerman. 1987. *Specific Absorbed Fractions of Energy at Various Ages from Internal Photon Sources: Parts I-VII*. Oak Ridge, TN: Oak Ridge National Laboratory.

Dillman, L. T. 1974. "Absorbed Gamma Dose Rate for Immersion in a Semi-Infinite Radioactive Cloud". *Health Physics* 27 (6):571–580.

Fritsch, F. N., and R. E. Carlson. 1980. "Monotone Piecewise Cubic Interpolation". *SIAM Journal on Numerical Analysis* 17 (2):238–246.

Haghighat, A. 2016. *Monte Carlo Methods for Particle Transport*. Boca Raton, FL: CRC Press.

Han, E. Y., W. E. Bolch, and K. F. Eckerman. 2006. "Revisions to the ORNL Series of Adult and Pediatric Computational Phantoms for Use with the MIRD Schema". *Health Phys.* 90 (4):337–56.

Hansen, G. E., and Sandmeier. 1965. "Neutron Penetration Factors Obtained by Using Adjoint Transport Calculations". *Nucle. Sci. Engr.* 22:315–320.

Helmke, P. A., and M. E. Sumner. 2000. "The Chemical Composition of Soils". In M. E. Sumner (ed), *Handbook of Soil Science*. Boca Raton, FL: CRC Press, B3–B24.

Hiller, M. M., K. Veinot, C. E. Easterly, N. E. Hertel, K. F. Eckerman, and M. B. Bellamy. 2016. "Reducing Statistical Uncertainties in Simulated Organ Doses of Phantoms Immersed in Water". *Radiation Protection Dosimetry* 174 (4):439–448.

Hough, M., P. Johnson, D. Rajon, D. Jokisch, C. Lee, and W. Bolch. 2011. "An Image-Based Skeletal Dosimetry Model for the ICRP Reference Adult Male—Internal Electron Sources". *Physics in Medicine & Biology* 56 (8):2309.

International Commission on Radiation Units. 2011. ICRU Report 85a: Fundamental Quantities and Units for Ionizing Radiation (Revised). *Journal of the ICRU* 11 (1).

International Commission on Radiation Units. 1994. ICRU Report 53: Gamma Ray Spectrometry in the Environment. *Journal of the ICRU* 27 (2).

International Commission on Radiation Units. 1998. "ICRU Report 57: Conversion Coefficients for Use in Radiological Protection against External Radiation". *Journal of the ICRU* 29 (2).

International Commission on Radiological Protection. 1971. *ICRP Publication 21: Data for Protection against Ionizing from External Sources: Supplement to ICRP Publication 15*. Oxford, U.K. Pergamon Press.

International Commission on Radiological Protection. 1977. ICRP Publication 26: Recommendations of the ICRP. *Ann ICRP* 1 (3).

International Commission on Radiological Protection. 1979. ICRP Publication 30 (Part 1): Limits for Intakes of Radionuclides by Workers. *Ann ICRP* 2 (3–4).

International Commission on Radiological Protection. 1985. ICRP Publication 45: Statement from the 1985 Paris Meeting of the International Commission on Radiological Protection: Quantitative Bases for Developing a Unified Index of Harm. *Ann ICRP* 15 (3).

International Commission on Radiological Protection. 1988. *ICRP Publication 51: Data for Use in Protection from External Radiation*. New York, NY: Pergamon Press.

International Commission on Radiological Protection. 1991. ICRP Publication 60: 1990 Recommendations of the International Commission on Radiological Protection. *Ann ICRP* 21 (1–3).

International Commission on Radiological Protection. 1994. "ICRP Publication 66: Human Respiratory Tract Model for Radiological Protection". *Ann ICRP* 24 (1–3):1–482.

International Commission on Radiological Protection. 1996. ICRP Publication 74: Conversion Coefficients for Use in Radiological Protection against External Radiation. *Ann ICRP* 26 (3–4).

International Commission on Radiological Protection. 2002. "ICRP Publication 89: Basic Anatomical and Physiological Data for Use in Radiological Protection - Reference Values". *Ann ICRP* 32 (3–4):1–277.

International Commission on Radiological Protection. 2007. ICRP Publication 103: The 2007 Recommendations of the International Commission on Radiological Protection. *Ann ICRP* 37 (2–4).

International Commission on Radiological Protection. 2008. "ICRP Publication 107: Nuclear Decay Data for Dosimetric Calculations." *Ann ICRP* 38 (3):1–26.

International Commission on Radiological Protection. 2009. "ICRP Publication 110: Adult Reference Computational Phantoms." *Ann ICRP* 39 (2):1–165.

International Commission on Radiological Protection. 2010. ICRP Publication 116: Conversion Coefficients for Radiological Protection Quantities for External Radiation Exposures. *Ann ICRP* 40 (2–5).

International Commission on Radiological Protection. 2015a. "ICRP Publication 128: Radiation Dose to Patients from Radiopharmaceuticals–A Compendium of Current Information Related to Frequently Used Substances". *Ann ICRP* 44 (2S):1–321.

International Commission on Radiological Protection. 2015b. "ICRP Publication 130: Occupational Intakes of Radionuclides: Part 1". *Ann ICRP* 44 (2):1–188.

International Commission on Radiological Protection. 2016a. "ICRP Publication 133: The ICRP Computational Framework for Internal Dose Assessment for Reference Adults: Specific Absorbed Fractions". *Ann ICRP* 45 (2):1–74.

International Commission on Radiological Protection. 2016b. "ICRP Publication 134: Occupational Intakes of Radionuclides: Part 2". *Ann ICRP* 45 (3–4):1–352.

International Commission on Radiological Protection. 2017. "ICRP Publication 137: Occupational Intakes of Radionuclides: Part 3". *Ann ICRP* 46 (3–4):1–486.

Jacob, P., H. G. Paretzke, H. Rosenbaum, and M. Zankl. 1986. "Effective Dose Equivalents for Photon Exposures from Plane Sources on the Ground". *Radiation Protection Dosimetry* 14 (4):299–310.

Jacob, P., R. Meckbach, H. G. Paretzke, I. Likhtarev, I. Los, L. Kovgan, and I. Komarikov. 1994. "Attenuation Effects on the Kerma Rates in Air after Cesium Depositions on Grasslands". *Radiation and Environmental Biophysics* 33 (3):251–267.

Johnson, P. B., A. A. Bahadori, K. F. Eckerman, C. Lee, and W. E. Bolch. 2011. "Response Functions for Computing Absorbed Dose to Skeletal Tissues from Photon Irradiation—an Update". *Physics in Medicine & Biology* 56 (8):2347.

Jokisch, D. W., D. A. Rajon, A. A. Bahadori, and W. E. Bolch. 2011. "An Image-Based Skeletal Model for the ICRP Reference Adult Male—Specific Absorbed Fractions for Neutron-Generated Recoil Protons". *Physics in Medicine & Biology* 56 (21):6857.

King, L. V. 1912. "Xx. Absorption Problems in Radioactivity". *The London, Edinburgh, and Dublin Philosophical Magazine and Journal of Science* 23 (134):242–250.

Kocher, D. C. 1980. "Effects of Indoor Residence on Radiation Doses from Routine Releases of Radionuclides to the Atmosphere". *Nuclear Technology* 48 (2):171–179.

Kocher, D. C. 1981. *Dose-Rate Conversion Factors for External Exposure to Photons and Electrons*. Oak Ridge, TN: Oak Ridge National Laboratory.

Lewis, E. E., and W. F. Miller. 1984. *Computational Methods of Neutron Transport*. New York, NY: John Wiley and Sons.

Loevinger, R., T. F. Budinger, and E. E. Watson. 1991. *MIRD Primer for Absorbed Dose Calculations (Revised)*. New York, NY: Society of Nuclear Medicine.

Lux, I., and L. Koblinger. 1991. *Monte Carlo Particle Transport Methods: Neutron and Photon Calculations*. Boca Raton, FL: Chemical Rubber Company.

Mayneord, W. V. 1940. "Energy Absorption - the Mathematical Theory of Integral Dose in Radium Therapy". *British Journal of Radiology* 17 (205):253–247.

National Council on Radiation Protection and Measurements. 1971. *NCRP Report 38: Protection against Neutron Radiation*. Bethesda, MD: National Council on Radiation Protection and Measurements.

O'Brien, K. 1980. "Lecture 2: The Physics of Radiation Transport". In W. R. Nelson and T. M. Jenkins (eds), *Computer Techniques in Radiation Transport and Dosimetry*. New York, NY: Plenum Press, 17–56.

O'Reilly, S. E., L. S. DeWeese, M R. Maynard, D. A. Rajon, M. B. Wayson, E. L. Marshall, and W. E. Bolch. 2016. "An Image-Based Skeletal Dosimetry Model for the ICRP Reference Adult Female—Internal Electron Sources". *Physics in Medicine & Biology* 61 (24):8794.

Pelowitz, D. B. 2013. *MCPN6 User's Manual Version 1.0*. Los Alamos, NM: Los Alamos National Laboratory.

Petoussi, N., P. Jacob, M. Zankl, and K. Saito. 1991. "Organ Doses for Foetuses, Babies, Children and Adults from Environmental Gamma Rays". *Radiation Protection Dosimetry* 37 (1):31–41.

Petoussi-Henss, N., H. Schlattl, M. Zankl, A. Endo, and K. Saito. 2012. "Organ Doses from Environmental Exposures Calculated Using Voxel Phantoms of Adults and Children". *Physics in Medicine & Biology* 57 (18):5679.

Poston, J. W., and W. S. Snyder. 1974. "A Model for Exposure to a Semi-Infinite Cloud of a Photon Emitter". *Health Physics* 26 (4):287–293.

Profio, A. E. 1979. *Radiation Shielding and Dosimetry*. New York, NY: Wiley.

Roesch, W. C. 1987. *US–Japan Joint Reassessment of Atomic Bomb Radiation Dosimetry in Hiroshima and Nagasaki. DS86 Dosimetry System 1986. Vol. 1*. Hiroshima, Japan: The Radiation Effects Research Foundation.

Ryman, J. C., R. E. Faw, and K. Shultis. 1981. "Air–Ground Interface Effect on Gamma-Ray Submersion Dose". *Health Physics* 41 (5):759–768.

Saito, K., N. Petoussi, M. Zankl, R. Veit, P. Jacob, and G. Drexler. 1990. *Calculation of Organ Doses from Environmental Gamma Rays Using Human Phantoms and Monte Carlo Methods. Pt. 1.* Munich: Gesellschaft fuer Strahlen-und Umweltforschung mbH.

Schaeffer, N. M. 1973. *Reactor Shielding for Nuclear Engineers.* Oak Ridge, TN: U.S. Atomic Energy Commission, Office of Information Services.

Shannon, M. P. 2009. *The Dosimetry of a Highly-Collimated Bremsstrahlung Source in Air.* Ph.D. thesis, George W. Woodruff School of Mechanical Engineering, Georgia Institute of Technology.

Shultis, J., and W. Dunn. 2012. *Exploring Monte Carlo Methods.* Boston, MA: Elsevier.

Shultis, J. K., and R. E. Faw. 2000. *Radiation Shielding.* La Grange Park, IL: American Nuclear Society.

Siebert, B. R. L., and R. H. Thomas. 1997. "Computational Dosimetry". *Radiation Protection Dosimetry* 70 (1–4):371–378.

Spencer, Lewis Van Clief, Arthur B. Chilton, and Charles Eisenhauer. 1980. *Structure Shielding against Fallout Gamma Rays from Nuclear Detonations.* Gaithersburg, MD: U.S. Dept. of Commerce, National Bureau of Standards.

Turner, J. E., H. A. Wright, and R. N. Hamm. 1985. "A Monte Carlo Primer for Health Physicists". *Health Physics* 48 (6):717–733.

U.S. Nuclear Regulatory Commission. 2014. *Code of Federal Regulations Title 10, Part 20: Standards for Protection against Radiation.* Washington, DC: Government Printing Office.

Veinot, K. G., K. F. Eckerman, and N. E. Hertel. 2015. "Organ and Effective Dose Coefficients for Cranial and Caudal Irradiation Geometries: Photons". *Radiation Protection Dosimetry* 168 (2):167–174.

Veinot, K. G., K. F. Eckerman, N. E. Hertel, and M. M. Hiller. 2016. "Organ and Effective Dose Coefficients for Cranial and Caudal Irradiation Geometries: Neutrons". *Radiation Protection Dosimetry* 175 (1):26–30.

Wayson, M., C. Lee, G. Sgouros, S. T. Treves, E. Frey, and Wy E. Bolch. 2012. "Internal Photon and Electron Dosimetry of the Newborn Patient—A Hybrid Computational Phantom Study". *Physics in Medicine & Biology* 57 (5):1433.

Young, R. W., and G. D. Kerr. 2002. *Reassessment of the Atomic Bomb Radiation Dosimetry for Hiroshima and Nagasaki: Dosimetry System 200 (DS02).* Vol. 2. Hiroshima, Japan: Radiation Effects Research Foundation.

Zankl, M., J. Becker, U. Fill, N. Petoussi-Henss, and K. F. Eckerman. 2005. "GSF Male and Female Adult Voxel Models Representing ICRP Reference Man—The Present Status". Paper presented at the conference *The Monte Carlo Method: Versatility Unbounded in a Dynamic Computing World*, Chattanooga, TN, April 17–21, 2005. La Grange Park, IL: American Nuclear Society.

Zankl, M., H. Schlattl, N. Petoussi-Henss, and C. Hoeschen. 2012. "Electron Specific Absorbed Fractions for the Adult Male and Female ICRP/ICRU Reference Computational Phantoms". *Physics in Medicine & Biology* 57 (14):4501.

Zankl, M., and A. Wittmann. 2001. "The Adult Male Voxel Model 'Golem' Segmented from Whole-Body Ct Patient Data". *Radiation and Environmental Biophysics* 40 (2):153–162.

CHAPTER 9

Cancer Risk Coefficients

David Pawel

CONTENTS

9.1 Purpose of a Cancer Risk Coefficient 395
    9.1.1 Limitations of Cancer Risk Coefficients 396
        9.1.1.1 Uncertainties Associated with Cancer Risk Coefficients 397
    9.1.2 Comparison with ICRP Risk Coefficients 399
    9.1.3 How Cancer Risk Coefficients Are Used: EPA vs. NRC 399
9.2 Computation of a Cancer Risk Coefficient for a Given Population and Exposure Pathway 400
9.3 Risk and Dose Coefficient Software 404
    9.3.1 Case Studies 405
        9.3.1.1 Ingestion of ^{90}Sr in Food 405
        9.3.1.2 Ingestion of ^{131}I in Milk 406
        9.3.1.3 External Exposure to ^{137}Cs on the Ground Surface 406
        9.3.1.4 Inhalation of ^{210}Po 408
        9.3.1.5 Inhalation of ^{239}Pu 409
Appendix 9-A: BEIR VII Risk Models and Formulas for Risk Coefficient Calculation 410
References 414

9.1 PURPOSE OF A CANCER RISK COEFFICIENT

In the field of radiation protection, a cancer risk coefficient is defined as an estimate of the probability, per unit intake or unit exposure, of cancer incidence or mortality. The purpose of radionuclide and pathway-specific cancer risk coefficients is to estimate the number of individuals expected to develop (or die from) cancer in a population as a result of exposures to radionuclides. The uses for risk coefficients include performing baseline site risk assessments, setting risk-based remediation goals, selecting among various remediation options (cost effectiveness), demonstrating compliance with risk-based regulations, and performing cost-benefit studies.

Cancer risk coefficients are tabulated in the U.S. Environmental Protection Agency's (EPA) Federal Guidance Report No. 13 (FGR 13) for over 800 radionuclides for the following modes of exposure: inhalation in air, ingestion of food, ingestion of tap water, external exposure from submersion in air, and external exposure from soil contaminated to an infinite depth

(Eckerman et al. 1999). The risk coefficients, when multiplied by activity intake (for internal exposures) or activity concentration (integrated over time for external exposures), provide estimates of the average probability of death or the development of a radiogenic cancer for the U.S. population. For the intake of a radionuclide, this is represented algebraically in FGR 13 by:

$$R = r \times I \tag{9.1}$$

In Equation (9.1), R is the average probability of radiogenic cancer incidence or death for a population, r is the cancer risk coefficient, and I is the per capita activity intake (Bq). The corresponding equation for external exposures is

$$R = r \times X \tag{9.2}$$

where X is the time-integrated activity in air (m^3/Bq-s) on the ground surface (m^2/Bq-s) or within the soil (kg/Bq-s). As explained in Section 9.2, risk coefficients can be used to approximate (1) cancer probabilities associated with a chronic lifelong exposure to a constant concentration of a radionuclide in the environmental medium, and (2) the average probability for members of a population acutely exposed to the radionuclide. Thus, from Equation (9.1), the number of individuals that would be expected to develop cancer, for example, from a lifelong exposure to a radionuclide at constant concentration in food, is $NR = N(r \times I)$. Section 9.2 describes the methodology for calculating the risk coefficients in FGR 13 and revisions to these coefficients that are underway for incorporating recent information on doses and risk associated with exposure to radiation and updated morbidity/mortality data. The revision to FGR 13, Federal Guidance Report 16 (FGR 16), promises to include cancer risk coefficients for both internal and external exposures for a much larger set of radionuclides than in FGR 13.

9.1.1 Limitations of Cancer Risk Coefficients

As stated in FGR 13 (Eckerman et al. 1999), "Analyses involving risk coefficients tabulated should be limited to estimation of prospective risks in hypothetical or large existing populations, or retrospective analyses of risks to large actual populations." The coefficients represent risk per unit intake or exposure averaged over gender and age(s) at which the exposure occurs; they were derived using a hypothetical (stationary) population defined by U.S. cancer and total mortality statistics for a fixed time period (1989–1991). The key characteristic that defines stationary populations (for our purposes) is that demographics (mortality and morbidity rates, gender and age distributions), do not change over time. For acute exposures, calculations based on the cancer risk coefficients are valid only for populations with similar gender and age distributions as the stationary one, and are not to be used to calculate risks for specific individuals. Calculations based on these coefficients for the risk of chronic lifelong exposures are only valid for populations with similar mortality and/or morbidity rates.

The coefficients for ingestion and inhalation were calculated, with few exceptions, using biokinetic and dosimetric models recommended by the International Commission on

Radiological Protection (ICRP). These models were designed for regulatory purposes to obtain absorbed doses to specific cancer sites for typical, or "reference," male and female members of the U.S. population and specified exposure scenarios. The absorbed doses from radionuclides distributed throughout the body are calculated using stylized anthropomorphic phantoms representing the newborn, other children of specified ages, and adults, and the calculations therefore do not account for variation in individual characteristics—other than age—that would affect dose.

For inhalation, the risk coefficients were calculated based on the assumption that the activity median aerosol size (AMAD) = 1 μm. In addition, the inhalation risk coefficients were calculated separately for ICRP defined categories for the rate of absorption from lungs to blood. For particulate form, there are three categories (fast, medium, and slow), with medium the ICRP default category for most radionuclides. An important cautionary note is given in FGR 13: "the information underlying the selection of an absorption type is often very limited and in many cases, reflects occupational rather than environmental exposures" (Eckerman et al. 1999). Thus, the user is advised to be careful in choosing an absorption type most appropriate for particular exposure situations. Drinking water consumption and food intake depend on age, so coefficients for ingestion of radionuclides should be applied with care for specific populations with diets that may differ from the population at large. In general, doses are calculated based on assumptions on intakes, human anatomy, and other factors that affect the distribution of radionuclides within the body, which for many exposure situations would not be expected to result in reasonable approximations of dose to individuals.

The coefficients are based on models for radiation risk for exposures involving "low" doses and/or low dose rates, and are not to be applied to exposures involving doses large enough to result in deterministic effects or to have a non-negligible effect on survival, for example, life expectancy. In FGR 13, low doses for low-LET radiation are defined as <0.2 Gy, and low dose rates are defined as <0.1 mGy min^{-1}. "Risks for high-let radiation are assumed to increase linearly with dose independent of dose rate" (Eckerman et al. 1999). For external exposures, the risk coefficients are based on specific exposure scenarios, for example, for a reference adult male, standing outdoors with no shielding [and] for which "activity distributions in air, on the ground surface, or in the soil are assumed to be of an infinite extent." The user must decide whether the calculations would need to be adjusted to account for shielding and other factors that may affect dose for the specific application.

9.1.1.1 Uncertainties Associated with Cancer Risk Coefficients

Health physicists should recognize that projections of cancer risk based upon the risk coefficients in FGR 13 are subject to substantial uncertainty. Analyses (Pawel et al. 2007) that considered some of the major sources of uncertainty indicate that, based on alternate models and input assumptions, plausible values for the risk coefficients for ingestion and/ or inhalation of radionuclides could differ from those published in FGR 13 by a factor of 3 or more; in some cases, the corresponding (uncertainty) factors exceed 10. Uncertainties can be especially large if—as is the case for ingestion of ^{239}Pu—the absorbed dose concentrates in individual tissues, such as the liver, for which the risk model is less reliable.

Uncertainties are also very large for many of the long-lived radionuclides for which doses would tend to be delivered remote from intake. In contrast, uncertainties are considered to be relatively small for intake of radionuclides, such as ^{137}Cs (by ingestion), and tritiated water, for which absorbed activity is more uniformly distributed.

Section 9.2 provides details on how cancer risk coefficients are calculated. In brief, the process involves the application of several different models including: biokinetic and dosimetric models for determining doses to specific tissues and radiation risk models for calculating risks that might be associated with the tissue specific doses. Characterization of uncertainties for cancer risk projections involves a variety of difficult problems (and often complex concepts) such as:

- Assessment of uncertainties ascribed to parameters in each of the models.

- Recognizing relationships among the parameters and resulting correlations in parameter estimates. For example, uncertainties in model parameters used to estimate internal doses in a study of cancer effects associated with the intake of a radionuclide will induce uncertainties in estimates of risk per unit dose. In such a study, errors in assigned dose might be inversely correlated with estimates of risk per unit dose.

- Uncertainties associated with the basic structure of models. For example, biokinetic models depend on the choice of hypothetical compartments used to characterize the movement of radionuclides within the body after intake.

- Consideration of the validity of underlying assumptions, which almost inevitably involves some degree of subjective judgment. It is arguably not feasible to quantify the uncertainty associated with some of these, e.g., the Linear Non-Threshold Hypothesis, which stipulates that at low doses and dose rates, the cancer-specific risk is approximately proportional to the absorbed dose to the corresponding tissue.

- Consideration of the validity of models that might have been derived for exposure conditions and to populations that are different in many respects to the types of exposures and U.S. populations for which the risk coefficients are to be applied.

- Consideration of the source of data used to derive the models. For example, for many radionuclides, biokinetic models have been derived from data on the behavior of chemically similar elements in animal subjects. Ideally, models would be derived directly from quantitative measurements on the same elements in human subjects.

Uncertainty analyses for both estimates of dose and risk inevitably involve some degree of subjective judgment. For an introduction to the subject of uncertainty analysis and a thorough description on how expert elicitations may be employed to quantify uncertainties involving subjective judgment, refer to the National Council on Radiation Protection and Measurements (NCRP) (1996). A comprehensive uncertainty analysis on doses and risks involving expert elicitations is given in a series of reports by the U.S.

Nuclear Regulatory Commission (NRC) and the Commission of European Communities (CEC), which conducted a joint study aimed at characterizing the uncertainties in predictions of the consequences of accidental releases of radionuclides into the environment (NRC and CEC 1997, 1998).

The published literature on uncertainties associated with risk and dose assessment also includes a chapter in a recent EPA technical report on revisions to that agency's risk models (EPA 2011), several NCRP reports (NCRP 2007, 2009a,b, 2012), Kocher et al. (2008), which describes methodology used for the Interactive Radio Epidemiological Program (IREP), and Puncher (2014), which describes an analysis—for the Environment Agency of England—of uncertainties of estimates of dose associated with the intake of selected radionuclides. See also the review article by Preston et al. (2013), which provides a more comprehensive list of the published reports and peer-reviewed journal articles on this topic.

9.1.2 Comparison with ICRP Risk Coefficients

The cancer risk coefficients described in this chapter are not to be confused with ICRP risk coefficients that have been used to determine tissue weighting factors. These ICRP risk coefficients represent the risk of a radiation-induced health effect "per unit dose" and are based on the concept of "detriment." Detriment was introduced in ICRP Publication 26 (ICRP 1977) as "a measure of the total harm that would eventually be experienced by an exposed group as a result of the group's exposure to a radiation source" (ICRP 1991). As originally defined, detriment was a weighted sum of the expected number of radiation-induced health effects, with weights representing the relative severity of each type of health effects. Detriment accounts for both cancers and severe genetic effects, and ICRP Publication 60 (ICRP 1991) presented a formula in which weights assigned to specific cancer types depend on their lethality fraction (the expected proportion that will be fatal) and the expected years of life loss per fatal cancer. For example, larger weights were assigned to cancers such as lung and liver cancer, with high lethality fractions, than more curable cancers such as skin and thyroid. However, ICRP (ICRP 1991) also concluded that there are many unsatisfactory aspects to using a single quantity, for example, there is no single best way to combine probabilities and severity for outcomes of a "multifarious nature." Therefore, ICRP reports include separate estimates of fatal and non-fatal risk (per unit dose), in addition to a measure which attempts to combine risks for different health outcomes into a single quantity. A very thorough discussion of the concepts underlying the terms "risk" and "detriment" is given in ICRP Publication 60 (ICRP 1991). The most notable advantage of the cancer risk coefficients in FGR 13, compared to the ICRP risk coefficients, is their relative ease of interpretation; cancer risk coefficients represent the probability of cancer incidence or death associated with exposure to radiation.

9.1.3 How Cancer Risk Coefficients Are Used: EPA vs. NRC

The mode in which the cancer risk coefficients in FGR 13 are used, or not used, depends upon the federal law which provides state governments and federal agencies such as the EPA, the NRC, and the DOE with the authority for regulating radionuclides. A concise yet comprehensive description of the principal federal laws which provide regulatory authority for radionuclides and chemicals is given by Kocher (2008). He states that although these laws are "varied and diverse," they all stem from one of two basic approaches for risk management: the "radiation paradigm" and the "chemical paradigm." Under the radiation paradigm, (1) "any practice that increases radiation exposure [must be *justified* in that it would] result in a positive net benefit to society," (2) limits for radiation dose are set that define exposures above which risks would be *intolerable*, and (3) for specific practices, *acceptable* exposures are often defined as those that do not exceed what ICRP refers to as dose constraints. The dose constraints are a fraction of the dose limits and are determined to ensure that *exposures are optimized or as low as reasonably achievable* (ALARA).

In contrast, the basic elements for the chemical paradigm are to define a risk range with the lower end of the range representing a point of departure for determining an acceptable level for exposure (such as a site cleanup goal). Risk management considerations, such as technical feasibility, cost effectiveness, and community acceptance, are factored into the final risk goal. This goal will typically be within the risk range (i.e., not exceeding the upper limit of the range), but achieving it may sometimes require reliance on additional controls, such as restrictions on future site use. In general, NRC regulations for radionuclides are based on the radiation paradigm, whereas EPA regulations for radionuclides can be based on either approach.

Risk coefficients are used to determine compliance with regulations that follow the chemical paradigm. These include regulations based on the Comprehensive Environmental Response, Compensation and Liability Act (CERCLA)—the law that established Superfund. CERCLA provides the EPA with the authority to regulate the cleanup of sites contaminated with hazardous substances and pollutants. According to Kocher (2008), CERCLA stipulates that one of the goals that "shall be developed" is to take "into account … an upper bound on lifetime risk of 10^{-6} to 10^{-4} from all substances and all exposure pathways combined at specific sites." The EPA recommends that, for compliance with CERCLA, risks should be calculated using "slope factors" which for the most part are based on the same type methodology and models used for FGR 13, as described in the next section (Section 9.2). The risk coefficients in FGR 13 have also been used by EPA to evaluate the protectiveness of proposed drinking water regulations. In contrast, risk coefficients are generally not used to comply with NRC regulations, although they can be used to provide insight as to their protectiveness.

9.2 COMPUTATION OF A CANCER RISK COEFFICIENT FOR A GIVEN POPULATION AND EXPOSURE PATHWAY

This section is an adaptation of material presented in FGR 13 (Eckerman et al. 1999) and the more recent EPA technical report on updates to that agency's radiogenic cancer risk models (Environmental Protection Agency 2011).

In FGR 13, cancer risk coefficients were computed using a process that combined results from several different types of models. For example, dosimetric and biokinetic models were used to convert activity intake to absorbed dose, and radiation risk models convert those absorbed doses to cancer risk. Since most of the models used for FGR 13 are more than two decades old, the cancer risk coefficients are undergoing revision to account for recent information on dose and the corresponding risks from exposure to radiation. For example, for most cancers, the risk models in FGR 16 will be those recommended by the National Academy of Sciences (NAS) BEIR VII Committee (NAS 2006). Nevertheless, the basic process for calculating risk coefficients remains largely unchanged. Steps involved in computing these risk coefficients for the ingestion or inhalation of a radionuclide are summarized next (see Appendix 9-A for more details):

(1) *Converting activity intake to absorbed dose:* Biokinetic and dosimetric models are applied to calculate, for each age of acute intake (x_i), the tissue-specific absorbed dose rates per unit activity intake for all times after intake. The biokinetic models determine how inventories of activity for the radionuclide are distributed within the body as a function of time. Dosimetric models are then used to convert these activities to absorbed dose rates to sensitive tissues of the body. For almost all radionuclides, the biokinetic and dosimetric models are the same as those used by ICRP.

(2) *Converting age-specific absorbed dose to risk:* For each of several cancer types (see Table 9.1) a radiation risk model is applied to calculate an estimate of the lifetime risk of cancer or cancer death per unit dose for each age (x) at which the tissues might be exposed to radiation.

For almost all cancer types, the risk models were derived from the large epidemiologic cohort of 120,000 atomic bomb survivors known as the Life Span Study (LSS). Besides its large size, the LSS has several important strengths, including a very long period of continuous follow-up (since 1950 for cancer mortality and 1958 for morbidity). For example, most of the BEIR VII risk models to be applied in FGR 16 are based on cancer incidence within the LSS cohort from 1958–1998. Other strengths include an exposed population of all ages and both genders; essentially uniform whole-body exposures for which all tissues were irradiated; accurate dose estimates compared to most other epidemiological studies; and a relatively healthy and homogeneous population. Thus, the LSS is a remarkable source of information that allows for reasonably accurate (direct) estimates of radiogenic risks for all types of cancers for acute low-LET exposures (at any age) for doses from less than 0.5 Sv to doses as large as 1–2 Sv.

TABLE 9.1 Radiation Risk Models in Federal Guidance Report (FGR) 13 and the Preliminary Version of FGR 16

Risk Model	Federal Guidance Report 13	Federal Guidance Report 16 (Preliminary Version)
Cancer sites with radiation risk models	Esophagus, stomach, colon, liver, lung, bone, skin, female breast, ovary, bladder, kidney, thyroid, leukemia	Same as for FGR 13 except FGR 16 includes a risk model for prostate cancer and pancreatic cancer, but none for esophageal cancer
Definition of residual site cancers	All solid cancers, e.g., of the prostate, not included above	All solid cancer, e.g., of the esophagus, not included above[a]

[a] The dose for residual cancers is calculated as the average (weights equal to 1/3 each) of doses to muscle, pancreas, and adrenal regions.

Nevertheless, because the risk estimates were derived from data on Japanese atomic bomb survivors, who received high, relatively acute doses of low-LET radiation, three different types of extrapolations are needed to project risks for low-level environmental and occupational exposures to the U.S. population:

First, to estimate risks at very low doses, the risk coefficients tabulated in FGR 13 incorporate a "Dose and Dose Rate Effectiveness Factor" (DDREF). Radiobiological data suggests that there would be a reduction in excess risk per unit dose at low dose and dose rates. Thus, it has long been common practice to divide risk estimates derived from the LSS by a DDREF >1 to project radiogenic risks at low doses and dose rates. In FGR 13, the DDREF was set to 2 for solid cancers other than breast. For FGR 16, the DDREF will be set to 1.5, the recommended value given in BEIR VII. Nevertheless, the risk estimates are valid only to the extent that at very low doses, radiogenic risks increase in proportion to risk. This so-called Linear Non-Threshold (LNT) Hypothesis is based on a conceptual model, in which (1) complex damage to DNA in a single cell can eventually result in cancer, and (2) the amount of damage to DNA done by radiation, as measured by the number of radiation-induced double strand breaks and sites of complex damage, is proportional to absorbed dose. Although there is research on low dose phenomena, e.g., adaptive response, genomic instability, and bystander effects, that casts doubt on the LNT assumption, the recent comprehensive NAS report on the Biological Effects of Ionizing Radiation (BEIR VII 2006) affirmed that "the balance of evidence from epidemiological, animal, and mechanistic studies tend to favor a simple proportionate relationship at low doses between radiation dose and cancer risk" (NAS 2006). Note that, to estimate risks for acute exposures, e.g., doses above 0.2 Gy, the division by the recommended DDREF values of 1.5 or 2 would not be appropriate. ("Ideal DDREFs" might be cancer specific and decrease continuously with dose.)

The second type of extrapolation concerns the problem of "risk transport," which refers to the projection of risk to a different population (U.S.) than the one the risk model was derived from (Japanese atomic bomb survivors). Two types of risk models are used to estimate excess risk in the LSS and to project risk to the U.S. population: an excess absolute risk (EAR) model, and an excess relative risk (ERR) model. ERR represents the ratio of the increase in cancer rates attributable to radiation divided by the baseline rate (the cancer rate for a non-exposed population), whereas EAR is simply the difference in rates associated with the radiation. An EAR model is appropriate for projecting risk for cancer sites for which radiogenic risk is independent of baseline rates. An ERR model is appropriate for cancer sites for which radiogenic risk is proportional to baseline rates. For cancers such as stomach and liver cancers, for which the baseline rates are higher in Japan than in the U.S., the EAR model will yield larger projections of risk to the U.S. population than the ERR model. In contrast, the ERR model yields the larger U.S. projection for bladder cancer, which is more common in the U.S. For most cancer sites, there is insufficient data to determine which of the two types of models would provide better approximations of risk. For both FGR 13 and its revision (FGR 16), a compromise between the two approaches is used, e.g., a weighted arithmetic mean of the ERR and EAR projections will be used for the revision.

Finally, the third type of extrapolation apples to radionuclides which emit high-LET radiation (α-particles). The radiation the atomic bomb survivors were exposed to was mostly in the form of γ-rays, and, except for radon, there are no human studies with comparable direct data on the effects of high-LET radiation. The preferred approach for estimating risks for high-LET radiation is to adjust risk estimates derived from the LSS by applying a "relative biological effectiveness" (RBE) factor. The RBE is generally defined as the relative effectiveness of a given type of radiation in producing a specified biological effect compared to some reference radiation. For most cancer sites, the estimated RBE is 20, based on laboratory experiments, i.e., the tissue-specific risks are set to equal 20 times that for γ-rays.

The BEIR VII risk models to be used in FGR 16 and formulas for calculating lifetime radiogenic risk from the LSS risk models are described in more detail in Appendix A. Essentially, the calculations involve a three-step process. First, the excess risk (excess probability of cancer) associated with the radiation are calculated for all ages after the age at which the tissues were exposed to radiation (x). Second, the excess risks at each age (after age x) are multiplied by a factor (equal to the conditional probability of survival given the person is alive at age x) to account for the possibility of dying from other causes beforehand. The last step is integrating these products over all ages after age x.

(3) Calculating lifetime cancer risk for intake at a specific age: Results from the two steps above are then combined to obtain the lifetime cancer risk per unit intake for each age of intake (x_i). First, for each age after intake ($x > x_i$), the absorbed dose per unit intake (determined from step (1) above) is multiplied by the probability that an individual at age x_i would survive to age x (to receive the absorbed dose) and the lifetime risk per unit absorbed dose (from step (2) above). These products are then integrated over all ages ($x > x_i$).

(4) Calculating lifetime cancer risk for chronic intakes: For chronic intakes, it is assumed that the concentration of the radionuclide in the environmental medium (for example air, tap water, food, cow's milk, etc.) is constant. Age-specific intakes are therefore assumed to be proportional to age-specific usage rates derived for the U.S. population from survey data, e.g., estimates of food intake used for FGR 13 were based on data from the Third National Health and Nutrition Examination Survey. As shown in Table 9.2, intake is often strongly dependent on age, e.g., on average about 4–5 times greater for young adults than at 1 y. For each age at intake (x_i), the lifetime risk associated with intake at that age is equal to the product of (a) the lifetime cancer risk per unit intake (from step (3)), (b) the activity intake at that age, and (c) the probability of survival to age x_i. The lifetime cancer risk for the chronic intake is then obtained by integrating these risks (associated with specific ages of intake) over all ages.

TABLE 9.2   Age and Gender Usage Rates for Food Intake (Other Than Milk)[a]

Age (y)	Food Energy[b] (kcal d^{-1})	
	Males	Females
0	478	470
1	791	752
5	1566	1431
10	1919	1684
15	2425	1828
20	2952	1927
50	2570	1758
75	1990	1508
Lifetime average	2418	1695

[a] From FGR 13, Table 3.1 (Eckerman et al. 1999).
[b] *Source*: Third National Health and Nutrition Examination Survey (McDowell et al. 1994).

(5) Lifetime cancer risk per unit activity intake: This is simply the ratio of the lifetime cancer risk (from step (4)) divided by the expected lifetime activity intake. The latter is obtained by integrating (over all ages) the expected age-specific intakes. The expected intake at age x_i is equal to the product of (a) the activity intake at that age (x_i), and (b) the probability of survival to age x_i.

9.3 RISK AND DOSE COEFFICIENT SOFTWARE

The risk coefficients tabulated in Federal Guidance Report 13 are appropriate for calculating age and sex-averaged risks for populations with demographic characteristics similar to the stationary population based on 1989–1991 mortality statistics. Software is available on the EPA's website (EPA 2016a) which allows the user to:

(1) Calculate tissue specific doses and cancer risks for specified age groups.

(2) Calculate risk and dose coefficients based on alternative assumptions, e.g., alternative age-specific values for f_1.

The software packages include:

(1) Dose and Risk Calculation Software (DCAL). DCAL is a versatile software package, which, as described on the EPA's website (EPA 2016a)

"consists of a series of computational modules, for the computation of dose and risk coefficients. The system includes extensive libraries of biokinetic and dosimetric data and models representing the current state of the art. It is intended for users familiar with the basic elements of computational radiation dosimetry. Components of DCAL have been used to prepare EPA *Federal Guidance Reports 12 and 13* and a number of publications of the ICRP. The dose and risk values calculated by [the first release of DCAL] are consistent with those published in *Federal Guidance Reports 12 and 13*."

For example, DCAL can be used to calculate dose and risk coefficients using alternative assumptions about the key dosimetric, biokinetic and risk model parameters used for FGR 13.

2. RiskTab. RiskTab is a software package for calculating lifetime radiogenic cancer risks from the intake of radionuclides for user-specified age ranges.

For further information about these and related software packages, see the EPA's website.

9.3.1 Case Studies

9.3.1.1 Ingestion of ^{90}Sr in Food

From 1954 through 1978, the average estimated concentration of ^{90}Sr per gram calcium in food is about 13 pCi in New York and 5 pCi in San Francisco (Leggett, Eckerman, and Williams 1982). Assume the average annual intake of calcium (g) during that period was 350 g (see Leggett et al., table 5). Estimate the average lifetime cancer risks to the populations in these two cities that can be associated with these exposures.

During the 25-y period from 1954 through 1978, the per capita activity intake in New York of ^{90}Sr from food was about:

$$25 \text{ y} \times 350 \text{ g y}^{-1} \times 13 \text{ pCi g}^{-1} \times 0.037 \text{ Bq/pCi} = 4.2 \times 10^3 \text{ Bq}$$

Similar calculations yield 1.6×10^3 Bq for San Francisco.

Strontium-90 ($T_{1/2} = 29.1$ y) decays into yttrium-90 ($T_{1/2} = 64$ h). Because of the relatively short half-life of ^{90}Y, it is reasonable to assume that ^{90}Sr is in equilibrium with ^{90}Y in diet. The mortality and morbidity coefficients for the ingestion of ^{90}Sr in food are 1.62×10^{-9} Bq^{-1} and 1.86×10^{-9} Bq^{-1}, respectively. For ^{90}Y, these coefficients are 3.96×10^{-10} Bq^{-1} and 7.16×10^{-10} Bq^{-1} Therefore, for New York, the estimated risks are:

Mortality: 4.2×10^3 Bq $\times (1.62 \times 10^{-9}$ Bq$^{-1} + 3.96 \times 10^{-10}$ Bq$^{-1})$
     $= 8.5 \times 10^{-6}$

Morbidity: 4.2×10^3 Bq $\times (1.86 \times 10^{-9}$ Bq$^{-1} + 7.16 \times 10^{-10}$ Bq$^{-1})$
     $= 1.1 \times 10^{-5}$

Similarly, the estimated risks for San Francisco are:

Mortality: 1.6×10^3 Bq $\times (1.62 \times 10^{-9}$ Bq$^{-1} + 3.96 \times 10^{-10}$ Bq$^{-1})$
     $= 3.2 \times 10^{-6}$

Morbidity: 1.6×10^3 Bq $\times (1.86 \times 10^{-9}$ Bq$^{-1} + 7.16 \times 10^{-10}$ Bq$^{-1})$
     $= 4.1 \times 10^{-6}$

Note that the estimates could (theoretically) be improved by using risk coefficients more appropriate for the specified time period (1958–1974). The risk coefficients in FGR 13 were calculated for a stationary population based on 1989–1991 mortality statistics.

9.3.1.2 Ingestion of ^{131}I in Milk

Background (Kahn, Straub, and Jones 1962): In 1961, about two months after the Russian nuclear test series was terminated, a study at Oregon State University's School of Agriculture was initiated to determine the effectiveness of sheltering cows to reduce ^{131}I in milk. For a period of just over two weeks, ^{131}I measurements were compared for composite milk samples taken daily from sheltered cows eating stored feed vs. from a herd on pasture. For the sheltered cows, the measurements provided no convincing evidence of ^{131}I. However, within two days after the study began, ^{131}I was measured at 270 pC/L in milk taken from cows on pasture. After that, ^{131}I decreased with an estimated half-life of about seven days.

> Suppose the average concentration of ^{131}I in cow's milk over a two-month period is 20pCi/L. Compute the lifetime risk (mortality and morbidity) associated with the ingestion of ^{131}I in this milk.

The decay during the relatively short time from milking to consumption "generally from 1 to 5 days" can be assumed to be small. The average intake of milk is 0.252 L/day (Eckerman et al. 1999, Table E-1). The mortality and morbidity coefficients for the ingestion of ^{131}I in milk (Eckerman et al. 1999, Table 2.2b) are 3.78×10^{-10} Bq^{-1} and 3.61×10^{-9} Bq^{-1}, respectively. Therefore, the estimated risks are:

$$\text{Mortality: } 20\frac{pCi}{L} \times 0.252L/d \times 60d \times 0.037\frac{Bq}{pCi}$$

$$\times \ 3.78 \times 10^{-10} Bq^{-1}$$

$$= 4.2 \times 10^{-9}$$

$$\text{Morbidity: } 20\frac{pCi}{L} \times 0.252L/d \times 60d \times 0.037\frac{Bq}{pCi}$$

$$\times \ 3.61 \times 10^{-9} Bq^{-1}$$

$$= 4.0 \times 10^{-8}$$

It is pertinent to note that lifetime risks from ingestion of ^{131}I are many times greater for children than adults. The estimated risks calculated above are age-averaged risks for the U.S. population; cancer risks from ingestion of ^{131}I would tend to be substantially larger (smaller) than suggested by these calculations for children (adults).

9.3.1.3 External Exposure to ^{137}Cs on the Ground Surface

The following example has been adapted from FGR 13 (Eckerman et al. 1999, F-4–F-5).

Suppose the ground surface was uniformly contaminated at time zero with ^{137}Cs at a level of 2 Bq m^{-2}. Assume that radioactive decay is the only mechanism by which contamination is reduced. (Reduction of the time-integrated exposure due to weathering is ignored here for simplicity.) Compute the average lifetime cancer risk (mortality and morbidity) resulting from external exposures during the first year following the initial deposition, assuming no shielding, and assuming that the age distribution of the exposed population is similar to that of the 1996 U.S. population.

Cesium-137 ($T_{1/2} = 30$ y) forms ^{137m}Ba ($T_{1/2} = 2.552$ m) in 94.6% of its decays. Due to the short half-life of ^{137m}Ba, the concentration of ^{137m}Ba on the ground surface will reach 1.89 Bq m^{-2} (0.946 × 2 Bq m^{-2}) within a half hour after time zero and will decline with the half-life of ^{137}Cs.

From Table 2.3 of FGR 13 (Eckerman et al. 1999), the mortality and morbidity risk coefficients for external exposure to ^{137}Cs distributed on the ground surface are 3.96 × 10^{-20} and 4.57 × 10^{-20} m^2 Bq^{-1} s^{-1} respectively. For ^{137m}Ba, the corresponding risk coefficients are 3.12 × 10^{-17} and 4.60 × 10^{-17} m^{2}Bq^{-1} s^{-1} respectively. The exposure (time-integrated concentration) for each radionuclide during the first year are:

$$\text{Exposure} = A_0 \int_0^T \exp\left(\frac{-\ln(2)t}{T_{1/2}}\right) dt = \frac{A_0 T_{\frac{1}{2}}}{\ln 2}\left(1 - \exp\left(\frac{-\ln 2 T}{T_{\frac{1}{2}}}\right)\right)$$

$$^{137}\text{Cs: } = \frac{2\text{Bq/m}^2 \times 30y \times 3.15 \times 10^7 \frac{s}{y}}{0.693}\left(1 - \exp\left(\frac{-0.6931y}{30y}\right)\right) = 6.23 \times 10^7 \frac{\text{Bq}-\text{s}}{\text{m}^2}$$

$$^{137m}\text{Ba: } \frac{1.89\text{Bq/m}^2 \times 30y \times 3.15 \times 10^7 \frac{s}{y}}{0.693}\left(1 - \exp\left(\frac{-0.6931y}{30y}\right)\right) = 5.89 \times 10^7 \frac{\text{Bq}-\text{s}}{\text{m}^2}$$

The lifetime risks resulting from external exposures during the first year are

$$\text{Mortality: } 6.23 \times 10^7 \frac{\text{Bq}-\text{s}}{\text{m}^2} \times 3.96 \times 10^{-20} \frac{\text{m}^2}{\text{Bq}-\text{s}}$$

$$+5.89 \times 10^7 \frac{\text{Bq}-\text{s}}{\text{m}^2} \times 3.12 \times 10^{-17} \frac{\text{m}^2}{\text{Bq}-\text{s}}$$

$$= 1.8 \times 10^{-9}$$

$$\text{Morbidity: } 6.23 \times 10^7 \frac{\text{Bq}-\text{s}}{\text{m}^2} \times 4.57 \times 10^{-20} \frac{\text{m}^2}{\text{Bq}-\text{s}}$$

$$+5.89 \times 10^7 \frac{\text{Bq}-\text{s}}{\text{m}^2} \times 4.60 \times 10^{-17} \frac{\text{m}^2}{\text{Bq}-\text{s}}$$

$$= 2.7 \times 10^{-9}$$

The radiations emitted by ^{137m}Ba are the main contributors to risk.

The discussion of this example in FGR 13 describes how these estimates might be refined by introducing a scaling factor to reflect differences in risk associated with the age distribution for the hypothetical 1996 population vs. the stationary population used to derive the risk coefficients published in FGR 13. The scaling factor used was 1.1, representing a relatively modest difference in the radiogenic risk that would be associated with the two populations.

9.3.1.4 Inhalation of ^{210}Po

Background (Henricsson and Persson 2012): ^{210}Po is an important carcinogen in tobacco smoke. Elevated levels of both ^{210}Po and ^{210}Pb found in tobacco leaves have been attributed to atmospheric fallout and absorption through the plants' roots. The average activity of ^{210}Po in smoke inhaled from a cigarette has been estimated to be 10 mBq.

> Discuss how one might attempt to estimate the lifetime risk (mortality and morbidity) associated with ^{210}Po and ^{210}Pb for smokers who on average smoke two packs (20 cigarettes) per day for 50 y.

From Table 2.1 of FGR 13, the mortality and morbidity coefficients for inhalation of ^{210}Po—based on default assumptions regarding speed of adsorption to blood and aerosol size (AMD = 1 μm) are 2.76×10^{-7} Bq^{-1} and 2.93×10^{-7} Bq^{-1}, respectively. For inhalation of ^{210}Pb, the corresponding coefficients—based on default assumptions regarding speed of adsorption to blood and aerosol size (AMD = 1 μm)—are 6.84×10^{-8} Bq^{-1} and 7.48×10^{-8} Bq^{-1}. Crude estimates of the risks, for example, they do not account for the fact that the vast majority of regular smokers are adults, are:

$$\text{Mortality}\,(^{210}\text{Po}): 20\,\frac{\text{cigarettes}}{d} \times \frac{365.25d}{y} \times 50\,y \times 0.01\,\frac{\text{Bq}}{\text{cigarette}}$$

$$\times 2.76 \times 10^{-7}\,\text{Bq}^{-1}$$

$$= 1.0 \times 10^{-3}$$

$$\text{Morbidity}\,(^{210}\text{Po}): 20\,\frac{\text{cigarettes}}{d} \times \frac{365.25d}{y} \times 50\,y \times 0.01\,\frac{\text{Bq}}{\text{cigarette}}$$

$$\times 2.93 \times 10^{-7}\,\text{Bq}^{-1}$$

$$= 1.1 \times 10^{-3}$$

$$\text{Mortality}\,(^{210}\text{Pb}): 20\,\frac{\text{cigarettes}}{d} \times \frac{365.25d}{y} \times 50\,y \times 0.01\,\frac{\text{Bq}}{\text{cigarette}}$$

$$\times 6.84 \times 10^{-8}\,\text{Bq}^{-1}$$

$$= 2.5 \times 10^{-4}$$

$$\text{Morbidity}\,(^{210}\text{Pb}): 20\,\frac{\text{cigarettes}}{d}\times\frac{365.25d}{y}\times 50\,y\times 0.01\,\frac{\text{Bq}}{\text{cigarette}}$$

$$\times\,7.48\times 10^{-8}\,\text{Bq}^{-1}$$

$$=2.7\times 10^{-4}$$

This example is given for illustrative purposes only; estimating radiogenic risks associated with tobacco smoke involves complex issues, which, in general, are beyond the scope of this chapter. In fact, radiogenic risks from tobacco smoke may be about two orders of magnitude smaller than these estimates indicate because the risk coefficients are based on default assumptions for speed of adsorption and aerosol size which are not valid for tobacco smoke. The amount of retention in the lungs for the radionuclides in tobacco smoke is likely to be far less than the biokinetic models would predict for the (default) Type M aerosol assumed for the calculations.

Other problems with these estimates are that they do not account for issues/facts such as: (1) the vast majority of regular smokers are adults, (2) smokers tend to die earlier than the rest of the U.S. population from a variety of cancer and non-cancer diseases, (3) radiation dose from tobacco smoke is distributed unevenly within respiratory epithelium and lung tissue, (4) radiation might act synergistically with other carcinogens in tobacco smoke to affect lung cancer risk, and (5) these risk coefficients are sex-averaged and thus do not provide any insight as to how risks might differ for males versus females.

As already mentioned in Section 9.3, the EPA provides software tools that can provide some insight as to how risks *might* depend on age-at-exposure. This includes the software program RiskTab, which provides risk coefficients for periods of chronic intake. For ages 18–68 y, mortality and morbidity coefficients for inhalation of ^{210}Po—based on the same methodology as used to calculate the risk coefficients in FGR 13—would be $1.79\times 10^{-7}\,\text{Bq}^{-1}$ and $1.90\times 10^{-7}\,\text{Bq}^{-1}$, respectively. For inhalation of ^{210}Pb, the corresponding coefficients are $4.37\times 10^{-8}\,\text{Bq}^{-1}$ and $4.79\times 10^{-8}\,\text{Bq}^{-1}$. Although these risk estimates are smaller than those published in Table 2.1 of FGR 13 (for exposures at all ages), it should be noted that the difference in these calculated risks—associated with age-at-exposure—are very small compared, for example, to uncertainties associated with aerosol type.

Radiogenic risk associated with tobacco smoke was a subject that generated a considerable amount of research 40–50 y ago. See, for example, Ferri and Baratta (1966), Ferri and Christiansen (1967), Cohen et al. (1979), and Holtzman and Ilcewicz (1966).

9.3.1.5 Inhalation of ^{239}Pu

From 1963–1977, the average concentration of ^{239}Pu in surface air in New York was about $6\times 10^{-6}\,\text{Bq m}^{-3}$; limited measurements indicated the respirable fraction was about 80% (Harley 1980; Volchok et al. 1974). Using risk coefficients given in FGR 13, approximate the associated risk from inhalation of Pu-239.

From FGR 13 (Table E.1), the average inhalation intake rate is 17.8 m³ d⁻¹; the mortality and morbidity coefficients (FGR 13, Table 2.1) for inhalation of ²³⁹Pu—assuming the default inhalation type (medium) are 7.94×10^{-7} Bq⁻¹ and 8.99×10^{-7} Bq⁻¹, respectively. Thus, estimates of risks associated with the inhalation over the 15-y period (1963–1977) are:

Mortality: 7.94×10^{-7} Bq⁻¹ × (0.8 × 6 × 10⁻⁶ Bq m⁻³) × (17.8 m³ d⁻¹ × 365.25 d/y × 15y)
= 3.7×10^{-7}

Morbidity: 8.99×10^{-7} Bq⁻¹ × (0.8 × 6 × 10⁻⁶ Bq m⁻³) × (17.8 m³ d⁻¹ × 365.25 d/y × 15y)
= 4.2×10^{-7}

Unlike the previous example for inhalation of ²¹⁰Po among smokers, the calculations are somewhat insensitive to particulate aerosol type. The mortality and morbidity coefficients are 1.26×10^{-6} Bq⁻¹ and 1.49×10^{-6} Bq⁻¹ for fast (F) and 8.45×10^{-7} Bq⁻¹ and 8.96×10^{-7} Bq⁻¹ for slow (S) particulate aerosol types. The estimates of risk for the three types of aerosols, given in Table 9.3 differ by less than a factor of 2. For ²³⁹Pu, an increase (decrease) in estimated risk for lung cancer associated with a faster (slower) absorption to the blood (for Type F and S aerosols) is at least partially compensated by a (decrease) increase in estimated risk for other types of cancer, for example, liver and bone cancer.

Finally, note that the estimates could in theory be improved by using risk coefficients more appropriate for the specified time period (1963–1977). The risk coefficients in FGR 13 were calculated for a stationary population based on 1989–1991 mortality statistics.

TABLE 9.3 Estimates of Mortality and Morbidity for Inhalation of ²³⁹Pu in New York for the Time Period 1963 through 1977 for Three Different Aerosol Types

Aerosol Type	Mortality	Morbidity
Fast (F)	5.9×10^{-7}	7.0×10^{-7}
Medium (M)	3.7×10^{-7}	4.2×10^{-7}
Slow (S)	4.0×10^{-7}	4.2×10^{-7}

APPENDIX 9-A: BEIR VII RISK MODELS AND FORMULAS FOR RISK COEFFICIENT CALCULATION

This Appendix includes abbreviated versions of the descriptions given in (1) the EPA technical report known as the "Blue Book" (Environmental Protection Agency 2011), which documents the new EPA radiogenic cancer risk models, and (2) FGR 13 (Eckerman et al. 1999) (EPA 2016b). The risk models that will be used to calculate risk per unit dose in FGR 16 are for the most part the EAR and ERR models recommended by the NAS BEIR VII Committee (National Academy of Sciences 2006). With the exception of leukemia, the BEIR VII models were derived from LSS cancer incidence data. This is a considered to be an improvement from previous models, which had been derived from LSS mortality data.

For the EAR models, excess cancer rate associated with radiation is a parametric function of: (1) the age at which an individual is exposed, (2) attained age—the age at which the individual develops (or dies from) cancer, (3) gender, and (4) cancer type. In the ERR, models, the excess cancer rate is equal to ERR (which depends on the same factors) multiplied by a baseline rate.

The EAR and ERR models for most solid cancers depend on dose (D), sex (s), age-at-exposure (x), and attained age (a) as follows:

$$\text{EAR}(D,s,x,a) \text{ or } \text{ERR}(D,s,x,a) = \beta_s D \exp(\gamma x^*)(a/60)^{\eta} \tag{A.1}$$

where

$$x^* = \frac{\min(x,30) - 30}{10}. \tag{A.2}$$

For ERR models, for most sites:

β, the ERR per Sv at age-at-exposure 30 y and attained age 60 y, tends to be larger for females than males;

$\gamma = -0.3$ implies the radiogenic risk of cancer falls by about 25% for every decade increase in age-at-exposure up to age 30 y; and

$\eta = -1.4$ implies the ERR is almost 20% smaller at attained age 70 than at age 60.

As a consequence, ERR decreases with age-at-exposure (up to age 30 y) and attained age. In contrast, for EAR models, $\gamma = -0.41$ and $\eta = 2.8$ for most sites. Thus, *EAR* decreases with age-at-exposure, but increases with attained age. (Thus, excess risks increase with attained age, but not in proportion to baseline rates.)

For an individual exposed to dose (D) at age (x), the tissue, age, and gender-specific values of ERR and EAR (from Equation A.1), are then used to calculate the lifetime attributable risk (LAR), which is a quantity that approximates the probability of a premature cancer from radiation exposure:

$$\text{LAR}(D,x) = \int_{x+L}^{110} \frac{M(D,x,a) \cdot S(a)}{S(x)} da \tag{A.3}$$

In Equation (A.3), $M(D, x, a)$ is the excess absolute risk at attained age a from an exposure at age x, $S(a)$ is the probability of surviving to age a, and L is the minimum latency period, that is, the minimum time from exposure to time of cancer diagnosis or death (assumed to be 5 y for solid cancers). LAR can most easily be thought of as a weighted sum (over attained ages up to 110 y) of the age-specific excess probabilities of radiation-induced cancer incidence or death, $M(D, x, a)$.

For any set of LAR calculations, the quantities $M(D, x, a)$ were obtained using either an EAR or ERR model. For cancer incidence, these were calculated using either:

$$M_I(D,x,a) = \text{EAR}_I(D,x,a) \qquad \left(\text{EAR model}\right) \tag{A.4}$$

$$M_I(D,x,a) = \text{ERR}_I(D,x,a) \cdot \lambda_I(a) \quad \left(\text{ERR model}\right) \tag{A.5}$$

where $\lambda_I(a)$ is the U.S. baseline cancer incidence rate at age a.

For mortality, the approach is very similar, but adjustments needed to be made to the equations since both ERR and EAR models were derived using incidence data. For example, it is assumed that the age-specific ERR is the same for both incidence and mortality, so that the ERR model-based excess risks are calculated using:

$$M_M(D,x,a) = \text{ERR}_M(D,x,a) \cdot \lambda_M(a) \tag{A.6}$$

that is, the age- and sex-specific mortality risks is the excess relative incidence risk times the baseline mortality rate. For EAR models, BEIR VII used essentially the same approach by assuming:

$$M_M(D,x,a) = \frac{\text{EAR}_I(D,x,a)}{\lambda_I(a)} \lambda_M(a) \tag{A.7}$$

Note that in Equation (A.7), the ratio of the age-specific EAR to the incidence rate is the ERR for incidence that would be derived from the EAR model. See the Blue Book (EPA 2011) for further details.

The age-specific "lifetime risk coefficient" (LRC), $r(x)$, is the risk per unit absorbed dose of a subsequent cancer death (Gy^{-1}) due to radiation received at age x. The LRC is closely related to the LAR, which is the primary risk measure used in the EPA Blue Book.

For purposes of this report, D is assumed to be sufficiently small that the linear component of the dose response model dominates and effects on the survival function are negligible. The risk coefficient is defined as the ratio of LAR(D, x) and dose D:

$$r(x) = \frac{\text{LAR}(D,x)}{D} \tag{A.8}$$

Following a unit intake of a radionuclide at age x_i, the absorbed dose rate $\dot{D}(x)$ to a given target tissue varies continuously with age $x \geq x_i$. The cancer risk $r_a(x_i)$ resulting from a unit intake of a radionuclide at age x_i is calculated from the continuously varying absorbed dose rate $\dot{D}(x)$ as follows:

$$r_a(x_i) = \frac{\int_{x_i}^{\infty} \dot{D}(x)r(x)S(x)dx}{S(x_i)} \tag{A.9}$$

where $r(x)$ is the cancer risk due to a unit absorbed dose (Gy^{-1}) at the site at age x. The absorbed dose rate is the absorbed dose rate for low-LET radiation, plus the product of the high-LET absorbed dose rate and the RBE applicable to the cancer type.

Age-specific male and female risk coefficients are combined by calculating a weighted mean:

$$r_a(x_i) = \frac{1.05 r_{ma}(x_i)u_{m(x_i)}S_m(x_i) + r_{fa}(x_i)u_f(x_i)S_f(x_i)}{1.05S_m(x_i)u_m(x_i) + S_f(x_i)u_f(x_i)} \tag{A.10}$$

where

$r_a(x_i)$ is the combined cancer risk coefficient for a unit intake of activity at age x_i,

1.05   is the presumed sex ratio at birth (male-to-female),

$r_{ma}(x_i)$ is the male risk per unit activity at age x_i,

$r_{fa}(x_i)$ is the female risk per unit activity at age x_i,

$S_m(x_i)$ is the male survival function at age x_i,

$S_f(x_i)$ is the female survival function at age x_i, and

$u_m(x_i)$ and $u_f(x_i)$ are the usage rates (see Eckerman et al. 1999) of the contaminated medium for males and females, respectively.

This formulation weights each sex-specific risk coefficient by the proportion of that sex in a stationary combined population at the desired age of intake.

The average lifetime risk coefficient for a radionuclide intake presumes that the intake rate is proportional to a constant environmental concentration (e.g., the radionuclide concentration in air). However, usage (e.g., the breathing rate) is also age- and sex-specific and therefore must be included in the averaging process. Defining the average lifetime risk as the quotient of the expected lifetime risk and the expected lifetime intake from exposure to a constant environmental concentration yields

$$\overline{r_a} = \frac{\int_0^\infty u(x)r_a(x)S(x)dx}{\int_0^\infty u(x)S(x)dx} \tag{A.11}$$

The radionuclide concentration in the environmental medium does not appear in the expression because it is a common factor in the numerator and denominator.

Lifetime risks for external radionuclide exposures are calculated in a manner similar to that for radionuclide intakes. Since the external exposure is not considered to be age dependent, the calculation is simpler. Given the age-specific cancer risk per unit dose, $r(x)$, and the corresponding dose per unit exposure coefficient, d_e, the lifetime risk is simply

$$r_e(x) = d_e r(x) \tag{A.12}$$

for an external exposure at age x. Age-specific male and female risk coefficients are combined by calculating a weighted mean as in Equation (A.11), but with the usage rates $u_m(x_i)$ and $u_f(x_i)$ removed from that equation. For lifetime external exposure at a constant exposure rate, d_e, the average lifetime risk is

$$\overline{r_e} = \frac{\int_0^\infty r_e(x)S(x)dx}{\int_0^\infty S(x)dx} \tag{A.13}$$

where $r_e(x)$ is given in Equation (A.12) and $S(x)$ is the sex-weighted survival function. This equation applies to a specific cancer site. The total risk is the sum over all cancer sites.

REFERENCES

Cohen, B. S., M. Eisenbud, M. E. Wrenn, and N. H. Harley. 1979. "Distribution of Polonium-210 in the Human Lung". *Radiation Research* 79 (1):162–168.

Eckerman, K. F., R. W. Leggett, C. B. Nelson, J. S. Pushkin, and A. C. B. Richardson. 1999. *Federal Guidance Report No. 13: Cancer Risk Coefficients for Environmental Exposure to Radionuclides*. Edited by Environmental Protection Agency. Washington, DC: U.S. Environmental Protection Agency.

Environmental Protection Agency. 2011. *EPA Radiogenci Cancer Risk Models and Projection for the U.S. Population*. EPA Report 402-R-001. Washington, DC: U.S. Environmental Protection Agency.

Environmental Protection Agency. 2016a. "DCAL Software and Resources". https://www.epa.gov/radiation/dcal-software-and-resources.

Environmental Protection Agency. 2016b. "Federal Guidance for Radiation Protection". https://www.epa.gov/radiation/federal-guidance-radiation-protection.

Ferri, E. S., and E. J. Baratta. 1966. "Polonium-210 in Tobacco Products and Human Tissues". *Radiological Health Data and Reports* 7 (9):485–488.

Ferri, E. S., and H. Christiansen. 1967. "Lead-210 in Tobacco and Cigarette Smoke". *Public Health Reports* 82 (9):828.

Harley, John H. "Plutonium in the environment—a review." *Journal of Radiation Research* 21, no. 1 (1980):83–104.

Henricsson, Fredrik, and Bertil R. R. Persson. 2012. "Polonium-210 in the Biosphere: Bio-Kinetics and Biological Effects". In Javier Guillén Gerada (ed), *Radionuclides: Sources, Properties and Hazards*. Hauppauge, NY: Nova Science Publishers, 1–39.

Holtzman, R. B., and F. H. Ilcewicz. 1966. "Lead-210 and Polonium-210 in Tissues of Cigarette Smokers". *Science* 153 (3741):1259–1260.

International Commission on Radiological Protection. 1977. ICRP Publication 26: Recommendations of the ICRP. *Ann ICRP* 1 (3).

International Commission on Radiological Protection. 1991. ICRP Publication 60: 1990 Recommendations of the International Commission on Radiological Protection. *Ann ICRP* 21 (1-3).

Kahn, B., C. P. Straub, and I. R. Jones. 1962. "Radioiodine in Milk of Cows Consuming Stored Feed and of Cows on Pasture". *Science* 138 (3547):1334–1335.

Kocher, D. C. 2008. "Regulations for Radionuclides in the Environment". In J. E. Till and H. R. Meyer (eds), *Radiological Risk Assessment and Environmental Analysis*. Oxford, U.K.: Oxford University Press.

Kocher, D. C., A. I. Apostoaei, R. W. Henshaw, F. O. Hoffman, M. K. Schubauer-Berigan, D. O. Stancescu, B. A. Thomas, J. R. Trabalka, E. S. Gilbert, and C. E. Land. 2008. "Interactive Radioepidemiological Program (Irep): A Web-Based Tool for Estimating Probability of Causation/Assigned Share of Radiogenic Cancers". *Health Physics* 95 (1):119.

Leggett, R. W., K. F. Eckerman, and L. R. Williams. 1982. "Strontium-90 in Bone: A Case Study in Age-Dependent Dosimetric Modeling". *Health Physics* 43 (3):307–322.

McDowell, M. A., R. D. Briefel, K. Alaimo, A. M. Bischof, C. R. Caughman, M. D. Carroll, C. M. Loria, and C. L. Johnson. 1994. *Energy and Macronutrient Intakes of Persons Ages 2 Months and over in the United States: Third National Health and Nutrition Examination Survey, Phase 1, 1989-1*. Advance Data 255. Edited by U.S. Department of Health and Human Services. Hyattsville, MD: U.S. Department of Health and Human Services.

National Academy of Sciences. 2006. *Health Risks from Exposure to Low Levels of Ionizing Radiation. BEIR VII Phase 2*. Washington, DC: The National Academies Press.

National Council on Radiation Protection and Measurements. 1996. *NCRP Commentary No 14: A Guide for Uncertainty Analysis in Dose and Risk Assessments Related to Environmental Contamination*. Bethesda, MD: National Council on Radiation Protection and Measurements.

National Council on Radiation Protection and Measurements. 2007. *NCRP Report No. 158: Uncertainties in the Measurement and Dosimetry of External Radiation*. Bethesda, MD: National Council on Radiation Protection and Measurements.

National Council on Radiation Protection and Measurements. 2009a. *NCRP Report No. 163: Radiation Dose Reconstruction: Principles and Practices*. Bethesda, MD: National Council on Radiation Protection and Measurements.

National Council on Radiation Protection and Measurements. 2009b. *NCRP Report No. 164: Uncertainties in Internal Radiation Dose Assessment*. Bethesda, MD: National Council on Radiation Protection and Measurements.

National Council on Radiation Protection and Measurements. 2012. *NCRP Report No. 171: Uncertainties in the Estimation of Radiation Risks and Probability of Disease Causation*. Bethesda, MD: National Council on Radiation Protection and Measurements.

Nuclear Regulatory Commission, and Commission of European Communities. 1997. *Probabilistic Accident Consequence Uncertainty Analysis. Late Health Effects Uncertainty Assessment: Volume 2*. NUREG/CR-6555, EUR 16774, SAND97-2322. Washington, DC: U.S. Nuclear Regulatory Commission.

Nuclear Regulatory Commission, and Commission of European Communities. 1998. *Probabilistic Accident Consequence Uncertainty Analysis:Volume 1*. NUREG/CR-6571, EUR 16773, SAND98-0119. Washington, DC: U.S. Nuclear Regulatory Commission.

Pawel, D. J., R. W. Leggett, K. F. Eckerman, and C. B. Nelson. 2007. *Uncertainties in Cancer Risk Coefficients for Environmental Exposure to Radionuclides. An Uncertainty Analysis for Risk Coefficients Reported in Federal Guidance Report No. 13*. ORNL/TM-2006/583. Edited by Oak Ridge National Laboratory. Oak Ridge, TN: Oak Ridge National Laboratory.

Preston, R. J. J., J. D. Boice Jr., A. B. Brill, R. Chakraborty, R. Conolly, F. O. Hoffman, R. W. Hornung, D. C. Kocher, C. E. Land, and R. E. Shore. 2013. "Uncertainties in Estimating Health Risks Associated with Exposure to Ionising Radiation". *Journal of Radiological Protection* 33 (3):573.

Puncher, M. 2014. "An Assessment of the Reliability of Dose Coefficients for Intakes of Radionuclides by Members of the Public". *Journal of Radiological Protection* 34 (3):625.

Volchok, H. L., R. Knuth, and M. T. Kleinman. "The respirable fraction of Sr-90, Pu-239, and Pb in surface air." *USAEC Report HASL-278 (Jan. 1974)* (1974)

CHAPTER 10

Interpretation of Bioassay Results to Assess the Intake of Radionuclides

David McLaughlin

CONTENTS

10.1 Introduction   418
10.2 Compartmental Analysis   418
     10.2.1 Closed System Catenary Transfer   420
     10.2.2 Open System Catenary Transfer   421
     10.2.3 Closed Recycling System   423
     10.2.4 Open Recycling System   424
     10.2.5 Summary for the Two-Compartment Model   425
     10.2.6 Matrix Solution   426
10.3 Retention Fractions   429
     10.3.1 Retention Functions for Chronic Intakes   429
     10.3.2 Excretion Functions for Acute Exposures   433
     10.3.3 Tabulations of Retention and Excretion Fractions   433
10.4 Limitations Affecting Incremental Excreta Bioassay Samples   434
     10.4.1 Best Estimate of Intake from Multiple Bioassay Measurements   434
10.5 Case Studies   441
     10.5.1 Case Study: Acute Inhalation of ^{137}Cs   441
     10.5.2 Case Study: Acute Ingestion of ^{32}P   446
10.6 Individual-Specific Modifications   450
     10.6.1 Case Study: Forced Fluids Following an Acute Uptake of ^{3}H   451
     10.6.2 Case Study: Modification of Respiratory Tract Absorption and
         Transfer Parameters Following an Acute Inhalation of ^{241}Am   454
     10.6.3 Modified Dose Coefficients   458
10.7 Regulatory Issues   459
10.8 Monitoring Intervals   461

Appendix 10-A: Laplace Transforms   462
Appendix 10-B: Heaviside Expansion Theorem   463
References   463

10.1 INTRODUCTION

Other than the medical administration of radiopharmaceuticals, the quantity of radioactive material inhaled, ingested, or otherwise taken into the body is unknown and must be estimated. Intake assessment requires an understanding of the transfer, retention, and excretion of radioactive material taken into the body. As discussed in the preceding chapters, mathematical models have been developed describing the metabolic processes acting upon materials taken into the body. Rather than focusing on a particular dosimetry system, such as those of the International Commission on Radiological Protection (ICRP) (that is, ICRP Publication 2, ICRP Publication 30, ICRP Publication 68, etc.), this chapter examines the process of interpreting metabolic models with the goal of estimating intake from bioassay data. This approach should enable the reader to transition between different dosimetry systems and adapt to future changes. Illustrative case studies are provided to enhance understanding.

10.2 COMPARTMENTAL ANALYSIS

For radiation protection purposes, biokinetic models are generally represented as a series of linearly connected compartments where material transfer is defined by fractional transfer coefficients (also referred to as translocation rate constants). Within this text, transfer coefficients are assigned the (lower case) symbol, k, with indices, $i \leftarrow j$, indicating the direction of travel. For example, $k_{\text{Liver} \leftarrow \text{Blood}}$ is interpreted as the transfer rate applied to the passage of material from blood to the liver.

For a quantity of material within a compartment (q), the product, qk, represents the instantaneous fractional transfer of material from the compartment. Similarly, $kd\,t$ is the probability that an atom will be transferred between the interval t to $t+dt$.

As shown in Figure 10.1, the time-dependent content of any ith compartment within the system is influenced by: (1) input from outside the system ($I_{\text{outside}}(t)$), (2) removal from the system ($k_{\text{outside} \leftarrow i}$), (3) transfer to another compartment ($k_{j \leftarrow i}$), and (4) receipt from another compartment ($k_{j \leftarrow i}$).

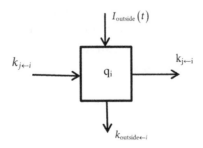

FIGURE 10.1   Compartmental model.

The rate of change within the ith compartment is given by the differential:

$$\frac{dq_i}{dt} = \text{Production} - \text{Removal} = \left[I_i(t) + \sum_{j \neq i}^{n} q_j k_{i \leftarrow j} \right] - \left[q_i k_{0 \leftarrow i} + \sum_{j \neq i}^{n} q_i k_{j \leftarrow i} \right] \quad (10.1)$$

A total removal rate constant, (K_i), accounting for all removal processes, can be defined as*:

$$K_i = \left[k_{0 \leftarrow i} + \sum_{j \neq i}^{n} k_{j \leftarrow i} \right] \quad (10.2)$$

Applying K_i, Equation (10.1) can be expressed as:

$$\frac{dq_i}{dt} = I_i(t) + \sum_{j \neq i}^{n} k_{i \leftarrow j} q_j - K_i q_i \quad (10.3)$$

Eliminating inputs from outside the system and accounting only for initial depositions (analogous to an acute exposure), Equation (10.3) simplifies to:

$$\frac{dq_i}{dt} = \sum_{j \neq i}^{n} k_{i \leftarrow j} q_j - K_i q_i \quad (10.4)$$

Regardless of the complexity of the model, Equation (10.4) applies to each member of the system (including excretion compartments). To aid understanding, a simple two compartmental model (Figure 10.2) is examined under four conditions: (1) catenary transfer in a closed system, (2) catenary transfer in an open system, (3) recycling within a closed system, and (4) recycling in an open system (Figure 10.2).

Differences between the examined conditions are defined by the value of the translocation rate constants. Parameters applied in this illustration are summarized in Table 10.1.

In general, all four cases assume that material is only deposited into *Compartment 1* ($q_1 = 100\%$) with no additional input from outside the system. In all cases, material is transported from *Compartment 1* to *Compartment 2* with a rate constant, $k_{2 \leftarrow 1} = 3 \cdot \text{time}^{-1}$ (an arbitrary time unit). Under conditions of recycling, material is transferred back to *Compartment 1* at a slower rate, $k_{1 \leftarrow 2} = 2 \cdot \text{time}^{-1}$. Open systems include radioactive decay

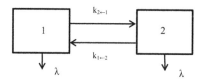

FIGURE 10.2   Two-compartment system.

---

* Note that within this text, the symbol upper case "K" with subscript "i" applies to the total removal rate constant for the ith compartment of the system.

TABLE 10.1   Parameters Applied to a Two-Compartment Model for Four Evaluated Conditions

Parameter	Closed System Catenary Transfer	Open System Catenary Transfer	Closed Recycling System	Open Recycling System
Initial quantity compartment 1 (q_1)	1	1	1	1
Initial quantity compartment 2 (q_2)	0	0	0	0
Transfer rate constant from 1 to 2 ($k_{2 \leftarrow 1}$)	3	3	3	3
Transfer rate constant from 2 to 1 ($k_{1 \leftarrow 2}$)	0	0	2	2
Decay constant (λ)	0	1	0	1
Total removal rate constant from Compartment 1 (K_1)	3	4	3	4
Total removal rate constant from Compartment 2 (K_2)	0	1	2	3

Note: Transfer rates given in arbitrary per time units.

which adds a removal mechanism (at an assigned rate of $1 \cdot \text{time}^{-1}$) to both compartments. Closed systems do not include radioactive decay and only consider transfer of material between system members.

10.2.1 Closed System Catenary Transfer

A system is closed if it neither receives nor transfers material outside the system. Catenary transfer is one-way passage without return. In this example, material is initially deposited in *Compartment 1* and transferred to *Compartment 2* where it is permanently retained. Under these conditions, the differential equation describing the time-dependent content of *Compartment 1* is:

$$\frac{dq_1}{dt} = -q_1 K_1 \tag{10.5}$$

The above differential equation may be solved using several methods. To aid the understanding of interdependencies between compartments, this text derives solutions using Laplace Transforms.* An alternate matrix approach is presented in Section 10.2.6.

Applying initial conditions and denoting the Laplace transform in terms of Y, Equation (10.5) may be re-expressed as:

$$sY_1 = -Y_1 K_1 + q_1(0) \tag{10.6}$$

Transformation yields:

$$Y_1 = \frac{q_1(0)}{(s + K_1)} \tag{10.7}$$

The inverse transformation gives:

$$q_1(t) = q_1(0)e^{-K_1 t} \tag{10.8}$$

---

* See Appendix 10-A for derivation of Laplace transforms.

which is recognized as the single exponential removal equation. Substitution of parameter values from Table 10.1 gives the solution as:

$$q_1(t) = 1e^{-3t} \tag{10.9}$$

For *Compartment 2*, the differential equation is:

$$\frac{dq_2}{dt} = -q_2 K_2 + q_1 k_{2 \leftarrow 1} \tag{10.10}$$

Recognizing that for this example, there is no removal from the second compartment, the above equation simplifies to:

$$\frac{dq_2}{dt} = q_1 k_{2 \leftarrow 1} \tag{10.11}$$

The Laplace transform is:

$$sY_2 = Y_1 k_{2 \leftarrow 1} \tag{10.12}$$

Substituting Equation (10.7) in Equation (10.12) gives:

$$Y_2 = \left[\frac{q_1(0)}{(s+K_1)} \right] \frac{k_{2 \leftarrow 1}}{s} \tag{10.13}$$

thus yielding,

$$q_2(t) = q_1(0) k_{2 \leftarrow 1} \left(\frac{1 - e^{-K_1 t}}{K_1} \right) \tag{10.14}$$

Recognizing in this example of a closed catenary system $K_1 = k_{1,2}$, the solution reduces to:

$$q_2(t) = q_1(0)\left(1 - e^{-K_1 t}\right) \tag{10.15}$$

Substitution of parameter values from Table 10.1 gives the solution as:

$$q_2(t) = 1\left(1 - e^{-3_1 t}\right) \tag{10.16}$$

10.2.2 Open System Catenary Transfer

A system is open if it receives or transfers material outside the system. In this example, radioactive decay (λ) serves as a transport mechanism removing material from the system. Therefore, the total removal rate constant for any compartment is the sum of the transfer coefficients plus the decay constant (i.e., $K_1 = k_{2\leftarrow 1} + \lambda$). Again, the differential equation describing the time-dependent content of *Compartment 1* is:

$$\frac{dq_1}{dt} = -q_1 K_1 \tag{10.17}$$

which yields the same solution:

$$q_1(t) = q_1(0) e^{-K_1 t} \tag{10.18}$$

except that the magnitude of K_1 has increased to include radioactive decay. Substitution of parameter values from Table 10.1 gives the solution as:

$$q_1(t) = 1 e^{-4t} \tag{10.19}$$

For *Compartment 2*, the differential equation now contains a removal component (due to radioactive decay):

$$\frac{dq_2}{dt} = -q_2 K_2 + q_1 k_{2\leftarrow 1} \tag{10.20}$$

The Laplace transform now gives:

$$sY_2 = -Y2K_2 + Y_1 k_{2\leftarrow 1} \tag{10.21}$$

Substituting Equation (10.7) into Equation (10.21) gives:

$$Y_2 = \left[\frac{q_1(0)}{(s + K_1)} \right] \frac{k_{2\leftarrow 1}}{(s + K_2)} \tag{10.22}$$

thus yielding:

$$q_2(t) = q_1(0) k_{2\leftarrow 1} \left(\frac{e^{-K_1 t}}{(K_2 - K_1)} + \frac{e^{-K_2 t}}{(K_1 - K_2)} \right) \tag{10.23}$$

which simplifies to:

$$q_2(t) = q_1(0) \frac{k_{2\leftarrow 1}}{(K_2 - K_1)} \left(e^{-K_2 t} - e^{-K_1 t} \right) \tag{10.24}$$

Substitution of parameter values from Table 10.1 gives the solution as:

$$q_2(t) = 1 \frac{3}{(4-1)} \left(e^{-1t} - e^{-4t} \right) \tag{10.25}$$

10.2.3 Closed Recycling System

Recycling adds feedback between compartments. With recycling, the differential equation for *Compartment 1* contains a dependency on material transfer from *Compartment 2*:

$$\frac{dq_1}{dt} = -q_1 K_1 + q_2 k_{1\leftarrow 2} \tag{10.26}$$

The above equation may be re-expressed as:

$$sY_1 = -Y_1 K_1 + Y_2 k_{1\leftarrow 2} + q_1(0) \tag{10.27}$$

The solution to Equation (10.27) requires the term Y_2 to be expressed in terms of Y_1. The solution for Y_2 is:

$$Y_2 = \frac{Y_1 k_{1\leftarrow 2}}{(s + K_2)} \tag{10.28}$$

Substitution and rearrangement gives:

$$Y_1(s + K_1) - \left[\frac{Y_1 k_{1\leftarrow 2}}{(s + K_2)}\right] k_{1\leftarrow 2} = q_1(0) \tag{10.29}$$

which yields:

$$Y_1 = \frac{q_1(0)}{\left[(s + K_1) - \left[\frac{k_{1\leftarrow 2}\, k_{2\leftarrow 1}}{(s + K_2)}\right]\right]} \tag{10.30}$$

Multiplying the numerator and denominator by the term $(s + K_2)$ gives:

$$Y_1 = \frac{(s + K_2) q_1(0)}{\left[(s + K_2)(s + K_1) - k_{1\leftarrow 2}\, k_{2\leftarrow 1}\right]} \tag{10.31}$$

which expands to:

$$Y_1 = \frac{(s + K_2) q_1(0)}{\left[s^2 + K_1 s + K_2 s + K_1 K_2 - k_{1\leftarrow 2}\, k_{2\leftarrow 1}\right]} \tag{10.32}$$

Equation (10.32) is recognized as the quotient of two polynomials. For the assigned parameter values (Table 10.1), the denominator has roots* $\gamma_1 = 0$ and $\gamma_1 = 5$. The above equation may therefore be re-expressed:

$$Y_1 = \frac{(s + K_2) q_1(0)}{(s + \gamma_1)(s + \gamma_2)} \tag{10.33}$$

---

* Roots are generally assigned the symbol λ. To avoid confusion with radioactive decay, the symbol γ is used in this text for roots and eigenvalues.

It is noted that the solution to Equation (10.33) is in the form $L^{-1}[p(s)/q(s)]$, where $p(s)$ and $q(s)$ are polynomials and the degree of $q(s)$ is greater than the degree of $p(s)$. Hence, the Heaviside Expansion Theorem* may be applied, giving a solution of:

$$q_1(t) = q_1(0) \left[\frac{(K_2 - \gamma_1)}{\gamma_2 - \gamma_1} e^{-\gamma_1 t} + \frac{(K_2 - \gamma_2)}{\gamma_1 - \gamma_2} e^{-\gamma_2 t} \right] \tag{10.34}$$

Substitution of parameter values from Table 10.1 gives the solution as:

$$q_1(t) = 1 \left[\frac{(2-0)}{5-0} e^{-0_1 t} + \frac{(2-5)}{0-5} e^{-5_2 t} \right] = 0.4e^{-0t} + 0.6e^{-5t} \tag{10.35}$$

Solving for *Compartment 2*:

$$Y_2 = Y_1 \left[\frac{k_{2 \leftarrow 1}}{(s + K_2)} \right] = \left[\frac{q_1(0)(s + K_2)}{(s + K_1)(s + K_2) - k_{2 \leftarrow 1} k_{1 \leftarrow 2}} \right] \left[\frac{k_{2 \leftarrow 1}}{(s + K_2)} \right] \tag{10.36}$$

Noting that the term $(s + K_2)$ cancels, the above expression becomes:

$$Y_2 = \left[= \frac{q_1(0) k_{2 \leftarrow 1}}{\left[(s^2 + sK_1 + sK_2 - K_2 K_1) - k_{2 \leftarrow 1} k_{1 \leftarrow 2} \right]} \right] \tag{10.37}$$

which yield the same roots as Equation (10.33) and a final solution of:

$$q_2(t) = q_1(0) \left[\frac{k_{2 \leftarrow 1}}{\gamma_2 - \gamma_1} e^{-\gamma_1 t} + \frac{k_{2 \leftarrow 1}}{\gamma_1 - \gamma_2} e^{-\gamma_2 t} \right] \tag{10.38}$$

Substitution of parameter values from Table 10.1 gives the solution as:

$$q_2(t) = 1 \left[\frac{3}{5-0} e^{-0_1 t} + \frac{3}{0-5} e^{-5_2 t} \right] = 0.6e^{-0t} - 0.6e^{-5t} \tag{10.39}$$

10.2.4 Open Recycling System

The solutions for *Compartments 1* and *2* are similar to that arrived at earlier (see Equations 10.33 and 10.37), with the exception that the total removal rate constants have increased to include radioactive decay, and the polynomial roots have also increased to $\gamma_1 = 1$ and $\gamma_2 = 6$. The solutions for *Compartments 1* and *2* now become:

$$q_1(t) = 1 \left[\frac{(3-1)}{6-1} e^{-1t} + \frac{(3-6)}{1-6} e^{-6t} \right] = 0.4e^{-1t} + 0.6e^{-6t} \tag{10.40}$$

---

* See Appendix 10-B for derivation of Heaviside Expansion Theorem.

TABLE 10.2   Solution Sets for a Two-Compartment Model Evaluated under Different Conditions

Modeling Condition	Solution for *Compartment 1*	Solution for *Compartment 2*
Closed catenary	$q_1(t) = 1e^{-3t}$	$q_2(t) = 1\left(1 - e^{-3t}\right)$
Open catenary	$q_1(t) = 1e^{-4t}$	$q_2(t) = 1\left(e^{-1t} - e^{-4t}\right)$
Closed recycling	$q_1(t) = 1\left[0.6e^{-5t} + 0.4e^{-0t}\right]$	$q_2(t) = 1\left[-0.6e^{-5t} + 0.6e^{-0t}\right]$
Open recycling	$q_1(t) = 1\left[0.6e^{-6t} + 0.4e^{-1t}\right]$	$q_2(t) = 1\left[-0.6e^{-6t} + 0.6e^{-1t}\right]$

and

$$q_2(t) = 1\left[\frac{3}{6-1}e^{-1t} + \frac{3}{1-6}e^{-6t}\right] = 0.6e^{-1t} - 0.6e^{-6t} \qquad (10.41)$$

10.2.5 Summary for the Two-Compartment Model

Compartmental solutions for the four evaluated conditions are summarized in Table 10.2, and graphically in Figure 10.3. A few observations are noted:

Closed Catenary System: Because the system is closed, material balance is maintained between the two compartments. While material is exponentially removed from one compartment, the receiving compartment increases exponentially. Transfer and ingrowth are both determined by the transfer rate constants between the compartments. In practice, similar dynamics may apply to tenaciously retained long-lived radionuclides.

Open Catenary System: Because the system is open, material balance is not maintained. Deposition sites are most rapidly cleared under these conditions, and the contents of receiving compartments generally increase, achieve a maximum, and then decrease. Such dynamics apply to material transported along the gastrointestinal tract.

Closed Recycling System: Material balance is again maintained between the compartments. In cases where material is only deposited in the first compartment, as in this example, the coefficients for the exponential terms for the first compartment must sum to unity. In contrast, for receiving compartments, the exponential terms must sum to zero. Values of the exponents no longer equal transfer or total removal rate coefficients, but are determined by the roots of polynomials derived from transfer and removal coefficients. In a closed system, one root should be zero (or nearly zero), as material is retained in the system. Compartmental contents achieve a steady-state condition based on the ratio of the transfer coefficients. In this example, under steady-state conditions, *Compartment 1* contains 40% of the initially deposited material (2/5), and *Compartment 2* contains 60% of the initial deposition (3/5).

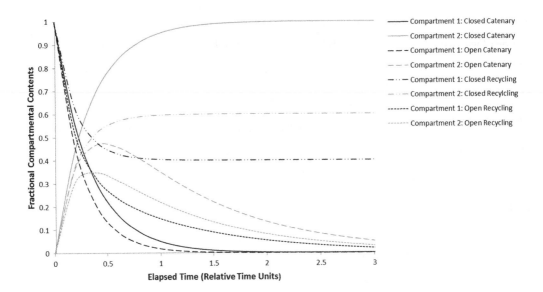

FIGURE 10.3 Comparison of compartmental contents under different conditions.

Open Recycling System: Material balance is not maintained between the compartments. However, because material is only deposited in the first compartment, the coefficients of the exponential terms for the solution for the first compartment must again sum to unity and the exponential terms for receiving compartments must sum to zero. Unlike a closed system, no root should be zero as material is removed from the system.

10.2.6 Matrix Solution

For a system without outside input, the differential equations for any compartment can be reduced to the following expression:

$$q' = kq \tag{10.42}$$

For a system of compartments this relationship can be expressed:

$$
\begin{bmatrix} q'_1 \\ q'_2 \\ \vdots \\ q'_n \end{bmatrix}
=
\begin{bmatrix}
k_{1,1} & k_{1,2} & \cdots & k_{1,n} \\
k_{2,1} & k_{2,2} & \cdots & \vdots \\
\vdots & \vdots & \ddots & \vdots \\
k_{n,1} & \vdots & \cdots & k_{n,n}
\end{bmatrix}
\begin{bmatrix} q_1 \\ q_2 \\ \vdots \\ q_n \end{bmatrix}
\tag{10.43}
$$

The above notation conforms with this text's convention for defining transfer rate constants* where $k_{1 \leftarrow 2}$ equates to $k_{(row=1, column=2)}$. The rate constants along the diagonal ($k_{1,1}$, $k_{2,2}$, etc.) are negative and represent the total removal rates constants for that particular compartment (see Equation 10.2).

---

* An alternate notation used by some authors to define transfer coefficients of requires matrix transposition.

As derived in Sections 10.2.1–10.2.4 above, the solution to this set of differential equations is a series of exponentials in the form of $q = ue^{\gamma t}$, the derivative of which is $\gamma e^{\gamma t}$, which leads to:

$$\gamma u e^{\gamma t} = k u e^{\gamma t} \tag{10.44}$$

Therefore, $\gamma u = ku$ and can be re-expressed as:

$$(k - \gamma I)u = 0 \tag{10.45}$$

The determinant of Equation (10.45) gives an nth degree polynomial in γ having roots $\gamma_1, \gamma_2, \ldots, \gamma_n$, which are defined to be the eigenvalues of k. The corresponding eigenvector $(ku_j = \gamma u)$ yields the solution $q = ue^{\gamma t}$. Thus, for a system of compartments:

$$q(t) = \sum_{j=1}^{n} C_j u_j e^{\gamma_j t} \tag{10.46}$$

where the constant values, C_j, are determined from initial conditions such that:

$$
\begin{bmatrix}
q_1(t) \\
q_2(t) \\
\vdots \\
q_n(t)
\end{bmatrix}
=
\begin{bmatrix}
C_1 u_{1,1} e^{\gamma_1 t} + C_2 u_{1,2} e^{\gamma_2 t} + \ldots + C_n u_{1,n} e^{\gamma_n t} \\
C_1 u_{2,1} e^{\gamma_1 t} + C_2 u_{2,2} e^{\gamma_2 t} + \ldots + C_n u_{2,n} e^{\gamma_n t} \\
\vdots \\
C_1 u_{n,1} e^{\gamma_1 t} + C_2 u_{n,2} e^{\gamma_2 t} + \ldots + C_n u_{n,n} e^{\gamma_n t}
\end{bmatrix}
\tag{10.47}
$$

Consider the two-compartment open recycling model addressed above. The differential equations and the corresponding matrix representing this model become:

$$\frac{dN_1}{dt} = -K_1 N_1 + N_2 k_{1 \leftarrow 2} \tag{10.48}$$

$$\frac{dN_2}{dt} = -K_2 N_2 + N_2 k_{2 \leftarrow 1} \tag{10.49}$$

$$q = \begin{bmatrix} 1 \\ 0 \end{bmatrix} \quad k = \begin{bmatrix} -4 & 2 \\ 3 & -3 \end{bmatrix} \tag{10.50}$$

The eigenvalues (or roots) are derived from the determinant:

$$|k - I\gamma| = 0 \quad \begin{bmatrix} -4 & 2 \\ 3 & -3 \end{bmatrix} - \begin{bmatrix} \gamma & 0 \\ 0 & \gamma \end{bmatrix} = \begin{bmatrix} (-4-\gamma) & 2 \\ 3 & (-3-\gamma) \end{bmatrix} = \gamma^2 + 7\gamma + 6 \tag{10.51}$$

which factors to $(\gamma+6)(\gamma+1)$ and has the same roots (−6 and −1) derived from previous polynomial expansion (see Equations 10.40 and 10.41). Therefore:

$$\gamma = \text{eigenvalues} = \begin{bmatrix} -6 \\ -1 \end{bmatrix} \tag{10.52}$$

Eigenvectors (u) for each eigenvalue are determined by substituting each root back into Equation (10.51). Applying the first root (-6) gives:

$$|\mathbf{k} - \mathbf{I}(-6)| = 0 \quad \begin{bmatrix} -4 & 2 \\ 3 & -3 \end{bmatrix} - \begin{bmatrix} -6 & 0 \\ 0 & -6 \end{bmatrix} = \begin{bmatrix} (-4+6) & 2 \\ 3 & (-3+6) \end{bmatrix} = \begin{bmatrix} 2 & 2 \\ 3 & 3 \end{bmatrix} \tag{10.53}$$

which yields:

$$\begin{bmatrix} 2 & 2 \\ 3 & 3 \end{bmatrix} \begin{bmatrix} u_1 \\ u_2 \end{bmatrix} = 2u_1 + 2u_2 = 0 \quad \text{and} \quad 3u_1 + 3u_2 = 0 \tag{10.54}$$

Therefore, $u_1 = u_2$ giving an eigenvector of:

$$\begin{bmatrix} 1 \\ -1 \end{bmatrix} \tag{10.55}$$

Similarly, for the second root (−1), $u_1 = 2/3u_2$ giving eigenvector:

$$\begin{bmatrix} 2 \\ 3 \end{bmatrix} \tag{10.56}$$

Combining the eigenvectors and eigenvalues produces the following general equation applying the initial boundary conditions of material only deposited in *Compartment 1*:

$$q(t) = C_1 \begin{bmatrix} 1 \\ -1 \end{bmatrix} e^{-6t} + C_2 \begin{bmatrix} 2 \\ 3 \end{bmatrix} e^{-1t} = \begin{bmatrix} 1 \\ 0 \end{bmatrix} \tag{10.57}$$

evaluated at time zero gives:

$$C_1 + 2C_2 = 1 \tag{10.58}$$

and

$$-C_1 + 3C_2 = 0 \tag{10.59}$$

the solution to which yields: $C_1 = 0.6$, $C_2 = 0.2$.

Substitution gives:

$$q_1(t) = (0.6)(1)e^{-6t} + (0.2)(2)e^{-1t} = 0.6e^{-6t} + 0.4e^{-1t} \tag{10.60}$$

$$q_2(t) = -(0.6)(1)e^{-6t} + (0.2)(3)e^{-1t} = -0.6e^{-6t} + 0.6e^{-1t} \tag{10.61}$$

These solutions are identical to those derived earlier (see Equations 10.40 and 10.41).

10.3 RETENTION FRACTIONS

The solutions derived above give the quantity of material contained in a compartment at an elapsed time post deposition ($q_i(t)$). A value $F(t)$ can be defined such that:

$$F(t) = \frac{q_i(t)}{q_1(0)} \tag{10.62}$$

where $F(t)$ is the fraction of the initial deposition retained in a compartment of interest at some elapsed time. For bioassay interpretation, $F(t)$ is referred to as the "retention fraction," as it provides an estimate of the fraction of the initial deposition retained in an anatomical or excretion compartment. Therefore, the deposition (or intake) can be estimated from a bioassay measurement as:

$$\text{Intake or deposition} = \frac{\text{Bioassay Measurement}}{F(t)} \tag{10.63}$$

10.3.1 Retention Functions for Chronic Intakes

A chronic exposure model, shown in Figure 10.4, is applied to assess intakes occurring over extended intervals where the rate of change within a single compartment is dependent upon the intake rate ($\dot{I}$) and the total removal rate (K).

As shown in Figure 10.5, the contents of the compartment increase throughout the exposure interval (T) and then decreases as time (t) elapses.

Conceptually, a chronic exposure can be treated as a series of acute exposures. As described above, for an acute exposure, the expected content of the compartment at an elapsed time post exposure ($q(t)$) is equal to the intake (I) multiplied by the time-dependent retention fraction ($F(t)$):

$$q(t) = I\,F(t) \tag{10.64}$$

Similarly, for a chronic exposure, the differential content ($dq(\tau)$) is equal to the differential intake (given as the product of the intake rate ($\dot{I}$) and the differential time element ($d\tau$))

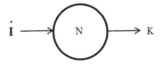

FIGURE 10.4   Chronic intake model for a single compartment with intake rate, $\dot{I}$, and removal rate, K.

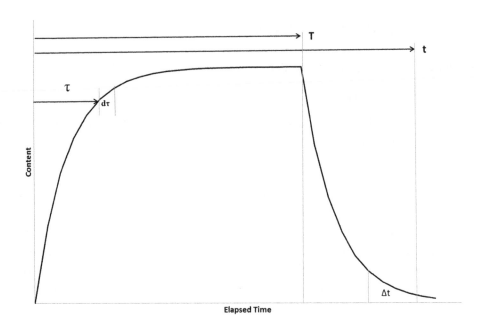

FIGURE 10.5 Compartmental buildup and removal for a chronic intake.

multiplied by the retention fraction evaluated at the elapsed time between the intake and the bioassay measurement of $(t - \tau)$:

$$dq(\tau) = \dot{I}\,dt\,F(t - \tau) \tag{10.65}$$

The intake retention fraction for a chronic exposure ($F_c(t)$) is therefore the integral of the above expression over the intake interval (0 to T) divided by the total intake, where the total intake is equal to the intake rate multiplied by the intake interval ($\dot{I}T$):

$$F_c(t) = \frac{q(t)}{I} = \frac{q(t)}{\dot{I}T} = \frac{\displaystyle\int_{\tau=0}^{\tau=T}\left(\dot{I}\,d\tau\right)\left(F(t-\tau)\right)}{\dot{I}T} \tag{10.66}$$

Note that in the right-hand expression, the intake rate ($\dot{I}$) cancels, reducing the expression to:

$$F_c(t) = \frac{q(t)}{I} = \frac{q(t)}{\dot{I}T} = \frac{\displaystyle\int_{\tau=0}^{\tau=T}\left(F(t-\tau)\right)d\tau}{T} \tag{10.67}$$

To integrate the above expression, the following substitution is helpful: let $u = (t - \tau)$, giving:

$$F_c(t) = \frac{q(t)}{\dot{I}T} = \frac{\displaystyle\int_{u=t-T}^{u=t}\left(F(u)\right)du}{T} \tag{10.68}$$

In the case of a single compartment having exponential removal, the retention function $F(u)$ is equal to e^{-Ku}, the integral to which is:

$$\int e^{-Ku} du = \frac{e^{-Ku}}{-K}$$

(10.69)

Applying the limits of integration gives:

$$F_c(t) = \frac{q(t)}{\dot{I}T} = \frac{1}{T}\left[\frac{e^{-Kt}}{-K} - \frac{e^{-K(t-T)}}{-K}\right]$$

(10.70)

Rearranging and combining terms gives:

$$F_c(t) = \frac{q(t)}{\dot{I}T} = \frac{1}{T}\left[\frac{e^{-K(t-T)} - e^{-Kt}}{K}\right]$$

(10.71)

which equals:

$$F_c(t) = \frac{q(t)}{\dot{I}T} = \left[\frac{\left(1 - e^{-KT}\right)e^{-K(t-T)}}{KT}\right]$$

(10.72)

This equation has a somewhat familiar construction, in that the term $\left(1 - e^{-KT}\right)/K$ is similar to the fraction of saturation activity analogous to that applied in neutron activation or activity collection on air filters, and the term $e^{K(t-T)}$ gives the exponential decay occurring after cessation of exposure. From the derivation, it was noted that the inclusion of the time parameter (T) in the denominator is due to the fact that an intake rate, rather than intake, is involved.

It is important to recognize that a bioassay measurement can be obtained during the ingrowth period or after the intake has ceased. The solution for the two conditions is different. The solution derived above applies to elapsed times greater than the exposure period ($t > T$). The intake retention fraction for measurements obtained during the intake interval ($t \leq T$) simplifies to the following expression:

$$F_c(t) = \frac{q(t)}{\dot{I}t} = \frac{\left(1 - e^{-kt}\right)}{Kt}$$

(10.73)

It should be noted that the above expression gives the fraction of the total intake contained in the compartment at time "t" during the buildup phase. The total intake continues to increase during the intake interval. As such, the above expression cannot be used to evaluate the total intake from multiple bioassay measurements. Rather, because the total intake is changing, the above expression is used to estimate the average intake rate ($\dot{I}$).

Applying the principles described for acute exposures, the average intake rate is calculated from the following two expressions:

For elapsed times $t \leq T$:

$$\dot{I} = \frac{q(t)}{\left[\dfrac{\left(1-e^{-kt}\right)}{Kt}\right]t} \tag{10.74}$$

For elapsed times $t > T$:

$$\dot{I} = \frac{q(t)}{\left[\dfrac{\left(1-e^{-kT}\right)e^{-k(t-T)}}{KT}\right]T} \tag{10.75}$$

For example, for a single compartment having a total removal rate constant of $K = 0.0693\,d^{-1}$, the average intake rate to maintain an activity burden of 1 µCi at 50 days into a chronic intake interval is calculated using the simplified expression:

$$F_c(t) = \frac{q(t)}{\dot{I}t} = \frac{\left(1-e^{-kt}\right)}{Kt} \tag{10.76}$$

into which the substitutions $q(50d) = 1\,\mu\text{Ci}$, $t = 50$ d, and $K = 0.0693$ d^{-1} are applied:

$$F_c(t) = \frac{1\,\mu\text{Ci}}{\left(\dot{I}\right)(50d)} = \frac{\left(1-e^{-\left(0.0693\,d^{-1}\right)(50d)}\right)}{\left(0.0693\,d^{-1}\right)(50d)} = 0.28 \tag{10.77}$$

giving:

$$\dot{I} = \frac{1\,\mu\text{Ci}}{(0.28)(50d)} = 0.07\frac{\mu\text{Ci}}{d} \tag{10.78}$$

The total intake over the 50-day interval is therefore 3.5 uCi.

For monitoring performed outside the intake interval, the expression containing the additional decay term is applied. For example, given a compartmental content of 0.5 µCi measured ten days after the cessation of a 50-day exposure interval (that is, $T = 50$ days and $t = 60$ days), the chronic intake retention fraction is calculated as:

$$F_c(t) = \frac{0.5\,\mu\text{Ci}}{\left(\dot{I}\right)(50d)} = \frac{\left(1-e^{-\left(0.0693\,d^{-1}\right)(50d)}\right)e^{-\left(0.0693\,d^{-1}\right)(10d)}}{\left(0.0693\,d^{-1}\right)(50d)} = 0.14 \tag{10.79}$$

giving an average intake rate of:

$$\dot{I} = \frac{0.5\,\mu\text{Ci}}{(0.14)(50d)} = 0.07\,\frac{\mu\text{Ci}}{d} \tag{10.80}$$

Therefore, the total intake over the 50-day interval is 3.5 μCi.

Though it is beyond the scope of this text, as described for acute intakes, the above discussion can be extended to evaluate incremental excretion. Unlike acute exposures, because time intervals (t and T) associated with chronic intakes are unique to the particular situation, retention/excretion fractions for chronic intakes have not been tabulated or published.

10.3.2 Excretion Functions for Acute Exposures

The retention functions described above are used to predict the contents of a continuously connected compartment within a chain where material is transferred to succeeding members of the chain. Such functions are not applicable to incremental excretion (urine or feces) where compartmental contents are collected over some interval and then removed from the chain. Conceptually, excretion functions give the fraction of the intake present in a sample collected over a defined interval, such as a 24-hour urine sample. Conceptually, the excretion fraction ($e_u(t)$) represents the change over a defined time interval in the systemic system adjusted for the fraction of material following that excretion pathway. For urinary excretion, this may be expressed:

$$e_u(t) = \left[F(t) - F(t - \Delta t)\right] f_u \tag{10.81}$$

where t is the elapsed time to the end of sample collection, Δt is the sample collection time, and f_u is the fraction of the systemic burden excreted via urine. Typically, for urinalysis programs, the collection interval Δt is set to 24-hours.

10.3.3 Tabulations of Retention and Excretion Fractions

Tabulations of acute intake retention/excretion factors for various dosimetry systems have been published. Most notable are NUREG-4884 (Nuclear Regulatory Commission 1987), which applied ICRP Publication 30 modeling assumptions (ICRP 1979) and the November 2002 issue of Health Physics (Potter 2002), in which ICRP Publication 68 (ICRP 1994) methods are applied.

Comparison of these tables shows that predicted retention/excretion values between the two systems can be remarkably similar, as is the case for whole-body retention of ^{137}Cs, where retention estimates differ by a constant factor related to initial respiratory tract deposition (see Figure 10.3). Conversely, the values can differ significantly, as is the case of ^{238}U urinary excretion, where modern models accounting for bone turnover and systemic recycling affect both the magnitude and pattern of urinary excretion (see Figure 10.3).

Care should be taken when interpreting tabulated retention/excretion values to ensure that radioactive decay corrections are applied. For example, the Potter tabulations are for stable elements for which nuclide-specific decay corrections are required (Tables 10.4 and 10.6).

10.4 LIMITATIONS AFFECTING INCREMENTAL EXCRETA BIOASSAY SAMPLES

As noted above, "incremental" excretion fractions are typically expressed in units of "fraction per day." As such, incremental excretion fractions published in documents, such as NUREG-4884, only apply to in vitro samples collected over a continuous 24-hour interval.

With respect to urine sample collection, to promote worker convenience, "approximate" or "simulated" 24-hour sampling, consisting of collecting urinary excretion over two consecutive nights has been employed (National Council on Radiation Protection and Measurements 1987). However, more typically, normalization is applied to samples deviating from the 24-hour collection protocol. Various normalization techniques have been used: (1) scale to 24-hour excretion based on the sample collection interval, (2) correct to a daily urine output volume of 1600 mL for males or 1200 mL for females (unless the individual-specific daily excretion rate is known), (3) adjust by creatinine, or (4) normalize by specific gravity.

Interpretation of incremental fecal samples is more problematic. The ICRP has cautioned that it

> should be recognized that the Publication 30 GI tract model was not specifically intended for bioassay purposes. There are considerable differences in average times between subjects, and for a given individual the transit times will depend on factors such as recent dietary intakes, physical and emotional stress, etc. (ICRP 2002).

Operational experience has shown that because material is rapidly transferred through the GI tract, fecal sampling is overly sensitive to chronic low-level (environmental) ingestion intakes (such as uranium in foodstuffs) which can interfere with the detection of occupational exposures. Fecal sampling programs also face increased hesitation and less cooperation from workers.

Despite the noted limitations, fecal sampling may be useful for incident response. For example, following the acute inhalation of insoluble uranium, as can be seen in Table 10.6, nearly half (47%) of the total intake is ultimately excreted through feces, and 75% of this total (36%) is excreted in the first three days following intake. As such, a large fraction of the intake can be captured through the collection of accumulated fecal excretion for several days following a significant exposure event.

10.4.1 Best Estimate of Intake from Multiple Bioassay Measurements

For a single bioassay measurement, the intake is estimated as the quotient of the bioassay measurement to the corresponding intake retention (or excretion) fraction:

$$I = \frac{q(t)}{F(t)}$$

TABLE 10.3 ^{137}Cs Retention Fraction Table from NUREG-4884 (Nuclear Regulatory Commission 1987)

Class D	AMAD = 1 Micron	Half-Life = 1.10E+04 Days			Cesium 137
Time after Single Intake		Fraction of Initial Intake in:			
Days	Systemic Organs	Lungs	Nasal Passages	GI Tract	Total Body
1.00E−01	3.90E−01	2.26E−01	3.03E−04	2.17E−02	6.38E−01
2.00E−01	4.33E−01	2.01E−01	2.96E−07	2.28E−03	6.36E−01
3.00E−01	4.55E−01	1.79E−01	0.00E+00	4.15E−04	6.35E−01
4.00E−01	4.73E−01	1.60E−01	0.00E+00	1.85E−04	6.33E−01
5.00E−01	4.88E−01	1.43E−01	0.00E+00	1.21E−04	6.31E−01
6.00E−01	5.02E−01	1.27E−01	0.00E+00	8.47E−05	6.29E−01
7.00E−01	5.14E−01	I.13E−01	0.00E+00	5.98E−05	6.28E−01
8.00E−01	5.25E−01	1.01E−01	0.00E+00	4.23E−05	6.26E−01
9.00E−01	5.34E−01	8.96E−02	0.00E+00	2.99E−05	6.24E−01
1.00E+00	5.43E−01	7.97E−02	0.00E+00	2.12E−05	6.22E−01
2.00E+00	5.81E−01	2.42E−02	0.00E+00	7.04E−07	6.06E−01
3.00E+00	5.85E−01	7.13E−03	0.00E+00	3.82E−08	5.92E−01
4.00E+00	5.79E−01	2.05E−03	0.00E+00	0.00E+00	5.81E−01
5.00E+00	5.71E−01	5.80E−04	0.00E+00	0.00E+00	5.72E−01
6.00E+00	5.64E−01	1.62E−04	0.00E+00	0.00E+00	5.64E−01
7.00E+00	5.58E−01	4.47E−05	0.00E+00	0.00E+00	5.58E−01
8.00E+00	5.53E−01	1.22E−05	0.00E+00	0.00E+00	5.53E−01
9.00E+00	5.48E−01	3.32E−06	0.00E+00	0.00E+00	5.48E−01
1.00E+0l	5.43E−01	8.96E−07	0. 00E+00	0.00E+00	5.43E−01
2.00E+01	5.08E−01	0.00E+00	0.00E+00	0.00E+00	5.08E−01
3.00E+0l	4.76E−01	0.00E+00	0.00E+00	0.00E+00	4.76E−01
4.00E+0l	4.47E−01	0.00E+00	0.00E+00	0.00E+00	4.47E−01
5.00E+0l	4.19E−01	0.00E+00	0.00E+00	0.00E+00	4.19E−01
6.00E+0l	3.94E−01	0.00E+00	0.00E+00	0.00E+00	3.94E−01
7.00E+0l	3.69E−01	0.00E+00	0.00E+00	0.00E+00	3.69E−01
8.00E+0l	3.47E−01	0.00E+00	0.00E+00	0.00E+00	3.47E−01
9.00E+0l	3.25E−01	0.00E+00	0.00E+00	0.00E+00	3.25E−01
1.00E+02	3.05E−01	0.00E+00	0.00E+00	0.00E+00	3.05E−01
2.00E+02	1.61E−01	0.00E+00	0.00E+00	0.00E+00	1.61E−01
3.00E+02	8.55E−02	0.00E+00	0.00E+00	0.00E+00	8.55E−02
4.00E+02	4.52E−02	0.00E+00	0.00E+00	0.00E+00	4.52E−02
5.00E+02	2.39E−02	0.00E+00	0.00E+00	0.00E+00	2.39E−02
6.00E+02	1.27E−02	0.00E+00	0.00E+00	0.00E+00	1.27E−02
7.00E+02	6.70E−03	0.00E+00	0.00E+00	0.00E+00	6.70E−03
8.00E+02	3.55E−03	0.00E+00	0.00E+00	0.00E+00	3.55E−03
9.00E+02	1.88E−03	0.00E+00	0.00E+00	0.00E+00	1.88E−03
1.00E+03	9.94E−04	0.00E+00	0.00E+00	0.00E+00	9.94E−04
2.00E+03	1.71E−06	0.00E+00	0.00E+00	0.00E+00	1.71E−06

TABLE 10.4   Cs Retention and Excretion Fraction Table from Health Physics (Potter 2002)

			Intake Retention Fractions for Class F Cesium			
Elapsed Time (days)	Whole Body w/o ET	Whole Body	Accumulated Urine	Accumulated Feces	Incremental Urine	Incremental Feces
0.25	4.80E−01	7.44E−01	9.00E−04	1.83E−06	9.00E−04	1.83E−06
0.5	4.78E−01	6.83E−01	3.14E−03	2.53E−05	3.14E−03	2.53E−05
0.75	4.75E−01	6.35E−01	5.93E−03	1.03E−04	5.93E−03	1.03E−04
1	4.72E−01	5.97E−01	8.90E−03	2.58E−04	8.90E−03	2.58E−04
1.25	4.69E−01	5.66E−01	1.18E−02	4.98E−04	1.09E−02	4.96E−04
1.5	4.66E−01	5.41E−01	1.47E−02	8.20E−04	1.15E−02	7.95E−04
1.75	4.43E−01	5.21E−01	1.73E−02	1.21E−03	1.14E−02	1.11E−03
2	4.60E−01	5.05E−01	1.99E−02	1.67E−03	1.10E−02	1.41E−03
2.25	4.57E−01	4.92E−01	2.22E−02	2.16E−03	1.04E−02	1.66E−03
2.5	4.54E−01	4.82E−01	2.44E−02	2.68E−03	9.79E−03	1.86E−03
2.75	4.51E−01	4.73E−01	2.65E−02	3.22E−03	9.19E−03	2.01E−03
3	4.49E−01	4.66E−01	2.85E−02	3.77E−03	8.62E−03	2.10E−03
4	4.40E−01	4.46E−01	3.52E−02	5.88E−03	6.73E−03	2.11E−03
5	4.33E−01	4.35E−01	4.06E−02	7.72E−03	5.37E−03	1.84E−03
6	4.27E−01	4.28E−01	4.50E−02	9.25E−03	4.41E−03	1.53E−03
7	4.22E−01	4.22E−01	4.87E−02	1.05E−02	3.72E−03	1.26E−03
8	4.18E−01	4.18E−01	5.20E−02	1.16E−02	3.24E−03	1.06E−03
9	4.14E−01	4.14E−01	5.48E−02	1.25E−02	2.89E−03	9.02E−04
10	4.10E−01	4.10E−01	5.75E−02	1.33E−02	2.64E−03	7.89E−04
20	3.84E−01	3.84E−01	7.87E−02	1.89E−02	1.95E−03	4.97E−04
30	3.60E−01	3.60E−01	9.75E−02	2.37E−02	1.82E−03	4.59E−04
40	3.38E−01	3.38E−01	1.15E−01	2.81E−02	1.71E−03	4.31E−04
50	3.17E−01	3.17E−01	1.32E−01	3.23E−02	1.60E−03	4.04E−04
60	2.98E−01	2.98E−01	1.47E−01	3.62E−02	1.50E−03	3.80E−04
70	2.80E−01	2.80E−01	1.62E−01	3.98E−02	1.41E−03	3.57E−04
80	2.63E−01	2.63E−01	1.75E−01	4.33E−02	1.33E−03	3.35E−04
90	2.47E−01	2.47E−01	1.88E−01	4.65E−02	1.25E−03	3.14E−04
100	2.32E−01	2.32E−01	2.00E−01	4.95E−02	1.17E−03	2.95E−04
200	1.23E−01	1.23E−01	2.86E−01	7.14E−02	6.22E−04	1.57E−04
300	6.57E−02	6.57E−02	3.32E−01	8.30E−02	3.32E−04	3.87E−05
400	3.50E−02	3.50E−02	3.57E−01	8.92E−02	1.77E−04	4.46E−05
500	1.86E−02	1.86E−02	3.70E−01	9.25E−02	9.40E−05	2.37E−05
600	9.92E−03	9.92E−03	3.77E−01	9.42E−02	5.01E−05	1.26E−05
700	5.28E−03	5.28E−03	3.81E−01	9.52E−02	2.67E−05	6.73E−06
800	2.81E−03	2.81E−03	3.83E−01	9.57E−02	1.42E−05	3.58E−06
900	1.50E−03	1.50E−03	3.84E−01	9.59E−02	7.56E−06	1.91E−06
1000	7.98E−04	7.98E−04	3.84E−01	9.61E−02	4.03E−06	1.02E−06
2000	1.46E−06	1.46E−06	3.85E−01	9.62E−02	7.38E−09	1.86E−09
3000	2.68E−09	2.68E−09	3.85E−01	9.62E−02	1.35E−11	3.42E−12
4000	4.92E−12	4.92E−12	3.85E−01	9.62E−02	2.48E−14	6.27E−15
5000	9.02E−15	9.02E−15	3.85E−01	9.62E−02	4.55E−17	1.15E−17
6000	1.65E−17	1.65E−17	3.85E−01	9.62E−02	8.35E−20	2.11E−20
7000	3.03E−20	3.03E−20	3.85E−01	9.62E−02	1.53E−22	3.87E−23
8000	5.56E−23	5.56E−23	3.85E−01	9.62E−02	2.81E−25	7.09E−26
9000	1.02E−25	1.02E−25	3.85E−01	9.62E−02	5.15E−28	1.30E−28
10000	1.87E−28	1.87E−28	3.85E−01	9.62E−02	9.44E−31	2.38E−31
20000	8.05E−56	8.05E−56	3.85E−01	9.62E−02	4.06E−58	1.03E−58
30000	3.46E−83	3.46E−83	3.85E−01	9.62E−02	1.75E−85	4.41E−86

TABLE 10.5 ^{238}U Excretion Fraction Table from NUREG-4884 (Nuclear Regulatory Commission 1987)

Class Y	AMAD = 1 Micron	Half-Life= 1.16E+12 Days		Uranium 238
Time after Single Intake		**Fraction of Initial Intake in:**		
Days	24-Hour Urine	Accumulated Urine	24-Hour Feces	Accumulated Feces
1.00E−01		3.80E−04		1.04E−05
2.00E−01		7.49E−04		2.10E−04
3.00E−01		1.05E−03		I.06E−03
4.00E−01		1.31E−03		3.07E−03
5.00E−01		1.53E−03		6.66E−03
6.00E−01		1.71E−03		1.21E−02
7.00E−01		I.87E−03		1.94E−02
8.00E−0I		2.01E−03		2.86E−02
9.00E−01		2.13E−03		3.95E−02
1.00E+00	2.23E−03	2.23E−03	5.20E−02	5.20E−02
2.00E+00	5.49E−04	2.78E−03	1.60E−01	2.13E−01
3.00E+00	2.30E−04	3.01E−03	1.31E−01	3.43E−01
4.00E+00	1.57E−04	3.17E−03	7.35E−02	4.17E−01
5.00E+00	1.31E−04	3.30E−03	3.63E−02	4.53E−01
6.00E+00	1.17E−04	3.41E−03	1.72E−02	4.70E−01
7.00E+00	1.07E−04	3.52E−03	8.11E−03	4.78E−01
8.00E+00	9.81E−05	3.62E−03	3.88E−03	4.82E−01
9.00E+00	9.07E−05	3.71E−03	1.91E−03	4.84E−01
1.00E+01	8.42E−05	3.79E−03	9.83E−04	4.85E−01
2.00E+01	4.69E−05	4.40E−03	1.35E−04	4.87E−01
3.00E+01	3.27E−05	4.78E−03	1.33E−04	4.89E−01
4.00E+01	2.65E−05	5.07E−03	1.31E−04	4.90E−01
5.00E+01	2.34E−05	5.32E−03	I.29E−04	4.91E−01
7.00E+01	2.04E−05	5.75E−03	1.26E−04	4.94E−01
8.00E+01	1.96E−05	5.95E−03	1.24E−04	4.95E−01
9.00E+01	1.91E−05	6.14E−03	1.22E−04	4.96E−01
1.00E+02	1.87E−05	6.33E−03	1.20E−04	4.97E−01
2.00E+02	1.81E−05	8.15E−03	1.05E−04	5.09E−01
3.00E+02	1.83E−05	9.97E−03	9.13E−05	5.18E−01
4.00E+02	1.82E−05	1.18E−02	7.95E−05	5.27E−01
5.00E+02	1.81E−05	1.36E−02	6.92E−05	5.34E−01
7.00E+02	1.74E−05	1.72E−02	5.24E−05	5.46E−01
8.00E+02	1.69E−05	1.89E−02	4.56E−05	5.51E−01
9.00E+02	1.64E−05	2.05E−02	3.97E−05	5.56E−01
I.00E+03	1.58E−05	2.21E−02	3.46E−05	5.59E−01
2.00E+03	9.76E−06	3.49E−02	8.65E−06	5.78E−01
3.00E+03	5.39E−06	4.23E−02	2.16E−06	5.83E−01
4.00E+03	2.86E−06	4.63E−02	5.40E−07	5.84E−01
5.00E+03	1.50E−06	4.84E−02	1.35E−07	5.84E−01
7.00E+03	4.36E−07	5.01E−02	0.00E+00	5.84E−01
8.00E+03	2.48E−07	5.04E−02	0.00E+00	5.84E−01
9.00E+03	1.50E−07	5.06E−02	0.00E+00	5.84E−01
1.00E+04	9.75E−08	5.07E−02	0.00E+00	5.84E−01
2.00E+04	1.31E−08	5.11E−02	0.00E+00	5.84E−01

TABLE 10.6 Uranium Retention and Excretion Fraction Table from Health Physics (Potter 2002)

	Intake Retention Fractions for Class S Uranium					
Elapsed Time (days)	Whole Body w/o ET	Whole Body	Accumulated Urine	Accumulated Feces	Incremental Urine	Incremental Feces
0.25	4.76E−01	7.40E−01	4.14E−04	4.14E−03	4.14E−04	4.14E−03
0.5	4.52E−01	6.58E−01	6.12E−04	2.78E−02	6.12E−04	2.78E−02
0.75	4.12E−01	5.72E−01	6.84E−04	6.82E−02	6.84E−04	6.82E−02
1	3.65E−01	4.89E−01	7.13E−04	1.16E−01	7.13E−04	1.16E−01
1.25	3.17E−01	4.14E−01	7.27E−04	1.63E−01	3.13E−04	1.59E−01
1.5	2.73E−01	3.49E−01	7.36E−04	2.07E−01	1.24E−04	1.79E−01
1.75	2.34E+02	2.93E−01	7.44E−04	2.46E−01	5.93E−05	1.78E−01
2	2.01E−01	2.47E−01	7.51E−04	2.79E−01	3.79E−05	1.63E−01
2.25	1.74E−01	2.10E−01	7.58E−04	3.06E−01	3.07E−05	1.43E−01
2.5	1.51E−01	1.79E−01	7.64E−04	3.29E−01	2.79E−05	1.22E−01
2.75	1.33E−01	1.55E−01	7.70E−04	3.47E−01	2.66E−05	1.01E−01
3	1.18E−01	1.35E−01	7.77E−04	3.62E−01	2.57E−05	8.30E−02
4	8.36E−02	9.00E−02	8.00E−04	3.97E−01	2.35E−05	3.46E−02
5	6.99E−02	7.24E−02	8.22E−04	4.10E−01	2.18E−05	1.37E−02
6	6.44E−02	6.54E−02	8.42E−04	4.16E−01	2.04E−05	5.52E−03
7	6.19E−02	6.24E−02	8.61E−04	4.18E−01	1.91E−05	2.42E−03
8	6.07E−02	6.10E−02	8.79E−04	4.19E−01	1.79E−05	1.25E−03
9	5.98E−02	6.01E−02	8.96E−04	4.20E−01	1.69E−05	8.14E−04
10	5.92E−02	5.94E−02	9.12E−04	4.21E−01	1.60E−05	6.44E−04
20	5.41E−02	5.43E−02	1.04E−03	4.26E−01	1.02E−05	4.39E−04
30	5.02E−02	5.04E−02	1.12E−03	4.30E−01	7.71E−06	3.50E−04
40	4.70E−02	4.72E−02	1.19E−03	4.33E−01	6.44E−06	2.81E−04
50	4.44E−02	4.46E−02	1.25E−03	4.35E−01	5.68E−06	2.28E−04
60	4.23E−02	4.25E−02	1.31E−03	4.37E−01	5.18E−06	1.86E−04
70	4.06E−02	4.08E−02	1.36E−03	4.39E−01	4.80E−06	1.53E−04
80	3.92E−02	3.94E−02	1.40E−03	4.40E−01	4.51E−06	1.27E−04
90	3.80E−02	3.82E−02	1.45E−03	4.42E−01	4.28E−06	1.07E−04
100	3.70E−02	3.72E−02	1.49E−03	4.43E−01	4.09E−06	9.06E−05
200	3.14E−02	3.16E−02	1.84E−03	4.48E−01	3.18E−06	2.30E−05
300	2.84E−02	2.85E−02	2.14E−03	4.50E−01	2.82E−06	2.43E−05
400	2.58E−02	2.60E−02	2.41E−03	4.53E−01	2.55E−06	2.12E−05
500	2.36E−02	2.38E−02	2.65E−03	4.55E−01	2.33E−06	1.89E−05
600	2.16E−02	2.18E−02	2.87E−03	4.57E−01	2.13E−06	1.70E−05
700	1.98E−02	1.99E−02	3.08E−03	4.58E−01	1.96E−06	1.53E−05
800	1.81E−02	1.83E−02	3.27E−03	4.60E−01	1.80E−06	1.37E−05
900	1.67E−02	1.68E−02	3.44E−03	4.61E−01	1.65E−06	1.23E−05
1000	5.30E−02	1.55E−02	3.60E−03	4.62E−01	1.52E−06	1.11E−05
2000	7.45E−03	7.61E−03	4.67E−03	4.69E−01	7.52E−07	3.88E−06
3000	4.43E−03	4.58E−03	5.25E−03	4.71E−01	4.53E−07	1.46E−06
4000	3.10E−03	3.24E−03	5.63E−03	4.72E−01	3.19E−07	6.14E−07
5000	2.40E−03	2.52E−03	5.91E−03	4.73E−01	2.47E−07	3.09E−07
6000	1.95E−03	2.06E−03	6.13E−03	4.73E−01	2.01E−07	1.87E−07
7000	1.62E−03	1.72E−03	6.32E−03	4.73E−01	1.67E−07	1.29E−07
8000	1.37E−03	1.45E−03	6.47E−03	4.73E−01	1.40E−07	9.71E−08
9000	1.16E−03	1.24E−03	6.60E−03	4.73E−01	1.19E−07	7.57E−08
10000	9.89E−04	1.06E−03	6.71E−03	4.73E−01	1.01E−07	6.00E−08
20000	2.59E−04	2.86E−04	7.24E−03	4.74E−01	2.56E−08	6.69E−09
30000	9.53E−05	1.05E−04	7.40E−03	4.74E−01	8.54E−09	7.79E−10

Bioassay results are, however, influenced by biological variation and measurement uncertainty. Consequently, multiple bioassay measurements connected with the same exposure yield different intake estimates. To resolve differences, statistical tests intended to achieve a "Best Estimate" of intake by minimizing the variation between measurement-specific intake estimates are typically applied. Three common approaches: (1) averaging, (2) unweighted least squares fitting, and (3) weighted least squares fitting are discussed below. It should be noted that intake estimation for replicate assays is not limited to the techniques presented below. For example, Bayesian methods have also been proposed. The methods discussed herein were selected because they are generally accepted, commonly applied, uncomplicated, and have been incorporated into commercially available software.

The fitting techniques described below minimize the chi-square statistic with respect to the intake. The primary difference is that different assumptions regarding the variance within the bioassay data are applied. Typically, fluctuation in bioassay data is considered to be dominated by biological variance as opposed to analytical variance; however, this may not always be true.

Averaging: Conventional averaging of measurement-specific intake estimates, also referred to as the "average of the slopes method," has been used to estimate intake. Under this approach, the best estimate of intake is simply the sum or the individual estimates divided by the number of measurements (see Table 10.7). The averaging technique assumes that each bioassay measurement is weighted inversely to the square of the expected value. That is, variance is proportional to the square of the expected result. This method is most appropriate when the variance is due primarily to biological factors rather than measurement uncertainty.

Unweighted Least Squares Fitting (ULSF): In unweighted fitting, each bioassay measurement is treated equally, and variance is assumed to be constant. The expression for estimating intake is provided in Table 10.7. This approach implies that variance is independent of the magnitude of the measurement. Unweighted fitting is appropriate if all measurements are equally significant. That is, the measurements are similarly accurate and are above detection limits.

Weighted Least Squares Fitting (WLSF): This approach assumes that bioassay measurements are weighted inversely proportional to their expected value. This assumption implies that variance is proportional to the magnitude of the expected value. The intake is estimated as the sum of the bioassay results divided by the sum of the retention (or excretion) fractions (see Table 10.7). Weighted least squares fitting is also referred to as the "ratio of the means," as it can be shown that the intake estimate is also equal to the quotient of the average bioassay result by the average retention (or excretion) fraction.

Method selection should be based on the quality of the bioassay data, the degree of confidence in the metabolic model, conditions associated with the exposure, and professional judgment. This text does not endorse a particular approach for fitting multiple bioassay

TABLE 10.7   Various Methods for Estimating Intake from Multiple Bioassay Measurements

Method	Intake Estimate	Variance in Intake Estimate
Weighted least squares Fit (WLSF) "ratio of means"	$$I = \frac{\sum \dfrac{F(t)_i q_i}{\sigma_i^2}}{\sum \dfrac{F(t)_i^2}{\sigma_i^2}} = \frac{\sum q_i}{\sum F(t)_i}$$	$$\sigma_i^2 = \frac{\sum \dfrac{\left(q_i - I \cdot F(t)_i\right)^2}{F(t)_i}}{(N-1)\sum F(t)_i}$$
Unweighted least squares fit (ULSF)	$$I = \frac{\sum F(t)_i q_i}{\sum F(t)_i^2}$$	$$\sigma_i^2 = \frac{\sum \left(q_i - I \cdot F(t)_i\right)^2}{(N-1)\sum F(t)_i^2}$$
Average (slopes)	$$I = \frac{\sum \dfrac{q_i}{F(t)_i}}{N}$$	$$\sigma_i^2 = \frac{\sum \left(\dfrac{q_i}{F(t)} - I\right)^2}{(N-1)}$$

measurements; however, the weighted least squares fit offers several advantages. As noted above, the intake is estimated as the sum of the bioassay results divided by the sum of the retention (or excretion) fractions. This relationship is intuitive, easy to remember, and analogous to that applied for single measurement. Mathematically, the weighted least squares fit also defines the intake parameter such that the sum of the bioassay measurements equals the sum of the expected measurements. This is a desirable feature in that the total activity collected (for example, urine sampling) equals the total activity predicted to be excreted.

It should be noted that for a long-lived radionuclide that is completely absorbed into the bloodstream and that is excreted via a single pathway, it is theoretically possible that the summation of the excretion fractions could equal unity $(\sum F(t) = 1)$. This simply means that the entire intake has been collected. Under this condition, it follows that the intake equals the sum of the bioassay measurements $(I = \sum \frac{q(t)}{1} = \sum q(t))$. This condition is not true for repetitive in vivo measurements where the sum of the retention fractions may exceed unity. For example, the retention fractions sum to 1.89 (assuming a respiratory tract deposition of 63%) for three whole-body counts performed in succession immediately following an intake of a long-lived radionuclide.

For well-behaved bioassay data, that is, data adhering to the correct model, the three fitting techniques provide essentially the same intake estimate. A hypothetical dataset associated with an intake of 100 arbitrary units is presented in Figure 10.7. As shown in *Panel A*, for measurements that trend as predicted, the three methods produce intake estimates that differ by less than 0.5%. *Panel B* illustrates the effect of late-term deviations where the elevation of the last two data points increases the average intake estimate by nearly a factor of five, increases the WLSF by 20%, and increases the uncertainty in the ULSF. Similarly, *Panel C* illustrates the effect of early-term deviations where elevation of the first two data points increases the average intake estimate by about 60%, increases the WLSF by more than a factor of 4, and increases the ULSF by more than a factor of 5. The effect of a single outlier is shown in *Panel D*.

Examination of Figure 10.7 shows, that in general: (1) the weighted least square fit tends to weight the later and smaller measurements more than earlier measurements, (2) the

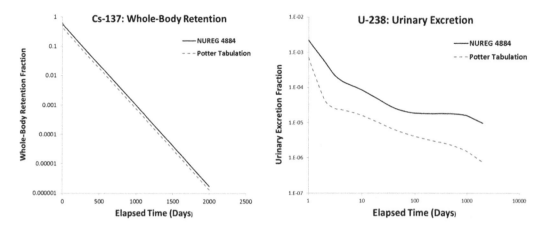

FIGURE 10.6   Comparison of predicted whole-body retention of ^{137}Cs and urinary excretion of ^{238}U using ICRP Publication 30 and ICRP publication 68 modeling.

unweighted fit tends to weight the earlier and larger measurements more than later and smaller measurements; and (3) the averaging method is equally influenced by variations in bioassay data throughout the dataset (Figure 10.7).

Table 10.7 also provides the expression for estimating the uncertainty in the respective intake estimates. Though tempting, declaring a "best fit" based solely on minimizing variance must be avoided. Variance is a measure of scatter about the predicted excretion or retention pattern. The curvature of the retention (excretion) pattern is dependent on the selected metabolic model. Consequently, a low variance could be calculated for an inappropriate model. For example, retention curves associated with the functions $1e^{-\frac{Ln(2)}{10}t}$ and $0.01e^{-\frac{Ln(2)}{10}t}$ fit the same bioassay data identically, but produce intake estimates differing by two orders of magnitude.

10.5 CASE STUDIES

10.5.1 Case Study: Acute Inhalation of ^{137}Cs

In the early 1960s, prior to the publication of ICRP 30, a healthy 37-year-old male worker was contaminated and experienced an acute inhalation intake of ^{137}Cs at the United Kingdom's Windscale Facility (Hesp 1964). Bioassay monitoring included whole-body counting and urinalysis. Results for 13 whole-body counts and 9 urine samples are summarized in Table 10.8. No other details about the exposure were reported.

Uncertainty estimates for the bioassay results were not reported; a uniform uncertainty of 10% is therefore assumed. For this example, a simplified two-compartment cesium systemic retention model is applied to evaluate the data. The model assumes that 10% of material transferred to soft tissues is retained with a two day half-time and the remaining 90% is retained with a 110 day half-time.

The intake assessment based on the whole-body count results in presented in Table 10.9. The first four columns of Table 10.9 give the elapsed time to the measurement, the bioassay result, the assumed uncertainty in the bioassay measurement, and the value of the acute

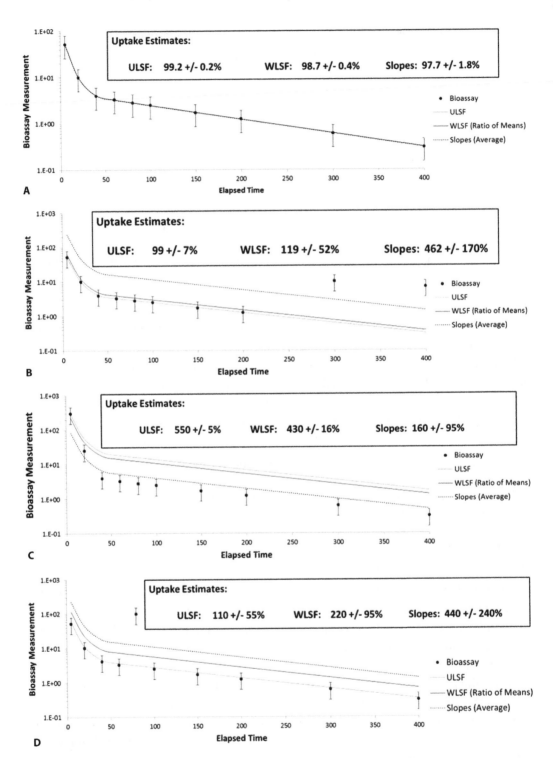

FIGURE 10.7 Weighting tendencies for three intake estimation and bioassay fitting methods. (a) Well-behaved data. (b) Elevated late data. (c) Elevated early data. (d) Single outlier.

TABLE 10.8 ^{137}Cs Bioassay Summary Following an Acute Inhalation Intake

Elapsed Time Post Intake (Days)	Whole-Body Content (nCi)	Urinary Excretion (nCi/d)
1	548	–
3	498	6.51
7	393	3.22
10	–	2.84
14	420	2.05
28	388	–
30	–	1.89
50	336	1.57
76	–	–
77	–	1.31
78	291	–
94	–	0.97
98	260	–
140	201	0.54
169	174	–
206	137	–
253	98	–
289	91	–

TABLE 10.9 ^{137}Cs Intake Estimate Based on Whole-Body Count Data

Elapsed Time (Days)	Whole-Body Content (nCi)	Measurement Uncertainty (nCi)	$F(t)$	$F(t)q$	$F(t)^2$	$q/F(t)$
1	548	54.8	6.08E−01	3.33E+02	3.70E−01	9.01E+02
3	498	49.8	5.79E−01	2.88E+02	3.35E−01	8.61E+02
7	393	39.3	5.48E−01	2.15E+02	3.00E−01	7.17E+02
14	420	42	5.19E−01	2.18E+02	2.70E−01	8.09E+02
28	388	38.8	4.74E−01	1.84E+02	2.25E−01	8.18E+02
50	336	33.6	4.12E−01	1.39E+02	1.70E−01	8.15E+02
78	291	29.1	3.45E−01	1.00E+02	1.19E−01	8.43E+02
98	260	26	3.04E−01	7.90E+01	9.23E−02	8.56E+02
140	201	20.1	2.33E−01	4.68E+01	5.41E−02	8.64E+02
189	174	17.4	1.70E−01	2.96E+01	2.90E−02	1.02E+03
206	137	13.7	1.53E−01	2.09E+01	2.34E−02	8.96E+02
253	98	9.8	1.13E−01	1.11E+01	1.28E−02	8.65E+02
289	81	8.1	9.01E−02	7.30E+00	8.12E−03	8.99E+02
SUM	3825		4.55E+00	1.67E+03	2.01E+00	1.12E+04
N	13					

Intake Estimates

WLSF	8.41E+02	+/−	2%
ULSF	8.33E+02	+/−	2%
Average	8.59E+02	+/−	8%

inhalation intake retention fraction. The Weighted Least Squares Fit (WLSF) methodology gives an intake estimate of:

$$I = \frac{\sum q_i}{\sum F(t)_i} = \frac{3825\,\text{nCi}}{4.55} = 841\,\text{nCi}$$

Columns 5 and 6 give the products of the bioassay measurement, and the retention fraction and the square of the retention fraction. The summation of these values, used in the Unweighted Least Squares Fit (ULSF), gives an intake estimate of:

$$I = \frac{\sum F(t)_i q_i}{\sum F(t)_i^2} = \frac{1670}{2.01} = 833\,\text{nCi}$$

The last column gives the quotient of each bioassay measurement to its respective retention fraction, which gives the measurement-specific intake estimate. The average of these values gives an intake estimate of:

$$I = \frac{\sum \dfrac{q_i}{F(t)_i}}{N} = \frac{11,200}{13} = 859\,\text{nCi}$$

All three fitting methods produce essentially the same intake estimate ranging between 833 and 859 nCi (~3% difference). As expected, a larger variance is associated with the averaging technique. A plot of the whole-body counting data compared to the body burden predicted by the WLSF is presented in Figure 10.8.

An intake evaluation based on the nine urine results and the simplified cesium excretion model is presented in Table 10.10.

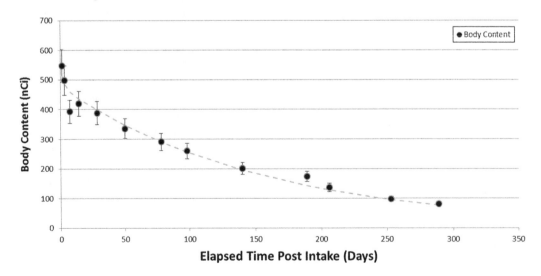

FIGURE 10.8 Comparison of ^{137}Cs whole-body count measurements to the body burden predicted by WLSF.

TABLE 10.10 ^{137}Cs Intake Estimate Based on Urinalysis Results

Elapsed Time (Days)	Urine Result (nCi/d)	Measurement Uncertainty (nCi/d)	$e_u(t)$	$e_u(t)q$	$e_u(t)^2$	$q/e_u(t)$
3	6.54	0.654	1.02E−02	6.69E−02	1.04E−04	6.40E+02
7	3.22	0.322	4.61E−03	1.49E−02	2.13E−05	6.98E+02
10	2.84	0.284	3.37E−03	9.57E−03	1.14E−05	8.43E+02
14	2.054	0.2054	2.81E−03	5.78E−03	7.91E−06	7.30E+02
30	1.89	0.189	2.39E−03	4.52E−03	5.73E−06	7.90E+02
50	1.57	0.157	2.11E−03	3.31E−03	4.44E−06	7.45E+02
77	1.31	0.131	1.77E−03	2.32E−03	3.15E−06	7.38E+02
94	0.97	0.097	1.59E−03	1.54E−03	2.54E−06	6.09E+02
140	0.64	0.064	1.19E−03	7.60E−04	1.41E−06	5.39E+02
SUM	21.034		3.01E−02	1.10E−01	1.62E−04	6.33E+03
N	9					

Intake Estimates

WLSF	6.99E+02	+/−	4%
ULSF	6.75E+02	+/−	3%
Average	7.04E+02	+/−	13%

Again, all three fitting methods produce essentially the same intake estimate ranging between 675 and 704 nCi (~4% difference). A larger variance is again associated with the averaging technique. A plot of the urinary excretion pattern based on the weighted least squares fit compared to the urinalysis results is shown in Figure 10.9.

Examination of the retention and excretion plots show that reasonable fits are achieved for both data sets. However, it is noted that the urine-based intake estimates are about 20% less than the body count-based results. Differences of this order of magnitude are not unexpected. Generally, agreement within a factor of two to three is considered good. The source for differences between intake estimates for different monitoring techniques is unknown, but is most likely related to measurement bias or modeling error. It is noted that in this example the difference between intake estimates could be accounted for by as little as a 15% bias in the whole-body measurements or a 15% difference in the fraction of cesium excreted via the urinary pathway ($f_u = 0.64$ rather than 0.80).

When differences exist and the intake is well below regulatory limits, the intake of record is typically set to the highest (or most conservative) assessment. For this example, based on the weighted least squares fitting results, an intake of 841 nCi would be assigned. An alternate approach is to take a weighted average where the individual intake estimates are weighted based on some measure of confidence.

In this example, weighting the WLSF intakes by the variance in the intake estimates gives a weighted intake of:

$$\bar{I} = \frac{\sum\left(\dfrac{I}{\sigma_I^2}\right)}{\sum\left(\dfrac{1}{\sigma_I^2}\right)} = \frac{\left(\dfrac{841}{17^2} + \dfrac{699}{28^2}\right)}{\left(\dfrac{1}{17^2} + \dfrac{1}{28^2}\right)} = 802\,\text{nCi}$$

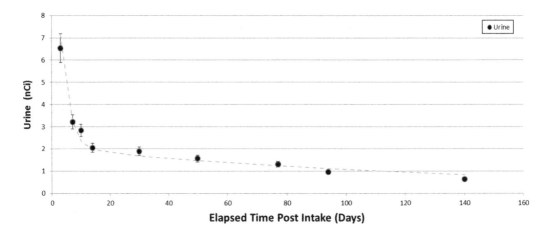

FIGURE 10.9   Comparison of ^{137}Cs urinalysis measurements to the urinary excretion pattern predicted by WLSF.

Note that in this case the weighted intake estimate favors the estimate derived from whole-body counting data as less uncertainty is associated with the fit to that dataset.

It is interesting to note that in the 1960s, the original investigators estimated the initial cesium deposition as 590 nCi (which, assuming 63% deposition, equates to an inhalation intake of 940 nCi). Recall that this event occurred before the publication of ICRP 30 during an era when internal doses were controlled by organ burdens rather than intakes. The original investigators also did not have the benefit of multi-compartmental retention and excretion models. The fact that exposure estimates made decades apart agree to within 20% is reassuring.

10.5.2 Case Study: Acute Ingestion of [32]P

The previous case study examined the use of multiple bioassay measurements from different monitoring techniques to estimate the intake following a known event. Achieving agreement between intake estimates for different monitoring techniques is also potentially useful for identifying when an unrecognized exposure may have occurred. This approach is illustrated below using whole-body counting and urine data collected in response to a suspected ingestion intake of [32]P (Nuclear Regulatory Commission 1995).

On August 19, 1995, a researcher self-identified the presence of ^{32}P contamination on clothing and in urine. An ensuing investigation, completed over a 2-month period, included the collection of 37 whole-body counts and 40 urine samples. Bioassay results for this event are summarized in Table 10.11 (Note: the uncertainty associated with the 15th urine sample is suspected to be incorrectly reported by a magnitude of order, but is retained in this text as published in Table 1, Appendix C of NUREG-1535 1995).

Figure 10.10 compares intake estimates derived from urinalysis and whole-body counting data for various assumed intake dates. The plot shows that it is unlikely that exposure occurred between May and July, as the intake estimates differ by about an order of magnitude.

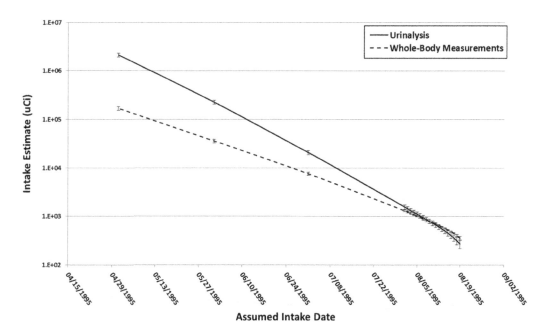

FIGURE 10.10 Comparison of ^{32}P intake estimates derived from urine and whole-body count measurements for various assumed exposure dates.

However, intake estimates derived from the two monitoring techniques converge during the second week of August.

This convergence is consistent with other information gathered during the investigation which suggested that the exposure likely occurred on August 14, 1995 (five days prior to discovery) due to the ingestion of food deliberately contaminated by a knowledgeable person.

Table 10.11 provides intake evaluations for both whole-body count and urine data using an assumed intake date of August 14, 1995. The retention/excretion fractions shown in Table 10.11 were calculated using contemporary ICRP models. The WLSF intake estimate for the two monitoring techniques is in good agreement with a conservative intake estimate of 570 µCi derived from the whole-body counting results. Fits for both data sets are presented in Figure 10.11.

The intake estimate of 570 µCi is in very good agreement with the final assessment of 579 µCi issued in 1995 using ICRP Publication 30-based modeling. The assessed intake was considered credible as a radioactive source inventory performed as part of the investigation found that nearly 500 µCi of ^{32}P could not be accounted for. It is assumed that the missing activity may have been diverted and combined with remnants of other sources to spike food consumed by the subject.

In addition to demonstrating the usefulness of harmonizing bioassay data, this case illustrates the importance of integrating workplace information into exposure assessments. Ideally, radiological indicators such as contamination levels, airborne concentrations, worker tasks and location, and radionuclide inventories should substantiate intake estimates. In this case, both the date and magnitude of the exposure were corroborated by workplace indicators.

TABLE 10.11 ^{32}P Intake Evaluation Based on Whole-Body Count Results and Urinalysis Results Following a Suspected Acute Ingestion

Elapsed Time (Days)	Whole-Body Content (uCi)	σ (uCi)	R(t)	F(t) B	F(t)²	B / F(t)	Elapsed Time (Days)	Urinalysis Result (uCi/d)	σ (uCi/d)	e_u(t)	F(t) B	F(t)²	B / F(t)
5	263	0.825	4.33E-01	1.14E+02	1.88E-01	6.07E+02	12	1.32	0.058	4.59E-03	6.06E-03	2.11E-05	2.87E+02
7	204	0.737	3.66E-01	7.46E+01	1.34E-01	5.58E+02	13	1.56	0.075	4.09E-03	6.37E-03	1.67E-05	3.82E+02
8	194	0.72	3.39E-01	6.57E+01	1.15E-01	5.73E+02	14	2.02	0.081	3.66E-03	7.40E-03	1.34E-05	5.51E+02
9	178	0.695	3.15E-01	5.60E+01	9.89E-02	5.66E+02	15	2.07	0.075	3.31E-03	6.84E-03	1.09E-05	6.26E+02
10	165	0.673	2.93E-01	4.83E+01	8.57E-02	5.64E+02	16	1.55	0.063	3.00E-03	4.64E-03	8.98E-06	5.17E+02
11	157	0.659	2.73E-01	4.29E+01	7.45E-02	5.75E+02	17	1.46	0.055	2.73E-03	3.98E-03	7.43E-06	5.36E+02
14	129	0.606	2.23E-01	2.87E+01	4.97E-02	5.79E+02	18	1.49	0.056	2.49E-03	3.71E-03	6.18E-06	5.99E+02
15	122	0.594	2.09E-01	2.54E+01	4.35E-02	5.85E+02	19	1.13	0.057	2.27E-03	2.57E-03	5.17E-06	4.97E+02
16	109	0.568	1.95E-01	2.13E+01	3.82E-02	5.58E+02	20	0.88	0.057	2.08E-03	1.83E-03	4.33E-06	4.23E+02
17	103	0.566	1.83E-01	1.89E+01	3.35E-02	5.63E+02	21	1.16	0.06	1.91E-03	2.21E-03	3.64E-06	6.08E+02
18	99	0.546	1.72E-01	1.70E+01	2.94E-02	5.77E+02	22	0.97	0.048	1.75E-03	1.70E-03	3.06E-06	5.55E+02
22	76	0.494	1.33E-01	1.01E+01	1.77E-02	5.71E+02	23	0.89	0.049	1.60E-03	1.43E-03	2.58E-06	5.55E+02
23	69	0.479	1.25E-01	8.62E+00	1.56E-02	5.52E+02	24	0.76	0.04	1.47E-03	1.12E-03	2.17E-06	5.16E+02
24	65	0.468	1.17E-01	7.63E+00	1.38E-02	5.54E+02	25	0.685	0.043	1.35E-03	9.27E-04	1.83E-06	5.06E+02
25	66	0.471	1.10E-01	7.28E+00	1.22E-02	5.98E+02	26	0.49	0.393	1.24E-03	6.09E-04	1.55E-06	3.94E+02
28	50	0.431	9.17E-02	4.59E+00	8.41E-03	5.45E+02	27	0.68	0.041	1.14E-03	7.77E-04	1.30E-06	5.95E+02
29	49	0.426	8.63E-02	4.23E+00	7.45E-03	5.68E+02	28	0.644	0.04	1.05E-03	6.76E-04	1.10E-06	6.14E+02
30	45	0.417	8.12E-02	3.66E+00	6.60E-03	5.54E+02	29	0.647	0.04	9.65E-04	6.24E-04	9.31E-07	6.71E+02
31	44	0.413	7.65E-02	3.36E+00	5.85E-03	5.75E+02	30	0.54	0.04	8.87E-04	4.79E-04	7.87E-07	6.09E+02
32	41	0.403	7.20E-02	2.95E+00	5.19E-03	5.69E+02	31	0.52	0.033	8.15E-04	4.24E-04	6.65E-07	6.38E+02
35	34	0.384	6.02E-02	2.05E+00	3.63E-03	5.64E+02	32	0.348	0.035	7.50E-04	2.61E-04	5.62E-07	4.64E+02
36	28.3	0.366	5.68E-02	1.61E+00	3.22E-03	4.98E+02	33	0.33	0.029	6.89E-04	2.27E-04	4.75E-07	4.79E+02
37	28.5	0.367	5.35E-02	1.53E+00	2.87E-03	5.32E+02	34	0.312	0.034	6.34E-04	1.98E-04	4.01E-07	4.92E+02
38	27.8	0.365	5.05E-02	1.40E+00	2.55E-03	5.50E+02	35	0.302	0.027	5.83E-04	1.76E-04	3.39E-07	5.18E+02
39	25.8	0.358	4.76E-02	1.23E+00	2.27E-03	5.41E+02	36	0.318	0.031	5.36E-04	1.70E-04	2.87E-07	5.94E+02

(Continued)

TABLE 10.11 (CONTINUED) ³²P Intake Evaluation Based on Whole-Body Count Results and Urinalysis Results Following a Suspected Acute Ingestion

Elapsed Time (Days)	Whole-Body Content (uCi)	σ (uCi)	$R(t)$	$F(t)B$	$F(t)^2$	$B/F(t)$
42	22.3	0.346	4.01E-02	8.94E-01	1.61E-03	5.57E+02
43	19.8	0.338	3.78E-02	7.49E-01	1.43E-03	5.23E+02
44	19.8	0.338	3.57E-02	7.08E-01	1.28E-03	5.54E+02
45	18.4	0.333	3.38E-02	6.21E-01	1.14E-03	5.45E+02
46	16.5	0.326	3.19E-02	5.27E-01	1.02E-03	5.17E+02
49	14.5	0.319	2.70E-02	3.91E-01	7.27E-04	5.38E+02
50	13.5	0.316	2.55E-02	3.44E-01	6.50E-04	5.29E+02
51	12.3	0.311	2.41E-02	2.97E-01	5.82E-04	5.10E+02
52	12.3	0.311	2.28E-02	2.81E-01	5.21E-04	5.39E+02
53	11.1	0.307	2.16E-02	2.40E-01	4.66E-04	5.14E+02
57	9.4	0.3	1.73E-02	1.63E-01	3.01E-04	5.42E+02
58	7.5	0.293	1.64E-02	1.23E-01	2.70E-04	4.56E+02
—	—	—	—	—	—	—
—	—	—	—	—	—	—
—	—	—	—	—	—	—
SUM	2549.8		4.50E+00	5.78E+02	1.01E+00	2.04E+04
N	37					

Elapsed Time (Days)	Urinalysis Result (uCi/d)	σ (uCi/d)	$e_u(t)$	$F(t)B$	$F(t)^2$	$B/F(t)$
37	0.367	0.031	4.93E-04	1.81E-04	2.43E-07	7.45E+02
38	0.341	0.034	4.53E-04	1.54E-04	2.05E-07	7.53E+02
39	0.396	0.026	4.17E-04	1.65E-04	1.74E-07	9.50E+02
40	0.252	0.029	3.83E-04	9.66E-05	1.47E-07	6.58E+02
41	0.222	0.03	3.52E-04	7.82E-05	1.24E-07	6.30E+02
42	0.231	0.03	3.24E-04	7.49E-05	1.05E-07	7.13E+02
43	0.206	0.026	2.98E-04	6.14E-05	8.89E-08	6.91E+02
44	0.174	0.022	2.74E-04	4.77E-05	7.52E-08	6.34E+02
45	0.211	0.029	2.52E-04	5.32E-05	6.37E-08	8.36E+02
46	0.17	0.023	2.32E-04	3.95E-05	5.39E-08	7.32E+02
47	0.177	0.025	2.14E-04	3.78E-05	4.56E-08	8.29E+02
48	0.149	0.021	1.97E-04	2.93E-05	3.86E-08	7.58E+02
49	0.149	0.024	1.81E-04	2.69E-05	3.27E-08	8.24E+02
50	0.142	0.022	1.66E-04	2.36E-05	2.77E-08	8.53E+02
51	0.122	0.016	1.53E-04	1.87E-05	2.35E-08	7.97E+02
SUM	26.385		5.00E-02	5.65E-02	1.17E-04	2.46E+04
N	40					

Intake Estimates-Whole-Body Data

WLSF	5.67E+02	+/-	1%
ULSF	5.74E+02	+/-	1%
Average	5.51E+02	+/-	5%

Intake Estimates-Urinalysis

WLSF	5.28E+02	+/-	4%
ULSF	4.82E+02	+/-	4%
Average	6.16E+02	+/-	23%

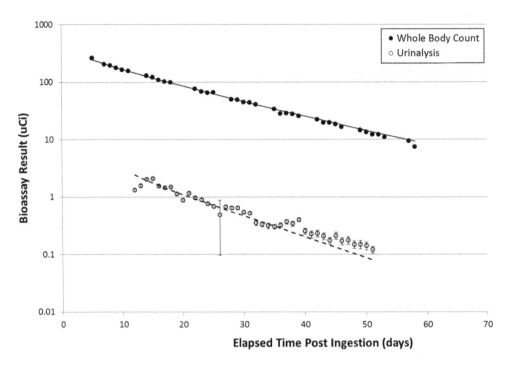

FIGURE 10.11 Comparison of predicted whole-body retention and urinary excretion of ^{32}P to bioassay measurements based on an acute ingestion intake occurring five days prior to discovery.

It is also interesting to note that although dosimetry systems have evolved, the contemporary assessment agrees with the earlier assessment. In this case, agreement is not unexpected as: (1) the intake was by ingestion which results in 100% deposition, (2) transfer of material through the gastrointestinal tract is unchanged between earlier and later models, (3) phosphorus is readily absorbed ($f_1 = 0.8$), and (4) the systemic model describing phosphorus retention has not been updated to include recycling. Therefore, ingestion modeling for phosphorus is essentially unchanged.

10.6 INDIVIDUAL-SPECIFIC MODIFICATIONS

Generally, published peer-reviewed metabolic models should be used to assess intake and dose. Revising models (i.e., "tweaking") is typically unsupported and unwarranted. Individual-specific modifications should only be attempted when adequate justification is available, the revision is supported by sufficient bioassay data, and the underlying mechanisms for the change are understood.

A classic example under which individual-specific modeling is supported is tritium exposure. The ICRP Publication 56 (as implemented in OCRP-78) systemic retention function for tritiated water is given as:

$$F(t) = 0.97 e^{-\left(\frac{0.693}{10}\right)t} + 0.03 e^{-\left(\frac{0.693}{40}\right)t}$$

Where 97% of systemically absorbed tritium is assumed to achieve equilibrium with free body water where it is retained with a biological half-time of 10 days and the remaining 3% is assumed to be incorporated into organic molecules where it is retained with a 40-day half-time. The dominant 10-day half-time is a function of the body's water balance. Given a free body water reservoir of 42 L and a daily intake/loss rate of 3 L/day, a biological removal rate constant (k) is estimated as:

$$k = \frac{3\,\text{L/d}}{42\,\text{L}} = 0.071\,\text{day}^{-1}$$

The above rate constant equates to a biological half-time of 9.7 days which, to one significant digit, is rounded to 10 days. Recognizing, that under homeostasis, the free water volume remains constant, tritium removal can be increased by increasing fluid intake (a plethoric hydrous diet).

10.6.1 Case Study: Forced Fluids Following an Acute Uptake of ^{3}H

Urinalysis data collected following an acute uptake of tritium, in which the subject increased fluid consumption up to 8 liters per day, was published as part of the 2007 IAEA Intercomparison Exercise on Internal Dose Assessment (IAEA 2007) and is presented in Table 10.12.

Because tritium reaches equilibrium with free body water, the activity concentration in urine equals the activity concentration within free body water, and the body burden (activity) can be estimated by scaling the urine concentration to the total free water body volume of 42 liters. Tritium is somewhat unique in that it is rapidly and completely absorbed. Therefore, because the initial urine sample was obtained shortly following exposure (but after equilibrium is established), the initial uptake (intake) is be estimated as:

$$I = \left(80.1\frac{\text{MBq}}{L}\right)42\,L = 3.4\times10^3\,\text{MBq}$$

However, as shown in Figure 10.12, the urinary excretion rate predicted from the standard retention model grossly exceeds subsequent measurements. The plot clearly indicates that increased fluid intake enhanced tritium removal through about 100 days post exposure, after which excretion slowed.

The subject's tritium excretion curve can be separated into at least two components: (1) a long-term portion defined by a 65-day half-time, and (2) a short-term component decreasing with a 6.5 day half-time (see Figure 10.13). The corresponding retention function is estimated by the function:

$$F(t) = 0.9993e^{-\left(\frac{0.693}{6.5}\right)t} + 0.0007e^{-\left(\frac{0.693}{65}\right)t}$$

TABLE 10.12 Tritium Urinalysis Results Influenced by a Plethoric Hydrous Diet Following an Acute Uptake

Time Post Intake (Days)	Tritium Concentration in Urine (MBq/L)	Time Post Intake (Days)	Tritium Concentration in Urine (MBq/L)	Time Post Intake (Days)	Tritium Concentration in Urine (MBq/L)
0	80.1	35	1.25	91	0.023
1	67.7	36	1.02	94	0.021
2	57.5	38	0.97	96	0.019
3	47.5	39	0.78	98	0.018
4	39.2	41	0.64	100	0.018
5	32	44	0.56	103	0.014
6	27.6	47	0.42	142	0.0087
7	24.2	49	0.36	149	0.0081
8	22.9	50	0.31	156	0.0074
9	19.5	54	0.23	163	0.0066
10	16.5	56	0.17	169	0.0064
11	14.3	58	0.15	177	0.0057
12	12.4	61	0.12	184	0.0063
13	11	63	0.11	191	0.0043
14	9.62	66	0.099	196	0.0048
15	8.23	68	0.078	216	0.004
16	7.81	70	0.064	219	0.0038
18	6.36	72	0.057	226	0.0041
20	5.25	75	0.05	233	0.0037
22	4.26	77	0.044	239	0.0033
24	3.52	79	0.044	246	0.0028
26	2.86	82	0.036	254	0.0025
28	2.8	84	0.034	268	0.002
30	2.08	87	0.029	270	0.0022
33	1.54	89	0.025	274	0.0021

In this particular case, refinement of the retention model does not markedly improve the intake estimate over that obtained from the first bioassay measurement. For example, the intake estimate from the urine sample collected at 84 days post intake is:

$$I = \frac{\left(0.034\dfrac{\text{MBq}}{L}\right)}{4.15\text{E}-04}\,42\,L = 3.4\text{E}+03\text{ MBq}$$

This estimate equals the original estimate which was obtained without knowledge of the enhanced excretion pattern. However, refinement of systemic retention has a significant impact on committed dose. Recall that committed dose is proportional to the number of nuclear transformations and that the number of transformations is equal to the integral of activity retention. As such, for this example, the committed dose is reduced by a factor

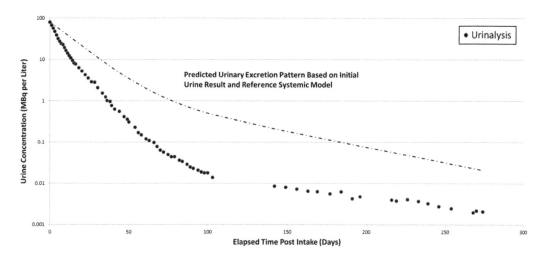

FIGURE 10.12   Comparison of the predicted ^{3}H urinary excretion pattern based on the initial urinalysis result and the default tritium model to collected urine results.

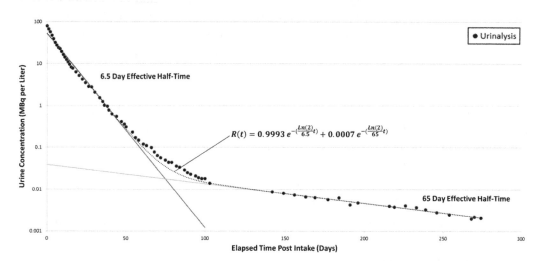

FIGURE 10.13   An individual-specific ^{3}H retention model based on observed urinary excretion.

equal to the ratio of the integral of the revised retention function to the default retention function:

$$\text{Reduction Factor} = \left(\frac{\int \left[0.9993e^{-\left(\frac{0.693}{6.5}\right)t} + 0.0007e^{-\left(\frac{0.693}{65}\right)t} \right]}{\int \left[0.97e^{-\left(\frac{0.693}{10}\right)t} + 0.03e^{-\left(\frac{0.693}{40}\right)t} \right]} \right) = 0.60$$

As such, forced fluids enhanced tritium removal and reduced the committed dose by 40%.

The above example is specific to tritium. Other situations where modeling revisions may be justified include changes in respiratory deposition based on particle size measurements, modified respiratory tract absorption rates based on material solubility studies, and altered respiratory tract transfer rates due to illness or smoking. An example of one such case, involving the inhalation of ^{241}Am, is discussed below.

10.6.2 Case Study: Modification of Respiratory Tract Absorption and Transfer Parameters Following an Acute Inhalation of ^{241}Am

In 1996, a healthy, non-smoking, 38-year-old male worker was performing radiological work in an area where loose transferable surface contamination of ^{241}Am at levels exceeding 1,000 Bq/cm^2 was detected. Bioassay measurements indicated that an inhalation intake occurred. The released americium was believed to be an oxide, and no particle size information was available. The individual enrolled in the United States Transuranium and Uranium Registries (USTUR) and was assigned Case Number 0855.

For the purposes of this case study, only the chest (or lung) and liver measurements are examined (see Table 10.13). It should be noted that "chest" counts may not provide a direct measure of lung burden, as activity contained in overlying skeletal tissue, thoracic lymph nodes, and the liver interfere. To account for the transfer of material to interfering tissues, a time-dependent cross-talk correction factor was applied to the chest measurements to estimate lung burden (Marsh, Bailey, and Birchall 2005). A uniform uncertainty of 30% was assumed for both data sets.

The right-hand portion of Table 10.13 gives acute inhalation retention fraction values for the lung and liver for default Type "M" and "S" materials. Revised values based on a modeling modification (discussed below) are also provided. Note that the organ-specific intake estimates differ by factors of 5 and 15 for default Type "M" and "S" materials, respectively. Modeling revisions improve agreement to within 3%.

Review of Table 10.13 shows a long-term retention pattern for americium retained in the lung. Liver content increased over a period of about 15 months and then decreased. Plots of the lung and liver measurements compared to retention patterns predicted using default ICRP Publication 68 modeling assumptions are shown in Figure 10.14. The dotted line (•••) applies to an assumption of moderately soluble Type "M" material, the dashed line (---) applies to insoluble Type "S" material, and the solid line (—) shows the retention pattern predicted after modeling adjustments are applied.

From Figure 10.14, it is noted that liver retention resembles Type M characteristics, whereas lung retention is more consistent with Type S behavior. Comparison of the intake estimates provided in Table 10.13 and the retention patterns shown in Figure 10.14 show that alternate solubility assumptions grossly underestimate measurement results for both organs. For example, applying an assumption of Type "S" material to the chest count results gives an intake estimate of 5,700 nCi. At this level of intake, the liver burden never exceeds 10 nCi, which is nearly an order of magnitude below observations. As such, the bioassay data is inconsistent with liver measurements favoring moderately soluble behavior and lung measurements favoring insoluble characteristics. This difference is irreconcilable using default modeling assumptions.

TABLE 10.13 ^{241}Am Chest and Liver Count Summary Following an Acute Inhalation and Associated Retention Fractions

Elapsed Time (Days)	Chest Measurement (nCi)	Cross-Talk Factor	Adjusted Lung Content (nCi)		Liver Measurement (nCi)		Lung R(t) Type M	Lung R(t) Type S	Lung R(t) Modified	Liver R(t) Type M	Liver R(t) Type S	Liver R(t) Modified
48	315	10%	284	+/- 85	40	+/- 12	3.15E-02	4.43E-02	4.19E-02	1.91E-02	3.72E-04	6.74E-03
84	241	10%	217	+/- 65	36	+/- 11	2.29E-02	3.83E-02	3.58E-02	2.12E-02	4.42E-04	7.35E-03
112	246	10%	221	+/- 66	38	+/- 11	1.85E-02	3.56E-02	3.31E-02	2.23E-02	4.88E-04	7.74E-03
140	226	10%	203	+/- 61	52	+/- 16	1.53E-02	3.38E-02	3.11E-02	2.32E-02	5.29E-04	8.06E-03
175	214	10%	193	+/- 58	64	+/- 19	1.23E-02	3.22E-02	2.92E-02	2.39E-02	5.77E-04	8.42E-03
203	195	15%	166	+/- 50	66	+/- 20	1.04E-02	3.11E-02	2.80E-02	2.42E-02	6.11E-04	8.67E-03
238	156	15%	133	+/- 40	64	+/- 19	8.40E-03	2.99E-02	2.67E-02	2.45E-02	6.52E-04	8.95E-03
315	146	15%	124	+/- 37	71	+/- 21	5.34E-03	2.77E-02	2.40E-02	2.44E-02	7.29E-04	9.45E-03
454	131	15%	111	+/- 33	82	+/- 25	2.37E-03	2.44E-02	2.01E-02	2.33E-02	8.39E-04	1.00E-02
988	109	20%	87	+/- 26	56	+/- 17	1.08E-04	1.52E-02	1.00E-02	1.68E-02	1.02E-03	1.02E-02
2135	82	20%	66	+/- 20	41	+/- 12	1.71E-07	6.63E-03	2.30E-03	8.84E-03	9.00E-04	7.07E-03
Sum			1805		610		1.27E-01	3.19E-01	2.82E-01	2.32E-01	7.15E-03	9.26E-02

Intake Estimates (nCi)

Technique	Type M	Type S	Modified
Lung	14,200	5,700	6,400
Liver	2,600	85,300	6,600

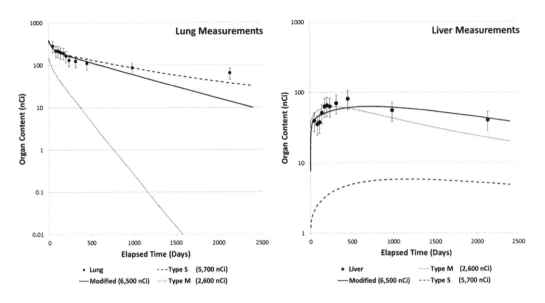

FIGURE 10.14 Comparison of ^{241}Am lung and liver measurements to retention predictions based on default modeling assumptions and an individual-specific modeling modification.

A detailed step-by-step guide for proposed modeling of USTUR Case 0855 has been published (Marsh, Bailey, and Birchall 2005). A partial synopsis of the recommended modifications is discussed below.

The modeling revision begins by recognizing that the bioassay data shows that: (1) long-term respiratory tract retention exists, (2) systemic transfer to the liver has occurred, and (3) default material types "M" and "S" do not apply. Thus, an "intermediate" clearance, sharing characteristics of both type "M" and "S" materials, must be crafted. This is accomplished by understanding that respiratory clearance involves both particle transport and systemic absorption.

Particle transfer pathways within the Human Respiratory Tract Model (HRTM) are shown in Figure 10.15. The rate constants defining these pathways are independent of particle solubility. Long-term retention within the lung is dominated by retention in the alveolar-interstitial region which is assigned three compartments (AI_1, AI_2, and AI_3) having, respectively, rapid, intermediate, and slow clearance rates. Because the majority of bioassay data is confined to "intermediate" times (50–100 days post exposure), the transfer rates associated with the rapidly cleared compartment AI_1 and the slowly cleared compartment AI_3 were retained. The intermediate rate constant associated with compartment AI_2 was decreased by a factor of three (0.001 d^{-1} → 0.00033 d^{-1}). This change prolongs lung retention during the intermediate time frame.

Absorption to blood occurs in all regions of the respiratory tract, except compartment ET_1 where removal occurs through nose blowing and wiping. Uptake to blood is modeled as a two-step process involving dissolution followed by absorption. Dissolution, which is time dependent, is modeled using both a rapid and slow component. Materials can also enter a "bound" state where absorption occurs at a different rate (see Figure 10.16).

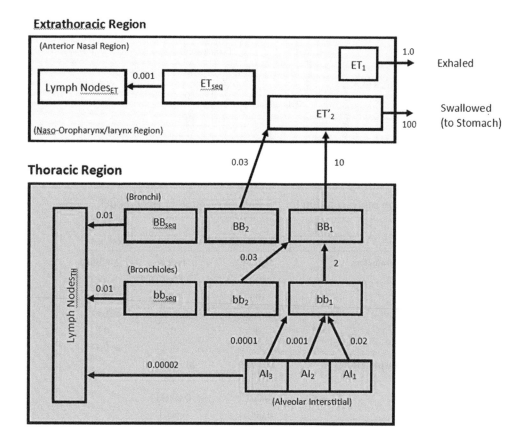

FIGURE 10.15 Particle transfer pathways and rate constants for the human respiratory tract model.

Because bioassay monitoring was not performed for 48 days, no justification exists to modify the rapid dissolution rate constant (S_r). The chest count results suggest that a bound condition may be present. However, the bounded fraction (f_b) and its associated absorption rate (S_b) cannot be discerned from the bioassay data. Therefore, experimentally determined values of $f_b = 0.87$ and $S_b = 0.15$ d^{-1} based on animal studies were applied. These values result in the rapid transfer of a large portion of slowly dissolved material to the blood (i.e., little hold-up in the bound state). Consequently, the fraction of material that is slowly dissolved was increased to 96% ($f_r = 0.04$), and the associated dissolution rate (S_S) was decreased, compared to type M material, by a factor of five (0.005 d^{-1} → 0.001 d^{-1}). The parameter changes are summarized in Table 10.14.

The combined effect of these modeling modifications is the formation of a hybrid "intermediate" material type that quickly transfers some material to blood (similar to Type "M" material) while retaining a portion of the intake in the respiratory tract (similar to Type "S" material). The modifications tend to resolve differences observed between lung and liver retention (see solid line in Figure 10.14) and improve agreement in the intake estimate associated with the two sets of measurements. Though improved, the modified model

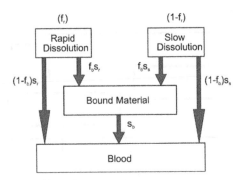

FIGURE 10.16   Absorption representation for the human respiratory tract model.

TABLE 10.14   Comparison of HRTM Parameter Values with Default Values

Absorption Parameters	Type M	Type S	Modified
f_r	0.1	0.001	0.04
S_r (d^{-1})	100	100	100
S_s(d^{-1})	0.005	0.0001	0.001
F_b	0	0	0.87
S_b	–	–	0.15
Particle Transport Rates (d−1)	**Default**	**Modified**	
AI$_1$ to bb$_1$	0.02	0.02	
AI$_2$ to bb$_1$	0.001	0.00033	
AI$_3$ to bb$_1$	0.0001	0.0001	

underestimates later-term lung measurements. This difference, along with the incorporation of other bioassay data, ultimately required additional changes to be applied.

The above discussion is not intended to be an exhaustive evaluation of USTUR Case
0855. Rather, it illustrates that unlike the simple modifications applied to the tritium example, a higher level of understanding is required to modify multi-compartmental models.
Such modification should only be attempted by skilled practitioners possessing a strong
understanding of modeling fundamentals.

10.6.3 Modified Dose Coefficients

The fact that under certain conditions, modification of metabolic models is justified raises
a question concerning the applicability of published dose coefficients. Three viewpoints
exist:

(1) *Never divert from published metabolic models or dose coefficients.* Advocates of this
approach argue that radiation protection standards (both internal and external) are
based on reference man and are uniformly applied across the workforce (for example,
external dosimeter calibrations and Derived Air Concentrations). Because individual-
specific adjustments are not applied to other radiation protection measurements, it is
not justified to deviate from reference man models for internal dose control.

(2) *Modify metabolic models based on bioassay observations, but apply published dose coefficients.* This position promotes the use of individual-specific modeling to best estimate intake. However, because parameters such as organ mass are unknown, the use of reference man dose coefficients is appropriate.

(3) *Modify metabolic models based on bioassay observations and revise dose coefficients accordingly.* Proponents of this tactic argue that, as shown in the forced fluid tritium case, dose is dependent upon retention, and that observed deviations in retention patterns should not be ignored.

This text does not endorse a particular approach. Rather, the reader should recognize that different positions exist and consider the validity of the various viewpoints for conditions that may be encountered.

10.7 REGULATORY ISSUES

At the time of writing, radiation protection regulations within the United States are controlled under the U.S. Nuclear Regulatory Commission (10 CFR 20) (Nuclear Regulatory Commission 1991), the NRC Agreement State Program, and the U.S. Department of Energy (10 CFR 835) (Department of Energy 2004). Though codified regulations establish conditions under which bioassay monitoring is to be performed, the Federal Code is silent with respect to how bioassay data is to be interpreted for the purpose of estimating intake and, ultimately, dose. This is advantageous, as it provides flexibility. Though the regulations are silent on assessment methods, examination of the tabulated Derived Air Concentration (DAC) values in the appendices of 10 CFR 20 and 10 CFR 835 reveals that the NRC implicitly endorses ICRP Publication 30 techniques, as its DAC values apply Class "D," "W," and "Y" assignments; whereas the DOE appears to favor ICRP Publication 68 (International Commission on Radiological Protection 1994) methods, as its DAC value use Type "F," "M," and "S" designations.

The conditions under which employee monitoring against the intake of radioactive materials are given in the following two requirements:

USNRC
10 CFR 20.1502(b)(1)   **CONDITIONS REQUIRING INDIVIDUAL**
MONITORING OF EXTERNAL AND INTERNAL
OCCUPATIONAL DOSE.

Each licensee shall monitor exposures to radiation and radioactive material at levels sufficient to demonstrate compliance with the occupational dose limits of this part. As a minimum—

  (b)   Each licensee shall monitor the occupational intake of radioactive material by and assess the committed effective dose equivalent to—

   (1)   Adults likely to receive, in 1 year, an intake in excess of 10 percent of the applicable ALI(s) in table 1, Columns 1 and 2, of appendix B to §§ 20.1001-20.2402;

USDOE
10 CFR 835.402(C)(1)   INDIVIDUAL MONITORING

(c)   For the purpose of monitoring individual exposures to internal radiation, internal dosimetry programs (including routine bioassay programs) shall be conducted for:

   (1)   Radiological workers who, under typical conditions, are likely to receive a committed effective dose of 0.1 rem (0.001 Sv) or more from all occupational radionuclide intakes in a year;

In essence, the threshold for requiring bioassay monitoring at NRC facilities is the likelihood of receiving 500 mrem committed effective dose equivalent, whereas a threshold of 100 mrem committed effective dose applies at DOE operations. For deterministically-controlled (non-stochastically controlled) radionuclides, a 5 rem committed dose equivalent to a controlling organ or tissue (typically, bone surfaces, thyroid, or other organ having a low weighting factor) may become limiting at NRC facilities, whereas DOE regulations only apply to the effective whole body.

Though, for an occupational setting, inhalation is the most likely mode of intake, federal monitoring requirements also apply to ingestion intakes. Assuming a situation where the ingestion of soluble ^{32}P is likely, the bioassay monitoring threshold for DOE and NRC facilities is calculated by applying the effective whole-body ingestion dose coefficient of 2.4×10^{-9} Sv/Bq (which is the same in both the ICRP Publication 30 and ICRP Publication 68 systems) against the regulatory dose thresholds. Therefore, the bioassay monitoring thresholds are:

$$\text{US-DOE}: \frac{0.001\,\text{Sv}}{2.4\times10^{-9}\,\dfrac{\text{Sv}}{\text{Bq}}} = 4.2\times10^5\,\text{Bq} = 11\,\mu\text{Ci}$$

$$\text{US-NRC}: \frac{0.005\,\text{Sv}}{2.4\times10^{-9}\,\text{Sv/Bq}} = 2.1\times10^6\,\text{Bq} = 56\,\mu\text{Ci}$$

As the same dose coefficient is applied, the NRC threshold is five times higher based on the dose limit (500 vs. 100 mrem). As noted earlier, difference of this magnitude may not apply to deterministically controlled radionuclides where 10% of the ALI corresponds to a 5 rem organ dose. Consider the inhalation of moderately soluble ^{239}Pu. The ICRP Publication 68 committed effective dose coefficient (as applied at DOE facilities) is 3.2×10^{-5} Sv/Bq. The ICRP Publication 30 committed dose equivalent coefficient for bone surfaces (as applied at NRC facilities) is 2.1×10^{-3} Sv/Bq. Applying these values give thresholds of:

$$\text{US-DOE}: \frac{0.001\,\text{Sv}}{3.2\times\dfrac{10^{-5}\,\text{Sv}}{\text{Bq}}} = 31\,\text{Bq} = 1900\,\text{dpm}$$

$$\text{US-NRC}: \frac{0.05\,\text{Sv}}{2.1\times10^{-3}\,\text{Sv/Bq}} = 23\,\text{Bq} = 1400\,\text{dpm}$$

In the case of moderately soluble plutonium, the threshold values are more comparable, with the NRC limit being more conservative.

It must be emphasized that the threshold values calculated above assume a single radionuclide is the sole source of exposure. Monitoring thresholds should be decreased if the likelihood of exposure to multiple radionuclides exists.

10.8 MONITORING INTERVALS

Both United States regulatory agencies require dosimetry programs be capable of "demonstrating" compliance with annual dose limits. Consequently, bioassay monitoring methods must provide detection levels that satisfy these requirements. Sensitivity requirements are therefore based on both the dosimetric significance and the retention (or excretion) of the radionuclide.

For example, consider the use of annual urine sampling to monitor against the inhalation intake of insoluble ^{232}U. The inhalation dose coefficient for Type "S" ^{232}U is 2.6×10^{-5} Sv/Bq (committed effective dose). The urinary excretion fraction at an elapsed time of 365 days for type "S" ^{232}U is 2.8×10^{-6}. Therefore, to detect an intake delivering an effective whole-body dose of 5 rem, the urine monitoring technique must provide a detection sensitivity of at least:

$$\text{Required Sensitivity} = \frac{0.05\,\text{Sv}}{2.6 \times 10^{-5}\,\text{Sv/Bq}} 2.82 \times 10^{-6} = 0.005\,\text{Bq} = 0.3\,\text{dpm}$$

Similar to the caution applied to the bioassay monitoring threshold, the above evaluation only considers dose from one source. Sensitivity requirements may need to be reduced if exposure to multiple radionuclides is possible or if external exposure is expected.

More often, detection capabilities are quantified (based upon detector efficiency, sample count time, background, etc.) and the appropriateness of the technique for bioassay monitoring is evaluated. This process gives the "minimum detectable dose" for the monitoring technique. Because retention (and excretion) varies with time, the minimum detectable dose is time dependent.

For example, the minimum detectable dose (MDD) for bi-monthly (60 days) and annual urine sampling of insoluble ^{232}U for a technique having a detection limit of 0.02 dpm is calculated as follows:

$$\text{Bi-monthly}: \quad \text{MDD} = \left[\frac{0.02\,\text{dpm}}{5.18 \times 10^{-6}}\right]\left[\frac{1\,\text{Bq}}{60\,\text{dpm}}\right]\left[2.6 \times 10^{-5}\,\frac{\text{Sv}}{\text{Bq}}\right]\left[10^{5}\,\frac{\text{mrem}}{\text{Sv}}\right] = 170\,\text{mrem}$$

$$\text{Annual}: \quad \text{MDD} = \left[\frac{0.02\,\text{dpm}}{2.82 \times 10^{-6}}\right]\left[\frac{1\,\text{Bq}}{60\,\text{dpm}}\right]\left[2.6 \times 10^{-5}\,\frac{\text{Sv}}{\text{Bq}}\right]\left[10^{5}\,\frac{\text{mrem}}{\text{Sv}}\right] = 310\,\text{mrem}$$

The first bracketed term in the above expressions gives the intake associated with a measurement at the detection level at the corresponding elapsed time, the second and fourth bracketed terms are unit conversions, and the third bracketed term is the dose coefficient. Tabulations of minimum detectable dose values can be generated for various times. Such tabulations are used to show compliance with regulatory limits and select appropriate monitoring intervals to satisfy regulatory requirements and other programmatic constraints.

APPENDIX 10-A: LAPLACE TRANSFORMS

The Laplace transform converts a function $f(t)$ into a function of a parameter s. Symbolically, the Laplace transform is denoted by $L\{f(t)\} = F(s)$. By definition, the Laplace transform is:

$$L\{f(t)\} = F(s) = \int_0^\infty e^{-st} f(t)dt = \lim_{b\to\infty} \int_0^b e^{-st} f(t)dt \tag{10.A.1}$$

For compartmental analysis, the Laplace transform of the exponential function e^{-kt} is of particular importance, where:

$$L\{e^{-kt}\} = \int_0^\infty e^{-st} e^{-kt}\, dt \tag{10.A.2}$$

$$= \int_0^\infty e^{-(s+k)t}\, dt \tag{10.A.3}$$

$$= \frac{-e^{-(s+k)t}}{s+k} \Bigg|_0^\infty \tag{10.A.4}$$

$$= \frac{1}{s+k}, \quad (s > -k) \tag{10.A.5}$$

The corollary, or inverse Laplace transform, follows that:

$$L^{-1}\left(\frac{1}{s+k}\right) = e^{-kt} \tag{10.A.6}$$

Other useful transforms include:

$$L^{-1}\left(\frac{1}{(s)(s+k)}\right) = \frac{1-e^{-kt}}{k} \tag{10.A.7}$$

$$L^{-1}\left(\frac{1}{(s+k_1)(s+k_2)}\right) = \frac{e^{-k_1 t}}{(k_2 - k_1)} + \frac{e^{-k_2 t}}{(k_1 - k_2)} \tag{10.A.8}$$

APPENDIX 10-B: HEAVISIDE EXPANSION THEOREM

The Heaviside Expansion Theorem states that if $P(s)$ and $Q(s)$ are polynomials of degree m and n, respectively, where $n > m$, and $Q(s)$ has simple distinct roots of $s_1, s_2, \ldots, s_n$, then the quotient $P(s)/Q(s)$ is the Laplace transform of the function $f(t)$ given by:

$$f(t) = L^{-1}\left(\frac{P(s)}{Q(s)}\right) = \sum_{k=1}^{n} \frac{P(s_k)}{Q'(s_k)} e^{s_k t} \qquad (10.B.1)$$

REFERENCES

Department of Energy. 2004. *Occupational Radiation Protection* 10 CFR 835. Washington, DC: U.S. Department of Energy.

Hesp, R. 1964. "The Retention and Excretion of Caesium-137 by Two Male Subjects". *International Atomic Energy Agency, Assessment of Radioactivity in Man. Vol. II. Proceedings of the Symposium on the Assessment of Radioactive Body Burdens*, Heidelberg, May 11–16, 1964.

International Atomic Energy Agency. 2007. *Intercomparison Exercise on Internal Dose Assessment* IAEA-TECDOC-1568. Vienna, Austria: International Atomic Energy Agency.

International Commission on Radiological Protection. 1979. ICRP Publication 30 (Part 1): Limits for Intakes of Radionuclides by Workers. *Ann ICRP* 2 (3–4).

International Commission on Radiological Protection. 1994. ICRP Publication 68: Dose Coefficients for Intakes of Radionuclides by Workers. *Ann ICRP* 24 (4).

International Commission on Radiological Protection. 2002. Supporting Guidance 3, Guide for the Practical Application of the ICRP Human Respiratory Tract Model. *Ann ICRP* 32 (1–2).

Marsh, James W., Michael R. Bailey, and Alan Birchall. 2005. "A Step-by-Step Procedure to Aid the Assessment of Intake and Doses from Measurement Data". *Radiation Protection Dosimetry* 114 (4):491–508.

National Council on Radiation Protection and Measurements. 1987. *NCRP Report 87: Use of Bioassay Procedures for Assessment of Internal Radionuclide Deposition*. Bethesda, MD: National Council on Radiation Protection and Measurements.

Nuclear Regulatory Commission. 1987. *Interpretation of Bioassay Measurements* NUREG/CR-4884. Washington, DC: U.S. Nuclear Regulatory Commission.

Nuclear Regulatory Commission. 1991. *Standards of Protection against Radiation* 10 CFR 20. Washington, DC: U.S. Nuclear Regulatory Commission.

Nuclear Regulatory Commission. 1995. *Ingestion of Phosphorus-32 at Massachusetts Institute of Technology, Cambridge, Massachusetts, Identified on August 19, 1995* NUREG-1535. Washington, DC: U.S. Nuclear Regulatory Commission.

Potter, Charles A. 2002. "Intake Retention Fractions Developed from Models Used in the Determination of Dose Coefficients Developed for ICRP Publication 68—Particulate Inhalation". *Health Physics* 83 (5):594–789.

Index

AAHP, *see* American Academy of Health Physics

Absorbed dose, 51, 56–57, 72
  for contaminated soil, 378–384
  converting activity intake to, 401
  converting age-specific, 401
  energy imparted, 56, 73
  for immersion in contaminated water, 379
  rate, 72, 154, 155
  for submersion in contaminated air, 379

Absorbed fraction; *see also* Specific Absorbed
      Fraction
  for charged particles, 330–332
  of energy, 312

Absorption types (F, M, S), 223–224

Active detectors, 132–134

Active dosimeters, 142–143
  electronic personal dosimeter, 142–143

Active marrow, 181, 185, 190, 379, 385

Active radiation-detecting instruments, 126

Acute ingestion of ^{32}P, 446–450

Acute inhalation of ^{241}Am, transfer parameters
      following, 454–458

Acute inhalation of ^{137}Cs, 441–446

Acute uptake of ^{3}H, forced fluids following, 451–454

ACXRP, *see* Advisory Committee on X-ray and
      Radium Protection

Adjoint Boltzmann transport equation, 344–346
  leakage, 346

Advisory Committee on X-ray and Radium
      Protection (ACXRP), 96–97, 99, 105

AEC, *see* Atomic Energy Commission

Age-dependent
  dose coefficients, 205, 206, 208
  element composition of skeleton, 190

Air composition, 374–375

Air kerma
  absorbed dose per, 328
  calibration, 153
  detection, 129
  dose rate coefficient, 379

dosimeter, 135
  rate, 143
  relationship to exposure, 127, 171

Air monitoring instruments, 129, 149

ALAP, *see* As Low As Practicable

ALARA, *see* As Low as Reasonably Achievable

ALATEP, *see* As Low As Technically and
      Economically Practical

Albedo neutron dosimeter, 144–146

ALGAMP code, 328

ALI, *see* Annual limit on intake

Alpha decay, 22–23, 27, 31–34, 43
  conservation of momentum, 32, 37
  Coulombic repulsion, 32
  neutron source, 27

Ambient dose equivalent, 15, 67–69, 152, 353–355
  detection, 124, 126, 132–133, 143
  H*(10), 353
  kerma approximation, 349
  rate, 126, 132–133, 143
  response, 148

American Academy of Health Physics (AAHP), 5

American Board of Health Physics (ABHP), 83

American National Standards Institute (ANSI), 4,
      82, 114, 149–150

American Nuclear Society (ANS), 5

American Roentgen Ray Society (ARRS), 91, 92

American Society for Testing and Materials
      (ASTM), 92, 114

Americium (^{241}Am)
  acute inhalation of, 454–458

Ampere, 14

AMU, *see* Atomic mass unit

Angular flux, 77

Annual limit on intake (ALI), 109, 367–368,
      384–385, 459–460

Annual Reference Levels of Intake (ARLI), 111

ANS, *see* American Nuclear Society

ANSI, *see* American National Standards Institute

Anthropomorphic models, *see* Phantom

Area monitoring instruments
  current-mode detectors
    ionization chambers, 126–128
  neutron and mixed field instruments, 131
    active detectors, 132–134
    high-energy neutron instruments, 134
    operational considerations, 134–135
    passive detectors, 132
  pulse-mode detectors
    Geiger–Mueller detectors, 129
    proportional counter detectors, 129
    scintillation detectors, 130
ARLI, *see* Annual Reference Levels of Intake
Arrhenius equation, 138
ARRS, *see* American Roentgen Ray Society
As Low As Practicable (ALAP), 107, 115
As Low as Reasonably Achievable (ALARA),
  2, 107
As Low As Technically and Economically Practical
  (ALATEP), 115
ASTM, *see* American Society for Testing and
  Materials
Atomic Energy Act, 101, 103, 115
Atomic Energy Commission (AEC),
  101–104, 106
  division, 113–115
Atomic mass
  fission products, 28
  unit (AMU), 16, 17, 19
Atomic number (Z), 16, 46
Atomic structure
  electron, 16–17
  electron orbital structure
    Bohr model, 22–24
    Sommerfeld model, 25
  excitation, 25
  ionization, 25–26
  liquid drop model, 21–22
  neutron, 16
  nucleus
    binding energy, 17–18
    binding energy per nucleon, 18–21
    mass defect, 18
  proton, 16
Attenuation coefficients, 47, 52–53, 136
  linear, 53
  total, 53
Auger electron, 53

Barn, 49, 72
Bateman equation, 42
BE, *see* Binding energy
BEAR, *see* Biological Effects of Atomic Radiation

Becquerel,
  Henri, 84–85
  unit (Bq), 39
BEIR, *see* Biological Effects of Ionizing Radiation
Beta decay, 34–35, 43
Beta particle calibrations, 154–156
Bias, measurement defined, 150–151, 160
Binding energy (BE), 17–18
  atomic mass energy, 18
  Coulombic repulsion, 17
  electron, 18
  mass energy, 18
  neutron separation, 21
  nuclear, 18
  nucleon, 18–21
  per nucleon, 18–21
  proton separation, 19–20
  Q-value, 21
Bioassay, interpretation of
  acute ingestion of ^{32}P, 446–450
  acute inhalation of ^{137}Cs, 441–446
  compartmental analysis, 418–420
    closed recycling system, 423–424
    closed system catenary transfer,
      420–421
    matrix solution, 426–429
    open recycling system, 424–425
    open system catenary transfer, 421–422
    two-compartment model, 425–426
  Heaviside expansion theorem, 463
  incremental excreta bioassay, limitations
    affecting, 434–441
  individual-specific modifications
    acute inhalation of ^{241}Am, transfer
      parameters following, 454–458
    acute uptake of ^{3}H, forced fluids following,
      451–454
    modified dose coefficients, 458–459
    respiratory tract absorption, modification of,
      454–458
  Laplace transforms, 462
  monitoring intervals, 461–462
  regulatory issues, 459–461
  retention fractions
    and excretion fractions, tabulations of,
      433–434
    excretion functions for acute exposures, 433
    retention functions for chronic intakes,
      429–433
  monitoring thresholds, 460–461
Biokinetic models, 216–217; *see also*
      International Commission on
      Radiological Protection

alimentary tract (*see* Human Alimentary
 Tract Model)
respiratory tract (*see* Human Respiratory
 Tract Model)
systemic, 237–295
 cesium, 271–283
 ICRP Publication 2, 237–238
 ICRP Publication 30, 238–239
 ICRP Publication 68 and 72 Series, 239–243
 iodine, 254–271
 plutonium, 283–295
 progeny, 295–302
 strontium, 243–254
Biological Effects of Atomic Radiation (BEAR), 106
Biological Effects of Ionizing Radiation (BEIR), 108
 VII risk models, 403
Biological half-life (time), 208, 230
 alimentary tract, 239
 plutonium, 284
 respiratory tract, 219, 221
 systemic, 238, 239, 274
 thyroid, 260–261, 268, 270–271
 tritium, 451
BIPM, *see* International Bureau of Weights
 and Measures
Blue Book, 410, 412
Blum, Theodore, 97
Bohr model, 22–24
 hydrogen, 23
 ionization energy, 24
 kinetic energy, 25
 orbital number, 24
 potential energy, 22
Boltzmann transport equation, 339–341
 Monte Carlo solution (*see* Monte Carlo method)
Bone remodeling rates, 170, 196, 200–201, 286
Bone-seeking elements, 239, 241
Bragg–Gray, 127, 134, 154–155
 equation, 127
Branching ratios, 43–44, 319, 321, 329
Bremsstrahlung, 45, 56, 57, 78, 322–323, 349, 353,
 374–379, 383
British Roentgen Society (BRS), 93
Bubble detectors, 136, 146–147
Bureau International des Poids et Mesures (BIPM),
 see International Bureau of Weights
 and Measures

CAA, *see* Clean Air Act
Cadaver imaging, 193–194
Calibration and testing in measurements,
 151–159
 frequency, 150

Cancer risk coefficient, 395
 BEIR VII risk models and formulas for, 410–414
 case studies, 405–410
 comparison with ICRP, 399
 dose and risk coefficient software, 404–405
 EPA *vs.* NRC, 399
 limitations of, 396–404
 for population and exposure pathway, 400–404
Candela, 14
Carbon–nitrogen–oxygen (CNO) cycle, 30
CDRH, *see* Center for Devices and Radiological
 Health
Center for Devices and Radiological Health
 (CDRH), 112
CERCLA, *see* Compensation and Liability Act
Certification, 82–83
 NRRPT, 83
Cesium (^{137}Cs), 271–283
 acute inhalation of, 441–446
 cancer risk on ground surface, 406–408
 ingestion of, 386–388
CFR, *see* Code of Federal Regulation
Chalk River Conference on Permissible Dose, 171
Charged particle equilibrium (CPE), 57, 72, 128,
 349, 354; *see also* Ionization chamber
Charged particles, tracking of, 348–350
Clean Air Act (CAA), 113
CNO cycle, *see* Carbon–nitrogen–oxygen cycle
Code of Federal Regulation (CFR), 4, 151
Codman, Ernest Amory, 91
Coherent derived units, 14–15
Collisional kerma, 56, 57, 72, 349
Collisional stopping power, 78
Committed effective dose coefficient, 364, 367, 385,
 459–461
Committed equivalent dose coefficient, 362–363,
 385–389
Compartmental analysis
 closed recycling system, 423–424
 closed system catenary transfer, 420–421
 Laplace transforms, 462
 matrix solution, 426–429
 open recycling system, 424–425
 open system catenary transfer, 421–422
 translocation rate constants, 418–419
 two-compartment model, 419
Comprehensive Environmental Response,
 Compensation and Liability Act
 (CERCLA), 113, 117, 400
Compton scatter, 46, 47–48, 54
 mass energy transfer coefficient, 52
Computation dosimetry
 defined, 336

Computed tomography (CT) imaging, 193, 208
CONCERT-European Joint Program, 5
Contaminated air, submersion in, 374–376
    calculation absorbed dose for, 383
Contaminated water
    calculation absorbed dose for, 384
    immersion in, 376–377
    submersion in, 384
Coolidge, William D., 91
Coulomb's law, 23
    repulsion effects, 17, 32
Council of Radiation Program Directors
    (CRCPD), 105
CPE, *see* Charged particle equilibrium
CRCPD, *see* Council of Radiation Program
    Directors
Cross section
    differential, 73
    elastic scatter, 48
    hydrogen, 131
    macroscopic, 72
    microscopic, 72
    neutron, 49, 130, 140
CT imaging, *see* Computed tomography imaging
Curie (Ci), 39
Current-mode detectors, 126–128

DAC, *see* Derived air concentration
Daily water balance, 196
Daughter radionuclide, 295, 321
DDREF, *see* Dose and Dose Rate Effectiveness
    Factor
Decay chain, 45, 295
    considerations, 366–367
Decay data, *see* Nuclear decay data
De-excitation, 26, 27
Department of Defense (DOD), 115, 117
Department of Energy (DOE), 113
    Title 10 Part 835, 114
Derived air concentration (DAC), 100, 109, 367, 368,
    384–385, 459
Derived units, 14–15
Detriment, 116, 399
Deuteron, 16, 21, 50
Directional dose equivalent, 15, 68–69
Direct-reading dosimeters, 142
Discrete ordinates method, 342–344
DOD, *see* Department of Defense
DOE, *see* Department of Energy
Dose and Dose Rate Effectiveness Factor
    (DDREF), 402
Dose calculation methodology, 143–144

Dose coefficients
    for external environmental radiation fields,
        208, 368
        contaminated soil, 368–374, 378–383
        immersion in contaminated water,
            376–377, 384
        submersion in contaminated air, 374–376, 383
    external irradiation
        adjoint transport equation, 344–346
        discrete ordinates method, 342–344
        Monte Carlo methods, 347–352
        neutron, 354–358
        photons, 352–354
        transport methods, 339–342
    for external occupational exposures, 207
    fluence-to-dose, 337, 356–357, 379
    ICRP Publication 60, 66
    ICRP Publication 74, 66
    ICRP Publication 107, 204–205
    ICRP Publication 116, 3, 66
    ICRU Report 57, 66
    for internal emitters
        calculation details, 364–367
        committed effective, 363, 364–367
        committed equivalent, 324, 362
        derived quantities, 367–368
        effective dose, 358–359
        equivalent dose, 359–360
        examples, 384–390
        number of nuclear transformations, 361–362
        S-coefficient, 360–361
        specific absorbed fraction, 363–364
    for internal occupational and environmental
        exposures, 205–207
    for medical exposures, 207–208
    for radionuclides, 384
    sample calculations based on
        ingestion of ^{137}Cs, 386–387
        ingestion of ^{131}I, 385–386
        inhalation of ^{210}Po, 387–388
        inhalation of ^{239}Pu, 388–390
        ingestion of ^{90}Sr, 384–385
Dose equivalent, 58, 68, 73
    rate, 157
Dosimeter
    accreditation, 151
    active, 142–143
    calibration, 143
    passive, 132–145
Dosimetric models
    absorbed fractions for charged particles, 330–332
    nuclear decay data, 317–321

source and target organs, 313–317

specific absorbed fraction (*see* Specific Absorbed Fraction)

    for neutrons, 329–330

    for photons, 173, 327–329

specific effective energy (SEE), 322–324, 330–332

EAR model, *see* Excess absolute risk model

Effective dose, 1, 62, 65, 68, 73, 358–359

Effective quality factor, 61, 77

Eisenhower, Dwight D., 105

Elastic scatter, 30–31, 51

    neutron, 51

Electrically powered device, 124

Electron, 16–17

    bremsstrahlung, 45

    capture, 35–36

    hard collisions, 44–45

    orbital structure

        Bohr model, 22–24

        Sommerfeld model, 25

    soft collisions, 44

    sources in water, 377

    trap, 137

Electronic personal dosimeters (EPD), 125, 142–143, 163

Electronics products, 112

Elemental tissue compositions/mass densities, 184–188

Element-specific systemic biokinetic models, 217

Embryo, 111, 188–191

    ICRP Publication 88, 174

Endogenous fecal excretion, 279

Energy fluence, 55, 73, 371

    rate, 73

Energy flux density, 73

Energy imparted, 73

Energy Research and Development Agency (ERDA), 113

Energy transferred, 74

ENSDF, *see* Evaluated Nuclear Structure Data Files

Environmental dose coefficients, 369

Environmental dose rate coefficients

    calculating absorbed dose for contaminated soil

        electrons, 383

        photons, 378–383

    calculation absorbed dose for submersion in contaminated air, 383

    calculation absorbed dose for submersion in contaminated water, 384

    for radionuclides, 384–385

Environmental exposures, 205–207, 261, 324, 397

Environmental Protection Agency (EPA), 1, 3–4, 112–113

    Clean Air Act, 113

    Comprehensive Environmental Response, Compensation and Liability Act, 113

    National Radon Action Plan, 113

    Safe Drinking Water Act, 113

    Uranium Mill Tailings Radiation Control Act, 113

EPA, *see* Environmental Protection Agency

EPD, *see* Electronic personal dosimeters

Equivalent dose, 1, 63, 70, 73, 74, 124, 359, 367, 368, 384

ERDA, *see* Energy Research and Development Agency

ERR model, *see* Excess relative risk model

Etched-track detector, 132, 135, 140, 141, 146

Evaluated Nuclear Structure Data Files (ENSDF), 320

Excess absolute risk (EAR) model, 402, 410, 411, 412

Excess relative risk (ERR) model, 402, 410, 411, 412

Exchangeable bone volume, 247–248, 252, 253

Excitation, 25

Excretion fractions, tabulations of, 433–434

Exponential transform, 351, 373

Exposure, 74

    acute *vs.* chronic, 2

    air kerma, 127

    dose relationship, 57–59, 348–350

    pathway, 400–404

    Rad, 58

    rate, 58, 125, 413

    rem, 25, 58

    roentgen, 58

External irradiation

    adjoint transport equation, 344–346

    discrete ordinates method, 342–344

    Monte Carlo methods, 347–352

    neutron, 355–358

    photons, 352–355

    transport methods, 339–342

External occupational exposures, 207

External radiation dosimetry, *see* Reference individuals, for external and internal radiation dosimetry

Extrapolation curves, 156

Extremity monitoring, 125

    dosimeters, 139

FDA, *see* Food and Drug Administration

Federal Guidance Report (FGR), 3–4

    FGR 12, 4, 404–405

    FGR 13, 5, 328, 395–397,399–400, 401, 404–410

    FGR 15, 4, 375, 377, 378, 380, 383–384, 395

    FGR 16, 396, 401

Federal Radiation Council (FRC), 111–112
Fetus, 111, 188–191; *see also* Embryo
FGR, *see* Federal Guidance Report
Film dosimeters, 135–136
First-order kinetics, 164, 216, 227, 230, 234–235,
    238, 260, 276
Fission, 27–29, 31, 50, 99, 103
    Atomic Energy Commission (AEC), 113–115
    average energy, 29
    energy, 29
    energy spectrum, 28
    fragments, 146, 329
    mass defect, 28
    Maxwell–Boltzmann distribution, 28
    products, 28
    spontaneous, 28, 36–37, 329, 384
    temperature, 28
Fluence, 50, 74
    rate, 75
Flux density, 75
Food and Drug Administration (FDA), 112
Force collisions, 351
Formerly Used Sites Remedial Action Program
    (FUSRAP), 117
Forward transport equation, 344
Fractional transfer coefficients, 418
Franklin, Milton, 92
FRC, *see* Federal Radiation Council
Free neutrons decay, 16
Fusion, 19, 30
    CNO cycle, 30
FUSRAP, *see* Formerly Used Sites Remedial
    Action Program

Gamma emissions, 37
Gamma ray sources, 128, 152
Gastrointestinal (GI) model, 217
    Publication 2, 227
    Publication 30, 227–228
    Publication 100, 228–235
Geiger–Mueller detectors
    P-10, 129
Glow curve, 137, 138
Gray, 71
Ground roughness effect, 370, 381

Hanford personal dosimeter, 140
Hard collisions, 44–45
HATM, *see* Human Alimentary Tract Model
Hazard, recognition of
    additional impetus, 90–91
    early reports of injury, 86–88
    protective measures, 88–90

Health Physics Society (HPS), 5, 101, 114
Heaviside expansion theorem, 424, 463
Heavy charged particles, 52, 170
High-energy neutron instruments, 134
Hole trap, 137
HPS, *see* Health Physics Society
HPSSC Committee, 114
HRTM, *see* Human Respiratory Tract Model
Human Alimentary Tract Model (HATM), 174, 217,
    231–232
    colon, 233–235, 264, 328
    ICRP Publication 2, 227
    ICRP Publication 30, 227
    ICRP Publication 100, 228–237
    lung mass (inclusive of blood) as a function of
        age, 205
    transit times of luminal content, 197–199
    urinary and fecal excretion rates, 199–200
Human Respiratory Tract Model (HRTM),
    217–227, 456
    ICRP Publication 2, 218–220
    ICRP Publication 10, 219, 220
    ICRP Publication 30, 220–221
    ICRP Publication 66, 221–224
    ICRP Publication 130, 224–227
    respiratory volumes and capacities, 196–197
    Task Group Lung Model (TGLM), 220–221
    ventilation rates, 196–198

IAEA, *see* International Atomic Energy Agency
ICRP, *see* International Commission on
    Radiological Protection
ICRU, *see* International Commission on Radiation
    Units and Measurements
ICXRP, *see* International Committee on X-Ray and
    Radium Protection
IEC, *see* International Electrotechnical Commission
Importance sampling, 351
Inactive marrow, 181, 185, 190
Incremental excreta bioassay samples, multiple
    bioassay measurements, 434–441
Individual organ systems, 176, 178–184
Individual-specific bioassay modifications,
    450–459
Inelastic scatter, 31, 51
    neutron, 51
Integration techniques, 366
Interlude, 101–102
Internal conversion, 36
Internal occupational exposures, 205–207
Internal radiation dosimetry, *see* Reference
    individuals, for external and internal
    radiation dosimetry

International Atomic Energy Agency (IAEA), 5,
105, 153, 201, 202–205, 451
International Bureau of Weights and Measures
(BIPM), 13–14
International Commission on Radiation Units and
Measurements (ICRU), 96, 126, 150, 337
Report 46, 2, 174, 185–190
Report 57, 66, 207, 353–358
Report 76, 159
Report 84, 207
Report 85a, 2, 349
slab, 69
sphere, 68
International Commission on Radiological
Protection (ICRP), 82, 95, 103, 108–110,
153, 170; see also Reference individuals,
for external and internal radiation
dosimetry
alimentary tract models (see Human alimentary
tract model)
biokinetic models for radionuclides in vivo,
295–302
cesium, 271–283
Committee 2, 103, 109, 172–173, 309
comparison radiation weighting factors in
Publication 60/103, 65–66
comparison tissue-weighting factors in
Publication 26/60/103, 64
data for use in radiological protection
Publication 70, 173
embryo fetus
Publication 88, 174
Environmental Intakes of Radionuclides, 207
gastrointestinal models
Publication 2, 227
Publication 30, 227–228
Publication 100, 228–235
International Committee on X-Ray and Radium
Protection, 96
International Congress on Radiology, 96
iodine, 254–271
colloid, 259
extrathyroidal T4 and T3, 260–261
iodide and organic iodine in thyroid, 257–260
requirements in humans, 254–256
Occupational Intakes of Radionuclides (OIR),
206–207
plutonium, 283–295
protection quantities, 61–67
Publication 1, 107, 109, 114
Publication 2, 218–220, 227, 229, 237–238, 249,
261, 274, 285
Publication 10, 219

Publication 19, 286
Publication 20, 236, 249
Publication 21, 60, 331
Publication 23, 172–173, 186, 192, 196, 198, 205,
311, 315, 317, 327
Publication 26, 62–63, 110, 353, 355, 399
Publication 30, 173, 197, 198, 205, 220–221,
227–228, 236, 238–239, 250, 261, 285, 286,
322, 324, 328, 433, 441, 447, 460
Publication 38, 205, 321
Publication 48, 285
Publication 51, 207, 329, 353
Publication 53, 207
Publication 54, 239, 285
Publication 56, 205, 287–288, 450
Publication 60, 63–64, 206, 288, 323, 399
Publication 66, 173, 196, 205, 221–224, 363
Publication 67, 205, 250–255, 288, 290, 292, 293,
294, 295, 298, 299
Publication 68, 205, 239–243, 261, 274,
282–283, 288–289, 299, 301–302, 433,
454, 459, 460
Publication 69, 205
Publication 70, 173
Publication 71, 186, 205
Publication 72, 205, 239–243, 261, 274
Publication 74, 2, 207, 328, 329, 353
Publication 80, 207
Publication 88, 173–174
Publication 89, 3, 174, 176–191, 201–204
comparison to Publication 23, 180–184
Publication 100, 174, 183, 228–235, 326
Publication 103, 3, 64–67, 174, 204, 206, 226, 332,
358–360, 362, 363, 371
Publication 106, 207
Publication 107, 3, 205, 206, 329, 360, 383, 384
Publication 110, 3, 192–195, 205, 207, 363
Publication 116, 3, 207, 338, 353–358
Publication 119, 206
Publication 123, 207
Publication 128, 207–208
Publication 130/134/137, 3, 366
Publication 133, 206, 363, 364, 365, 366–367
respiratory tract models (see Human respiratory
tract model)
specific individuals, 174–175
strontium, 245–256
systemic biokinetic models
Publication 2, 237–238
Publication 30, 238–239
Publication 68 and 72 Series, 239–243
uncertainty, 174
voxel phantoms, 192–195

International Committee on X-Ray and Radium
    Protection (ICXRP), 96–97
International committees and organizations,
    governmental and non-governmental
    organizations, 5
International Electrotechnical Commission (IEC),
    82, 114
International Radiation Protection Association
    (IRPA), 5
International standard units (SI)
    base units
        units with special names and
            symbols, 14–15
    and corresponding system of quantities, 13–14
    quantities and units, 13
    traditional units for radiation protection, 15–16
International X-ray Unit Committee (IXRUC), 96
Interpolation techniques, 365–366
Iodine (^{131}I), 262–271, 385–386
    absorption and distribution of inorganic iodide,
        256–257
    extrathyroidal T4 and T3, 260–261
    iodide and organic iodine in thyroid, 257–260
    in milk, 406
    requirements in humans, 254–256
Ionization, 25–26
    chamber, 58, 96, 126–128
        Bragg–Gray, 127
        charged particle equilibrium, 128
        direct-reading dosimeter, 142
        Geiger–Mueller, 129
        pulse-mode, 128–129
        survey meter, 132–135
        tissue-equivalent, 139
    energy, 1–2, 24
Ionizing radiation, 1–2, 100, 104, 107, 108
    radioactivity, discovery of, 84–85
    X-rays, discovery of, 83–84
IRPA, *see* International Radiation Protection
    Association
Irradiation duration, 158
Isobars, 17
Isomeric transitions, 37
Isotones, defined, 17
Isotopes, defined, 17
IXRUC, *see* International X-ray Unit Committee

JCAE, *see* Joint Committee on Atomic Energy
Joint Committee on Atomic Energy (JCAE), 106

Kaye, G.W.C., 96
Kelvin, 14

Kerma, 54–56, 75
    approximation, 348–350
    coefficient, 55
    collisional, 56
    energy transfer coefficient, 56
    mass energy transfer coefficient, 55
    neutron, 55
    photon, 55
    radiative, 56
    rate, 75
Kilogram, 14
Kinematics
    of Compton scatter events, 47
    of photoelectric events, 47
Klein–Nishina equation, 48

Langham's equation, 286
Laplace transforms, 420, 462, 463
LAR, *see* Lifetime attributable risk
Laser-based counting techniques, 132
LBM, *see* Lean body mass
Lean body mass (LBM), 177–178
Legal bases of radiation protection guidance, 81
Lens of the eye, 66, 68, 69, 109, 125, 135, 143, 153
Leonard, Charles Lester, 90
LET, *see* Linear energy transfer
Licensure, 82–83, 90, 92, 104, 108, 113
Life Span Study (LSS), 401
Lifetime attributable risk (LAR), 411–412
Lifetime risk coefficient (LRC), 412
Light pulses, 130
Lineal energy, 75, 129, 133
Linear energy transfer (LET), 59–61, 62, 75
    unrestricted, 60, 61, 62, 75–78, 320
Linear Non-Threshold (LNT), 105, 106
    hypothesis, 402
Liquid drop model, 21–22
    binding energy per nucleon, 18, 21
    magic numbers, 21–22
LNT, *see* Linear Non-Threshold
Lower limit of detection, 162
LRC, *see* Lifetime risk coefficient
LSS, *see* Life Span Study

MADE, *see* Maximum Absorbed Dose Equivalent
Magic numbers, 22
Manhattan Engineer District (MED), 98–99,
    102, 105
Manhattan Project, 6, 80, 98
Marrow
    active marrow, 181, 185, 190, 379, 385
    inactive marrow, 181, 185, 190

MARSSIM, *see* Multi-Agency Radiation Survey and Site Investigation Manual
Mass attenuation coefficients, 76, 136
  plots of, 47
Mass defect, 18, 20–21, 28, 31–32, 34
  alpha decay, 31
Mass energy absorption coefficient, 54, 56, 58, 59, 76
Mass energy transfer coefficient, 53–54, 76
  Auger electron, 53
  Compton scatter, 54
  pair production, 54
  photoelectric, 54
Matrix solution, 426–429
Maturation [of radiation protection policy]
  immediate postwar period, 99–101
  organizing for radiation protection, 96–97
  radium rears, 97–98
  roentgen, 95–96
  tolerance dose, 95
  World War, 98–99
Maximum Absorbed Dose Equivalent (MADE), 344, 355
Maximum Permissible Body Burden, 109
Maximum Permissible Concentration (MPC), 100
Maximum permissible dose (MPD), 106
Maxwell–Boltzmann distribution, 28
  fission, 28
Mean absorbed dose, 66, 73, 76, 359
Mean energy imparted, 71, 73, 76–77
Mean quality factor, 61, 63
Mean transit time, 230
Mean work function, 57, 77
Measurement methods/procedures
  calibration and testing
    beta particle calibrations, 154–156
    neutron calibrations, 156–158
    photon calibrations, 153–154
    surface contamination monitors, 158–159
  measurement traceability, 159–160
  national standards and reports, 149–151
  regulatory guidance, 151
  statistics of radiation measurements
    uncertainty analysis, 162–163
Measurement traceability, 159–160
MED, *see* Manhattan Engineer District
Medical exposures, 101, 170, 207–209
Medical Internal Radiation Dose (MIRD) Committee, 4, 173
Mesons, 17
Metastable energy states, 37, 43
Meter, 14

MIRD Committee, *see* Medical Internal Radiation Dose Committee
Mixed field dosimeters
  etched track detectors, 146
  nuclear emulsion dosimeters, 144
  personal neutron accident dosimeters, 147–148
  superheated drop (bubble) detectors, 146–147
  thermoluminescent detectors, 144–146
Mixed field instruments
  active detectors, 132–134
  high-energy neutron instruments, 134
  operational considerations, 134–135
  passive detectors, 132
Modified dose coefficients, 458–459
Mole, 14
Monitoring intervals, 461–462
Monte Carlo method, 337, 347–352
Morgan, Karl Z., 100
Morton, William, 86
MPC, *see* Maximum Permissible Concentration
MPD, *see* Maximum permissible dose
Multi-Agency Radiation Survey and Site Investigation Manual (MARSSIM), 117
Multi-element personal dosimeters, 144
Multiple bioassay measurements, 434–441
Multi-sphere spectrometers, 132
Mutscheller, Arthur, 95

National Council on Radiation Protection and Measurements (NCRP), 1, 82, 96, 99, 110–111
  Advisory Committee on X-ray and Radium Protection, 96
  Report 17 (1954), 99
  reports and commentaries of, 2, 3
  X-ray protection, 111
National Environmental Policy Act, 112
National Fire Protection Association, 92
National Institute of Standards and Technology (NIST), 82, 125
National Physical Laboratory (NPL), 94
National Radiological Protection Board (NRPB), 288
National Registry of Radiation Protection Technologists (NRRPT), 83
National Voluntary Laboratory Accreditation Program (NVLAP), 151
NCRP, *see* National Council on Radiation Protection and Measurements
Negligible Individual Risk Level (NIRL), 111
Net leakage, 341
Neutron, 16, 49–50, 49–51, 144–148, 354–358
  absorption, 27, 39, 50
  calibrations, 156–158

capture, 35–36
cross sections, 49–50
detection
  ^{3}He, 132
  ^{6}Li, 132
  active detectors, 132–134
  BF_3, 132
  cadmium filter, 132
  high-energy, 134
  moderator, 134
  passive, 135–142
  spectrometry, 147
  tissue equivalent proportional counter, 134
elastic scatter, 51
fluence, 50
inelastic scatter, 51
mean free path, 49
measurement, 29
and mixed field dosimeters
  etched track detectors, 146
  nuclear emulsion dosimeters, 144
  personal neutron accident dosimeters, 147–148
  superheated drop (bubble) detectors, 146–147
  thermoluminescent detectors, 144–146
and mixed field instruments
  active detectors, 132–134
  high-energy neutron instruments, 134
  operational considerations, 134–135
  passive detectors, 132
moderation, 28
reaction rate, 30
separation energy, 19–20, 21
sources, 152
specific absorbed fractions for, 329–330
spectrum, 28
in tissue, 50
Neutron number (N), 17
Nicoloff and Dowling model, 264
NIRL, see Negligible Individual Risk Level
NIST, see National Institute of Standards and Technology
Non-exchangeable bone volume, 247–253
Non-governmental organizations, 5
Non-governmental standards bodies, 108
Non-uniform rational basis-spline (NURBS), 194–195
NPL, see National Physical Laboratory
NRC, see Nuclear regulatory commission
NRPB, see National Radiological Protection Board
NRRPT, see National Registry of Radiation Protection Technologists

Nuclear decay data
  applications and uncertainties of, 321
  ICRP Publication 38, 318–321
  ICRP Publication 107, 321
  sources of, 318–321
  uncertainties, 321
  updates, 321
Nuclear emulsion dosimeters, 144
Nuclear reactions
  absorption, 26–27
  elastic scatter, 30–31
  endothermic, 27
  exothermic, 27
  fission, 27–29
  fusion, 30
  inelastic scatter, 31
  Q-value, 21
Nuclear Regulatory Commission (NRC), 4, 104–105, 113–114
  10 CFR 20, 114
Nuclear transformations, 322, 324, 361–362, 365, 384–387, 389, 452
Nucleus
  binding energy, 17–18
  binding energy per nucleon, 18–21
  mass defect, 18
Nuclide, 17
NURBS, see Non-uniform rational basis-spline
NVLAP, see National Voluntary Laboratory Accreditation Program

Occupational Intakes of Radionuclides (OIR), 206–207
Occupational Safety and Health Administration (OSHA), 112, 115
OIR, see Occupational Intakes of Radionuclides
Open catenary system, 425
Open recycling system, 424–425, 426
Open system catenary transfer, 421–422
Operational quantities
  ambient dose equivalent, 67–68
  dose equivalent, 67–68
  ICRU, 67–68
  personal dose equivalent, 68–70
Optical density, 136
Optically stimulated luminescence (OSL) dosimeters, 135, 140–141, 143
Oral cavity, 228, 229, 230
Organ dose from electrons, in air, 375–376
Organ equivalent dose, 67
OSHA, see Occupational Safety and Health Administration
OSL dosimeters, see Optically stimulated luminescence dosimeters

Pacific Northwest National Laboratory (PNNL), 114
PADC, *see* Plastic polyallyl diglycol carbonate
Pair production, 46, 48–49, 52–54, 130
    mass energy transfer coefficient, 76
Parker, Herbert M., 99–101
Particle radiance, 77, 337
Particle range, 52
Particle transfer pathways, 456–457
Passive detectors, 132
Passive dosimeters, 142
    direct-reading dosimeters, 142
    optically stimulated luminescence dosimeters,
        140–141
    photographic dosimeters, 135–136
    thermoluminescent dosimeters, 137–140
Performance quotient, 150, 160, 161
Permissible dose, 96–99, 103, 109–110, 309
Personal dose equivalent, 68–70, 124, 135, 143–147,
        152, 156, 160
Personal dosimeters, 6, 124, 125, 135–148, 150,
        160, 163
    passive dosimeters
        direct-reading dosimeters, 142
        optically stimulated luminescence
            dosimeters, 140–141
        photographic dosimeters, 135–136
        thermoluminescent dosimeters, 137–140
Personal neutron accident dosimeters (PNAD),
        147–148
Personal neutron dosimeters, 143
Phantom models, 307–308, 310; *see also* ICRP
        Publication 110; Reference Man;
        Standard Man
    complex humanoid phantom models, 67
    geometry, 149
    heterogeneous, 313, 315–317
    hybrid computational phantoms, 194–195
    ICRU, 68, 69, 152
    MIRD, 316–317
    NCAT, 194–195
    NUBS, 194–195
    pillar phantoms, 69
    PMMA, 152–154, 157, 158
    similitude phantoms, 315
    slab, 152–154, 343–344
    Snyder–Fisher adult human phantom, 312
    sphere, 126, 135
    stylized computational phantoms, 173, 191–193,
        310–312, 315, 317–319
        ORNL, 194, 312–313, 317, 318–320, 328, 378
    voxel computational phantoms, 193–195
Phase flux, 77
Phosphorus (^{32}P), acute ingestion of, 446–450

Photoelectric (PE) effect, 46–47, 53–54, 130, 372
    mass energy transfer coefficient, 53
Photographic dosimeters, 135–136
Photographic films, 136
Photon, 25, 45–46, 352–354
    absorption, 139
    calibrations, 153–154
    Compton scatter, 47–48
    emissions, 26, 34, 37, 53
    energy, 25, 37, 46, 49, 50, 56, 58, 130, 139, 173
    interactions, 52
    mass energy absorption coefficient, 76
    mass energy transfer coefficient, 76
    pair production, 48–49
    photoelectric effect, 46–47
    photonuclear, 49
    photonuclear reactions, 49
    sources in water, 376–377
    specific absorbed fractions for, 327–329
    submersion dose due to, 374–375
Photonuclear reactions, 49
PHS, *see* U.S. Public Health Service
Planck's constant, 46
Plastic polyallyl diglycol carbonate (PADC), 132
Plutonium (^{239}Pu), 98, 99–100, 103, 105, 235, 239,
        283–295, 409–410
    inhalation of, 388–390
PMMA, *see* Polymethyl-methacrylate
PNAD, *see* Personal neutron accident dosimeters
PNNL, *see* Pacific Northwest National Laboratory
Poincaré, Henri, 84–85
Polar plot of fraction of photons, 48
Polonium (^{210}Po), 85
    inhalation of, 385, 387–389, 408–409
Polymethyl-methacrylate (PMMA), 68
Positron decay, 35
Precision, 152
Pregnant female, 174, 188–191, 195, 196, 201–
        202, 205
Progeny, 7, 295
    radionuclide, 297
Prompt radiations, 28
Proportional counter detectors, 129
Prospective dose assessment, 176
Protection pioneers
    hazard, recognition of
        additional impetus, 90–91
        early reports of injury, 86–88
        protective measures, 88–90
Protection quantities, 61–67, 67
    International Commission on Radiological
        Protection
        Publication 26, 62–63

Publication 60, 63–64
Publication 103, 64–67
Protection standards, status of, 92–93
Protective measures, 88–90
Proton, 16, 30, 35–36, 50, 131–132, 144
   separation energy, 19, 20
Public Law, 117
PHS, *see* U.S. Public Health Service
U.S. Public Health Service (PHS), 104
Pulse-mode detectors
   Geiger–Mueller detectors, 129
   proportional counter detectors, 129
   scintillation detectors, 130

Quality factor, 1, 59–61, 77–78
   biological effectiveness, 59
   effective, 61
   function, Q(L), 59
   ICRP Publication 21, 60
   ICRP Publication 26, 62–63
   ICRP Publication 60, 63–64
   linear energy transfer, 73, 75
   mean, 75
Quantitative risk factors, 109
Quiescence
   first efforts, 93–94
   protection standards, status of, 92–93
Q-value, 21
   absorption, 26

Radiation absorption, 126
Radiation detection and measurement; *see also*
      Radiation measurement
   area monitoring instruments, 126–135
   measurement methods and procedures, 148–163
   personal dosimeters, 135–148
Radiation dose, regulation of, *see* Radiation
      protection guidance evolution, in United
   International Commission on Radiation Units
      and Measurements, reports of, 2
   International Commission on Radiological
      Protection, reports of, 3
   International committees and organizations
      governmental and non-governmental
         organizations, 5
   National Council on Radiation Protection,
      reports and commentaries of, 3
   U.S. regulations
      Environmental Protection Agency, 3–4
      Nuclear Regulatory Commission, 4
      Standards and Guidelines, 4
Radiation dosimetry, 1–2, 5, 6, 13, 50
Radiation equivalent man (REM), 100

Radiation equivalent physical (REP), 100
Radiation exposure, 1, 6, 89, 171, 188, 328,
      400, 411
Radiation measurement, 159
   air kerma, 171
   calibration, 458
   Geiger–Mueller, 129
   lower limit of detection, 162
   performance criterion, 161
   proportional counter, 129
   relative error, 161
   scintillation, 130
   standard deviation, 161, 162
   standard error of mean, 162
   statistics, 160–163
   uncertainty analysis, 162–164
   uncertainty standards, 163
   variance, 161
Radiation protection guidance evolution, in United
      States, 80–118
Radiation protection professional societies, 5
Radiation protection quantities/units
   absorbed dose
      charged particle equilibrium, 57
      exposure–dose relationship, 57–59
      kerma, 54–56
      linear energy transfer and quality factor, 59–61
Radiation regulations and standards
      Atomic Energy Commission Fissions,
         113–115
      electronics products, 112
      Environmental Protection Agency, 113
      Federal Radiation Council, 111–112
      International Commission on Radiological
         Protection, 108–110
      National Council on Radiation Protection
         and Measurements, 110–111
      proliferation of, 107–108
      regulatory agencies, 115
      tolerance dose to LNT paradigm, 105–107
      watershed, 102–105
Radiation Research Society (RRS), 5
Radiation risk models, 400, 401
Radiation-transport theory, 77
Radiation weighted absorbed dose, 63
Radiation weighting factor, 63, 64, 67, 69, 74, 78, 143,
      317, 324, 329, 332, 355, 359, 360, 364
   ICRP Publication 26, 62–63
   ICRP Publication 60, 63–64
   ICRP Publication 103, 64–67
   neutron, 64
Radiation interactions with matter
   attenuation coefficients, 52–53

electron
    bremsstrahlung, 45
    hard collisions, 44–45
    soft collisions, 44
  heavy charged particles, 50
  mass energy absorption coefficient, 54
  mass energy transfer coefficient, 53–54
  neutrons
    absorption, 50
    cross sections, 49–50
    elastic scatter, 51
    inelastic scatter, 51
  photon
    Compton scatter, 47–48
    pair production, 48–49
    photoelectric effect, 46–47
    photonuclear reactions, 49
  range, 52
Radiative capture, 132
Radiative kerma, 56
Radioactive decay
  alpha decay, 31–34
  beta decay, 34–35
  branching ratios, 43–44
  decay constant, 38
  electron capture, 35–36
  gamma emissions, 37
  half-life, 38–39
  internal conversion, 36
  isomeric transitions (metastable energy
    states), 37
  law, 37–38
  mean lifetime, 38
  positron decay, 35
  production and decay, 39–40
  radioactive decay law, 37–38
  radioactive half-life and decay constant, 38–39
  secular equilibrium, 42–43
  serial decay, 40–42
  specific activity, 40
  spontaneous fission, 36–37
  transient equilibrium, 43
  units, 40
Radioactive half-life/decay constant, 38–39
Radioactive progeny, 7, 295–298, 366
  produced in vivo, 298–302
Radioactivity, discovery of, 84–85
Radionuclides, *see individual entries*
Radium, 80
  Byers, Eban, 98
  dial painting, 97
  Evans, Robley D., 97
  Keene, Mae, 97

  Martland, Harrison, 97
  U.S. Radium Corporation, 97
Range, 52
  half-value layer, 52
  straggling, 52
RBE, *see* Relative biological effectiveness
Reciprocity method, 351–352
Reference geometries, 338
Reference individuals, for external and internal
    radiation dosimetry
  forms of dose assessment and role of, 174–176
  historical development, 171–174
  ICRP anatomical aspects of
    compared to ICRU Report 46, 186–188
    elemental tissue compositions and mass
      densities, 174, 184–188
    embryo, fetus, and pregnant female, 174,
      188–191
    individual organ systems, 178–184
    publication 89, 174
    total-body measurements, 176–178
  ICRP computational phantoms
    hybrid computational phantoms, 194–196
    stylized computational phantoms, 192
    voxel computational phantoms, 192–194
  ICRP in external and internal dosimetry
    external environmental exposures, dose
      coefficients for, 207
    external occupational exposures, dose
      coefficients for, 207
    internal occupational and environmental
      exposures, dose coefficients for, 205–207
    medical exposures, dose coefficients for, 207–208
  ICRP physiological aspects of
    bone remodeling rates, 200–201
    daily water balance, 196
    developing fetus and mother, 201
    lung mass (inclusive of blood) as a function
      of age, 205
    respiratory volumes and capacities, 196
    time budgets and ventilation rates, 196
    transit times of luminal content in alimentary
      tract, 197–199
    urinary and fecal excretion rates, 199–200
  ICRP reference data with Asian populations,
    201–204
  ICRP technical basis for, 170–171
Reference Man, 172–173, 202, 312, 315, 458; *see also*
    Standard Man
  ICRP Publication 23, 172
Reference values
  for age-depdendent element composition of the
    skeleton, 190

for blood flow to organs of non-pregnant and pregnant adult female, 201

for body mass of fetus, 190

for bone remodeling rates, 201

for daily time budgets and ventilation parameters, 198

for daily urinary excretion, 200

for daily ventilation rates for adult workers, 198

for density of skeletal components, 185

for division of bone mass in adult male or female, 184

for elemental composition of body tissue constituents, 186

for eye lens depth and size in adult males and females, 186

for gender-specific tissues for children and adults, 188

for gender-specific tissues for newborns, 190

for height, mass, and surface area of the total body, 177

for masses of body fat, 178

for masses of organs/tissues, 179–180

for masses of skeletal tissues and skeletal calcium (g), 185

for mass of epidermis, dermis, and total skin (g), 185

for mass of feces excreted per day, 200

for organ mass in developing fetus, 191

for regional blood volumes and blood flow rates in adults, 182–183

for respiratory volumes/capacities, 196

for soft tissue composition for children and adults, 187

for soft tissue composition for newborns, 189

for transit times of luminal contents, 199

for volume and surface area of bone in adult male, 184

for water balance in adults, 197

Regulation of Radiation Exposure by Legislative Means, 104

Regulations, regulatory guides, 81, 114

Regulatory agencies, 110, 115

Regulatory guides, 81

Regulatory issues, 459–461

Relative biological effectiveness (RBE), 67, 69, 78, 99, 317, 320, 403, 412

Relative error, 161

REM, *see* Radiation equivalent man

REP, *see* Radiation equivalent physical

Respiratory models, *see* Human Respiratory Tract Model

Respiratory tract absorption, modification of, 454–458

Respiratory tract model, 217; *see also* HRTM

Respiratory volumes/capacities, 196

Retention fraction, 429–432, 444, 454

excretion functions for acute exposures, 433

functions for chronic intakes, 429–433

retention and excretion fractions, tabulations of, 433–434

Retrospective dose assessment, 176

Riggs model, 261–262

Röntgen, Wilhelm Konrad, 58, 59, 74, 83–84, 95–96

ROT, *see* Rotational geometry

Rotational geometry (ROT), 338–339

RRS, *see* Radiation Research Society

Russian roulette, 351

SAF, *see* Specific absorbed fractions

Safe Drinking Water Act (SDWA), 113

Saturability, 17

Scintillation detectors, 6, 130

BGO, 130

CsI (Tl), 130

CZT, 130

gamma ray, 130

LaBr$_3$, 130

^{6}Li, 130

LiI(Eu), 130

NaI(Tl), 130

organic, 130

S-coefficient, defined, 360–361

SDWA, *see* Safe Drinking Water Act

Second, 14

Secular equilibrium, 42–43

SED, *see* Skin erythema dose

SEE, *see* Specific effective energy

Serial decay, 40–42

Shell model, 22

SI, *see* Systemè Internationale

Single bioassay measurement, 434

Site remediation, 116–117

Skeletal tissues, 193

Skin erythema dose (SED), 95

Small intestine, 232–233

SMR, *see* Standard mortality ratio

Soft collisions, 44

Soft X-rays, 90

Soil composition, 372, 373

Sommerfeld model, 25

Specific absorbed fraction (SAF), 206, 207, 315, 330, 360, 363–364, 366, 387

for charged particles, 330–332

for neutrons, 329–330

for photons, 173, 327–329

stylized, 192

Specific effective energy (SEE), 321–327, 330–332

for charged particles, 330–332

Spontaneous fission, 28, 36–37, 329, 384
Standard deviation, 150–151, 161–162
Standard error of mean, 162
Standard Man, 100, 171–172, 308, 309, 311; *see also* Reference Man
Standard mortality ratio (SMR), 94
Standards/standards setting bodies, 82
Stannard, J. Newell, 98
Static burns, 88
Statistics
    bias, 161
    lower limit of detection, 162
    performance criterion, 161
    radiation measurement, 160–163
    relative error, 161
    standard deviation, 161, 162
    standard error of mean, 162
    type A and type B, 162
    uncertainty analysis, 162–163
    variance, 161
Statutory Law, 81
Stomach, 231–232
Stopping power, 78
Strontium (^{90}Sr), 243–254
    ingestion of, 384–385
        in food, 405–416
Superheated drop (bubble) detectors, 146–147
Supplemental dosimeter, 125, 142
Supralinearity, 139
Surface contamination monitors, 158–159
Survey meters, 124, 133, 135, 142–143, 149, 160, 163
Systemè Internationale (SI), 95
Systemic biokinetic models, 237–295
    cesium, 271–283
    ICRP Publication 2, 237–239
    ICRP Publication 30, 238–239
    ICRP Publication 68 and 72 Series, 239–243
    iodine, 254–271
    plutonium, 283–295
    progeny, 295–302
    Riggs model, 261–262
    strontium, 243–254

Task Group Lung Model (TGLM), 220–221
Taylor, Lauriston, 96
TGLM, *see* Task Group Lung Model
Thermal neutrons, 50, 131, 132, 143, 144
Thermoluminescent dosimeters (TLDs), 6, 50, 139, 144–146
    dopants, 137
    glow curves, 138
    materials, 139
Thomson, Elihu, 86, 90

Thyroidal uptake of iodine, *see* Systemic biokinetic models, iodine
Time budgets, 196–197
Time-dependent absorption, model of, 223
Tissue-equivalent ionization chambers, 134
Tissue weighting factor, 64, 73, 78, 153, 164, 325–326, 332, 359
    ICRP Publication 26, 62–63
    ICRP Publication 60, 63–64
    ICRP Publication 103, 64–67
TLDs, *see* Thermoluminescent dosimeters
Tolerance dose, 95, 97, 99, 102, 110, 117
    to LNT paradigm, 105–107
Total-body measurements, 176–178
Traditional units, for radiation protection, 15–16
Transient equilibrium, 43
Transit times of luminal content, in alimentary tract, 197–199
Translocation rate constants, 418–419
Transport equation, 339–342
Tripartite Conferences on Internal Dosimetry, 100
Tritium (^{3}H), 451–454
Triton, 16, 21
Truman, Harry S., 101
Tunneling, 32

ULSF, *see* Unweighted least squares fitting
UMTRCA, *see* Uranium Mill Tailings Radiation Control Act
Uncertainty analysis, 162–163; *see also* Statistics
    type A and type B, 162
Unit
    of amount of substance, 14
    of electric current, 14
    of length, 14
    of luminous intensity, 14
    of mass, 14
    of thermodynamic temperature, 14
    of time, 14
United Nations Scientific Committee on the Effects of Atomic Radiation (UNSCEAR), 4, 106, 108
United States Transuranium and Uranium Registries (USTUR), 295, 454
UNSCEAR, *see* United Nations Scientific Committee on the effects of Atomic Radiation
Unweighted least squares fit (ULSF), 439, 444
Uranium (^{238}U)
    excretion fraction table, 437
    retention and excretion fraction table, 438
Uranium Mill Tailings Radiation Control Act (UMTRCA), 113
Urinary excretion of cesium, 279

Urinary/fecal excretion rates, 199–200
Urine monitoring technique, 461
U.S. Public Health Service (PHS), 104
    Radiological Health Handbook, 104
    Yellow Book, 104
U.S. regulations
    Environmental Protection Agency, 3–4
    Nuclear Regulatory Commission, 4
    Standards and Guidelines, 4
USTUR, *see* United States Transuranium and
    Uranium Registries

Variance, 161
Variance reduction, 344, 373
    techniques, 350–351
Ventilation rates, 196–197
Volumization, 286

Voluntary standards, 80, 82, 94, 98, 101, 107–108,
    114, 116, 117, 118
Voxel computational phantoms, 193–194

Watershed, 102–105
Weighted least squares fit (WLSF), 439, 440, 444–445
WLSF, *see* Weighted least squares fitting
World War
    Manhattan Project, 98
    WWII, 80, 94, 98–99, 101, 106, 116, 117–118, 171, 313

X-rays
    discovery of, 83–84
    photons, 88, 152
    Röntgen, Wilhelm Konrad, 1

Yellow Book, 104

Printed and bound by CPI Group (UK) Ltd, Croydon, CR0 4YY

24/10/2024

01778295-0014